FRACTALS
EVERYWHERE

SECOND EDITION

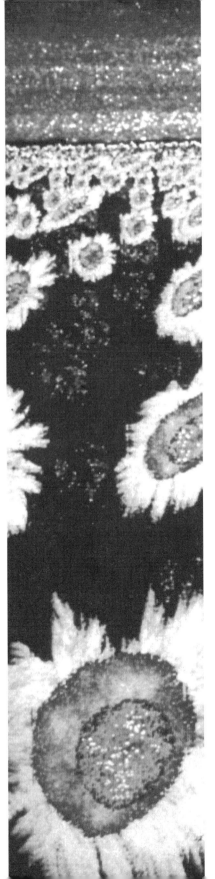

FRACTALS EVERYWHERE

SECOND EDITION

MICHAEL F. BARNSLEY

Iterated Systems, Inc. Atlanta, Georgia

Revised with the assistance of Hawley Rising III
Answer key by Hawley Rising III

Academic Press Professional

A Division of Harcourt Brace & Company

Boston • San Diego • New York
London • Sydney • Tokyo • Toronto

Figure credits and other acknowledgments appear at
the end of the book.

ACADEMIC PRESS PROFESSIONAL
955 Massachusetts Avenue, Cambridge, MA 02139

An imprint of ACADEMIC PRESS, INC.
A division of HARCOURT BRACE & COMPANY

United Kingdom Edition published by
ACADEMIC PRESS LIMITED
24–28 Oval Road, London NW1 7DX

Library of Congress Cataloging-in-Publication Data

Barnsley, M. F. (Michael Fielding), date.
    Fractals Everywhere/Michael Barnsley, with the assistance of
  Hawley Rising III.
        p. cm.
    Includes bibliographical references and index.
    ISBN 0-12-079061-0
    1. Fractals. I. Rising, Hawley. II. Title.
  QA614.86.B37   1993   93-15985
  514'.74—dc20   CIP

Printed in the United States of America
93 94 95 96 97 MV 9 8 7 6 5 4 3 2 1

I dedicate the second edition of this book to my daughter
Diana Gabriel Barnsley

Contents

| Chapter VII | Julia Sets | 246 |

| Chapter VIII | Parameter Spaces and Mandelbrot Sets | 294 |

| Chapter IX | Measures on Fractals | 330 |

Foreword to the Second Edition

Much has changed in the world of fractals, computer graphics and modern mathematics since the first edition of *Fractals Everywhere* appeared. The company Iterated Systems, Inc., founded by Michael Barnsley and Alan Sloan, is now competing in the image compression field with both hardware and software products that use fractal geometry to compress images. Indeed, there is now a plethora of texts on subjects like fractals and chaos, and these terms are rapidly becoming "household words."

The fundamental approach to fractal geometry through iterated function systems remains sound as an introduction to the subject. This edition of *Fractals Everywhere* leaves this approach largely as it stands. One still needs a grounding in concepts in metric space theory and eventually (see Chapter IX) measure theory to get a working understanding of the subject. However, there have been several additions to help ease and broaden the reader's development.

Primary to these is the addition of answers to the mathematical problems. These were done largely by starting at one end of the book writing the answers until the other cover was reached. Most of the answers found in the key have been worked over at least twice, in hopes improving the accuracy of the key. Every effort has been make to rely solely on the material presented ahead of each problem, although in a few of the harder problems some concepts have been introduced in the answers themselves. These are not considered necessary to the development of the main thread of the text; however, if the reader finds some areas of mathematics touched on in looking at the presented solutions which extend the feeling for the subject, the key has served its purpose.

In addition the the answer key, there have been some other changes as well. In Chapter III, section 11, the main theorem has been qualified. The reader with more mathematical background will recognize that the additional Lipshitz condition satisfies the need for equicontinuity in Theorem 11.1. This is not the only way to satisfy it, just the clearest in terms of the presumed mathematical background.

There have been problems added to several chapters to develop the idea of Cartesian products of code spaces. This was done because it helps bridge the gap between IFS theory and the reversible systems found in physical chaos, and because it presents an interesting way of looking the the Random Iteration Algorithm in Chapter IX. The thread of these problems begins in Chapter II, leads up to the baker's transformation in Chapter IV, and is completed as an example in Chapter IX. Additional problems were added in Chapter III to develop some basic properties of eigenvalues and eigenvectors, which can be useful in examining dynamics both from the point of view described in the text, and elsewhere. It is hoped that with these additional tools those readers whose goals are application-oriented will come away with more at their disposal, while the text itself will retain its readable style.

I would like to thank Lyman Hurd for many useful discussions about the topological nature of nonempty compact sets, and John Elton for his patience while I ran many of my new examples and problems past him to check them and to check the "excitement level" of the additional material.

Hawley Rising

It seems now that deterministic fractal geometry is racing ahead into the serious engineering phase. Commercial applications have emerged in the areas of image compression, video compression, computer graphics, and education. This is good because it authenticates once again the importance of the work of mathematicians. However, sometimes mathematicians lose interest in wonderful areas once scientists and engineers seem to have the subject under control. But there is so much more mathematics to be done. What is a useful metric for studying the contractivity of the vector recurrent IFS of affine maps in $\mathbb{R}^2$? What is the information content of a picture? Measures, pictures, dreams, chaos, flowers and information theory—the hours of the days keep rushing by: do not let the beauty of all these things pass us by too.

Michael Fielding Barnsley

Acknowledgments

I acknowledge and thank many people for their help with this book. In particular I thank Alan Sloan, who has unceasingly encouraged me, who wrote the first Collage software, and who so clearly envisioned the application of iterated function systems to image compression and communications that he founded a company named *Iterated Systems Incorporated*. Edward Vrscay, who taught the first course in deterministic fractal geometry at Georgia Tech, shared his ideas about how the course could be taught, and suggested some subjects for inclusion in this text. Steven Demko, who collaborated with me on the discovery of iterated function systems, made early detailed proposals on how the subject could be presented to students and scientists, and provided comments on several chapters. Andrew Harrington and Jeffrey Geronimo, who discovered with me orthogonal polynomials on Julia sets. My collaborations with them over five years formed for me the foundation on which iterated function systems are built. Watch for more papers from us!

Les Karlovitz, who encouraged and supported my research over the last nine years, obtained the time for me to write this book and provided specific help, advice, and direction. His words can be found in some of the sentences in the text. Gunter Meyer, who has encouraged and supported my research over the last nine years. He has often given me good advice. Robert Kasriel, who taught me some topology over the last two years, corrected and rewrote my proof of Theorem 7.1 in Chapter II and contributed other help and warm encouragement. Nathanial Chafee, who read and corrected Chapter II and early drafts of Chapters III and IV. His apt constructive comments have increased substantially the precision of the writing. John Elton, who taught me some ergodic theory, continues to collaborate on exciting research into iterated function systems, and helped me with many parts of the book. Daniel Bessis and Pierre Moussa, who are filled with the wonder and mystery of science, and taught me to look for mathematical events that are so astonishing that they may be called miracles. Research work with Bessis and Moussa at Saclay during 1978, on the Diophantine Moment Problem and Ising Models, was the seed that grew into this book. Warren Stahle, who provided some of his experimental research results for inclusion in Chapter VI.

Graduate students John Herndon, Doug Hardin, Peter Massopust, Laurie Reuter, Arnaud Jacquin, and François Malassenet, who have contributed in many ways to this book. They helped me to discover and develop some of the ideas. Els Withers and Paul Blanchard, who supported the writing of this book from the start and suggested some good ideas that are used. The research papers by Withers on iterated functions are deep. Edwina Barnsley, my mother, whose house was always full of flowers. Her encouragement and love helped me to write this book. Thomas Stelson, Helena Wisniewski, Craig Fields, and James Yorke who, early on, supported the development of applications of iterated function systems. Many of the pictures in this text were produced in part using software and hardware in the DARPA/GTRC funded Computergraphical Mathematics Laboratory within the School of Mathematics at Georgia Institute of Technology.

George Cain, James Herod, William Green, Vince Ervin, Jamie Good, Jim Osborne, Roger Johnson, Li Shi Luo, Evans Harrell, Ron Shonkwiler, and James Walker who contributed by reading and correcting parts of the text, and discussing research. Thomas Morley, who contributed many hours of discussion of research and never asks for any return. William Ames who encouraged me to write this book and introduced me to Academic Press. Annette Rohrs, who typed the first drafts of Chapters II, III, and IV. William Kammerer, who introduced me to EXP, the technical word processor on which the manuscript was written, and who has warmly supported this project.

This book owes its deepest debt to Alan Barnsley, my father, who wrote novels and poems under the *nom-de-plume* Gabriel Fielding. I learnt from him care for precision, love of detail, enthusiasm for life, and an endless amazement at all that God has made.

Michael Barnsley

Chapter I

Introduction

Fractal geometry will make you see everything differently. There is danger in reading further. You risk the loss of your childhood vision of clouds, forests, galaxies, leaves, feathers, flowers, rocks, mountains, torrents of water, carpets, bricks, and much else besides. Never again will your interpretation of these things be quite the same.

The observation by Mandelbrot [Mandelbrot 1982] of the existence of a "Geometry of Nature" has led us to think in a new scientific way about the edges of clouds, the profiles of the tops of forests on the horizon, and the intricate moving arrangement of the feathers on the wings of a bird as it flies. Geometry is concerned with making our spatial intuitions objective. Classical geometry provides a first approximation to the structure of physical objects; it is the language that we use to communicate the designs of technological products and, very approximately, the forms of natural creations. Fractal geometry is an extension of classical geometry. It can be used to make precise models of physical structures from ferns to galaxies. Fractal geometry is a new language. Once you can speak it, you can describe the shape of a cloud as precisely as an architect can describe a house.

This book is based on a course called "Fractal Geometry," which has been taught in the School of Mathematics at the Georgia Institute of Technology for two years. The course is open to all students who have completed two years of calculus. It attracts both undergraduate and graduate students from many disciplines, including mathematics, biology, chemistry, physics, psychology, mechanical engineering, electrical engineering, aerospace engineering, computer science, and geophysical science. The delight of the students with the course is reflected in the fact there is now a second course, entitled "Fractal Measure Theory." The courses provide a compelling vehicle for teaching beautiful mathematics to a wide range of students.

Here is how the course in Fractal Geometry is taught. The core is Chapter II, Chapter III, sections 1–5 of Chapter IV, and sections 1–3 of Chapter V. This is followed by a collection of delightful special topics, chosen from Chapters VI, VII, and VIII. The course is taught in 30 one-hour lectures.

Chapter II introduces the basic topological ideas that are needed to describe subsets to spaces such as $\mathbb{R}^2$. The framework is that of metric spaces; this is adopted because metric spaces are both rigorously and intuitively accessible, yet full of suprises. They provide a suitable setting for fractal geometry. The concepts introduced include openness, closedness, compactness, convergence, completeness, connectedness, and equivalence of metric spaces. An important theme concerns properties that are preserved under equivalent metrics. Chapter II concludes by presenting the most exciting idea: a metric space, denoted $\mathcal{H}$, whose elements are the nonempty compact subsets of a metric space. Under the right conditions this space is complete, sequences converge, and fractals can be found!

Chapter III deals with transformations on metric spaces. First, the goal is to develop intuition and practical experience with the actions of elementary transformations on subsets of spaces. Particular attention is devoted to affine transformations and Möbius transformations in $\mathbb{R}^2$. Then the contraction mapping principle is revealed, followed by the construction of contraction mappings on $\mathcal{H}$. Fractals are discovered as the fixed points of certain set maps. We learn how fractals are generated by the application of "simple" transformations on "simple" spaces, and yet they are geometrically complicated. We explain what an iterated function system (IFS) is, and how it can define a fractal. Iterated function systems provide a convenient framework for the description, classification, and communication of fractals. Two algorithms, the "Chaos Game" and the Deterministic Algorithm, for computing pictures of fractals are presented. Attention is then turned to the inverse problem: given a compact subset of $\mathbb{R}^2$, fractal, how do you go about finding a fractal approximation to it? Part of the answer is provided by the Collage Theorem. Finally, the thought of the wind blowing through a fractal tree leads to discovery of conditions under which fractals depend continuously on the parameters that define them.

Chapter IV is devoted to dynamics on fractals. The idea of addresses of points on certain fractals is developed. In particular, the reader learns about the metric space to which addresses belong. Nearby addresses correspond to nearby points on the fractal. This observation is made precise by the construction of a continuous transformation from the space of addresses to the fractal. Then dynamical systems on metric spaces are introduced. The ideas of orbits, repulsive cycles, and equivalent dynamical systems are described. The concept of the shift dynamical system associated with an IFS is introduced and explored. This is a visual and simple idea in which the author and the reader are led to wonder about the complexity and beauty of the available orbits. The equivalence of this dynamical system with a corresponding system on the space of addresses is established. This equivalence takes no account of the geometrical complexity of the dance of the orbit on the fractal. The chapter then moves towards its conclusion, the definition of a chaotic dynamical system and the realization that "most" orbits of the shift dynamical system on a fractal are chaotic. To this end, two simple and delightful ideas are shown to the reader. The Shadow Theorem illustrates how apparently random orbits may actually be the "shadows" of deterministic motions in higher-dimensional spaces. The Shadow*ing* Theorem demonstrates how a rottenly inaccurate orbit may be trailed by a precise orbit, which clings like a secret agent. These ideas are used to make an explanation of why the "Chaos Game" computes fractals.

Chapter V introduces the concept of fractal dimension. The fractal dimension of a set is a number that tells how densely the set occupies the metric space in which it lies. It is invariant under various stretchings and squeezings of the underlying space. This makes the fractal dimension meaningful as an experimental observable; it possesses a certain robustness and is independent of the measurement units. Various theoretical properties of the fractal dimension, including some explicit formulas, are developed. Then the reader is shown how to calculate the fractal dimension of real-world data, and an application to a turbulent jet exhaust is described. Lastly the Hausdorff-Besicovitch dimension is introduced. This is another number that can be associated with a set. It is more robust and less practical than the fractal dimension. Some mathematicians love it; most experimentalists hate it; and we are intrigued.

Chapter VI is devoted to fractal interpolation. The aim of the chapter is to teach the student practical skill in using a new technology for making complicated curves and fitting experimental data. It is shown how geometrically complex graphs of continuous functions can be constructed to pass through specified data points. The functions are represented by succinct formulas. The main existence theorems and computational algorithms are provided. The functions are known as fractal interpolation functions. It is explained how they can be readily computed, stored, manipulated and communicated. "Hidden variable" fractal interpolation functions are introduced and illustrated; they are defined by the shadows of the graphs of three-dimensional fractal paths. These geometrical ideas are extended to introduce space-filling curves.

Chapter VII gives an introduction to Julia sets, which are deterministic fractals that arise from the iteration of analytic functions. The objective is to show the reader how to understand these fractals, using the ideas of Chapters III and IV. In so doing we have the pleasure of explaining and illustrating the Escape Time Algorithm. This algorithm is a means for computergraphical experimentation on dynamical systems that act on two-dimensional spaces. It provides illumination and coloration, a seachlight to probe dynamical systems for fractal structures and regions of chaos. The algorithm relies on the existence of "repelling sets" for continuous transformations which map open sets to open sets. The applications of Julia sets to biological modelling and to understanding Newton's method are considered.

Chapter VIII is concerned with how to make maps of certain spaces, known as parameter spaces, where every point in the space corresponds to a fractal. The fractals depend "smoothly" on the location in the parameter space. How can one make a picture that provides useful information about what kinds of fractals are located where? If both the space in which the fractals lie and the parameter space are two-dimensional, the parameter space can sometimes be "painted" to reveal an associated Mandelbrot set. Mandelbrot sets are defined, and three different examples are explored, including the one discovered by Mandelbrot. A computergraphical technique for producing images of these sets is described. Some basic theorems are proved.

Chapter IX is an introduction to measures on fractals and to measures in general. The chapter is an outline that can be used by a professor as the basis of a course in fractal measure theory. It can also be used in a standard measure theory course as a source of applications and examples. One goal is to demonstrate that measure theory is a workaday tool in science and engineering. Models for real-world images can be made using measures. The variations in color and brightness, and the complex textures in a color picture, can be successfully modelled by measures that can be written down explicitly in terms of succinct "formulas." These measures are desirable for image engineering applications, and have a number of advantages over nonnegative "density" functions. Section 1 provides an intuitive description of measures and motivates the rest of the chapter. The context is that of Borel measures on compact metric spaces. Fields, sigma-fields, and measures are defined. Carathéodory's extension theorem is introduced and used to explain what a Borel measure is. Then the integral of a continuous real-valued function, with respect to a measure, is defined. The reader learns to evaluate some integrals. Next the space $\mathcal{P}$ of normalized Borel measures on a compact metric space is defined. With an appropriate metric, $\mathcal{P}$ becomes a compact metric space. Succinctly defined contraction mappings on this space lead to measures that live on fractals. Integrals with respect to these measures can be evaluated with the aid of Elton's ergodic theorem. The book ends with a description of the application of these measures to computer graphics.

This book teaches the tools, methods, and theory of *deterministic* geometry. It is useful for describing *specific* objects and structures. Models are represented by succinct "formulas." Once the formula is known the model can be reproduced. We do not consider statistical geometry. The latter aims at discovering general statistical laws that govern families of similar-looking structures, such as *all* cumulus clouds, *all* maple leaves, or *all* mountains.

In deterministic geometry, structures are defined, communicated, and analyzed, with the aid of elementary transformations such as affine transformations, scalings, rotations, and congruences. A fractal set generally contains infinitely many points whose organization is so complicated that it is not possible to describe the set by specifying directly where each point in it lies. Instead, the set may be defined by "the relations between the pieces." It is rather like describing the solar system by quoting the law of gravitation and stating the initial conditions. Everything follows from that. It appears always to be better to describe in terms of relationships.

Chapter II

Metric Spaces; Equivalent Spaces; Classification of Subsets; and the Space of Fractals

1 Spaces

In fractal geometry we are concerned with the structure of subsets of various very simple "geometrical" spaces. Such a space is denoted by $\mathbf{X}$. It is the space on which we think of drawing our fractals; it is the place where fractals live. What is a fractal? For us, for now, it is just a subset of a space. Whereas the space is simple, the fractal subset may be geometrically complicated.

Definition 1.1 *A space* $\mathbf{X}$ *is a set. The* points *of the space are the elements of the set.*

Although this definition does not say it, the nomenclature "space" implies that there is some structure to the set, some sense of which points are close to which. We give some examples to show the sort of thing this may mean. Throughout this text $\mathbb{R}$ denotes the set of real numbers, and "$\in$" means "belongs to."

Examples

1.1. $\mathbf{X} = \mathbb{R}$. Each "point" $x \in \mathbf{X}$ is a real number, or a dot on a line.

1.2. $\mathbf{X} = C[0, 1]$, the set of continuous functions that take the real closed interval $[0, 1] = \{x \in \mathbb{R} : 0 \le x \le 1\}$ into the real line $\mathbb{R}$. A "point" $f \in \mathbf{X}$ is a function $f : [0, 1] \overset{cts.}{\to} \mathbb{R}$. f may be represented by its graph.

Figure II.1. A point x in $\mathbb{R}$.

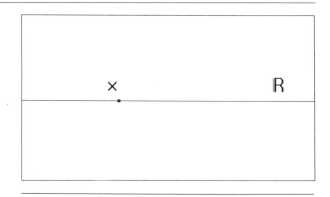

Figure II.2. A point f in the space of continuous functions on $[0, 1]$.

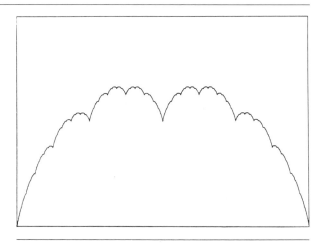

Notice that here $f \in \mathbf{X}$ is not a point on the x-axis; it is the whole function. A continuous function on an interval is characterized by the fact that its graph is unbroken; as a picture it contains no rips or tears; it can be drawn without removing the pencil from the paper.

1.3. $\mathbf{X} = \mathbb{R}^2$, the Euclidean plane, the coordinate plane of calculus. Any pair of real numbers $x_1, x_2 \in \mathbb{R}$ determines a single point in $\mathbb{R}^2$. A point $x \in \mathbf{X}$ is represented in several equivalent ways:

$$x = (x_1, x_2) = \begin{pmatrix} x_1 \\ x_2 \end{pmatrix} = \text{a point in a figure such as Figure II.3.}$$

The spaces in examples 1.1, 1.2, and 1.3 are each *linear spaces:* there is an obviously defined way, in each case, of adding two points in the space to obtain a new one in the same space. In 1.1 if x and $y \in \mathbb{R}$, then $x + y$ is also in $\mathbb{R}$; in 1.2 we define $(f + g)(x) = f(x) + g(x)$; and in 1.3 we define

$$x + y = \begin{pmatrix} x_1 \\ x_2 \end{pmatrix} + \begin{pmatrix} y_1 \\ y_2 \end{pmatrix} = \begin{pmatrix} x_1 + y_1 \\ x_2 + y_2 \end{pmatrix}.$$

Similarly, in each of the above examples, we can multiply members of **X** by a scalar, that is, by a real number $\alpha \in \mathbb{R}$. For example, in 1.2 $(\alpha f)(x) = \alpha f(x)$ for any $\alpha \in \mathbb{R}$, and $\alpha f \in C[0, 1]$ whenever $f \in C[0, 1]$. Example 1.1 is a one-dimensional linear space; 1.2 is an ∞-dimensional linear space (can you think why the dimension is infinite?); and 1.3 is a two-dimensional linear space. A linear space is also called a vector space. The scalars may be complex numbers instead of real numbers.

1.4. The complex plane, $\mathbf{X} = \mathbb{C}$, where any point $x \in \mathbf{X}$ is represented

$$x = x_1 + ix_2, \qquad \text{where } i = \sqrt{-1},$$

for some pair of real numbers $x_1, x_2 \in \mathbb{R}$. Any pair of numbers $x_1, x_2 \in \mathbb{R}$ determines a point of $\mathbb{C}$. It is obvious that $\mathbb{C}$ is essentially the same as $\mathbb{R}^2$, but there is an implied distinction. In $\mathbb{C}$ we can multiply two points x, y and obtain a new point in $\mathbb{C}$. Specifically, we define

$$x \cdot y = (x_1 + ix_2)(y_1 + iy_2) = (x_1 y_1 - x_2 y_2) + i(x_2 y_1 + x_1 y_2)$$

1.5. $\mathbf{X} = \hat{\mathbb{C}}$, the Riemann sphere. Formally, $\hat{\mathbb{C}} = \mathbb{C} \cup \{\infty\}$; that is, all the points of $\mathbb{C}$ together with the "point at infinity." Here is a way of constructing and thinking about $\hat{\mathbb{C}}$. Place a sphere on the plane $\mathbb{C}$, with the South Pole on the origin, and the North Pole N vertically above it.

To a given point $x \in \mathbb{C}$ we associate a point x' on the sphere by constructing the straight line from N to x and marking where this line intersects the sphere. This associates a unique point $x' = h(x)$ with each point $x \in \mathbb{C}$. The transformation $h : \mathbb{C} \to$ sphere is clearly continuous in the sense that nearby points go to nearby points. Points farther and farther away from 0 in the plane $\mathbb{C}$ end up closer and closer to N. $\hat{\mathbb{C}}$ consists of the completion of the range of h by including N on the sphere: The "point at infinity (∞)" can be thought of as a giant circle, infinitely far out in $\mathbb{C}$, whose image under h is N. It is easier to think of $\hat{\mathbb{C}}$ being the whole of the sphere, rather than as the plane together with ∞. It is of interest that $h : \mathbb{C} \to$ sphere

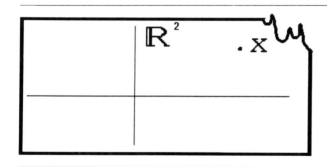

Figure II.3.   A point x in the space $\mathbb{R}^2$.

Figure II.4. Construction of a geometrical representation for the Riemann sphere. N is the North Pole and corresponds to the "point at infinity."

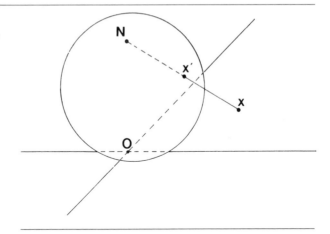

is *conformal*: it preserves angles. The image under h of a triangle in the plane is a curvaceous triangle on the sphere.

Although the sides of the triangle on the sphere are curvaceous they meet in well-defined angles, as one can visualize by imagining the globe to be magnified enormously. The angles of the curvaceous triangle are the same as the corresponding angles of the triangle in the plane.

Examples & Exercises

1.6. $X = \Sigma$, the *code space* on N symbols. Usually the symbols are the integers $\{0, 1, 2, \ldots, N - 1\}$. A typical point in X is a semi-infinite word such as

$$x = 2\ 17\ 0\ 0\ 1\ 21\ 15\ (N - 1)\ 3\ 0 \ldots.$$

There are infinitely many symbols in this sequence. In general, for a given element $x \in X$, we can write

Figure II.5. A triangle in the plane corresponds to a curvaceous triangle on the sphere.

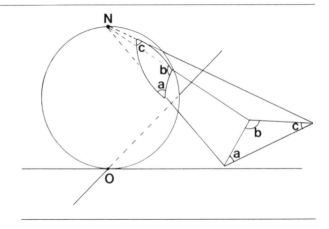

$$x = x_1 x_2 x_3 x_4 x_5 x_6 x_7 x_8 \ldots, \qquad \text{where each } x_i \in \{0, 1, 2, \ldots, N - 1\}.$$

There are many names attached to this space because of its importance in a variety of branches of mathematics and physics. When each symbol is intended to represent a random choice from N possibilities, each point in this space represents a particular sequence of events, from a set of N possible events. In this case, the space is sometimes called the space of Bernoulli trials. When there are several code spaces being referred to, it is customary to write the code space on N symbols as Σ_N.

1.7. A few other favorite spaces are defined as follows.

(a) A disk in the plane with center at the origin and with finite radius $R > 0$:

$$\bullet = \{x \in \mathbb{R}^2 : x_1^2 + x_2^2 \leq R^2\}.$$

(b) A "filled" square:

$$\blacksquare = \{x \in \mathbb{R}^2 : 0 \leq x_1 \leq 1, 0 \leq x_2 \leq 1\}.$$

(c) An interval:

$$[a, b] = \{x \in \mathbb{R} : a \leq x \leq b\}, \text{ where } a \text{ and } b \text{ are real numbers with } a < b.$$

(d) Body space:

$$\text{\includegraphics{}} = \{x \in \mathbb{R}^3 : \text{coordinate points implied by a cadaver frozen in } \mathbb{R}^3\}.$$

(e) Sierpinski space

$$\Delta = \{x \in \mathbb{R}^2 : x \text{ is a point on a certain fixed Sierpinski triangle}\}.$$

Sierpinski triangles occur often in this text. See, for example, Figure IV.94.

1.8. Show that the examples in 1.5, 1.6, and 1.7 are not vector spaces, at least if addition and multiplication by reals are defined in the usual way.

1.9. The notation $A \subset \mathbf{X}$ means A is a *subset* of $\mathbf{X}$; that is, if $x \in A$ then $x \in \mathbf{X}$, or $x \in A$ implies $x \in \mathbf{X}$. The symbol $\emptyset$ means the empty set. It is defined to be the set such that the statement "$x \in \emptyset$" is always false. We use the notation $\{x\}$ to denote the set consisting of a single point $x \in \mathbf{X}$. Show that if $x \in \mathbf{X}$, then $\{x\}$ is a subset of $\mathbf{X}$.

1.10. Any set of points makes a space, if we care to define it as such. The points are what we choose them to be. Why, do you think, have the spaces defined above been picked out as important? Describe other spaces that are equally important.

1.11. Let $\mathbf{X}_1$ and $\mathbf{X}_2$ be spaces. These can be used to make a new space denoted $\mathbf{X}_1 \times \mathbf{X}_2$, called the Cartesian product of $\mathbf{X}_1$ and $\mathbf{X}_2$. A point in $\mathbf{X}_1 \times \mathbf{X}_2$ is represented by the ordered pair (x_1, x_2), where $x_1 \in \mathbf{X}_1$ and $x_2 \in \mathbf{X}_2$. For example, $\mathbb{R}^2$ is the Cartesian product of $\mathbb{R}$ and $\mathbb{R}$.

1.12. As another example of a Cartesian product let

$$\mathbf{X} = \{(x, y) : x, y \in \Sigma\} = \Sigma \times \Sigma,$$

where Σ is the code space on N symbols. This has an interpretation in terms of the random choices mentioned in exercise 1.6. We call y the past and x the future. Then each element of the space represents a sequence of "coin tosses" (the coins are really more like N-sided dice); y represents the tosses that have already happened, beginning with the latest one, and x represents the tosses to come (beginning with the next one). If we rewrite the point (x, y) with a dot marking the "present,"

$$\ldots y_3 y_2 y_1 \ . \ x_1 x_2 x_3 \ldots$$

then the act of moving the dot to the right moves one future coin toss to a past coin toss; the obvious interpretation is that it represents flipping the coin. Moving the dot is called a *shift*, and the space is called *the space of shifts on N symbols*. It is also denoted Σ, whether it is this space or code space that is being referred to is usually clear from context. In this book Σ will always be code space unless specifically mentioned.

2   Metric Spaces

We use the notation "$\forall$" to mean "for all." We also introduce the notation $A \setminus B$ to mean the set A "take away" the set B. That is, $A \setminus B = \{x \in A : x \notin B\}$. We use "$\Rightarrow$" to mean "implies."

Definition 2.1   *A metric space* $(\mathbf{X}, d)$ *is a space* $\mathbf{X}$ *together with a real-valued function* $d : \mathbf{X} \times \mathbf{X} \to \mathbb{R}$, *which measures the* distance *between pairs of points x and y in* $\mathbf{X}$. *We require that d obeys the following axioms:*

(1) $d(x, y) = d(y, x) \ \forall x, y \in \mathbf{X}$
(2) $0 < d(x, y) < \infty \ \forall x, y \in \mathbf{X}, x \neq y$
(3) $d(x, x) = 0 \ \forall x \in \mathbf{X}$
(4) $d(x, y) \leq d(x, z) + d(z, y) \ \forall x, y, z \in \mathbf{X}$.

Such a function d is called a metric.

The concept of shortest paths between points in a space, *geodesics*, is dependent on the metric. The metric may determine a *geodesic structure* of the space. Geodesics on a sphere are great circles; in the plane with the Euclidean metric they are straight lines.

Examples & Exercises

2.1. Show that the following are all metrics in the space $\mathbf{X} = \mathbb{R}$:

(a) $d(x, y) = |x - y|$ (Euclidean metric)
(b) $d(x, y) = 2 \cdot |x - y|$
(c) $d(x, y) = |x^3 - y^3|$

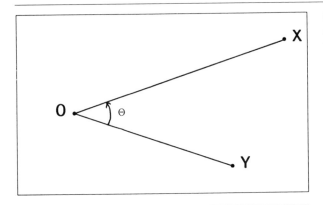

Figure II.6. (The angle θ, and the distances r_1, r_2 used to construct a metric on the punctured plane.) Acute angle subtended by two straight lines.

2.2. Show that the following are metrics in the space $\mathbf{X} = \mathbb{R}^2$:

(a) $d(x, y) = \sqrt{(x_1 - y_1)^2 + (x_2 - y_2)^2}$ (Euclidean metric)
(b) $d(x, y) = |x_1 - y_1| + |x_2 - y_2|$ (Manhattan metric)

Why is the name "Manhattan" used in connection with (b)?

2.3. Show that $d(x, y) = |xy|$ does not define a metric in $\mathbb{R}$.

2.4. Let $\mathbb{R}^2 \setminus \{O\}$ denote the punctured plane. Define $d(x, y)$ as follows:

$$d(x, y) = |r_1 - r_2| + |\theta|,$$

where $r_1 =$ Euclidean distance from x to O, $r_2 =$ Euclidean distance from y to O, O is the origin, and θ is the smallest angle subtended by the two straight lines connecting x and y to the origin. Show that d is a metric.

2.5. On the code space Σ define

$$d(x, y) = d(x_1 x_2 x_3 \ldots, y_1 y_2 y_3 \ldots) = \sum_{i=1}^{\infty} \frac{|x_i - y_i|}{(N + 1)^i}$$

Show that every pair of points in Σ is a *finite* distance apart. That is, d is indeed a function that takes $\Sigma \times \Sigma$ into $\mathbb{R}$. Verify that (Σ, d) is a metric space. Try to envisage a possible geometry for Σ. (Do not confuse the possible meanings of the symbol Σ; from its context it should be clear when it refers to code space and when it refers to summation.)

*2.6. Define $\mathbf{X} = \{(x, y) : x, y \in \Sigma\}$, the space of shifts on N symbols as in exercise 1.12. We can define a *Euclidean* distance by treating each coordinate as a base $N + 1$ number between 0 and 1. That is, we make the distance between (x, y) and (u, v) equal to

$$\sqrt{\left(\sum_{i=1}^{\infty} \frac{x_i - u_i}{(N + 1)^i}\right)^2 + \left(\sum_{i=1}^{\infty} \frac{y_i - v_i}{(N + 1)^i}\right)^2}$$

Show that this is indeed a metric space.

2.7. In $X =$ [figure] define $d(x, y)$ to be the Euclidean length of the shortest path lying entirely within **X** which connects x and y. Show that this is a metric. Discuss the utility of this metric in anatomy. The distance from a toenail to a fingertip does not depend much on the configuration of the body, whereas the usual spatial distance would.

2.8. Invent a function $d : \blacksquare \times \blacksquare \to \mathbb{R}$ which is not a metric. Define a metric for the space [figure], namely an annulus, which makes it seem like the curved wall of a cylinder: [figure].

*2.9. Show that a metric on $X = \hat{\mathbb{C}}$ is defined by shortest great circle distances on the sphere. Compare the distances from 0, and from $1 + i$, to ∞.

Definition 2.2 *Two metrics d_1 and d_2 on a space* **X** *are equivalent if there exist constants $0 < c_1 < c_2 < \infty$ such that*

$$c_1 d_1(x, y) \leq d_2(x, y) \leq c_2 d_1(x, y), \qquad \forall (x, y) \in \mathbf{X} \times \mathbf{X}.$$

Examples & Exercises

2.10. Definition 2.2 looks unsymmetrical; it does not appear to make the same requirements on d_1 as it does on d_2. Show that this is an illusion by establishing that if the definition holds then there are constants $0 < e_1 < e_2 < \infty$ so that

$$e_1 d_2(x, y) \leq d_1(x, y) \leq e_2 d_2(x, y), \qquad \forall (x, y) \in \mathbf{X} \times \mathbf{X}.$$

2.11. Are the Manhattan and Euclidean metrics equivalent on $\blacksquare \subset \mathbb{R}^2$? What about on $\mathbb{R}^2$?

2.12. Show that the metric in exercise 2.4 is *not* equivalent to the Euclidean metric on [figure] $\setminus \{0\}$.

One notion underlying the concept of equivalent metrics is that any pair of equivalent metrics gives the same notion of which points are close together and which are far apart. It is as though there were a standard way for boundedly deforming the space, whereby distances are determined both before and after deformation.

For example, consider a pair of points x and y in $\blacksquare \subset \mathbb{R}^2$. Let the Euclidean distance between these points be $d_1(x, y)$. Think of a thin rubber sheet lying over $\blacksquare$. This sheet is stretched in some repeatable fashion, carrying copies of the points x and y to new locations, as illustrated in Figure II.7. The Euclidean distance between these moved points is called $d_2(x, y)$. The condition of equivalence is the requirement that there is no extreme (infinite) stretching or compression of the space.

This leads us to the idea of equivalent metric spaces:

Definition 2.3 *Two metric spaces $(\mathbf{X}_1, d_1)$ and $(\mathbf{X}_2, d_2)$ are equivalent if there is a function $h \colon \mathbf{X}_1 \to \mathbf{X}_2$ that is one-to-one and onto (i.e., it is invertible), such that*

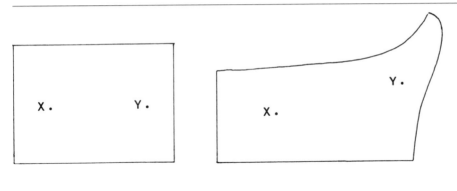

Figure II.7. A thin rubber sheet lies over the ■ in the plane and is stretched. The Euclidean distances between points are determined before and after deformation, yielding two metrics. These metrics may be equivalent if the deformation leads to no rips, tears, or infinite stretching.

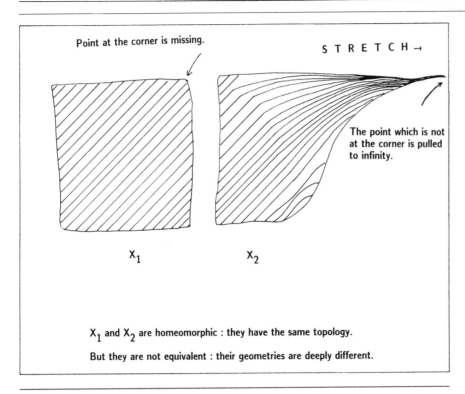

Point at the corner is missing.

STRETCH →

The point which is not at the corner is pulled to infinity.

X_1

X_2

X_1 and X_2 are homeomorphic : they have the same topology.

But they are not equivalent : their geometries are deeply different.

Figure II.8. This picture suggests two metric spaces X_1 and X_2 that have the same topology, but that are not metrically equivalent: their "geometries" are deeply different.

the metric $\tilde{d}_1$ on X_1 defined by

$$\tilde{d}_1(x, y) = d_2(h(x), h(y)), \qquad \forall x, y \in X_1$$

is equivalent to d_1.

One can think of Definition 2.3 as requiring that X_1 and X_2 are related to one another by a bounded deformation, and nowhere is there an arbitrarily large compression or stretching; also, there is no overlapping, folding, or ripping.

Definition 2.4 *A function $f : X_1 \to X_2$ from a metric space (X_1, d_1) into a metric space (X_2, d_2) is* continuous *if, for each $\epsilon > 0$ and $x \in X_1$, there is a $\delta > 0$ so that*

$$d_1(x, y) < \delta \Rightarrow d_2(f(x), f(y)) < \epsilon.$$

If f is also one-to-one and onto, and thus invertible, and if also the inverse f^{-1} of f is continuous, then we say that f is a homeomorphism *between X_1 and X_2. In such a case we say that X_1 and X_2 are* homeomorphic.

The assertion that two spaces are equivalent metrically is much stronger than the statement that they are homeomorphic: to be equivalent there must be a bounded relationship between ϵ and δ independent of x. Homeomorphism is the equivalence relationship for topological properties; two spaces that are homeomorphic are identical *topological* spaces. Two metrics d_1 and d_2 on a given space X are identical topologically (define the same topological space) if the *identity map* $\iota : (X, d_1) \to (X, d_2)$ given by $\iota(x) = x$ is a homeomorphism.

Examples & Exercises

2.13. Let $X_1 = [1, 2]$ and $X_2 = [0, 1]$. Let d_1 denote the Euclidean and let $d_2(x, y) = 2 \cdot |x - y|$ in X_2. Show that (X_1, d_1) and (X_2, d_2) are equivalent metric spaces.

2.14. Show that (■, Euclidean) and (■, Manhattan) are equivalent metric spaces.

2.15. Show that ($\mathbb{C}$, Euclidean) and ($\mathbb{R}^2$, Manhattan) are equivalent metric spaces.

2.16. Define two different metrics on the space $X = (0, 1] = \{x \in \mathbb{R} : 0 < x \leq 1\}$ by

$$d_1(x, y) = |x - y| \quad \text{and} \quad d_2(x, y) = \left|\frac{1}{x} - \frac{1}{y}\right|.$$

Show that (X, d_1) and (X, d_2) are not equivalent metric spaces.

2.17. Figure II.9 suggests a subset (black) of (■, Euclidean). It also shows the space and set deformed by a metric equivalence. Discuss the properties of the image that would be invariant under (a) any metric equivalence, and (b) any homeomorphism. To what extent might one be able to "see" these invariances? Think about how much deformation an image can withstand while remaining recognizably the same image. Look at reflections of sets and images in the back of a shiny spoon.

2.18. Show that if two metric spaces are metrically equivalent then there is a homeomorphism between them.

∗2.19. We can define metrics on Σ, our code space, by

$$d_k(x, y) = \sum_{i=1}^{\infty} \frac{|x_i - y_i|}{k^i},$$

where $k > 1$. The most important choices for k are $N + 1$ and N, but other choices (mostly real numbers between these two) are useful. Show that the functions $d_k :$ $\Sigma \times \Sigma \to \mathbb{R}$ are metrics, and that these metrics are identical topologically.

3   Cauchy Sequences, Limit Points, Closed Sets, Perfect Sets, and Complete Metric Spaces

Fractal geometry is concerned with the description, classification, analysis and observation of subsets of metric spaces $(\mathbf{X}, d)$. The metric spaces are usually, but not always, of an inherently "simple" geometrical character; the subsets are typically geometrically "complicated." There are a number of general properties of subsets of metric spaces, which occur over and over again, which are very basic, and which form part of the vocabulary for describing fractal sets and other subsets of metric spaces. Some of these properties, such as openness and closedness, which we are going to introduce, are of a *topological* character. That is to say, they are invariant under homeomorphism.

Figure II.9. What features of the set (black) are invariant under a metric equivalence transformation? Two sets that are metrically equivalent to (a) are shown in (b) and (c).

(a)

Figure II.9. (b)

Figure II.9. (c)

For us what is important, however, is that there is another class of properties that are invariant under metric space equivalence. These include openness, closedness, boundedness, completeness, compactness, and perfection; these properties are introduced in this and the next section. Later we will discover another such property, the fractal dimension of a set. If a subset of a metric space has one of these properties, and the space is deformed with bounded distortion, then the corresponding subsets in the deformed space still have that same property.

We are also about another business in this section. In our search for fractals we are always going to look in a certain type of metric space known as "complete." We need to understand this concept.

Definition 3.1 *A sequence* $\{x_n\}_{n=1}^{\infty}$ *of points in a metric space* $(\mathbf{X}, d)$ *is called a* Cauchy sequence *if, for any given number* $\epsilon > 0$, *there is an integer* $N > 0$ *so that*

$$d(x_n, x_m) < \epsilon \qquad \text{for all } n, m > N.$$

In other words, the further along the sequence one goes, the closer together become the points in the sequence. Mentally one pictures something like the image in Figure II.10.

However, just because a sequence of points moves closer together as one goes

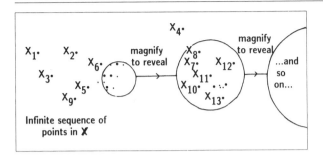

Figure II.10. Image representing successive magnifications on a Cauchy sequence, an infinite sequence of points in **X**. Just because the points are getting closer and closer together as one looks in at higher magnification does not mean that there is a point x to which the sequence is converging!

along the sequence, we must not infer that they are approaching a point. Perhaps they are trying to approach a point that is not there?

Definition 3.2 *A sequence $\{x_n\}_{n=1}^{\infty}$ of points in a metric space $(\mathbf{X}, d)$ is said to* converge *to a point $x \in \mathbf{X}$ if, for any given number $\epsilon > 0$, there is an integer $N > 0$ so that*

$$d(x_n, x) < \epsilon \qquad \text{for all } n > N.$$

In this case the point $x \in \mathbf{X}$, to which the sequence converges, is called the limit *of the sequence, and we use the notation*

$$x = \lim_{n \to \infty} x_n.$$

The limit x of a convergent sequence $\{x_n\}_{n=1}^{\infty}$ has this property: let

$$B(x, \epsilon) = \{y \in \mathbf{X} : d(x, y) \leq \epsilon\}$$

denote a closed ball of radius $\epsilon > 0$ centered at x, as illustrated in Figure II.11.

Any such ball centered at x contains all of the points x_n after some index N, where N typically becomes larger and larger as ϵ becomes smaller and smaller. See Figure II.12.

Theorem 3.1 *If a sequence of points $\{x_n\}_{n=1}^{\infty}$ in a metric space $(\mathbf{X}, d)$ converges to a point $x \in \mathbf{X}$, then $\{x_n\}_{n=1}^{\infty}$ is a Cauchy sequence.*

Figure II.11. Uncelebrated small ball $B(x, \epsilon)$ with its center at x and radius ϵ. Beware! Balls do not usually look like balls. It depends on the metric and on the space. Balls (a)–(c) represent balls (marked in black) in spaces **X** that are subsets of $\mathbb{R}^2$, with the Euclidean metric. In (a) **X** has a ragged boundary, viewed as a subset of $\mathbb{R}^2$. In (b) the point $x \in \mathbf{X}$ is isolated. In (c) **X** is a curvaceous Sierpinski triangle. The ball depicted in (d) is in $\mathbb{R}^2$, but the metric is $d(x, y) = \max\{|x_1 - y_1|, |x_2 - y_2|\}$.

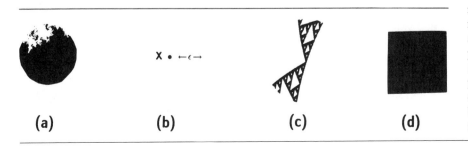

(a)　　　　(b)　　　　(c)　　　　(d)

Figure II.12. Magnifying glass looking at a magnifying glass near a limit point.

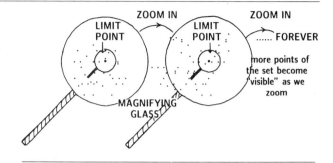

Definition 3.3 *A metric space* $(\mathbf{X}, d)$ *is* complete *if every Cauchy sequence* $\{x_n\}_{n=1}^{\infty}$ *in* $\mathbf{X}$ *has a limit* $x \in \mathbf{X}$.

In other words, there actually exists, in the space, a point x to which the Cauchy sequence is converging. This point x is of course the limit of the sequence. If $\{x_n\}_{n=1}^{\infty}$ is a Cauchy sequence of points in $\mathbf{X}$ and if $\mathbf{X}$ is complete, then there is a point $x \in \mathbf{X}$ such that, for each $\epsilon > 0$, $B(x, \epsilon)$ contains x_n for infinitely many integers n.

We will sometimes use the notation $\{x_n\}$ in place of $\{x_n\}_{n=1}^{\infty}$ and lim in place of $\lim_{n \to \infty}$ when it is clear from the context what the domain of the index is.

Examples & Exercises

3.1. Prove that if $\{x_n\}_{n=1}^{\infty}$ is a Cauchy sequence of points in $\mathbf{X}$ and if $\mathbf{X}$ is complete, then there is a point $x \in \mathbf{X}$ such that, for each $\epsilon > 0$, $B(x, \epsilon)$ contains x_n for infinitely many integers n.

∗3.2. Show that $(\mathbb{R}, \text{Euclidean metric})$ is a complete metric space.

3.3. Show that $(\mathbb{R}^2, \text{Euclidean metric})$ is a complete metric space.

3.4. Show that $(\blacksquare, \text{Euclidean metric})$ is a complete metric space.

3.5. Show that $(\hat{\mathbb{C}}, \text{metric on sphere})$ is a complete metric space.

3.6. Show that $(\Sigma, \text{code space})$ is a complete metric space.

3.7. $(C[0, 1], D)$ is a complete metric space, where the metric D is defined by

$$D(f, g) = \max\{|f(s) - g(s)| : s \in [0, 1]\}.$$

3.8. Let $(\mathbf{X}_1, d_1)$ and $(\mathbf{X}_2, d_2)$ be equivalent metric spaces. Suppose $(\mathbf{X}_1, d_1)$ is complete. Show that $(\mathbf{X}_2, d_2)$ is complete.

3.9. Show that there are many different "shortest paths" between most pairs of points in $(\blacksquare, \text{Manhattan})$.

3.10. Prove Theorem 3.1.

3.11. Prove that any sequence in a metric space can have at most one limit.

Definition 3.4 *Let $S \subset \mathbf{X}$ be a subset of a metric space $(\mathbf{X}, d)$. A point $x \in \mathbf{X}$ is called a* limit point *of S if there is a sequence $\{x_n\}_{n=1}^{\infty}$ of points $x_n \in S \setminus \{x\}$ such that $\lim_{n \to \infty} x_n = x$.*

Definition 3.5 *Let $S \subset \mathbf{X}$ be a subset of a metric space $(\mathbf{X}, d)$. The* closure *of S, denoted $\overline{S}$, is defined to be $\overline{S} = S \cup \{Limit\ points\ of\ S\}$. S is* closed *if it contains all of its limit points, that is, $S = \overline{S}$. S is* perfect *if it is equal to the set of all its limit points.*

Exercises & Examples

3.12. Show that 0 is a limit of the sequence $\{x_n = \frac{1}{n}\}_{n=1}^{\infty}$ in the metric space ([0, 1], Euclidean) but not in the metric space ((0, 1], Euclidean).

3.13. A metric space $(\mathbf{X}, d)$ consists of a single point $\mathbf{X} = \{a\}$, together with a metric defined by $d(a, a) = 0$. Show that $\mathbf{X}$ contains a Cauchy sequence and the limit of the Cauchy sequence, but that it possesses no limit points. Hence show that $\mathbf{X}$ is closed and complete but not perfect.

3.14. Show that the sequence $\{x_n = n\}_{n=1}^{\infty}$ has no limit in ($\mathbb{R}$, Euclidean), but that it does when the points are treated as belonging to ($\hat{\mathbb{C}}$, spherical).

3.15. Show that if $h : \mathbf{X}_1 \to \mathbf{X}_2$ makes the metric spaces $(\mathbf{X}_1, d_1)$ and $(\mathbf{X}_2, d_2)$ metrically equivalent, then the statements "$x \in \mathbf{X}_1$ is a limit point of $S \subset \mathbf{X}_1$" and "$h(x) \in \mathbf{X}_2$ is a limit point of $h(S) \subset \mathbf{X}_2$" are equivalent. Here we use the notation $h(S) = \{h(s) : s \in S\}$.

3.16. Find all of the limit points of the set $\{x_n = (\frac{1}{n} + (-1)^n, \frac{1}{n} + (-1)^{2n}) : n = 1, 2, 3, \ldots\}$ in the metric space ($\blacksquare$, Euclidean).

3.17. Show that the subset $S = \{x = \frac{1}{n} : n = 1, 2, 3, \ldots\}$ is closed in ((0, 1], Euclidean) .

3.18. Show that $S = [0, 1]$ is a perfect subset of ($\mathbb{R}$, Euclidean).

3.19. Show that $S = \{\frac{1}{n} : n = 1, 2, 3, \ldots\} \cup \{0\}$ is not a *perfect* subset of ($\mathbb{R}$, Euclidean), but that $S = \overline{S}$.

3.20. Show that $S = \Sigma$ is a perfect subset of (Σ, code space metric).

3.21. Let S be a subset of a complete metric space $(\mathbf{X}, d)$. Then (S, d) is a metric space. Show that (S, d) is complete if, and only if, S is closed in $\mathbf{X}$.

---

4   Compact Sets, Bounded Sets, Open Sets, and Boundaries

We continue the description of basic properties that are to be used to describe sets and subsets of metric spaces. Where are the fractals? What are they? They are everywhere and soon you will be able to see them. Not just the pictures, which are shadows of fractals, but in your mind's eye you will find what and where they really *are*.

Definition 4.1 *Let $S \subset X$ be a subset of a metric space (X, d). S is* compact *if every infinite sequence $\{x_n\}_{n=1}^{\infty}$ in S contains a subsequence having a limit in S.*

Definition 4.2 *Let $S \subset X$ be a subset of a metric space (X, d). S is* bounded *if there is a point $a \in X$ and a number $R > 0$ so that*

$$d(a, x) < R \forall x \in X.$$

Definition 4.3 *Let $S \subset X$ be a subset of a metric space (X, d). S is* totally bounded *if, for each $\epsilon > 0$, there is a finite set of points $\{y_1, y_2, \ldots, y_n\} \subset S$ such that whenever $x \in X$, $d(x, y_i) < \epsilon$ for some $y_i \in \{y_1, y_2, \ldots, y_n\}$. This set of points $\{y_1, y_2, \ldots, y_n\}$ is called an ϵ-net.*

Theorem 4.1 *Let (X, d) be a complete metric space. Let $S \subset X$. Then S is compact if and only if it is closed and totally bounded.*

Proof Suppose that S is closed and totally bounded. Let $\{x_i \in S\}$ be an infinite sequence of points in S. Since S is totally bounded we can find a finite collection of closed balls of radius 1 such that S is contained in the union of these balls. By the Pigeon-Hole Principle (a huge number of pigeons laying eggs in two letter boxes $\Rightarrow$ at least one letter box contains a huge number of angry pigeons), one of the balls, say B_1, contains infinitely many of the points x_n. Choose N_1 so that $x_{N_1} \in B_1$. It is easy to see that $B_1 \cap S$ is totally bounded. So we can cover $B_1 \cap S$ by a finite set of balls of radius $1/2$. By the Pigeon-Hole Principle, one of the balls, say B_2, contains infinitely many of the points x_n. Choose N_2 so that $x_{N_2} \in B_2$ and $N_2 > N_1$. We continue in this fashion to construct a *nested* sequence of balls,

$$B_1 \supset B_2 \supset B_3 \supset B_4 \supset B_5 \supset B_6 \supset B_7 \supset B_8 \supset B_9 \supset \cdots \supset B_n \supset \cdots$$

where B_n has radius $\frac{1}{2^{n-1}}$ and a sequence of integers $\{N_n\}_{n=1}^{\infty}$ such that $x_{N_n} \in B_n$. It is easy to see that $\{x_{N_n}\}_{n=1}^{\infty}$, which is a subsequence of the original sequence $\{x_n\}$, is a Cauchy sequence in S. Since S is closed, and X is a complete metric space, S is complete as well; see exercise 4.2. So $\{x_n\}$ converges to a point x in S. (Notice that x is exactly $\bigcap_{n=1}^{\infty} B_n$.) Thus, S is compact.

Conversely, suppose that S is compact. Let $\epsilon > 0$. Suppose that there does not exist an ϵ-net for S. Then there is an infinite sequence of points $\{x_n \in S\}$ with $d(x_i, x_j) \geq \epsilon$ for all $i \neq j$. But this sequence must possess a convergent subsequence $\{x_{N_i}\}$. By Theorem 3.1 this sequence is a Cauchy sequence, and so we can find a pair of integers N_i and N_j with $N_i \neq N_j$ so that $d(x_{N_i}, x_{N_j}) < \epsilon$. But $d(x_{N_i}, x_{N_j}) \geq \epsilon$, so we have a contradiction. Thus there *does* exist an ϵ-net. This completes the proof.

Definition 4.4 *Let $S \subset X$ be a subset of a metric space (X, d). S is* open *if for each $x \in S$ there is an $\epsilon > 0$ such that $B(x, \epsilon) = \{y \in X : d(x, y) \leq \epsilon\} \subset S$.*

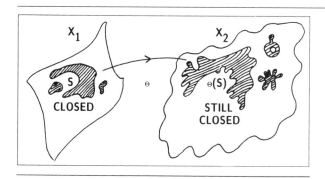

Figure II.13. A transformation θ between two metric spaces, establishing the equivalence of the spaces and carrying the closed set S onto a closed set $\theta(S)$.

Examples & Exercises

4.1. Show that if $(\mathbf{X}, d)$ is a metric space then $\mathbf{X}$ is closed. Give an example of a metric space that is closed but not complete.

4.2. Let S be a closed subset of a complete metric space $(\mathbf{X}, d)$. Show that (S, d) is a complete metric space.

4.3. Let $(\mathbf{X}_1, d_1)$ and $(\mathbf{X}_2, d_2)$ be equivalent metric spaces, and let a transformation $\theta : \mathbf{X}_1 \to \mathbf{X}_2$ provide this equivalence. Let $S \subset \mathbf{X}_1$ be closed. Show that $\theta(S) = \{\theta(s) : s \in S\}$ is closed. This idea is illustrated in Figure II.13.

4.4. If $(\mathbf{X}, d)$ is a metric space, then $\mathbf{X}$ is open.

Proof Let $x \in \mathbf{X}$. Clearly $B(x, 1) \subset \mathbf{X}$.

4.5. If $(\mathbf{X}, d)$ is a metric space, then "$S \subset \mathbf{X}$ is open" is the same as "$\mathbf{X} \setminus S$ is closed."

Proof Suppose "$S \subset \mathbf{X}$ is open." Suppose $\{x_n\}$ is a sequence in $\mathbf{X} \setminus S$ with a limit $x \in \mathbf{X}$. We must show that $x \in \mathbf{X} \setminus S$. Assume that $x \in S$. Then every ball $B(x, \epsilon)$ with $\epsilon > 0$ contains a point $x_n \in \mathbf{X} \setminus S$, which means that S is not open. This is a contradiction. The assumption is false. Therefore $x \in \mathbf{X} \setminus S$. Therefore "$\mathbf{X} \setminus S$ is closed."

Suppose "$\mathbf{X} \setminus S$ is closed." Let $x \in S$. We want to show there is a ball $B(x, \epsilon) \subset S$. Assume there is no ball $B(x, \epsilon) \subset S$. Then for every integer $n = 1, 2, 3, \ldots$, we can find a point $x_n \in B(x, \frac{1}{n}) \cap (\mathbf{X} \setminus S)$. Clearly $\{x_n\}$ is a sequence in $\mathbf{X} \setminus S$, with limit $x \in \mathbf{X}$. Since $\mathbf{X} \setminus S$ is closed we conclude that $x \in \mathbf{X} \setminus S$. This contradicts $x \in S$. The assumption that there is no ball $B(x, \epsilon) \subset S$ is false. Therefore there is a ball $B(x, \epsilon) \subset S$. Therefore "S is open."

4.6. Every bounded subset S of $(\mathbb{R}^2, \text{Euclidean})$ has the Bolzano-Weierstrass property: "Every infinite sequence $\{x_n\}_{n=1}^\infty$ of points of S contains a subsequence which is a Cauchy sequence." The proof is suggested by the picture in Figure II.14.

We deduce that every closed bounded subset of $(\mathbb{R}^2, \text{Euclidean})$ is compact. In particular, every metric space of the form (closed bounded subset of $\mathbb{R}^2$, Euclidean)

Figure II.14. Demonstration of the Bolzano-Weierstrass Theorem. (Government warning: This is not a proof.)

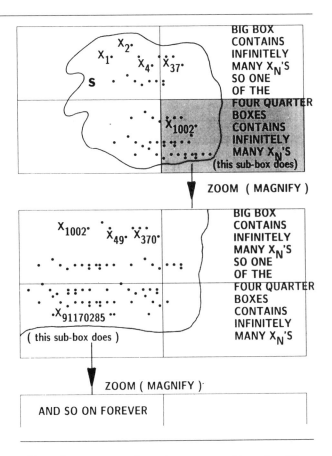

is a complete metric space. Show that we can make a rigorous proof by using Theorem 4.1. Begin by proving that any bounded subset of $\mathbb{R}^n$ is totally bounded.

4.7. Let $(\mathbf{X}, d)$ be a metric space. Let $f : \mathbf{X} \to \mathbf{X}$ be continuous. Let A be a compact nonempty subset of $\mathbf{X}$. Show that $f(A)$ is a compact nonempty subset of $\mathbf{X}$. (This result is proved later as Lemma 7.2 of Chapter III.)

4.8. Let $S \subset (\mathbf{X}_1, d_1)$ be open, and let $(\mathbf{X}_2, d_2)$ be a metric space equivalent to $(\mathbf{X}_1, d_1)$, the equivalence being provided by a function $h : \mathbf{X}_1 \to \mathbf{X}_2$. Show that $h(S)$ is an open subset of $\mathbf{X}_2$.

4.9. Let $(\mathbf{X}, d)$ be a metric space. Let $C \subset \mathbf{X}$ be a compact subset of $\mathbf{X}$. Let $\{C_n : n = 1, 2, 3, \ldots\}$ be a set of open subsets of $\mathbf{X}$ such that "$x \in C$" implies "$x \in C_n$ for some n." $\{C_n\}$ is called a countable open cover of C. Show that there is a finite integer N so that "$x \in C$" implies "$x \in C_n$ for some integer $n < N$."

   Proof Assume that an integer N does not exist such that "$x \in C$" implies "$x \in C_n$ for some $n < N$." Then for each N we can find

$$x_N \in C \setminus \bigcup_{n=1}^{N} C_n.$$

Since $\{x_N\}_{N=1}^{\infty}$ is in C it possesses a subsequence with a limit $y \in C$. Clearly y does not belong to any of the subsets C_n. Hence "$y \in C$" does not imply "$y \in C_n$ for some integer n." We have a contradiction. This completes the proof.

The following even stronger statement is true. Let $(\mathbf{X}, d)$ be a metric space. Let $C \subset \mathbf{X}$ be compact. Let $\{C_i : i \in I\}$ denote *any* collection of open sets such that whenever $x \in C$, it is true that $x \in C_i$ for some index $i \in I$. Then there is a *finite* sub-collection, say $\{C_1, C_2, \ldots, C_n\}$ such that $C \subset \bigcup_{i=1}^{N} C_i$. The point is that the original collection of open sets need not even be countably infinite. A good discussion of compactness in metric spaces can be found in [Mendelson 1963], Chapter V.

4.10. Let $\mathbf{X} = (0, 1) \cup \{2\}$. That is, $\mathbf{X}$ consists of an open interval in $\mathbb{R}$, together with an "isolated" point. Show that the subsets $(0, 1)$ and $\{2\}$ of $(\mathbf{X}, \text{Euclidean})$ are open. Show that $(0, 1)$ is closed in $\mathbf{X}$. Show that $\{2\}$ is closed in $\mathbf{X}$. Show that $\{2\}$ is compact in $\mathbf{X}$ but $(0, 1)$ is not compact in $\mathbf{X}$.

Definition 4.5 *Let $S \subset \mathbf{X}$ be a subset of a metric space $(\mathbf{X}, d)$. A point $x \in \mathbf{X}$ is a* boundary point *of S if for every number $\epsilon > 0$, $B(x, \epsilon)$ contains a point in $\mathbf{X} \setminus S$ and a point in S. The set of all boundary points of S is called the* boundary *of S and is denoted ∂S.*

Definition 4.6 *Let $S \subset \mathbf{X}$ be a subset of a metric space $(\mathbf{X}, d)$. A point $x \in S$ is called an* interior point *of S if there is a number $\epsilon > 0$ such that $B(x, \epsilon) \subset S$. The set of interior points of S is called the* interior *of S and is denoted S^0.*

Examples & Exercises

4.11. Let S be a subset of a metric space $(\mathbf{X}, d)$. Show that $\partial S = \partial(\mathbf{X} \setminus S)$. Deduce that $\partial \mathbf{X} = \emptyset$.

4.12. Show that the property of being a boundary of a set is invariant under metric equivalence.

4.13. Let $(\mathbf{X}, d)$ be the real line with the Euclidean metric. Let S denote the set of all rational points in $\mathbf{X}$ (i.e., real numbers that can be written $\frac{p}{q}$ where p and q are integers with $q \neq 0$). Show that $\partial S = \mathbf{X}$.

4.14. Find the boundary of $\mathbb{C}$ viewed as a subset of $(\hat{\mathbb{C}}, \text{spherical metric})$.

4.15. Let S be a closed subset of a metric space. Show that $\partial S \subset S$.

4.16. Let S be an open subset of a metric space. Show that $\partial S \cap S = \emptyset$.

4.17. Let S be an open subset of a metric space. Show that $S^0 = S$. Conversely, show that if $S^0 = S$ then S is open.

4.18. Let S be a closed subset of a metric space. Show that $S = S^0 \cup \partial S$.

Figure II.15. How well can topological concepts such as open, boundary, etc., be used to model land, sea, and coastlines?

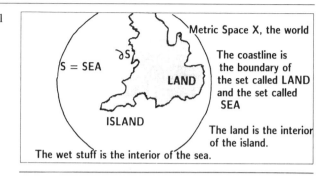

Metric Space X, the world

The coastline is the boundary of the set called LAND and the set called SEA

The land is the interior of the island.

The wet stuff is the interior of the sea.

4.19. Show that the property of being the interior of a set is invariant under metric equivalence.

4.20. Show that the boundary of a set S in a metric space always divides the space into two disjoint open sets whose union, with the boundary ∂S, is the whole space. Illustrate this result in the following cases, in the metric space ($\mathbb{R}^2$, Euclidean): (a) $S = \{(x, y) \in \mathbb{R}^2 : x^2 + y^2 < 1\}$; (b) $S = \mathbb{R}^2$.

4.21. Show that the boundary of a set is closed.

✳4.22. Let S be a subset of a compact metric space. Show that ∂S is compact.

4.23. Figure II.15 shows how we think of boundaries and interiors. What features of the picture are misleading?

4.24. To what extent does Mercator's projection provide a metric equivalence to a Cartesian map of the world?

4.25. Locate the boundary of the set of points marked in black in Figure II.16.

4.26. Prove the assertion made in the caption to Figure II.17.

5   Connected Sets, Disconnected Sets, and Pathwise-Connected Sets

Definition 5.1 *A metric space* $(\mathbf{X}, d)$ *is* connected *if the only two subsets of* $\mathbf{X}$ *that are simultaneously open and closed are* $\mathbf{X}$ *and* $\emptyset$. *A subset* $S \subset \mathbf{X}$ *is* connected *if the metric space* (S, d) *is connected.* S *is* disconnected *if it is not connected.* S *is* totally disconnected *provided that the only nonempty connected subsets of* S *are subsets consisting of single points.*

Definition 5.2 *Let* $S \subset \mathbf{X}$ *be a subset of a metric space* $(\mathbf{X}, d)$. *Then* S *is* pathwise-connected *if, for each pair of points* x *and* y *in* S, *there is a continuous function* $f : [0, 1] \to S$, *from the metric space* $([0, 1]$, *Euclidean) into the metric space* (S, d), *such that* $f(0) = x$ *and* $f(1) = y$. *Such a function* f *is called a* path *from* x *to* y *in* S. S *is* pathwise-disconnected *if it is not pathwise-connected.*

Figure II.16. Should the black part be called open and the white part closed? Locate the boundary of the set of points marked in black.

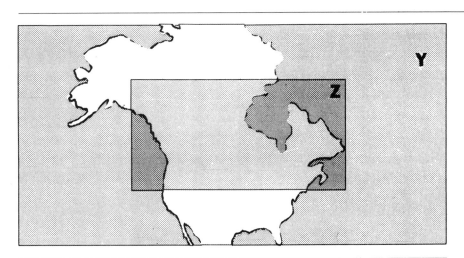

Figure II.17. The interior of the "land" set is an open set in the metric space ($Y =$ ▨ , Euclidean). The smaller filled rectangle denotes a subset $Z =$ ■ of Y. The intersection of the interior of the land with Z is an open set in the metric space (Z, Euclidean), despite the fact that it includes some points of the "border" of ■.

One can also define *simply connected* and *multiply connected*. Let S be pathwise-connected. A pair of points x, $y \in S$ is *simply connected* in S if, given any two paths f_0 and f_1 connecting x, y in S, we can continuously deform f_0 to f_1 without leaving the subset S. What does this mean?

Let there be the two points x, $y \in S$ and the two paths f_0, f_1 connecting x, y in S. In other words, f_0, f_1 are two continuous functions mapping the unit interval $[0, 1]$ into S so that $f_0(0) = f_1(0) = x$ and $f_0(1) = f_1(1) = y$. By a *continuous deformation* of f_0 into f_1 within S we mean a function g continuously mapping the Cartesian product $[0, 1] \times [0, 1]$ into S, so that

(a) $g(s, 0) = f_0(s)(0 \le s \le 1)$
(b) $g(s, 1) = f_1(s)(0 \le s \le 1)$
(c) $g(0, t) = x(0 \le t \le 1)$
(d) $g(1, t) = y(0 \le t \le 1)$

Thus, we say that two points x, y in S are *simply connected* in S if, given any two paths f_0, f_1 going from x to y in S, there exists a function g as just described. This idea is illustrated in Figure II.18.

If x, y are not simply connected in S, then we say that x, y are *multiply connected* in S. S itself is called simply connected if every pair of points x, y in S is simply connected in S. Otherwise, S is called multiply connected. In the latter case we can imagine that S contains a "hole," as illustrated in Figure II.19.

Examples & Exercises

∗5.1. Show that the properties of being (pathwise) connected, disconnected, simply connected, and multiply connected are invariant under metric equivalence.

5.2. Show that the metric space (■, Euclidean) is simply connected.

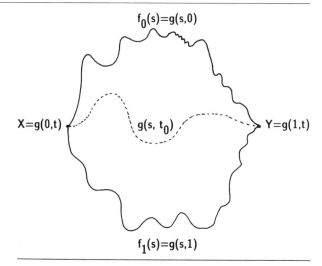

Figure II.18. A path f_0 which connects the points x and y is continuously deformed, while remaining "attached" to x and y, to become a second path f_1.

$f_0(s)=g(s,0)$

$X=g(0,t)$ $g(s, t_0)$ $Y=g(1,t)$

$f_1(s)=g(s,1)$

5.3. Show that the metric space ($\mathbf{X} = (0, 1) \cup \{2\}$, Euclidean) is disconnected.

5.4. Show that the metric space (Σ, code space metric) is totally disconnected.

∗5.5. Show that the metric space (⬤, Manhattan) is multiply connected.

5.6. Suppose $S_1 \supset S_2 \supset \cdots \supset S_n \supset \cdots$ is a nested sequence of nonempty connected subsets. Is $\bigcap_{n=1}^{\infty} S_n$ necessarily connected?

5.7. Identify pathwise connected subsets of the metric space suggested in Figure II.20.

5.8. Is (🧍, Euclidean) simply or multiply connected?

5.9. Discuss which set-theoretic properties (open, closed, connected, compact, bounded, etc.) would be best suited for a model of a cloud, treated as a subset of $\mathbb{R}^3$.

5.10. The property that $\{x_n\}_{n=1}^{\infty}$ is a Cauchy sequence in the metric space ($\mathbf{X}, d$) is not invariant under homeomorphism but is invariant under metric equivalence, as illustrated in Figure II.21.

---

6   The Metric Space ($\mathcal{H}(X), h$): The Space Where Fractals Live

We come to the ideal space in which to study fractal geometry. To start with, and always at the deepest level, we work in some complete metric space such as ($\mathbb{R}^2$, Euclidean) or ($\hat{\mathbb{C}}$, spherical), which we denote by ($\mathbf{X}, d$). But then, when we wish to discuss pictures, drawings, "black-on-white" subsets of the space, it becomes natural to introduce the space $\mathcal{H}$.

Definition 6.1 *Let* ($\mathbf{X}, d$) *be a complete metric space. Then* $\mathcal{H}(X)$ *denotes the space whose points are the compact subsets of* $\mathbf{X}$, *other than the empty set.*

Examples & Exercises

6.1. Show that if x and $y \in \mathcal{H}(X)$ then $x \cup y$ is in $\mathcal{H}(X)$. Show that $x \cap y$ need not be in $\mathcal{H}(X)$. A picture of this situation is given in Figure II.22.

6.2. What is the difference between a subset of $\mathcal{H}(X)$ and a compact nonempty subset of $\mathbf{X}$?

Definition 6.2 *Let* ($\mathbf{X}, d$) *be a complete metric space,* $x \in \mathbf{X}$, *and* $B \in \mathcal{H}(X)$. *Define*

$$d(x, B) = \min\{d(x, y) : y \in B\}.$$

$d(x, B)$ *is called the* distance from *the point* x *to the set* B.

Figure II.19. In a multiply connected space there exist paths that cannot be continuously deformed from one to another. There is some kind of "hole" in the space.

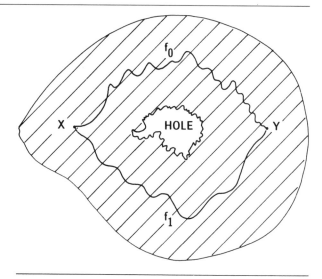

Figure II.20. Locate the largest connected subsets of this subset of $\mathbb{R}^2$.

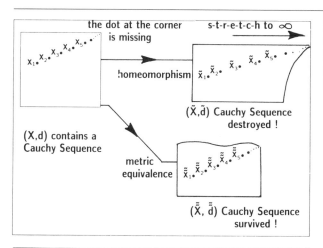

Figure II.21. A Cauchy sequence being preserved by a metric equivalence and destroyed by a certain homeomorphism.

How do we know that the set of real numbers $\{d(x, y) : y \in B\}$ contains a minimum value, as claimed in the definition? This follows from the compactness and nonemptyness of the set $B \in \mathcal{H}(\mathbf{X})$. Consider the function $f : B \to \mathbb{R}$ defined by

$$f(y) = d(x, y) \qquad \text{for all } y \in B.$$

From the definition of the metric it follows that f is continuous, viewed as a transformation from the metric space (B, d) to the metric space $(\mathbb{R}, \text{Euclidean})$. Let $P = \inf\{f(y) : y \in B\}$, where "inf" is defined in exercise 6.19, and also in Definition 6.2 in Chapter III. Since $f(y) \geq 0$ for all $y \in B$, it follows that P is finite. We claim there is a point $\hat{y} \in B$ such that $d(x, \hat{y}) = P$. We can find an infinite sequence of points $\{y_n : n = 1, 2, 3, \ldots\} \subset B$ such that $f(y_n) - P < \frac{1}{n}$ for each positive integer n. Using the compactness of B, we find that $\{y_n : n = 1, 2, 3 \ldots\}$ has a limit $\hat{y} \in B$. Using the continuity of f we discover that $f(\hat{y}) = P$, which is what we needed to show.

Color Plate 2 shows a picture of the metric space ($\blacksquare$, Manhattan). It has been colored as follows. Let $\mathcal{F}$ denote a certain subset of $\blacksquare$ whose "geometry" is that of a piece of a fern. Then the color of each point $a \in \blacksquare$ is fixed by the value of $d(a, \mathcal{F})$.

Definition 6.3 *Let* $(\mathbf{X}, d)$ *be a complete metric space. Let* $A, B \in \mathcal{H}(\mathbf{X})$. *Define*

$$d(A, B) = \max\{d(x, B) : x \in A\}.$$

$d(A, B)$ *is called the* distance *from* the set $A \in \mathcal{H}(\mathbf{X})$ *to the set* $B \in \mathcal{H}(\mathbf{X})$.

Just as above, using the compactness of A and B, we can prove that this definition is meaningful. In particular, there are points $\hat{x} \in A$ and $\hat{y} \in B$ such that $d(A, B) = d(\hat{x}, \hat{y})$.

Examples & Exercises

6.3. Show that $B, C \in \mathcal{H}(\mathbf{X})$, with $B \subset C$, implies $d(x, C) \leq d(x, B)$.

6.4. Calculate $d(x, B)$ if $(\mathbf{X}, d)$ is the space $(\mathbb{R}^2, \text{Euclidean})$, $x \in \mathbb{R}^2$ is the point $(1, 1)$, and B is a closed disk of radius $\frac{1}{2}$ centered at the point $(\frac{1}{2}, 0)$.

6.5. Same as 6.4 above, but use the Manhattan metric.

6.6. Calculate $d(x, B)$ if $(\mathbf{X}, d)$ is $(\mathbb{R}, \text{Euclidean})$, $x = \frac{1}{2}$, and

$$B = \{x_n = 3 + (-1)^n \frac{n}{n^2 + 1} : n = 1, 2, 3, \ldots\} \cup \{3\}.$$

6.7. Let $A, B \in \mathcal{H}(\mathbf{X})$, where $(\mathbf{X}, d)$ is a metric space. Show that, in general, $d(A, B) \neq d(B, A)$. Conclude that d does not provide a metric on $\mathcal{H}(\mathbf{X})$. It is not symmetrical: the distance from A to B need not equal the distance from B to A.

6.8. Figure II.23 shows two subsets A and B of ($\blacksquare \subset \mathbb{R}^2$, Euclidean). A is the white part and B is the black part. (a) Estimate the location of a pair of points, $x \in A$ and $y \in B$, such that $d(x, y) = d(A, B)$. (b) Estimate the location of a pair of points, $x \in A$ and $y \in B$, such that $d(x, y) = d(B, A)$.

6.9. Figure II.24 shows two fern-like subsets, A and B, of $(\mathbb{R}^2, \text{Manhattan})$. Locate points $x \in A$ and $y \in B$ such that: (a) $d(x, y) = d(A, B)$; (b) $d(x, y) = d(B, A)$.

Figure II.22. Points in the space $\mathcal{H}(\mathbb{R}^2)$ may be interpreted as black-and-white images. Unions of points yield new points. Be careful with intersections, however.

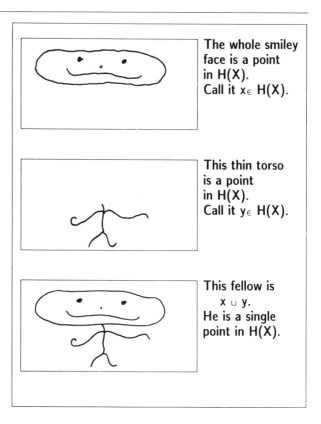

The whole smiley face is a point in H(X). Call it x∈ H(X).

This thin torso is a point in H(X). Call it y∈ H(X).

This fellow is x ∪ y. He is a single point in H(X).

Figure II.23. This fractal image contains a pair of disjoint subsets of $\blacksquare \subset \mathbb{R}^2$, "black" and "white." Let A denote the closure of the set in black and let B denote the closure of its complement. Find a pair of points $x \in A$ and $y \in B$, such that $d(x, y) = d(A, B)$. Find a pair of points $\tilde{x} \in A$ and $\tilde{y} \in B$ such that $d(\tilde{x}, \tilde{y}) = d(B, A)$. Why do we "close" the sets before we begin?

6.10. Find d(France, U.S.A.) and d(U.S.A., France) on ($\hat{\mathbb{C}}$, spherical metric). Which is larger? Also compare d(Georgia, U.S.A.) to d(U.S.A., Georgia).

6.11. Let ($\mathbf{X}$, d) be a complete metric space. Let A and B be points in $\mathcal{H}(\mathbf{X})$ such that $A \neq B$. Show that either $d(A, B) \neq 0$ or $d(B, A) \neq 0$. Show that if $A \subset B$ then $d(A, B) = 0$.

6.12. Let ($\mathbf{X}$, d) be a complete metric space. Show that if A, B and $C \in \mathcal{H}(\mathbf{X})$ then $B \subset C \Rightarrow d(A, C) \leq d(A, B)$. (Hint: Use 6.3.)

6.13. Let ($\mathbf{X}$, d) be a complete metric space. Show that if A, B, and $C \in \mathcal{H}(\mathbf{X})$ then

Figure II.24. Find a pair of points $\hat{x}$ and $\hat{y}$, one in the dark fern and one in the pale fern, such that the Hausdorff distance between the two fern images is the same as the distance between the points.

$$d(A \cup B, C) = d(A, C) \vee d(B, C).$$

We use the notation $x \vee y$ to mean the maximum of the two real numbers x and y.

Proof $d(A \cup B, C) = \max\{d(x, C) : x \in A \cup B\} = \max\{d(x, C) : x \in A\} \vee \max\{d(x, C) : x \in B\}$.

6.14. Let A, B, and C belong to $\mathcal{H}(\mathbf{X})$, where $(\mathbf{X}, d)$ is a metric space. Show that

$$d(A, B) \leq d(A, C) + d(C, B).$$

Also determine whether or not the inequality

$$d(A, B) \leq d(C, A) + d(C, B)$$

is true in general.

Definition 6.4 *Let* $(\mathbf{X}, d)$ *be a complete metric space. Then the Hausdorff distance between points A and B in* $\mathcal{H}(\mathbf{X})$ *is defined by*

$$h(A, B) = d(A, B) \vee d(B, A).$$

Examples & Exercises

6.15. Show that h is a metric on the space $\mathcal{H}(\mathbf{X})$.

Proof Let $A, B, C \in \mathcal{H}(\mathbf{X})$. Clearly $h(A, A) = d(A, A) \vee d(A, A) = d(A, A) = \max\{d(x, A) : x \in A\} = 0$. $h(A, B) = d(a, b)$ for some $a \in A$ and $b \in B$, using the compactness of A and B. Hence $0 \leq h(A, B) < \infty$. If $A \neq B$ we can assume there is an $a \in A$ so that $a \notin B$. It follows that $h(A, B) \geq d(A, B) > 0$. To show that $h(A, B) \leq h(A, C) + h(C, B)$ we first show that $d(A, B) \leq d(A, C) + d(C, B)$. We have for any $a \in A$

$$d(a, B) = \min\{d(a, b) : b \in B\}$$
$$\leq \min\{d(a, c) + d(c, b) : b \in B\} \forall c \in C,$$
$$= d(a, c) + \min\{d(c, b) : b \in B\} \forall c \in C, \text{ so}$$
$$d(a, B) \leq \min\{d(a, c) : c \in C\} + \max\{\min\{d(c, b) : b \in B\} : c \in C\},$$
$$= d(a, C) + d(C, B), \text{ so}$$
$$d(A, B) = d(A, C) + d(C, B).$$

Similarly,

$$d(B, A) \leq d(B, C) + d(C, A), \text{ whence}$$
$$h(A, B) = d(A, B) \vee d(B, A) \leq d(B, C) \vee d(C, B) + d(A, C) \vee d(C, A)$$
$$= h(B, C) + h(A, C), \text{ as desired.}$$

6.16. Show that $h(A \cup B, C \cup D) \leq h(A, C) \vee h(B, D)$, for all $A, B, C,$ and $D \in \mathcal{H}(\mathbf{X})$.

6.17. Show that $h(A, B) = d(a, b)$ for some $a \in B$ and $b \in B$.

6.18. The same situation as in 6.9, but this time locate a pair of points $\hat{x} \in A$ and $\hat{y} \in B$ such that $d(\hat{x}, \hat{y}) = h(A, B)$, the Hausdorff distance from A to B.

6.19. Let $S \subset \mathbb{R}$, with $S \neq \emptyset$. The *supremum* of S is denoted by sup S. The *infinum* of S is denoted by inf S. If there is no real number greater than all of the numbers in S, then sup $S = +\infty$; otherwise sup $S = \min\{x \in \mathbb{R} : x \geq s \forall s \in S\}$. If there is no real number less than all of the numbers in S, then inf $S = -\infty$; otherwise inf $S = \max\{x \in \mathbb{R} : x \leq s \forall s \in S\}$. Show that sup S and inf S are well defined. Show that if S is compact then sup $S = \max S$ and inf $S = \min S$. Further exercises on sup and inf are given following Definition 6.2 in Chapter III.

By replacing max by sup and min by inf, respectively, throughout the definition of the Hausdorff metric, define a "distance" between arbitrary pairs of subsets of a metric space. Give several reasons why this "distance" is not usually a metric.

7 The Completeness of the Space of Fractals

We refer to $(\mathcal{H}(\mathbf{X}), h)$ as the "space of fractals." It is too soon to be formal about the exact meaning of a "fractal." At the present stage of development of science and mathematics, the idea of a fractal is most useful as a broad concept. Fractals are not defined by a short legalistic statement, but by the many pictures and contexts that refer to them. For the first eight chapters of this book, any subset of $(\mathcal{H}(\mathbf{X}), h)$ is a fractal. However, as with the concept of a "space," more meaning is suggested than is formalized.

In this section our principal goal is to establish that the space of fractals $(\mathcal{H}(\mathbf{X}), h)$ is a complete metric space. We also want to characterize convergent sequences in $\mathcal{H}(\mathbf{X})$. To achieve these goals using only the tools introduced so far is quite difficult. Indeed, at this juncture, we want to introduce another notion; namely, the idea of *extending* certain Cauchy subsequences.

Definition 7.1 *Let $S \subset \mathbf{X}$ and let $\Gamma \geq 0$. Then $S + \Gamma = \{y \in \mathbf{X} : d(x, y) \leq \Gamma$ for some $x \in S\}$. $S + \Gamma$ is sometimes called, for example, in the theory of set morphology, the* dilation *of S by a ball of radius Γ.*

Lemma 7.1 *Let A and B belong to $\mathcal{H}(\mathbf{X})$ where $(\mathbf{X}, d)$ is a metric space. Let $\epsilon > 0$. Then*

$$h(A, B) \leq \epsilon \Leftrightarrow A \subset B + \epsilon \quad \text{and} \quad B \subset A + \epsilon.$$

Proof Begin by showing that $d(A, B) \leq \epsilon \Leftrightarrow A \subset B + \epsilon$. Suppose $d(A, B) \leq \epsilon$. Then $\max\{d(a, B) : a \in A\} \leq \epsilon$ implies $d(a, B) \leq \epsilon$ for all $a \in A$. Hence for each $a \in A$ we have $a \in B + \epsilon$. Hence "$A \subset B + \epsilon$." Suppose "$A \subset B + \epsilon$." Consider $d(A, B) = \max\{d(a, B) : a \in A\}$. Let $a \in A$. Since $A \subset B + \epsilon$, there is $a, b \in B$ so that $d(a, b) \leq \epsilon$ for all $a \in A$. Hence $d(a, B) \leq \epsilon$. This is true for each $a \in A$. So "$d(A, B) \leq \epsilon$." This completes the proof.

Let $\{A_n : n = 1, 2, \ldots, \infty\}$ be a Cauchy sequence of sets in $(\mathcal{H}(\mathbf{X}), h)$. That is, given $\epsilon > 0$, there is N so that $n, m \geq N$ implies

$$A_n + \epsilon \supset A_m \text{ and } A_m + \epsilon \supset A_n,$$

i.e., $h(A_n, A_m) \leq \epsilon$. We are concerned with Cauchy sequences $\{x_n\}_{n=1}^{\infty}$ in $\mathbf{X}$ with the property that $x_n \in A_n$ for each n. In particular, we need the following property that allows the *extension* of a Cauchy *subsequence* $\{x_{n_j} \in A_{n_j}\}_{j=1}^{\infty}$, with the property that $x_{n_j} \in A_{n_j}$ for each j, to a Cauchy sequence $\{x_n \in A_n\}_{n=1}^{\infty}$.

Lemma 7.2 The Extension Lemma. *Let $(\mathbf{X}, d)$ be a metric space. Let $\{A_n : n = 1, 2, \ldots, \infty\}$ be a Cauchy sequence of points in $(\mathcal{H}(\mathbf{X}), h)$. Let $\{n_j\}_{j=1}^{\infty}$ be an infinite sequence of integers*

$$0 < n_1 < n_2 < n_3 < \cdots.$$

Suppose that we have a Cauchy sequence $\{x_{n_j} \in A_{n_j} : j = 1, 2, 3, \ldots\}$ in $(\mathbf{X}, d)$. Then there is a Cauchy sequence $\{\tilde{x}_n \in A_n : n = 1, 2, \ldots\}$ such that $\tilde{x}_{n_j} = x_{n_j}$, for all $j = 1, 2, 3, \ldots$.

Proof We give the construction of the sequence $\{\tilde{x}_n \in A_n : n = 1, 2, \ldots\}$. For each $n \in \{1, 2, \ldots, n_1\}$, choose $\tilde{x}_n \in \{x \in A_n : d(x, x_{n_1}) = d(x_{n_1}, A_n)\}$. That is, $\tilde{x}_n$ is the closest point (or one of the closest points) in A_n to x_{n_1}. The existence of such a

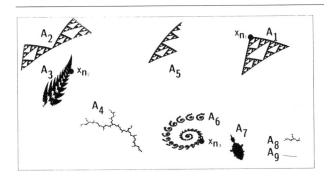

Figure II.25. The beginning of a Cauchy sequence $\{A_n\}$ of sets in $\mathcal{H}(\mathbb{R}^2)$ is shown. A Cauchy subsequence of points $\{x_{n_i}\}$ belonging to a subsequence of the sets is also indicated. Make a photocopy of the figure, and mark on it the extension of the subsequence of points to the visible sets in $\{A_n\}$.

closest point is ensured by the compactness of A_n. Similarly, for each $j \in \{2, 3, \ldots\}$ and each $n \in \{n_j + 1, \ldots, n_{j+1}\}$, choose $\tilde{x}_n \in \{x \in A_n : d(x, x_{n_1}) = d(x_{n_j}, A_n)\}$.

Now we show that $\{\tilde{x}_n\}$ has the desired properties, that it is indeed an extension of $\{x_{n_j}\}$ to $\{A_n\}$. Clearly $\tilde{x}_{n_j} = x_{n_j}$ and $x_n \in A_n$, by construction. To show that it is a Cauchy sequence let $\epsilon > 0$ be given. There is N_1 so that $n_k, n_j \geq N_1$ implies $d(x_{n_k}, x_{n_j}) \leq \epsilon/3$. There is N_2 so that $m, n \geq N_2$ implies

$$d(A_m, A_n) \leq \epsilon/3.$$

Let $N = \max\{N_1, N_2\}$ and note that, for $m, n \geq N$,

$$d(\tilde{x}_m, \tilde{x}_n) \leq d(\tilde{x}_m, x_{n_j}) + d(x_{n_j}, x_{n_k}) + d(x_{n_k}, \tilde{x}_n),$$

where $m \in \{n_{j-1} + 1, n_{j-1} + 2, \ldots, n_j\}$ and $n \in \{n_{k-1} + 1, n_{k-1} + 2, \ldots, n_k\}$. Since $h(A_m, A_{n_j}) < \epsilon/3$ there exists $y \in A_m \cap (\{x_{n_j}\} + \epsilon/3)$ so that $d(\tilde{x}_m, x_{n_j}) \leq \epsilon/3$. Similarly $d(x_{n_k}, \tilde{x}_n) \leq \epsilon/3$. Hence $d(\tilde{x}_n, \tilde{x}_n) \leq \epsilon$ for all $m, n > N$. This completes the proof.

Examples & Exercises

7.1. A Cauchy sequence $\{A_n\}$ of sets in $(\mathcal{H}(\mathbb{R}^2), h)$ is sketched in Figure II.25. The underlying metric space is $(\mathbb{R}^2, \text{Euclidean})$. A Cauchy subsequence $\{x_{n_j} \in A_{n_j}\}$ is also shown. Sketch, in the same figure, the extension $\{\tilde{x}_n\}$ of this subsequence to $\{A_n\}$.

7.2. Repeat 7.1 but this time with reference to Figure II.26.

The central result toward which we have been driving is this:

Theorem 7.1 The Completeness of the Space of Fractals. *Let* $(\mathbf{X}, d)$ *be a complete metric space. Then* $(\mathcal{H}(\mathbf{X}), h)$ *is a complete metric space. Moreover, if* $\{A_n \in \mathcal{H}(\mathbf{X})\}_{n=1}^{\infty}$ *is a Cauchy sequence, then*

$$A = \lim_{n \to \infty} A_n \in \mathcal{H}(\mathbf{X})$$

can be characterized as follows:

$$A = \{x \in \mathbf{X} : \text{there is a Cauchy sequence } \{x_n \in A_n\} \text{ that converges to } x\}.$$

Proof Let $\{A_n\}$ be a Cauchy sequence in $\mathcal{H}(\mathbf{X})$ and let A be defined as in the statement of the theorem. We break the proof up into the following parts:

(a) $A \neq \emptyset$;

(b) A is closed and hence complete since $\mathbf{X}$ is complete;

(c) for $\epsilon > 0$ there is N such that for $n \geq N$, $A \subset A_n + \epsilon$;

(d) A is totally bounded and thus by (b) is compact;

(e) $\lim A_n = A$.

Proof of (a): We shall prove this part by proving the existence of a Cauchy sequence $\{a_i \in A_i\}$ in $\mathbf{X}$. Toward this end find a sequence of positive integers $N_1 < N_2 < N_3 < \cdots < N_n < \cdots$ such that

$$h(A_m, A_n) < \frac{1}{2^i} \text{ for } m, n > N_i.$$

Choose $x_{N_1} \in A_{N_1}$. Then since $h(A_{N_1}, A_{N_2}) \leq \frac{1}{2}$, we can find an $x_{N_2} \in A_{N_2}$ such that $d(x_{N_1}, x_{N_2}) \leq \frac{1}{2}$. Assume that we have selected a finite sequence $x_{N_i} \in A_{N_i}$; $i = 1, 2, \ldots, k$ for which $d(x_{N_{i-1}}, x_{N_i}) \leq \frac{1}{2^{i-1}}$. Then since $h(A_{N_k}, A_{N_{k+1}}) \leq \frac{1}{2^k}$, and $x_{N_k} \in A_{N_k}$, we can find $x_{N_{k+1}} \in A_{N_{k+1}}$ such that $d(x_{N_k}, x_{N_{k+1}}) \leq \frac{1}{2^k}$. For example let $x_{N_{k+1}}$ be the point in $A_{N_{k+1}}$ that is closest to x_{N_k}. By induction we can find an infinite sequence $\{x_{N_i} \in A_{N_i}\}$ such that $d(x_{N_i}, x_{N_{i+1}}) \leq \frac{1}{2^i}$. To see that $\{x_{N_i}\}$ is a Cauchy sequence in $\mathbf{X}$, let $\epsilon > 0$ and choose N_ϵ such that $\sum_{i=N_\epsilon}^{\infty} \frac{1}{2^i} < \epsilon$. Then for $m > n \geq N_\epsilon$ we have

$$d(x_{N_m}, x_{N_n}) \leq d(x_{N_m}, x_{N_{m+1}}) + d(x_{N_{m+1}}, x_{N_{m+2}}) + \cdots + d(x_{N_{n-1}}, x_{N_n})$$

$$< \sum_{i=N_\epsilon}^{\infty} \frac{1}{2^i} < \epsilon.$$

By the Extension Lemma, there exists a convergent subsequence $\{a_i \in A_i\}$ for which $a_{N_i} = x_{N_i}$. Then $\lim a_i$ exists and by definition is in A. Thus $A \neq \emptyset$.

Proof of (b): To show that A is closed, suppose $\{a_i \in A_i\}$ is a sequence that converges to a point a. We will show that $a \in A$, hence making A closed. For each positive integer i, there exists a sequence $\{x_{i,n} \in A_n\}$ such that $\lim_{n \to \infty} x_{i,n} = a_i$. There exists an *increasing* sequence of positive numbers $\{N_i\}_{i=1}^{\infty}$ such that $d(a_{N_i}, a) < \frac{1}{i}$.

Figure II.26. The same problem as for Figure II.25. The sets $\{A_n\}$ look very different here.

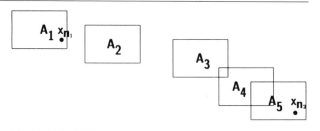

Furthermore, there is a sequence of integers $\{m_i\}$ such that $d(x_{N_i,m_i}, a_{N_i}) \leq \frac{1}{i}$. Thus $d(x_{N_i,m_i}, a) \leq 2/i$. If we let $y_{m_i} = x_{N_i,m_i}$ we see that $y_{m_i} \in A_{m_i}$ and $\lim i \to \infty y_{m_i} = a$. By the Extension Lemma, $\{y_{m_i}\}$ can be extended to a convergent sequence $\{z_i \in A_i\}$, and so $a \in A$. Thus we have shown A is closed.

Proof of (c): Let $\epsilon > 0$. There exists an N such that for $m, n \geq N$, $h(A_m, A_n) \leq \epsilon$. Now let $n \geq N$. Then for $m \geq n$, $A_m \subset A_n + \epsilon$. We need to show that $A \subset A_n + \epsilon$. To do this, let $a \in A$. There is a sequence $\{a_i \in A_i\}$ that converges to a. We may assume N is also large enough so that for $m \geq N$, $d(a_m, a) < \epsilon$. Then $a_m \in A_n + \epsilon$ since $A_m \subset A_n + \epsilon$. Since A_n is compact, it can be shown that $A_n + \epsilon$ is closed. Then since $a_m \in A_n + \epsilon$ for all $m \geq N$, a must also be in $A_n + \epsilon$. This completes the proof that $A \subset A_n + \epsilon$ for n large enough.

Proof of (d): Suppose A were not totally bounded. Then for some $\epsilon > 0$ there would not exist a finite ϵ-net. We could then find a sequence $\{x_i\}_{i=1}^{\infty}$ in A such that $d(x_i, x_j) \geq \epsilon$ for $i \neq j$. We shall show that this gives a contradiction. By (c) there exists an n large enough so that $A \subset A_n + \frac{\epsilon}{3}$. For each x_i, there is a corresponding $y_i \in A_n$ for which $d(x_i, y_i) \leq \frac{\epsilon}{3}$. Since A_n is compact, some subsequence $\{y_{n_i}\}$ of $\{y_i\}$ converges. Then we can find points in the sequence $\{y_{n_i}\}$ as close together as we wish. In particular we can find two points y_{n_i} and y_{n_j} such that $d(y_{n_i}, y_{n_j}) < \frac{\epsilon}{3}$. But then

$$d(x_{n_i}, x_{n_j}) \leq d(y_{n_i}, y_{n_i}) + d(y_{n_i}, y_{n_j}) + d(y_{n_j}, x_{n_j}) < \frac{\epsilon}{3} + \frac{\epsilon}{3} + \frac{\epsilon}{3},$$

and we have a contradiction to the way $\{x_{n_i}\}$ was chosen. Thus A is totally bounded and by part (b) compact.

Proof of (e): From part (d), $A \in \mathcal{H}(\mathbf{X})$. Hence by part (c) and Lemma 7.1 the proof that $\lim A_i = A$ will be complete if we show that for $\epsilon > 0$, there exists an N such that for $n \geq N$, $A_n \subset A + \epsilon$. To show this let $\epsilon > 0$ and find N such that for $m, n \geq N$, $h(A_m, A_n) \leq \frac{\epsilon}{2}$. Then for $m, n \geq N$, $A_m \subset A_n + \frac{\epsilon}{2}$. Let $n \geq N$. We will show that $A_n \subset A + \epsilon$. Let $y \in A_n$. There exists an increasing sequence $\{N_i\}$ of integers such that $n < N_1 < N_2 < N_3 < \cdots < N_k < \cdots$ and for $m, n \geq N_j$, $A_m \subset A_n + \frac{\epsilon}{2^{j+1}}$. Note that $A_n \subset A_{N_1} + \frac{\epsilon}{2}$. Since $y \in A_n$, there is an $x_{N_1} \in A_{N_1}$ such that $d(y, x_{N_1}) \leq \frac{\epsilon}{2}$. Since $x_{N_1} \in A_{N_1}$, there is a point $x_{N_2} \in A_{N_2}$ such that $d(x_{N_1}, x_{N_2}) \leq \frac{\epsilon}{2^2}$. In a similar manner we can use induction to find a sequence $x_{N_1}, x_{N_2}, x_{N_3}, \ldots$, such that $x_{N_j} \in A_{N_j}$ and $d(x_{N_j}, x_{N_{j+1}}) < \frac{\epsilon}{2^{j+1}}$. Using the triangle inequality a number of times we can show that

$$d(y, x_{N_j}) \leq \frac{\epsilon}{2} \qquad \text{for all } j$$

and also that $\{x_{N_j}\}$ is a Cauchy sequence. From the way n was chosen, each $A_{N_j} \subset A_n + \frac{\epsilon}{2}$. $\{x_{N_j}\}$ converges to a point x and since $A_n + \frac{\epsilon}{2}$ is closed, $x \in A_n + \frac{\epsilon}{2}$ also. Moreover, $d(y, x_{N_j}) \leq \epsilon$ implies that $d(y, x) \leq \epsilon$. We have thus shown that $A_n \subset A + \epsilon$ for $n \geq N$. This completes the proof that $\lim A_n = A$ and consequently that $(\mathcal{H}(\mathbf{X}), h)$ is a complete metric space.

Examples & Exercises

7.3. A tree waves in the wind. A special camera photographs the tree at times $t_n = (l - \frac{1}{n})$ sec, $n = 1, 2, 3, \ldots$. Show that with reasonable assumptions the sequence of pictures thus obtained form a Cauchy sequence $\{A_n\}_{n=1}^{\infty}$ in $\mathcal{H}(\mathbb{R}^2)$. What does $A = \lim_{n \to \infty} A_n$ look like?

7.4. The Sierpinski triangle $\triangle$ is a compact subset of $(\mathbb{R}^2$, Euclidean). Hence $(\triangle$, Euclidean) is a compact metric space. Given an example of an infinite set in $(\mathcal{H}(\triangle), h)$, demonstrate a Cauchy sequence $\{A_n \in \mathcal{H}(\triangle)\}$ that is contained in your set and describe its limit.

7.5. Figure II.27 shows a convergent sequence of sets in $\mathcal{H}(\blacksquare)$ converging to a fern. Pick a point in A. Find a Cauchy sequence $\{x_n \in A_n\}$ that converges to it.

*7.6.   Let $(\mathbf{X}, d)$ be a compact metric space. Show that $(\mathcal{H}(\mathbf{X}), h)$ is a compact metric space, where h is the Hausdorff metric on the space $\mathcal{H}(\mathbf{X})$.

It's a good idea to get familiar with $\mathcal{H}(\mathbf{X})$, by trying to see what properties of the space $\mathbf{X}$ are also true of $\mathcal{H}(\mathbf{X})$. We have seen that completeness is one of these, and exercise 7.6 shows that compactness is as well. So are some forms of connectedness, and by way of example we prove this for path-connectedness of $\mathcal{H}(\mathbb{R})$ as follows:

Theorem 7.2 *The function* $f : \mathbb{R} \to \mathcal{H}$ *given by* $f(x) = \{x\}$ *is continuous.*

Proof Let $\{x_n\}$ be a sequence in $\mathbb{R}$ that converges to a point x. Then given $\epsilon > 0$ there is an N such that for $n > N$, $d(x_n, x) < \epsilon$. By definition, $h(\{x_n\}, \{x\}) = d(x_n, x)$, since there is only one element in each set. Hence $\{\{x_n\}\}$ converges to $\{x\}$ in $\mathcal{H}(\mathbb{R})$. So the function $f(x)$ is continuous. Thus the image of $\mathbb{R}$ in $\mathcal{H}(\mathbb{R})$ is path-connected. Notice that a function like this guarantees that there will be a copy of any space in its associated space of non-empty compact sets.

Theorem 7.3 *The functions* $f_x : [0, 1] \to \mathcal{H}$ *given by* $f_x(a) = [x, x + a], 0 \le a \le 1$ *are continuous, thereby showing that there is a path in $\mathcal{H}$ from an interval to one of its endpoints.*

Proof As before, let $\{a_n\}$ be a convergent sequence in $[0, 1]$ with limit a. Suppose $d(a_n, a) < \epsilon$. Then the distance $h([x, x + a_n], [x, x + a])$ is given by

$$h([x, x + a_n], [x, x + a]) = d([x, x + a_n], [x, x + a]) \vee d([x, x + a], [x, x + a_n])$$
$$= d(a, a_n) < \epsilon.$$

Hence each of the functions f_x is continuous.

Theorem 7.4 *If A is a compact subset of $\mathbb{R}$ then the function $f_A : [0, b] \to \mathcal{H}$ given by $f_A(a) = \bigcup [x, x + a]$ such that $x \in A$ is continuous.*

Proof By part (b), the function f_A is continuous at each point in A, when $b = 1$. Hence the statement is true for $b = 1$. To get the whole statement, we note that if $g : [0, b] \to [0, 1]$ is given by $g(x) = (1/b)x$, g is continuous. Hence f_A can be written as the composition of two continuous functions and is continuous.

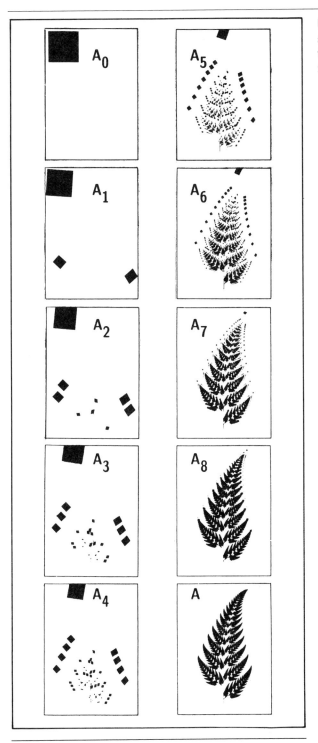

Figure II.27. A Cauchy sequence of sets $\{A_n\}$ in the space $\mathcal{H}(\mathbb{R}^2)$ converging to a fern-like set.

Theorem 7.5 *If A is a compact subset of $\mathbb{R}$ then the set $\bigcup[x, x + b]$ such that $x \in A$ is an interval for b large enough.*

Proof A is compact, hence it is bounded. If we take b to be the length of any interval $[c, d]$ such that $A \subset [c, d]$, then the set $[x, x + b]$ with x the least element of A will overlap the set $[y, y + b]$ where y is the greatest element of A. Thus all the sets of this form will overlap, and their union forms one interval, which is path-connected.

Theorem 7.6 *If A and B are compact subsets of $\mathbb{R}$ then there is a path in $\mathcal{H}$ connecting them.*

Proof For any space $\mathbf{X}$ and three points $a, b, c \in \mathbf{X}$, if there is a path from a to b and one from b to c then there is one from a to c. We construct a path between A and B in $\mathcal{H}(\mathbb{R})$ by taking first a path from A to the interval formed in part (d), namely choose b as in (d) and construct the path

$$f : [0, 1] \rightarrow [0, b] \rightarrow \{f_A(x) : 0 \leq x \leq b\},$$

which is continuous by part (c). Thus there is a path between every point in $\mathcal{H}(\mathbb{R})$ and a point that is an interval. Similarly, there is a path from some interval to every point in $\mathcal{H}(\mathbb{R})$. By part (b) there is a path from every interval to one of its endpoints. By part (a) there is a path (the image of $\mathbb{R}$) between any two of these endpoints. Therefore a path from A to B may be constructed by taking A to an interval, taking the interval to an endpoint, moving to the a new point of $\mathbb{R}$, making a path to an interval, and taking the path from the interval to B. Hence $\mathcal{H}(\mathbb{R})$ is pathwise-connected. It is true, though somewhat more complicated, that if $\mathbf{X}$ is connected then so is $\mathcal{H}(\mathbf{X})$.

8   Additional Theorems about Metric Spaces

We state here a number of theorems that we shall use later on. Full proofs are not provided. They can be found in most introductory topology texts. We particularly recommend [Kasriel 1971] and [Mendelson 1963]. These theorems may be treated as exercises in metric space theory.

Theorem 8.1 *Let $(\mathbf{X}, d)$ be a metric space. Let $\{x_n\}$ be a Cauchy sequence convergent to $x \in \mathbf{X}$ (or equivalently let $\{x_n\}$ be a sequence and x be a point, such that $\lim_{n \rightarrow \infty} d(x, x_n) = 0$). Let $f : \mathbf{X} \rightarrow \mathbf{X}$ be continuous. Then*

$$\lim_{n \rightarrow \infty} f(x_n) = f(x).$$

Proof See your first calculus book.

Theorem 8.2 *Let $(\mathbf{X}_1, d_1)$ and $(\mathbf{X}_2, d_2)$ be metric spaces. Let $f : \mathbf{X}_1 \rightarrow \mathbf{X}_2$ be continuous. Let $E \subset \mathbf{X}_1$ be compact. Then $f : E \rightarrow \mathbf{X}_2$ is uniformly continuous: that* *is, given $\epsilon > 0$ there is a number $\delta > 0$ so that*

$$d_2(f(x), f(y)) < \epsilon \text{ whenever } d_1(x, y) < \delta \text{ for all } x, y \in E.$$

Proof Use the fact that any open cover of E contains a finite subcover.

Theorem 8.3 *Let $(\mathbf{X}_i, d_i)$ be metric spaces for $i = 1, 2, 3$. Let $f : \mathbf{X}_1 \times \mathbf{X}_2 \to \mathbf{X}_3$ have the following property. For each $\epsilon > 0$ there exists $\delta > 0$ such that*

$$\text{(i) } d_1(x_1, y_1) < \delta \Rightarrow d_3(f(x_1, x_2), f(y_1, y_2)) < \epsilon,$$
$$\forall x_1, y_1 \in \mathbf{X}_1$$
$$\forall x_2 \in \mathbf{X}_2$$

and

$$\text{(ii) } d_2(x_2, y_2) < \delta \Rightarrow d_3(f(y_1, x_2), f(y_1, y_2)) < \epsilon,$$
$$\forall y_1 \in \mathbf{X}_1$$
$$\forall x_2, y_2 \in \mathbf{X}_2$$

Then f is continuous on the metric space $(\mathbf{X} = \mathbf{X}_1 \times \mathbf{X}_2, d)$, where $d((x_1, x_2), (y_1, y_2)) = \max\{d_1(x_1, y_1), d_2(x_2, y_2)\}$.

Proof Use

$$d(f(x_1, x_2), f(y_1, y_2)) \leq d(f(x_1, x_2), f(y_1, x_2)) + d(f(y_1, x_2), f(y_1, y_2)),$$

but check first that d is a metric.

Theorem 8.4 *Let $(\mathbf{X}_i, d_i)$ be metric spaces for $i = 1, 2$ and let the metric space $(\mathbf{X}, d)$ be defined as in Theorem 8.3. If $K_1 \subset \mathbf{X}_1$ and $K_2 \subset \mathbf{X}_2$ are compact, then $K_1 \times K_2 \subset \mathbf{X}$ is compact.*

Proof Deal with the component in K_1 first.

Theorem 8.5 *Let $(\mathbf{X}_i, d_i)$ be compact metric spaces for $i = 1, 2$. Let $f : \mathbf{X}_1 \to \mathbf{X}_2$ be continuous, one-to-one, and onto. Then f is a homeomorphism.*

Chapter III

Transformations on Metric Spaces; Contraction Mappings; and the Construction of Fractals

1 Transformations on the Real Line

Fractal geometry studies "complicated" subsets of geometrically "simple" spaces such as $\mathbb{R}^2$, $\mathbb{C}$, $\mathbb{R}$, and $\hat{\mathbb{C}}$. In deterministic fractal geometry the focus is on those subsets of a space that are generated by, or possess invariance properties under, simple geometrical transformations of the space into itself. A simple geometrical transformation is one that is easily conveyed or explained to someone else. Usually it can be completely specified by a small set of parameters. Examples include affine transformations in $\mathbb{R}^2$, which are expressed using 2×2 matrices and 2-vectors, and rational transformations on the Riemann Sphere, which require the specification of the coefficients in a pair of polynomials.

Definition 1.1 *Let $(\mathbf{X}, d)$ be a metric space. A* transformation *on $\mathbf{X}$ is a function $f : \mathbf{X} \to \mathbf{X}$, which assigns exactly one point $f(x) \in \mathbf{X}$ to each point $x \in \mathbf{X}$. If $S \subset \mathbf{X}$ then $f(S) = \{f(x) : x \in S\}$. f is* one-to-one *if $x, y \in \mathbf{X}$ with $f(x) = f(y)$ implies $x = y$. f is* onto *if $f(\mathbf{X}) = \mathbf{X}$. f is called* invertible *if it is one-to-one and onto: in* *this case it is possible to define a transformation $f^{-1} : \mathbf{X} \to \mathbf{X}$, called the inverse of f, by $f^{-1}(y) = x$, where $x \in \mathbf{X}$ is the unique point such that $y = f(x)$.*

Definition 1.2 *Let f: $\mathbf{X} \to \mathbf{X}$ be a transformation on a metric space. The* forward iterates *of f are transformations $f^{\circ n} : \mathbf{X} \to \mathbf{X}$ defined by $f^{\circ 0}(x) = x$, $f^{\circ 1}(x) = f(x)$, $f^{\circ (n+1)}(x) = f \circ f^{(n)}(x) = f(f^{(n)}(x))$ for $n = 0, 1, 2, \ldots$. If f is invertible then the* backward iterates *of f are transformations $f^{\circ (-m)}(x) : \mathbf{X} \to \mathbf{X}$ defined by $f^{\circ(-1)}(x) = f^{-1}(x)$, $f^{\circ(-m)}(x) = (f^{\circ m})^{-1}(x)$ for $m = 1, 2, 3, \ldots$.*

In order to work in fractal geometry one needs to be familiar with the basic families of transformations in $\mathbb{R}$, $\mathbb{R}^2$, $\mathbb{C}$, and $\hat{\mathbb{C}}$. One needs to know well the relationship between "formulas" for transformations and the geometric changes, stretchings, twistings, foldings, and skewings of the underlying fabric, the metric space upon which they act. It is more important to understand what the transformations do to sets than how they act on individual points. So, for example, it is more useful to know how an affine transformation in $\mathbb{R}^2$ acts on a straight line, a circle, or a triangle, than to know to where it takes the origin.

Examples & Exercises

1.1. Let f: $\mathbf{X} \to \mathbf{X}$ be an invertible transformation. Show that

$$f^{\circ m} \circ f^{\circ n} = f^{\circ(m+n)} \qquad \text{for all integers } m \text{ and } n.$$

1.2. A transformation $f : \mathbb{R} \to \mathbb{R}$ is defined by $f(x) = 2x$ for all $x \in \mathbb{R}$. Is f invertible? Find a formula for $f^{\circ n}(x)$ that applies for all integers n.

1.3. A transformation $f : [0, 1] \to [0, 1]$ is defined by $f(x) = \frac{1}{2}x$. Is this transformation one-to-one? Onto? Invertible?

1.4. The mapping $f : [0, 1] \to [0, 1]$ is defined by $f(x) = 4x \cdot (1 - x)$. Is this transformation one-to-one? Onto? Is it invertible?

1.5. Let C denote the *Classical Cantor Set*. This subset of the metric space $[0, 1]$ is obtained by successive deletion of middle-third open subintervals as follows. We construct a nested sequence of closed intervals

$$I_0 \supset I_1 \supset I_2 \supset I_3 \supset I_4 \supset I_5 \supset I_6 \supset I_7 \ldots \supset I_n \supset \ldots,$$

where

Figure III.28. Construction of the Classical Cantor Set $\mathcal{C}$.

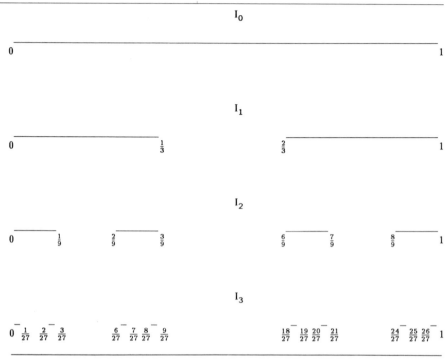

$I_0 = [0, 1],$

$I_1 = [0, \frac{1}{3}] \cup [\frac{1}{3}, \frac{2}{3}],$

$I_2 = [0, \frac{1}{9}] \cup [\frac{2}{9}, \frac{3}{9}] \cup [\frac{6}{9}, \frac{7}{9}] \cup [\frac{8}{9}, \frac{9}{9}],$

$I_3 = [0, \frac{1}{27}] \cup [\frac{2}{27}, \frac{3}{27}] \cup [\frac{6}{27}, \frac{7}{27}] \cup [\frac{8}{27}, \frac{9}{27}]$

$\qquad \cup [\frac{18}{27}, \frac{19}{27}] \cup [\frac{20}{27}, \frac{21}{27}] \cup [\frac{24}{27}, \frac{25}{27}] \cup [\frac{26}{27}, \frac{27}{27}],$

$I_4 = I_3$ take away the middle open third of each interval in I_3,

$\vdots$

$I_n = I_{N-1}$ take away the middle open third of each interval in I_{N-1}.

This construction is illustrated in Figure III.28. We define

$$\mathcal{C} = \cap_{n=0}^{\infty} I_n.$$

$\mathcal{C}$ contains the point $x = 0$, so it is nonempty. In fact $\mathcal{C}$ is a perfect set that contains uncountably many points, as discussed in Chapter IV. $\mathcal{C}$ is an official fractal and we will often refer to it.

We are now able to work in the metric space $(\mathcal{C}, \text{Euclidean})$. A transformation $f : \mathcal{C} \to \mathcal{C}$ is defined by $f(x) = \frac{1}{3}x$. Show that this transformation is one-to-one but

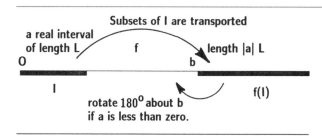

Figure III.29. The action of the affine transformation $f : \mathbb{R} \rightarrow \mathbb{R}$ defined by $f(x) = ax + b$.

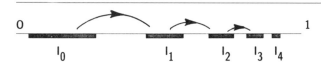

Figure III.30. This figure suggests a sequence of intervals $\{I_n\}_{n=0}^{\infty}$. Find an affine transformation $f : \mathbb{R} \rightarrow \mathbb{R}$ so that $f^{\circ n}(I_0) = I_n$ for $n = 0, 1, 2, 3, \ldots$. Use a straight-edge and dividers to help you.

not onto. Also, find another affine transformation (see example 1.7, which maps $\mathcal{C}$ one-to-one into $\mathcal{C}$).

1.6. $f : \mathbb{R}^2 \rightarrow \mathbb{R}^2$ is defined by $f(x_1, x_2) = (2x_1, x_2^2 + x_1)$ for all $(x_1, x_2) \in \mathbb{R}^2$. Show that f is not invertible. Give a formula for $f^{\circ 2}(x)$.

1.7. *Affine transformations* in $\mathbb{R}^1$ are transformations of the form $f(x) = a \cdot x + b$, where a and b are real constants. Given the interval $I = [0, 1]$, $f(I)$ is a new interval of length $|a|$, and f rescales by a. The left endpoint 0 of the interval is moved to b, and $f(I)$ lies to the left or right of b according to whether a is positive or negative, respectively (see Figure III.29).

We think of the action of an affine transformation on all of $\mathbb{R}$ as follows: the whole line is stretched away from the origin if $|a| > 1$, or contracted toward it if $|a| < 1$; flipped through $180°$ about $\mathcal{O}$ if $a < 0$; and then *translated* (shifted as a whole) by an amount b (shift to the left if $b < 0$, and to the right if $b > 0$).

1.8. Describe the set of affine transformations that takes the real interval $\mathbf{X} = [1, 2]$ into itself. Show that if f and g are two such transformations then $f \circ g$ and $g \circ f$ are also affine transformations on $[1, 2]$. Under what conditions does $f \circ g(\mathbf{X}) \cup g \circ f(\mathbf{X}) = \mathbf{X}$?

1.9. A sequence of intervals $\{I_n\}_{n=0}^{\infty}$ is indicated in Figure III.30. Find an affine transformation $f : \mathbb{R} \rightarrow \mathbb{R}$ so that $f^{\circ n}(I_0) = I_n$ for $n = 0, 1, 2, 3, \ldots$. Use a straight-edge and dividers to help you. Also show that $\{I_n\}_{n=1}^{\infty}$ is a Cauchy sequence in $(\mathcal{H}(\mathbb{R}), h)$, where h is the Hausdorff distance on $\mathcal{H}(\mathbb{R})$ induced by the Euclidean metric on $\mathbb{R}$. Evaluate $I = \lim_{n \rightarrow \infty} I_n$.

Figure III.31. Picture of a convergent geometric series in $\mathbb{R}^1$ (see exercise 1.10).

$I_0 \qquad I_1 \qquad I_2 \quad I_3 \quad I_4 \quad I_5 \, I_6$

1.10. Consider the geometric series $\sum_{n=0}^{\infty} b \cdot a^n = b + a \cdot b + a^2 b + a^3 b + a^4 b + \cdots > 0, 0 < b < 1$. This is associated with a sequence of intervals $I_0 = [0, b]$, $I_n = f^{\circ n}(I_0)$, where $f(x) = ax + b$, $n = 1, 2, 3, \ldots$, as illustrated in Figure III.31.

Let $I = \cup_{n=0}^{\infty} I_n$ and let l denote the total length of I. Show that $f(I) = I \setminus I_0$, and hence deduce that $al = l - b$ so that $l = b/(1 - a)$. Deduce at once that

$$\sum_{n=0}^{\infty} b \cdot a^n = b/(1 - a).$$

Thus we see from a geometrical point of view a well-known result about *geometric series*. Make a similar geometrical argument to cover the case $-1 < a < 0$.

Definition 1.3 *A transformation* $f : \mathbb{R} \to \mathbb{R}$ *of the form*

$$f(x) = a_0 + a_1 x + a_2 x^2 + a_3 x^3 + \cdots + a_n x^n,$$

where the coefficients a_i *(*$i = 0, 1, 2, \ldots, N$*) are real numbers,* $a_n \neq 0$*, and* N *is a nonnegative integer, is called a* polynomial transformation. N *is called the* degree *of the transformation.*

Examples & Exercises

1.11. Show that if $f : \mathbb{R} \to \mathbb{R}$ and $g : \mathbb{R} \to \mathbb{R}$ are polynomial transformations, then so is $f \circ g$. If f is of degree N, calculate the degree of $f^{\circ m}(x)$ for $m = 1, 2, 3, \ldots$.

1.12. Show that for $n > 1$ a polynomial transformation $f : \mathbb{R} \to \mathbb{R}$ of degree n is not generally invertible.

1.13. Show that far enough out (i.e., for large enough $|x|$), a polynomial transformation $f : \mathbb{R} \to \mathbb{R}$ always stretches intervals. That is, view f as a transformation from $(\mathbb{R}, \text{Euclidean})$ into itself. Show that if I is an interval of the form $I = \{x : |x - a| \leq b\}$ for fixed $a, b \in \mathbb{R}$, then for any number $M > 0$ there is a number $\beta > 0$ such that if $b > \beta$, then the ratio (length of $f(I)$)/(length of I) is larger than M. This idea is illustrated in Figure III.32.

1.14. A polynomial transformation $f : \mathbb{R} \to \mathbb{R}$ of degree n can produce at most $(n - 1)$ *folds*. For example $f(x) = x^3 - 3x + 1$ behaves as shown in Figure III.33.

1.15. Find a family of polynomial transformations of degree 2 which map the interval $[0, 2]$ into itself, such that, with one exception, if $y \in f([0, 2])$ then there exist two distinct points x_1 and x_2 in $[0, 2]$ with $f(x_1) = f(x_2) = y$.

1.16. Show that the one-parameter family of polynomial transformations f_λ :

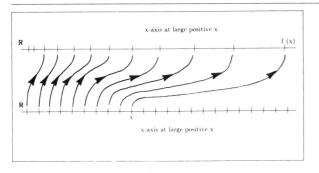

Figure III.32. A polynomial transformation $f : \mathbb{R} \to \mathbb{R}$ of degree > 1 stretches $\mathbb{R}$ more and more the farther out one goes.

$[0, 2] \to [0, 2]$, where

$$f_\lambda(x) = \lambda \cdot x \cdot (2 - x),$$

and the parameter λ belongs to $[0, 1]$, indeed takes the interval $[0, 2]$ into itself. Locate the value of x at which the fold occurs. Sketch the behavior of the family, in the spirit of Figure III.33.

1.17. Let $f : \mathbb{R} \to \mathbb{R}$ be a polynomial transformation of degree n. Show that values of x that are transformed into fold points are solutions of

$$\frac{df}{dx}(x) = 0, x \in \mathbb{R}.$$

Solutions of this equation are called (real) *critical points* of the function f. If c is a critical point then $f(c)$ is a *critical value*. Show that a critical value need not be a fold point.

1.18. Find a polynomial transformation such that Figure III.34 is true.

1.19. Recall that a polynomial transformation of an interval $f : I \subset \mathbb{R} \to I$ is normally represented as in Figure III.35. This will be useful when we study iterates $f^{\circ n}(x)_{n=1}^\infty$. However, the folding point of view helps us to understand the idea of the deformation of space.

1.20. Polynomial transformations can be lifted to act on subsets of $\mathbb{R}^2$ in a simple

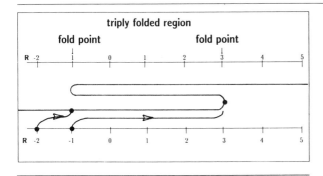

Figure III.33. The polynomial transformation $f(x) = x^3 - 3x + 1$.

Figure III.34. Find a polynomial transformation $f : \mathbb{R} \to \mathbb{R}$, so that this figure correctly represents the way it folds on the real line.

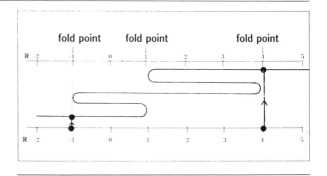

Figure III.35. The usual way of picturing a polynomial transformation.

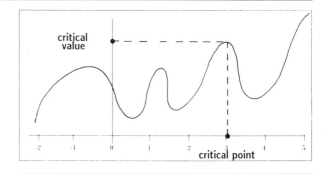

way: we can define, for example, $F(x) = (f_1(x_1), f_2(x_2))$, where f_1 and f_2 are polynomial transformations in $\mathbb{R}$, so that $F : \mathbb{R}^2 \to \mathbb{R}^2$. Desired foldings in two orthogonal directions can be produced; or shrinking in one direction and folding in another. Show that the transformation $F(x_1, x_2) = (\frac{8}{5}x_1^3 - \frac{36}{5}x_1^2 + \frac{48}{5}x_1, x_2)$ acts on the triangular set S in Figure III.36 as shown.

The real line can be extended to a space which is topologically a circle by including the point at infinity. One way to do this is to think of $\mathbb{R}$ as a subset of $\hat{\mathbb{C}}$, and

Figure III.36. A polynomial transformation acting on a set S in the plane.

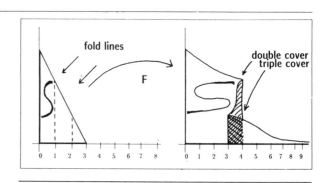

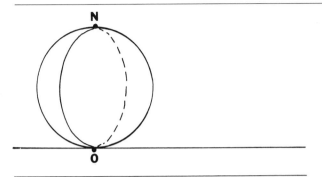

Figure III.37. $\mathbb{R} \cup \{\infty\}$ becomes a circle on a sphere.

then include the North Pole on $\hat{\mathbb{C}}$. We define this space to be $\hat{\mathbb{R}} = \mathbb{R} \cup \{\infty\}$ and will usually give it the spherical metric.

Definition 1.4 *A transformation* $f : \hat{\mathbb{R}} \to \hat{\mathbb{R}}$ *defined in the form*

$$f(x) = \frac{ax + b}{cx + d}, \qquad a, b, c, d \in \mathbb{R}, ad \neq bc,$$

is called a linear fractional transformation *or a* Möbius transformation. *If $c \neq 0$, then $f(-d/c) = \infty$, and $f(\infty) = a/c$. If $c = 0$, then $f(\infty) = \infty$.*

Examples & Exercises

1.21. Show that a Möbius transformation is invertible.

1.22. Show that if f_1 and f_2 are both Möbius transformations then so is $f_1 \circ f_2$.

1.23. What does $f(x) = 1/x$ do to $\hat{\mathbb{R}}$ on the sphere?

1.24. Show that the set of Möbius transformations f such that $f(\infty) = \infty$ is the set of affine transformations.

1.25. Find a Möbius transformation $f : \hat{\mathbb{R}} \to \hat{\mathbb{R}}$ so that $f(1) = 2$, $f(2) = 0$, $f(0) = \infty$. Evaluate $f(\infty)$.

1.26. Figure III.38 shows a Sierpinski triangle before and after the polynomial transformation $x \mapsto ax(x - b)$ has been applied to the x-axis. Evaluate the real constants a and b. Notice how well fractals can be used to illustrate how a transformation acts.

2 Affine Transformations in the Euclidean Plane

Definition 2.1 *A transformation* $w : \mathbb{R}^2 \to \mathbb{R}^2$ *of the form*

$$w(x_1, x_2) = (ax_1 + bx_2 + e, cx_1 + dx_2 + f),$$

where $a, b, c, d, e,$ and f are real numbers, is called a (two-dimensional) affine transformation.

We will often use the following equivalent notations

$$w(x) = w \begin{pmatrix} x_1 \\ x_2 \end{pmatrix} = \begin{pmatrix} a & b \\ c & d \end{pmatrix} \begin{pmatrix} x_1 \\ x_2 \end{pmatrix} + \begin{pmatrix} e \\ f \end{pmatrix} = Ax + t.$$

Here $A = \begin{pmatrix} a & b \\ c & d \end{pmatrix}$ is a two-dimensional, 2×2 real matrix and t is the column

vector $\begin{pmatrix} e \\ f \end{pmatrix}$, which we do not distinguish from the coordinate pair $(e, f) \in \mathbb{R}^2$.

Such transformations have important geometrical and algebraic properties. From this point on, we shall assume that the reader is familiar with matrix multiplication.

The matrix A can always be written in the form

$$\begin{pmatrix} a & b \\ c & d \end{pmatrix} = \begin{pmatrix} r_1 \cos \theta_1 & -r_2 \sin \theta_2 \\ r_1 \sin \theta_1 & r_2 \cos \theta_2 \end{pmatrix},$$

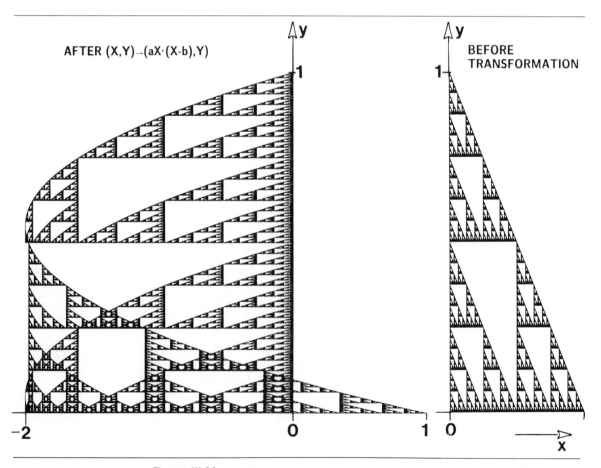

Figure III.38.   A Sierpinski triangle before and after the polynomial transformation $x \mapsto ax(x - b)$ is applied to the x-axis. Evaluate the real constants a and b.

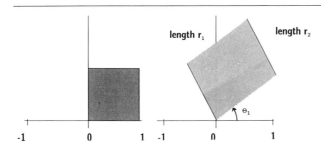

Figure III.39. An affine transformation takes parallelograms into parallelograms.

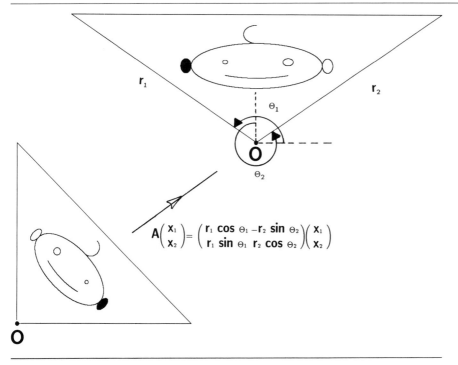

Figure III.40. A linear transformation can turn pictures over.

$$A\begin{pmatrix} x_1 \\ x_2 \end{pmatrix} = \begin{pmatrix} r_1 \cos \theta_1 & -r_2 \sin \theta_2 \\ r_1 \sin \theta_1 & r_2 \cos \theta_2 \end{pmatrix} \begin{pmatrix} x_1 \\ x_2 \end{pmatrix}$$

where (r_1, θ_1) are the polar coordinates of the point (a, c) and $(r_2, (\theta_2 + \pi/2))$ are the polar coordinates of the point (b, d). The *linear* transformation

$$\begin{pmatrix} x_1 \\ x_2 \end{pmatrix} \rightarrow A \begin{pmatrix} x_1 \\ x_2 \end{pmatrix}$$

in $\mathbb{R}^2$ maps any parallelogram with a vertex at the origin to another parallelogram with a vertex at the origin, as illustrated in Figure III.39. Notice that the parallelogram may be "turned over" by the transformation, as illustrated in Figure III.40.

The general affine transformation $w(x) = Ax + t$ in $\mathbb{R}^2$ consists of a linear transformation, A which deforms space relative to the origin, as described above, followed by a *translation* or *shift* specified by the vector t (see Figure III.41).

Figure III.41. An affine transformation consists of a linear transformation followed by a translation.

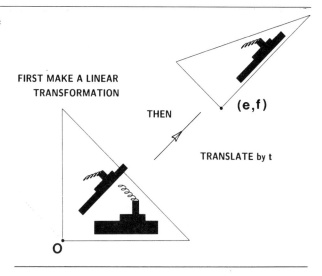

FIRST MAKE A LINEAR
TRANSFORMATION

THEN

(e,f)

TRANSLATE by t

O

Figure III.42. Two ivy leaves lying on the Euclidean Plane determine an affine transformation.

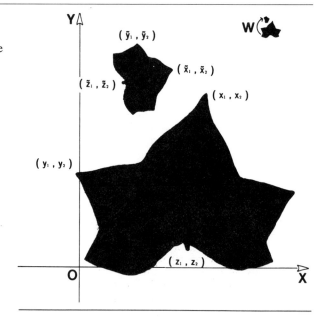

How can one find an affine transformation that approximately transforms one given set into another given set in $\mathbb{R}^2$? Let's show how to find the affine transformation that almost takes the big leaf to the little leaf in Figure III.42. This figure actually shows a photocopy of two real ivy leaves. We wish to find the numbers $a, b, c, d, e,$ and f defined above, so that

$$w(\text{BIG LEAF}) \text{ approximately equals LITTLE LEAF.}$$

Begin by introducing x and y coordinate axes, as already shown in Figure III.42. Mark three points on the big leaf (we've chosen the leaf tip, a side spike, and the point where the stem joins the leaf) and determine their coordinates (x_1, x_2), (y_1, y_2), and (z_1, z_2). Mark the corresponding points on the little leaf, assuming that a caterpillar hasn't eaten them, and determine their coordinates; say $(\tilde{x}_1, \tilde{x}_2)$, $(\tilde{y}_1, \tilde{y}_2)$, and $(\tilde{z}_1, \tilde{z}_2)$, respectively.

Then a, b, and e are obtained by solving the three linear equations

$$x_1 a + x_2 b + e = \tilde{x}_1,$$

$$y_1 a + y_2 b + e = \tilde{y}_1,$$

$$z_1 a + z_2 b + e = \tilde{z}_1;$$

while c, d, and f satisfy

$$x_1 c + x_2 d + f = \tilde{x}_2,$$

$$y_1 c + y_2 d + f = \tilde{y}_2,$$

$$z_1 c + z_2 d + f = \tilde{z}_2.$$

Examples & Exercises

2.1. Find an affine transformation in $\mathbb{R}^2$ that takes the triangle with vertices at $(0, 0)$, $(0, 1)$, $(1, 0)$ to the triangle with vertices at $(4, 5)$, $(-1, 2)$, and $(3, 0)$. Show what this transformation does to a circle inscribed in the first triangle.

2.2. Show that a necessary and sufficient condition for the affine transformation

$$\begin{pmatrix} a & b \\ c & d \end{pmatrix} \begin{pmatrix} x_1 \\ x_2 \end{pmatrix} + \begin{pmatrix} e \\ f \end{pmatrix} = Ax + t$$

to be invertible is $\det A \neq 0$, where $\det A = (ad - bc)$ is the determinant of the 2×2 matrix A.

2.3. Show that if $f_1 : \mathbb{R}^2 \to \mathbb{R}^2$ and $f_2 : \mathbb{R}^2 \to \mathbb{R}^2$ are both affine transformations, then so is

$$f_3 = f_1 \circ f_2.$$

If $f_i(x) = A_i x + t_i$, $i = 1, 2, 3$, where A_i is a 2×2 real matrix, express A_3 in terms of A_1 and A_2.

2.4. Let A and B be 2×2 matrices, with determinants $\det A$ and $\det B$, respectively. Show that the determinant of the product is the product of the determinants, i.e.,

$$\det(AB) = \det A \cdot \det B.$$

Definition 2.2 *A transformation* $w : \mathbb{R}^2 \to \mathbb{R}^2$ *is called a* similitude *if it is an affine transformation having one of the special forms*

$$w \begin{pmatrix} x_1 \\ x_2 \end{pmatrix} = \begin{pmatrix} r \cos \theta & -r \sin \theta \\ r \sin \theta & r \cos \theta \end{pmatrix} \begin{pmatrix} x_1 \\ x_2 \end{pmatrix} + \begin{pmatrix} e \\ f \end{pmatrix}$$

$$w \begin{pmatrix} x_1 \\ x_2 \end{pmatrix} = \begin{pmatrix} r \cos \theta & r \sin \theta \\ r \sin \theta & -r \cos \theta \end{pmatrix} \begin{pmatrix} x_1 \\ x_2 \end{pmatrix} + \begin{pmatrix} e \\ f \end{pmatrix}$$

for some translation $(e, f) \in \mathbb{R}^2$, *some real number* $r \neq 0$, *and some angle* $\theta, 0 \leq \theta < 2\pi$. θ *is called the rotation angle while* r *is called the* scale factor *or* scaling. *The linear transformation*

$$R_\theta \begin{pmatrix} x_1 \\ x_2 \end{pmatrix} = \begin{pmatrix} r \cos \theta & -r \sin \theta \\ r \sin \theta & r \cos \theta \end{pmatrix} \begin{pmatrix} x_1 \\ x_2 \end{pmatrix}$$

is a rotation. *The linear transformation*

$$R \begin{pmatrix} x_1 \\ x_2 \end{pmatrix} = \begin{pmatrix} 1 & 0 \\ 0 & -1 \end{pmatrix} \begin{pmatrix} x_1 \\ x_2 \end{pmatrix}$$

is a reflection.

Figure III.43 shows some of the things a similitude can do. Notice that a similitude preserves angles.

Examples & Exercises

2.5. Find the scaling ratios r_1, r_2 and the rotation angles θ_1, θ_2 for the affine transformation that takes the triangle $(0, 0)$, $(0, 1)$, $(1, 0)$ onto the straight-line segment from $(1, 1)$ to $(2, 2)$ in $\mathbb{R}^2$ in such a way that both $(0, 1)$ and $(1, 0)$ go to $(1, 1)$.

2.6. Let S be a region in $\mathbb{R}^2$ bounded by a polygon or other "nice" boundary. Let $w : \mathbb{R}^2 \to \mathbb{R}^2$ be an affine transformation, $w(x) = Ax + t$. Show that

$$(\text{area of } w(S)) = |\det A| \cdot (\text{area of } S);$$

see Figure III.44. Show that $\det A < 0$ has the interpretation that S is "flipped over" by the transformation. (Hint: suppose first that S is a triangle.)

2.7. Show that if $w : \mathbb{R}^2 \to \mathbb{R}^2$ is a similitude, $w(x) = Ax + t$, where t is the translation and A is a 2×2 matrix, then A can always be written either $A = r R_\theta$ or $A = r R R_\theta$.

2.8. View the railway tracks image in Figure III.45 as a subset S of $\mathbb{R}^2$. Find a similitude $w : \mathbb{R}^2 \to \mathbb{R}^2$ such that $w(S) \subset S, w(S) \neq S$.

2.9. We use the notation introduced in Definition 2.2. Find a nonzero real number r, an angle θ, and a translation vector t such that the similitude $wx = r R_\theta x + t$ on $\mathbb{R}^2$ obeys

$$w(\triangle) \subset \triangle, \qquad \text{with } w(\triangle) \neq \triangle,$$

where $\triangle$ denotes a Sierpinski triangle with vertices at $(0, 0)$, $(1, 0)$, and $(\frac{1}{2}, 1)$.

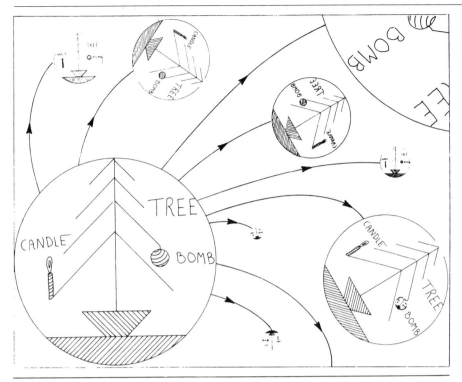

Figure III.43. Some of the things that a similitude can do.

2.10. Show that if $w : \mathbb{R}^2 \to \mathbb{R}^2$ is affine, $w(x) = Ax + t$, then it can be reexpressed

$$w(x) = \begin{pmatrix} r_1 & 0 \\ 0 & r_2 \end{pmatrix} R_\theta \begin{pmatrix} r_3 & 0 \\ 0 & r_4 \end{pmatrix} \begin{pmatrix} x_1 \\ x_2 \end{pmatrix} + t,$$

where $r_i \in \mathbb{R}$ and $0 \leq \theta < 2\pi$. We call a transformation of the form

$$w \begin{pmatrix} x_1 \\ x_2 \end{pmatrix} = \begin{pmatrix} r_1 & 0 \\ 0 & r_2 \end{pmatrix} \begin{pmatrix} x_1 \\ x_2 \end{pmatrix}$$

a coordinate rescaling.

2.11. Let S denote the two-dimensional orchard subset of $\mathbb{R}^2$ shown in Figure III.46. Find two fundamentally different affine transformations that map S into S but not onto S. Define the transformations by specifying how they act on three points.

2.12. Show that if A is a 2×2 matrix such that $\det A \neq 0$, with

$$A = \begin{pmatrix} a & b \\ c & d \end{pmatrix},$$

then the inverse of A, denoted A^{-1}, is given by

$$A^{-1} = \frac{1}{\det A} \begin{pmatrix} d & -b \\ -c & a \end{pmatrix} = \begin{pmatrix} \frac{d}{\det A} & \frac{-b}{\det A} \\ \frac{-c}{\det A} & \frac{a}{\det A} \end{pmatrix}.$$

Figure III.44. The scaling factor by which an affine transformation changes area is determined by the determinant of its linear part.

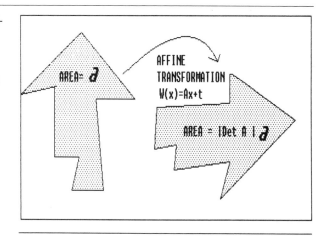

Figure III.45. Railway to infinity. Can you find an affine transformation that nearly maps the track ties into themselves?

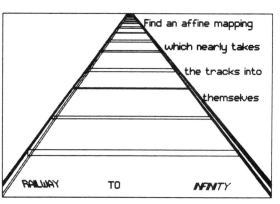

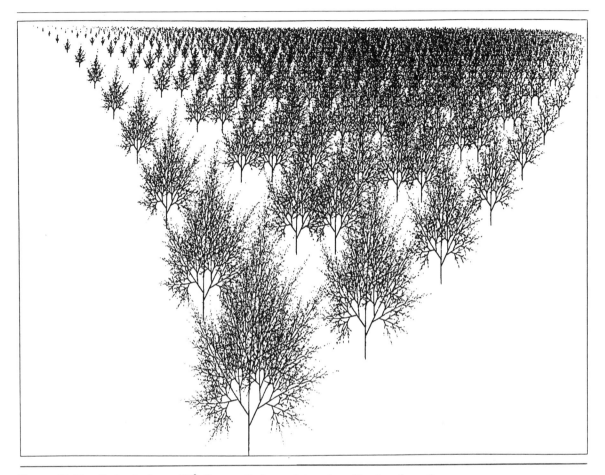

Figure III.46. Orchard subset of $\mathbb{R}^2$. Can you find some interesting affine transformations that map this set into itself?

2.13. The *trace* of a matrix A is the sum of the elements along the diagonal, that is

$$\operatorname{tr} A = \sum a_{ii}.$$

Let A be a 2×2 matrix, and let B be another 2×2 matrix such that $\det B \neq 0$. Show that

$$\operatorname{tr}(BAB^{-1}) = \operatorname{tr} A$$

and

$$\det(BAB^{-1}) = \det A.$$

2.14. Let $w(x) = Ax$ denote a linear transformation in the metric space $(\mathbb{R}^2, D)$ where

$$A = \begin{pmatrix} a & b \\ c & d \end{pmatrix}.$$

Define the *norm* of a point $x \in \mathbb{R}^2$ to be $|x| = D(x, O)$, where O denotes the origin. Define the norm of the linear transformation A by

$$|A| = \max \left\{ \frac{|Ax|}{|x|} : x \in \mathbb{R}^2, x \neq 0 \right\}$$

when this maximum exists. Show that $|A|$ is defined when D is the Euclidean metric and when it is the Manhattan metric. Find an expression for $|A|$ in terms of a, b, c, and d in each case. Make a geometrical interpretation of $|A|$. Show that when $|A|$ exists we have

$$|Ax| \leq |A| \cdot |x| \qquad \text{for all } x \in \mathbb{R}^2.$$

3   Möbius Transformations on the Riemann Sphere

Definition 3.1 *A transformation* $f : \hat{\mathbb{C}} \to \hat{\mathbb{C}}$ *defined by*

$$f(z) = \frac{(az + b)}{(cz + d)},$$

where $a, b, c,$ *and* $d \in \mathbb{C}$, $ad - bc \neq 0$, *is called a* Möbius transformation *on* $\hat{\mathbb{C}}$. *If* $c \neq 0$ *then* $f(-d/c) = \infty$, *and* $f(\infty) = a/c$. *If* $c = 0$, *then* $f(\infty) = \infty$.

As shown by the following exercises and examples, one can think of a Möbius transformation as follows. Map the whole plane $\mathbb{C}$, together with the point at infinity, onto the sphere $\hat{\mathbb{C}}$, as described in Chapter II. A sequence of operations is then applied to the sphere. Each operation is elementary and has the property that it takes circles to circles. The possible operations are rotation about an axis, rescaling (uniformly expand or contract the sphere), and translation (the whole sphere is picked up and moved to a new place on the plane, without rotation). Finally, the sphere is mapped back onto the plane in the usual way. Since the mappings back and forth from the plane to the sphere take straight lines and circles in the plane to circles on the sphere, we see that a Möbius transformation transforms the set of straight lines and circles in the plane onto itself. We also see that a Möbius transformation is invertible. It is wonderful how the quite complicated geometry of Möbius transformations is handled by straightforward complex algebra, where we simply manipulate expressions of the form $(az + b)/(cz + d)$.

Examples & Exercises

3.1. Show that the most general Möbius transformation, which maps ∞ to ∞, is of the form $f(z) = az + b, a, b \in \mathbb{C}, a \neq 0$, and that this is a similitude. Show that any

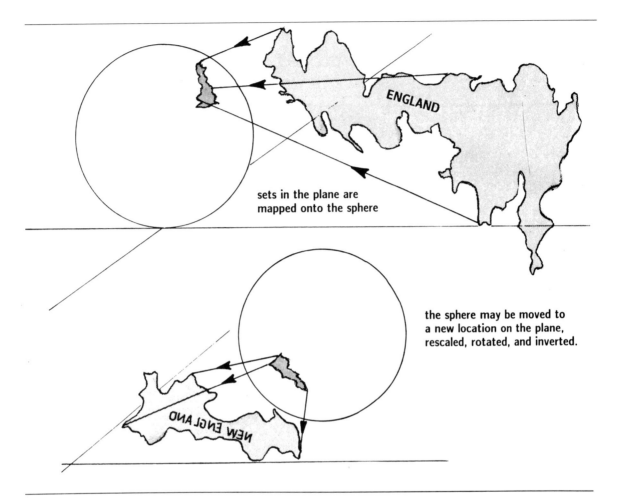

sets in the plane are
mapped onto the sphere

the sphere may be moved to
a new location on the plane,
rescaled, rotated, and inverted.

Figure III.47.   A Möbius transformation acting on England to produce a new country.

two-dimensional similitude that does not involve a reflection can be written in this
form. That is, disregarding changes in notation,

$$f(z) = f(x_1 + ix_2) = (a_1 + ia_2)(x_1 + ix_2) + (b_1 + ib_2)$$
$$= re^{i\theta}(x_1 + ix_2) + (b_1 + ib_2), \qquad (i = \sqrt{-1})$$
$$= \begin{pmatrix} r\cos\theta & -r\sin\theta \\ r\sin\theta & r\cos\theta \end{pmatrix} \begin{pmatrix} x_1 \\ x_2 \end{pmatrix} + \begin{pmatrix} b_1 \\ b_2 \end{pmatrix}.$$

Find r and θ in terms of a_1 and a_2. Show that the transformation can be achieved as
illustrated in Figure III.48.

3.2. Show that the Möbius transformation $f(z) = 1/z$ corresponds to first mapping
the plane to the sphere in such a way that the unit circle $\{z \in \mathbb{C} : |z| = 1\}$ goes to
the equator, followed by an inversion of the sphere (turn it upside down by rotating

Figure III.48. The mechanism of the similitude $f(z) = re^{i\theta}z + b$ in terms of the sphere.

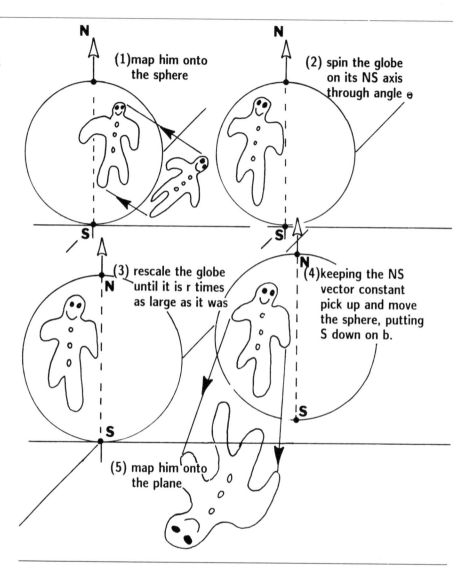

about an axis through $+1$ and -1 on the equator), and finally mapping back to the plane.

3.3. Show that any Möbius transformation that is not a similitude may be written

$$f(z) = e + \frac{f}{z + g} \qquad \text{for some } e, f, g \in \mathbb{C}, f \neq 0.$$

3.4. Sketch what happens to the picture in Figure III.49 under the Möbius transformation $f(z) = \frac{1}{z}$.

3.5. What happens to Figure III.49 under the Möbius transformation $f(z) = 1 + iz$?

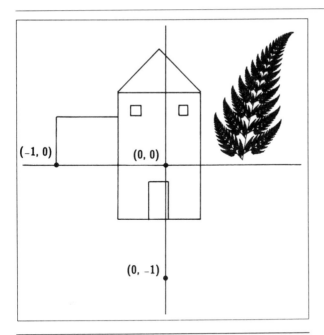

Figure III.49. Up the Garden Path. What does the Möbius transformation $z \mapsto 1 + iz$ do to this picture?

3.6. Show algebraically that a Möbius transformation $f : \hat{\mathbb{C}} \to \hat{\mathbb{C}}$ is always invertible.

3.7. Show that if f_1 and f_2 are Möbius transformations, then $f_1 \circ f_2$ is a Möbius transformation.

3.8. Find a Möbius transformation that takes the real line to the unit circle centered at the origin.

3.9. Evaluate $f^{\circ n}(z)$ if $f(z) = 1/(1 + z), n \in \{-2, -1, 0, 1, 2, 3, \ldots\}$.

3.10. Interpret the Möbius transformation $f(z) = i + 1/(z - i)$ in terms of operations on the sphere.

4 Analytic Transformations

In this section we continue the discussion of transformations on the metric spaces $(\mathbb{C}, \text{Euclidean})$ and $(\hat{\mathbb{C}}, \text{Spherical})$. We introduce a generalization of the Möbius transformations, called analytic transformations. We concentrate on the behavior of quadratic transformations. It is recommended that during a first reading or first course the reader obtains a good mental picture of how the quadratic transformation acts on the sphere. The reader may then want to study this section more closely after reading about Julia sets in Chapter VII.

The similitude $f : \hat{\mathbb{C}} \to \hat{\mathbb{C}}$ defined by the formula $f(z) = 3z + 1$ is an example of an analytic transformation. It maps circles to circles magnified by a factor of three. A disk with center at z_0 is taken to a disk with center at $f(z_0) = 3z_0 + 1$. The tranformation is continuous, and it maps open sets to open sets. Nowhere does it "fold back along the dotted line."

The similitude $f : \hat{\mathbb{C}} \to \hat{\mathbb{C}}$ defined by $f(z) = (3 + 3i)z + (1 - 2i)$ is similarly described. The circles and disks are now rotated by 45° in addition to being magnified and translated.

Loosely a transformation on $\hat{\mathbb{C}}$ is analytic if it is continuous and it locally "behaves like" a similitude. If you take a very small region indeed (How small? Small enough! There is a smallness such that what is about to be said is true!) and you watch what the transformation does to that tiny region, you will typically find that it is magnified or shrunk, rotated, and translated, in almost exactly the same manner that some similitude would do the job. The similitude will always be of the special type discussed in exercise 3.1 above.

We make this description more precise. Let us decide to look at what our transformation does in the vicinity of a point $z_0 \in \hat{\mathbb{C}}$. Assume that z_0 is not a critical point, defined below. Let T denote a tiny region, a disk for example, which contains the point z_0. Let $f(T)$ be its image under the transformation. Then one can rescale T by a factor that makes it roughly the size of the unit square, and one can rescale $f(T)$ by the same factor. The assertion of the previous paragraph is that the action of the transformation, viewed as taking T, rescaled, onto $f(T)$, rescaled, can be described more accurately by a similitude. If you like, one could consider a picture P drawn in T and examine the transformed image $f(P)$: if P and $f(P)$ are rescaled by the same factor so that P is the size of the unit square, then $f(P)$ looks more and more like a similitude applied to P. This description becomes more and more precise the tinier the region under discussion.

Consider the quadratic transformation $f : \hat{\mathbb{C}} \to \hat{\mathbb{C}}$ defined by

$$f(z) = z^2 = (x_1 + ix_2)^2 = (x_1^2 - x_2^2) + 2x_1x_2i = f_1(x_1, x_2) + f_2(x_1, x_2)i,$$

where $f_1(x_1, x_2) = (x_1^2 - x_2^2)$ is called the real part of $f(z)$, and $f_2(x_1, x_2) = 2x_1x_2$ is called the imaginary part of f. Pictures of what this transformation does to some Sierpinski triangles in $\mathbb{C}$ are illustrated in Figure III.50.

Two features are to be noticed. (I) Provided that we stay away from the origin, the transformation behaves locally like a similitude: for points z close to z_0, $f(z)$ is approximated by the similitude

$$w(z) = az + b \qquad \text{where } a = 2z_0 \text{ and } b = -z_0^2.$$

This fact shows up in Figure III.50: upon close examination (we suggest the use of a magnifying glass here) of the transformed Sierpinski triangles, one sees that they are built up out of small triangles whose shapes are only slightly different from that of their preimages. The only place where this is not true is at the forward image of the

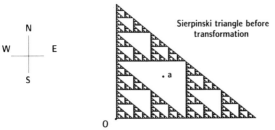

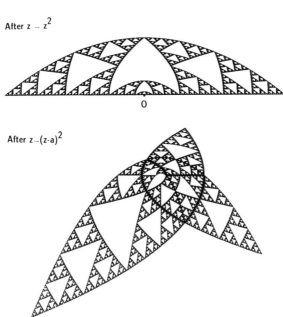

Figure III.50. Quadratic transformations are described by showing how they act on a Sierpinski triangle. Use a magnifying glass to check that the transformations behave locally like similitudes.

origin, which is a critical point. (II) The transformation maps the space twice around the origin.

One can track analytically what happens to the point

$$z = R \cos t + i R \sin t,$$

where $R > 0$. As the time parameter t goes from zero to 2π, z moves anticlockwise once around the circle of radius R. The transformed point $f(z)$ is given by

$$f(z) = R^2 \cos 2t + i R^2 \sin 2t.$$

As the time parameter t goes from 0 to 2π, $f(z)$ moves twice around the circle of radius R^2.

On the Riemann sphere the transformation $z \mapsto z^2$ can be described as follows. Let us say that the Equator corresponds to the circle of unit radius in the plane, that the South Pole corresponds to the Origin, and that the North Pole corresponds to the

Point at Infinity. Then the transformation leaves both Poles fixed. The Line of Longitude L connecting the Poles, which corresponds to the positive real axis, is mapped into itself, and the Equator is mapped into itself. Here is what we must picture. First, points that lie above the Equator are moved closer to the North Pole; points that lie below the Equator are moved closer to the South Pole; and the Equator is not shifted. Second, the skin of the sphere is cut along the Line of Longitude L. One side of the cut is held fixed while the other side is pulled around the sphere (following the terminator when the Sun is high above the Equator), uniformly stretching the space, until the edge of the cut is back over L. The two lips of the the cut are rejoined. The sphere has been mapped twice over itself. The Poles are the critical points of the transformation; they are the points about which wrapping occurs. This description is illustrated in Figure III.51.

The most general quadratic transformation on the sphere is expressible by a formula of the form $f(z) = Az^2 + Bz + C$, where A, B, and C are complex numbers.

Figure III.51. The action of the quadratic transformation $z \mapsto z^2$ in terms of the sphere.

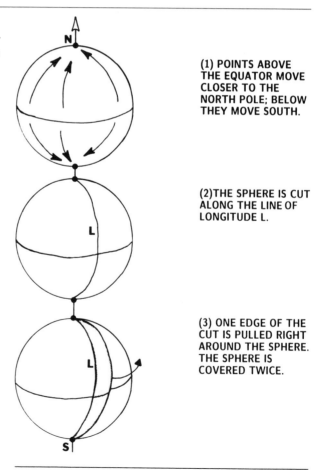

(1) POINTS ABOVE THE EQUATOR MOVE CLOSER TO THE NORTH POLE; BELOW THEY MOVE SOUTH.

(2) THE SPHERE IS CUT ALONG THE LINE OF LONGITUDE L.

(3) ONE EDGE OF THE CUT IS PULLED RIGHT AROUND THE SPHERE. THE SPHERE IS COVERED TWICE.

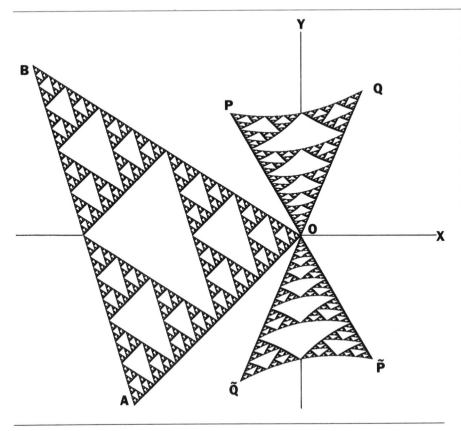

Figure III.52. The set valued inverse, f^1, of the quadratic transformation $f(z) = z^2$, maps the Sierpinski triangle AOB into $POQ \cup \tilde{P}O\tilde{Q}$. More generally, f^{-1} maps the Space of Fractals into itself. Look carefully at this image! Several important features of analytic transformations are illustrated here.

One can show there is a change of coordinates, $z \mapsto \theta(z)$, where θ is a similitude, such that $f(z)$ becomes expressible in the special form $f(z) = z^2 + \tilde{C}$ for some complex number $\tilde{C}$; see Exercise 5.20 in the following section. Hence the description of the most general quadratic transformation on the sphere can be made in the same terms as above, except that at the end there is a translation by some constant amount $\tilde{C}$. This translation leaves the Point at Infinity fixed.

The quadratic transformation $f(z) = z^2$ maps the punctured plane $\mathbb{C}$ onto itself twice. Each point on $z \in \mathbb{C}\{0\}$ has two preimages. Hence $f : \hat{\mathbb{C}} \to \hat{\mathbb{C}}$ is not an invertible transformation. In such situations we can define a set-valued inverse function.

Definition 4.1 *Let $f : \hat{\mathbb{C}} \to \hat{\mathbb{C}}$ be an analytic transformation such that $f(\hat{\mathbb{C}}) = \hat{\mathbb{C}}$. Then the* set-valued inverse *of f is the mapping $f^{-1} : \mathcal{H}(\hat{\mathbb{C}}) \to \mathcal{H}(\hat{\mathbb{C}})$ defined by*

$$f^{-1}(A) = \{w \in \hat{\mathbb{C}} : f(w) \in A\} \qquad \text{for all } A \in \mathcal{H}(X).$$

In Figure III.52 we illustrate the transformation f^{-1} acting on the Space of Fractals, in the case of the quadratic transformation $f(z) = z^2$.

One can obtain explicit formulas for $f^{-1}(z)$ when f is a quadratic transformation. For example for $f(z) = z^2$, $f^{-1}(O) = O$, $f^{-1}(\infty) = \infty$, and $f^{-1}(z) = \{w_1(z), w_2(z)\}$ for $z \in \hat{\mathbb{C}} \setminus \{0, \infty\}$. Here $w_1(x_1 + ix_2) = a(x_1, x_2) + ib(x_1, x_2)$, and $w_2(x_1, x_2) = -a(x_1, x_2) - ib(x_1, x_2)$, where

$$a(x_1, x_2) = \sqrt{\frac{\sqrt{x_1^2 + x_2^2} + x_1}{2}} \qquad \text{when } x_2 \geq 0,$$

$$a(x_1, x_2) = -\sqrt{\frac{\sqrt{x_1^2 + x_2^2} + x_1}{2}} \qquad \text{when } x_2 < 0,$$

$$b(x_1, x_2) = \sqrt{\frac{\sqrt{x_1^2 + x_2^2} - x_1}{2}}.$$

Each of the two functions $w_1(z)$ and $w_2(z)$ is itself analytic on $\mathbb{C} \setminus \{0, \infty\}$.

The following definition formalizes what is meant by an analytic transformation on the complex plane. We recommend further reading, for example [Rudin, 1966].

Definition 4.2 *Let $(\mathbb{C}, d)$ denote the complex plane with the Euclidean metric. A transformation $f : \mathbb{C} \to \mathbb{C}$ is called* analytic *if for each $z_0 \in \mathbb{C}$ there is a similitude of the form*

$$w(z) = az + b, \qquad \text{for some pair of numbers } a, b \in \mathbb{C},$$

such that $d(f(z), w(z))/d(z, z_0) \to 0$ as $z \to z_0$. The numbers a and b depend on z_0. If, corresponding to a certain point $z_0 = c$, we have $a = 0$, then c is called a critical point *of the transformation, and $f(c)$ is called a* critical value.

If the analytic transformation $f(z)$ is a rational transformation, which means that it is expressible as a ratio of two polynomials in z, such as

$$\text{(i)} f(z) = 1 + 2i + 27z^2 - 9z^3,$$

$$\text{(ii)} f(z) = \frac{1 + z}{1 - z},$$

$$\text{(iii)} f(z) = \frac{1 + z + z^2}{1 - z + z^2};$$

then the numbers a and b in the similitude $w(z)$ in Definition 4.2 are given by the formulas

$$a = f'(z_0) \text{ and } b = f(z_0) - az_0.$$

The derivative $f'(z)$ of the rational function $f(z)$ can be calculated by treating z as though it were the real variable x and applying the standard differentiation rules of calculus. The critical points $c \in \mathbb{C}$ are the solutions of the equation $f'(c) = 0$.

For example, close enough to any point $z_0 \in \mathbb{C}$ such that $f'(z_0) \neq 0$, the cubic transformation (i) is well described by the similitude

$$w(z) = (54z_0 - 27z_0^2)z + (1 + 2i - 27z_0^2 + 18z_0^3).$$

The finite critical points associated with (i) may be obtained by solving

$$54c - 27c^2 = 0$$

and are accordingly $c = 0 + i0$ and $c = 2 + i0$. By making the change of coordinates $z' = 1/z$ (see section 5), one can also analyze the behavior near the point at infinity. It turns out that $c = \infty$ is always a critical point for a polynomial transformation $f(z)$ on $\hat{\mathbb{C}}$. The space is "wrapped" an integral number of times about the image of critical point. For example, the cubic transformation (i) wraps space twice about each of the points $f(0 + i0) = 1 + 2i$, and $f(2 + i0) = 37 + 2i$, and it wraps it three times about $f(\infty) = \infty$.

Examples & Exercises

4.1. Sketch a globe representing $\hat{\mathbb{C}}$, including a subset that looks like Africa, and show what happens to the subset under the quadratic transformation $f(z) = z^2$.

4.2. Verify the following explicit formulas for $f^{-1}(z)$, corresponding to $f(z) = z^2 - 1$: $f^{-1}(-1) = 0$; $f^{-1}(\infty) = \infty$; and $f^{-1}(z) = \{w_1(z), w_2(z)\}$ for $z \in \hat{\mathbb{C}} \setminus \{-1, \infty\}$, where $w_1(x_1 + ix_2) = a(x_1, x_2) + ib(x_1, x_2)$, and $w_2(x_1, x_2) = -a(x_1, x_2) - ib(x_1, x_2)$. Here

$$a(x_1, x_2) = \sqrt{\frac{\sqrt{(1 + x_1)^2 + x_2^2} + 1 + x_1}{2}} \qquad \text{when } x_2 \geq 0,$$

$$a(x_1, x_2) = -\sqrt{\frac{\sqrt{(1 + x_1)^2 + x_2^2} + 1 + x_1}{2}} \qquad \text{when } x_2 < 0,$$

and

$$b(x_1, x_2) = \sqrt{\frac{\sqrt{(1 + x_1)^2 + x_2^2} - 1 - x_1}{2}}.$$

Both $w_1(z)$ and $w_2(z)$ are analytic on $\mathbb{C} \setminus \{-1\}$.

4.3. Locate the critical points and critical values of the quadratic transformation $f(z) = z^2 + 1$.

4.4. Draw a side view of a man with an arm stretched out in front of him, holding a knife. The blade should point down. Choose the origin of coordinates to be his navel. Draw another picture to explain how *hara-kiri* can be achieved by applying the inverse of the quadratic transformation $f(z) = z^2$ to your image.

4.5. Find a similitude that approximates the behavior of the given analytic transformation in the vicinity of the given point: (a) $f(z) = z^2$ near $z_0 = 1$; (b) $f(z) = 1/z$ near $z_0 = 1 + i$; (c) $f(z) = (z - 1)^3$ near $z_0 = 1 - i$.

---

5 How to Change Coordinates

In describing transformations on spaces we usually make use of an underlying coordinate system. Most spaces have a coordinate system by means of which the points in the space are located. This underlying coordinate system is implied by the specification of the space: for example, $\mathbf{X} = [1, 2]$ provides a collection of points together with the natural coordinate x restricted by $l \leq x \leq 2$. We can think of either the space, made of points $x \in \mathbf{X}$, or equivalently the system of coordinates. If the space $\mathbf{X}$ is $\mathbb{R}^2$ or $\mathbb{C}$, then the underlying coordinate system may be Cartesian coordinates. If $\mathbf{X} = \hat{\mathbb{C}}$, then the coordinate system may be angular coordinates on the sphere.

In each case the underlying coordinate system is itself a subset of a metric space. We denote this metric space by $\mathbf{X}_C$. Usually we do not consciously distinguish between a point $x \in \mathbf{X}$ and its coordinate $x \in \mathbf{X}_C$. Notice, however, that the space $\mathbf{X}_C$ may contain points (coordinates) that do not correspond to any point in the space $\mathbf{X}$. For example, in the case of the space $\mathbf{X} = \blacksquare$ it is natural to take $\mathbf{X}_C = \mathbb{R}^2$; then points $x \in \mathbf{X}$ in the space correspond to coordinates $x = (x_1, x_2) \in \mathbf{X}_C$ restricted by $0 \leq x_1 \leq 1$ and $0 \leq x_2 \leq 1$. However, the coordinates $(3, 5) \in \mathbf{X}_C$ do not correspond to a point in $\mathbf{X}$. We would like the reader to think of the space itself as "lying above" its coordinate system, as suggested in Figure III.53.

A change of coordinate system may be described by a transformation $\theta : \mathbf{X}_C \to \mathbf{X}_C$. We can think of a change of coordinates being effected by physically moving each point $x \in \mathbf{X}$ so that it no longer lies above $x \in \mathbf{X}_C$ but instead above the coordinate $x' = \theta(x) \in \mathbf{X}_C$. Thus we must now distinguish between a point x lying in the space, $\mathbf{X}$, from its coordinate $x \in \mathbf{X}_C$. Then we want to think of the change of coordinates $\theta : \mathbf{X}_C \to \mathbf{X}_C$ as moving $\mathbf{X}$ relative to the underlying coordinate space $\mathbf{X}_C$, as illustrated in Figure III.54.

Example

5.1. Let $\mathbf{X} = [1, 2]$ and take $\mathbf{X}_C$ to be $\mathbb{R}$. Let $\theta : \mathbb{R} \to \mathbb{R}$ be defined by $\theta(x) = 2x + 1$. Then the coordinate of the point $x = 1.5$ becomes changed to 4. We want to think of the space $\mathbf{X}$ as being moved relative to the coordinate space $\mathbf{X}_C$, which is held fixed , as illustrated in Figure III.55.

Let $\theta : \mathbf{X}_C \to \mathbf{X}_C$ denote a change of coordinates. In order that the new coordinate system be useful, it is usually necessary that θ, treated as a transformation from $\mathbf{X}$ to $\theta(\mathbf{X})$, be one-to-one and onto, and hence invertible. Let $f : \mathbf{X} \to \mathbf{X}$ be a transformation on a metric space $\mathbf{X}$. We want to consider how the transformation f should

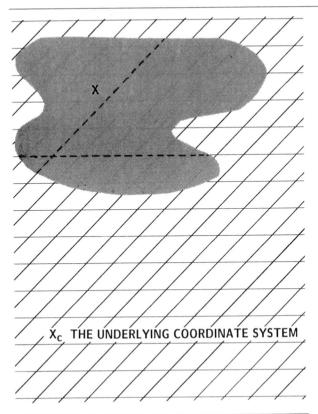

Figure III.53.   The underlying coordinate system $\mathbf{X}_C$ for the space $\mathbf{X}$.

$\mathbf{X}_C$ THE UNDERLYING COORDINATE SYSTEM

be expressed after the change of coordinates. Let x denote simultaneously a point in $\mathbf{X}$ and the coordinates of that point. Let $f(x)$ denote simultaneously the point to which x is transformed by f, and the coordinates of that point. Let x' denote the point $x \in \mathbf{X}$ in the new coordinate system. That is, $x' = \theta(x) \in \mathbf{X}_C$ denotes the new coordinates of the point x. Let $f'(x')$ denote the same transformation $f : \mathbf{X} \to \mathbf{X}$, but expressed in the new coordinate system. Then the relation between the two coordinate systems is expressed by the commutative diagram in Figure III.57, and is illustrated in Figure III.56.

Theorem 5.1 *Let $\mathbf{X}$ be a space and let $\mathbf{X}_C \supset \mathbf{X}$ be a coordinate space for $\mathbf{X}$. Let a change of coordinates be provided by a transformation $\theta : \mathbf{X}_C \to \mathbf{X}_C$. Let θ be invertible when treated as a transformation from $\mathbf{X}$ to $\theta(\mathbf{X})$. Let the coordinates of a point $x \in \mathbf{X}$ be denoted by x before the change of coordinates, and by x' after the change of coordinates, so that*

$$x' = \theta(x).$$

Let $f : \mathbf{X} \to \mathbf{X}$ be a transformation on the space $\mathbf{X}$. Let $x \mapsto f(x)$ be the formula for f expressed in the original coordinates. Let $x' \mapsto f'(x')$ be the formula for f expressed in the new coordinates. Then

Figure III.54. A change of coordinates in terms of **X** and $\mathbf{X}_C$. We think of **X** as being removed relative to the underlying coordinate space $\mathbf{X}_C$.

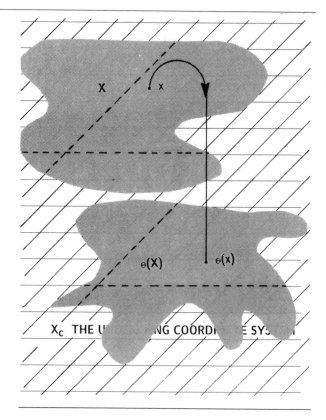

$$f(x) = (\theta^{-1} \circ f' \circ \theta)(x),$$
$$f'(x') = (\theta \circ f \circ \theta^{-1})(x').$$

Examples & Exercises

5.2. Consider an affine transformation $f(x) = ax + b$, $a \neq 0$, $a \neq 1$, $a, b \in \mathbb{R}$. This has a fixed point $x_f \in \mathbb{R}$ defined by $f(x_f) = x_f$. We find $x_f = b/(1-a)$. x_f is clearly the interesting point in the action of an affine transformation on $\mathbb{R}$. Accordingly let us change coordinates to move x_f to the origin: that is $x' = \theta(x) = x - x_f$. What does f look like in this new coordinate system?

$$f'(x') = (\theta \circ f \circ \theta^{-1})(x') = \theta \circ (x' + x_f) = a(x' + x_f) + b - x_f;$$

$f'(x') = ax'$, which is simply a rescaling! Now using the first formula we get

Figure III.55. A change of coordinates for the space $[1, 2]$ given by the transformation $x' = \theta(x) = 2x + 1$.

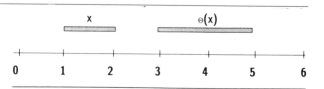

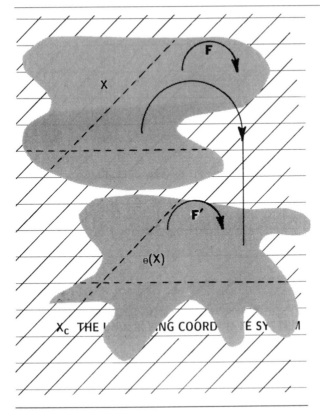

Figure III.56. The transformation F acting on $\mathbf{X}$ is equivalent to F' acting on $\theta(\mathbf{X})$.

$$f(x) = a(x - x_f) + x_f$$

and

$$f^{\circ n}(x) = a^n(x - x_f) + x_f \qquad \text{for all } n \in \{0, 1, 2, 3, \ldots\}.$$

We now see a new way of visualizing an affine transformation on $\mathbb{R}$: for example, if $a > 1$, we see the image in Figure III.58.

5.3. Show that for any affine transformation $f(x) : \mathbb{R}^2 \to \mathbb{R}^2$ given by $f(x) = Ax + t$, with fixed point x_f, that the coordinate transformation $\theta(x) = x - x_f$ transforms the function $f'(x') = Ax'$.

5.4. Let $\mathbf{X} = [1, 2]$ and let a change of coordinates be defined by $x' = 2x - 1$. Let a transformation $f : \mathbf{X} \to \mathbf{X}$ be defined by $f(x) = (x - 1)^2 + 1$. Express f in the new coordinate system.

Definition 5.1 *Let $f : \mathbf{X} \to \mathbf{X}$ be a transformation on a metric space. A point $x_f \in X$ such that $f(x_f) = x_f$ is called a* fixed point *of the transformation.*

The fixed points of a transformation are very important. They tell us which parts of the space are pinned in place, not moved, by the transformation. The fixed points of a transformation restrict the motion of the space under nonviolent, nonripping transformations of bounded deformation.

Examples & Exercises

5.5. Find the fixed points x_1 and x_2 of the Möbius transformation

$$f(z) = \frac{(z+2)}{(4-z)}$$

on $\hat{\mathbb{C}}$. Make a change of coordinates so that x_1 becomes the origin and x_2 becomes the point at infinity. Hence interpret the action of $f(z)$ on the sphere in geometrical terms.

5.6. Let $W(x) = Ax + t$ where $\det A \neq 0$ is a two-dimensional affine transformation acting on the space $\mathbf{X} = \mathbb{R}^2$. Find the fixed point x_f. Change coordinates so that x_f becomes the origin of coordinates. Hence describe the action geometrically of a two-dimensional, nondegenerate affine transformation. What can happen if $\det A = 0$?

5.7. Suppose we can find a coordinate transformation $BAB^{-1} = D$, where D is a diagonal matrix we denote by

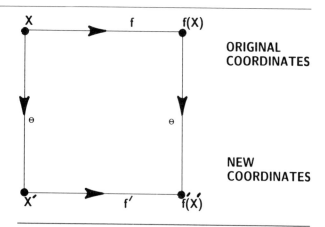

Figure III.57. Commutative diagram for the coordinate change $\theta : \mathbf{X}_C \to \mathbf{X}_C$.

ORIGINAL COORDINATES

NEW COORDINATES

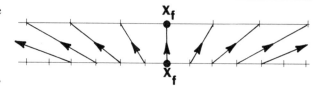

Figure III.58. An affine transformation on $\mathbb{R}$. We see rescaling (magnification or diminution) centered at the fixed point, together with a flip of 180° if $a < 0$.

$$D = \begin{pmatrix} \lambda_1 & 0 \\ 0 & \lambda_2 \end{pmatrix}.$$

Show that λ_1 and λ_2 satisfy the equation

$$\det \begin{pmatrix} e-\lambda & f \\ g & h-\lambda \end{pmatrix} = \begin{vmatrix} e-\lambda & f \\ g & h-\lambda \end{vmatrix} = \lambda^2 - \operatorname{tr} A\lambda + \det A = 0.$$

5.8. Analyze the behavior of the affine transformation $w(z) = 7z + 1$ on $\hat{\mathbb{C}}$ near the point at infinity by making the change of coordinates $h(z) = 1/z$.

5.9. Two one-parameter families of transformations on $\mathbb{R}$ are $f_\mu(x) = x^2 - \mu$ and $g_\lambda(x) = \lambda x(1-x)$, where μ and λ are real parameters. Find a change of coordinates and a function $\mu = \mu(\lambda)$ so that $f'_{\mu(\lambda)}(x') = g_\lambda(x')$ is valid for an appropriate interval on the λ-axis.

5.10. Find the real fixed points of $g(x) = x^2 - \frac{1}{2}$. Analyze the behavior of g near each of its fixed points by changing coordinates so as to move first one then the other to the origin. Another method for looking at the behavior of g near a fixed point is to approximate $g(x)$ by the first two terms of its Taylor series expansion about the fixed point. Compare these methods.

5.11. Suppose that the 2×2 matrix

$$A = \begin{pmatrix} e & f \\ g & h \end{pmatrix}$$

satisfies the condition $(\operatorname{tr} A)^2 - 4 \det A > 0$. Show that there is a B such that

$$BAB^{-1} = D,$$

where D is a diagonal matrix. Furthermore, show that one choice for B is given by

$$B = \begin{pmatrix} 1 & \frac{f}{\lambda_1 - h} \\ 1 & \frac{f}{\lambda_2 - h} \end{pmatrix}.$$

What do you think happens if $(\operatorname{tr} A)^2 - 4 \det A < 0$?

5.12. Let $w : \mathbb{R}^2 \to \mathbb{R}^2$ denote the affine transformation

$$w \begin{pmatrix} x_1 \\ x_2 \end{pmatrix} = \begin{pmatrix} 1 & 2 \\ 2 & 3 \end{pmatrix} \begin{pmatrix} x_1 \\ x_2 \end{pmatrix} + \begin{pmatrix} 1 \\ 1 \end{pmatrix}.$$

Make a change of coordinates so that the transformation is simply a coordinate rescaling. What are the rescaling factors?

Definition 5.2 *Let F denote a set of transformations on a metric space $\mathbf{X}$. F is called a* semigroup *if $f, g \in F$ implies $f \circ g \in F$. F is called a* group *if it is a semigroup of invertible transformations, and $f \in F$ implies $f^{-1} \in F$.*

We introduce this definition because we will use semigroups (and groups) of transformations both to characterize and to compute fractal subsets of $\mathbf{X}$. However, we do not use any deep theorems from group theory.

Examples & Exercises

5.13. Let $f : \mathbf{X} \to \mathbf{X}$ be a transformation on a metric space. Show that the set of transformations $\{f^{\circ n} : n=0, 1, 2, 3, \ldots\}$ forms a semigroup.

5.14. A transformation $T : \Sigma \to \Sigma$ on code space is defined by

$$T(x_1 x_2 x_3 x_4 x_5 \ldots) = x_2 x_3 x_4 x_5 x_6 \ldots$$

and is called a *shift operator*. Describe the semigroup of transformations $\{T^{\circ n} : n = 0, 1, 2, 3, \ldots\}$. What are the *fixed points* of $T^{\circ 3}$ if the code space is built up from the two symbols $\{0, 1\}$?

5.15. Show that the set of Möbius transformations on $\hat{\mathbb{R}}$ forms a group.

5.16. Show that the set of Möbius transformations on $\hat{\mathbb{C}}$ forms a group.

5.17. Show that the set of invertible affine transformations on $\mathbb{R}^2$ forms a group.

5.18. Show that the set of transformations $f : \mathbb{R}^2 \to \mathbb{R}^2$ such that $f(\Delta) \subset \Delta$ forms a semigroup.

5.19. Show that a group of transformations is provided by the set of affine transformations of the form $w(x) = Ax + t$, where $A = \begin{pmatrix} a & 0 \\ b & c \end{pmatrix}$ for $a, b, c \in \mathbb{R}$, with $ac \neq 0$, and the translation vector t is arbitrary.

5.20. The most general analytic quadratic transformation $f : \hat{\mathbb{C}} \to \hat{\mathbb{C}}$ can be expressed by a formula of the form $f(z) = Az^2 + Bz + C$, where A, B, and C are complex numbers, and $A \neq 0$. Show that by means of a suitable change of coordinates, $z' = \theta(z)$, where θ is a similitude, show that $f(z)$ can be reexpressed as a quadratic transformation of the special form $f'(z) = (z')^2 + \tilde{C}$ for some complex number $\tilde{C}$.

6   The Contraction Mapping Theorem

Definition 6.1 *A transformation $f : \mathbf{X} \to \mathbf{X}$ on a metric space $(\mathbf{X}, d)$ is called* contractive *or a* contraction mapping *if there is a constant $0 \leq s < 1$ such that*

$$d(f(x), f(y)) \leq s \cdot d(x, y) \forall x, \qquad y \in \mathbf{X}.$$

Any such number s is called a contractivity factor *for f.*

It would be convenient to be able to talk about the largest number and the smallest number in a set of real numbers. However, a set such as $S = (-\infty, 3)$ does not possess either. This difficulty is overcome by the following definition.

Definition 6.2 *Let S denote a set of real numbers. Then the* infimum *of S is equal to $-\infty$ if S contains negative numbers of arbitrarily large magnitude. Otherwise the infimum of $S = \max\{x \in \mathbb{R} : x \leq s$ for all $s \in S\}$. The infimum of S always* *exists because of the nature of the real number system, and it is denoted by* inf *S. The* supremum *of S is similarly defined. It is equal to* $+\infty$ *if S contains arbitrarily large numbers; otherwise it is the minimum of the set of numbers that are greater than or equal to all of the numbers in S. The supremum of S always exists, and it is denoted by* sup *S.*

Examples & Exercises

6.1. Find the supremum and the infimum of the following sets of real numbers: (a) $(-\infty, 3)$; (b) $\mathcal{C}$, the Classical Cantor Set; (c) $\{1, 2, 3, 4, \ldots\}$; (d) the positive real numbers.

6.2. Let $f : \mathbf{X} \to \mathbf{X}$ be a contraction mapping on a compact metric space $(\mathbf{X}, d)$. Show that $\inf\{s \in \mathbb{R} : s$ is a contractivity factor for $f\}$ is a contractivity factor for f.

6.3. Show that if $f : \mathbf{X} \to \mathbf{X}$ and $g : \mathbf{X} \to \mathbf{X}$ are contraction mappings on a space $(\mathbf{X}, d)$, with contractivity factors s and t, respectively, then $f \circ g$ is a contraction mapping with contractivity factor st.

Theorem 6.1 *[(The Contraction Mapping Theorem).] Let $f : \mathbf{X} \to \mathbf{X}$ be a contraction mapping on a complete metric space $(\mathbf{X}, d)$. Then f possesses exactly one fixed point $x_f \in \mathbf{X}$ and moreover for any point $x \in \mathbf{X}$, the sequence $\{f^{\circ n}(x) : n = 0, 1, 2, \ldots\}$ converges to x_f. That is,*

$$\lim_{n \to \infty} f^{\circ n}(x) = x_f, \qquad \text{for each } x \in \mathbf{X}.$$

Figure III.59 illustrates the idea of a contractive transformation on a compact metric space.

Proof Let $x \in \mathbf{X}$. Let $0 \le s < 1$ be a contractivity factor for f. Then

$$d(f^{\circ n}(x), f^{\circ m}(x)) \le s^{m \wedge n} d(x, f^{\circ |n-m|})(x) \tag{1}$$

for all $m, n = 0, 1, 2, \ldots$, where we have fixed $x \in \mathbf{X}$. The notation $u \wedge v$ denotes the minimum of the pair of real numbers u and v. In particular, for $k = 0, 1, 2, \ldots$, we have

$$d(x, f^{\circ k}(x)) \le d(x, f(x)) + (f(x), f^{\circ 2}(x)) + \cdots + d(f^{\circ (k-1)}(x), f^{\circ k}(x))$$
$$\le (1 + s + s^2 + \cdots + s^{k-1})d(x, f(x))$$
$$\le (1 - s)^{-1}d(x, f(x)),$$

so substituting into equation (1) we now obtain

$$d(f^{\circ n}(x), f^{\circ m}(x)) \le s^{m \wedge n} \cdot (1 - s)^{-1} \cdot (d(x, f(x)),$$

from which it immediately follows that $\{f^{\circ n}(x)\}_{n=0}^{\infty}$ is a Cauchy sequence. Since $\mathbf{X}$ is complete this Cauchy sequence possesses a limit $x_f \in \mathbf{X}$, and we have

$$\lim_{n \to \infty} f^{\circ n}(x) = x_f.$$

Figure III.59. (a) Illustrates the idea of a contractive transformation on a metric space. (b) A contraction mapping doing its work, drawing all of a compact metric space **X** toward the fixed point.

(a)

(b)

Now we shall show that x_f is a fixed point of f. Since f is contractive it is continuous and hence

$$f(x_f) = f(\lim_{n \to \infty} f^{\circ n}(x)) = \lim_{n \to \infty} f^{\circ(n+1)}(x) = x_f.$$

Finally, can there be more than one fixed point? Suppose there are. Let x_f and y_f be two fixed points of f. Then $x_f = f(x_f)$, $y_f = f(y_f)$, and

$$d(x_f, y_f) = d(f(x_f), f(x_f)) \le s d(x_f, y_f),$$

where $(1 - s)d(x_f, y_f) \le 0$, which implies $d(x_f, y_f) = 0$ and hence $x_f = y_f$. This completes the proof.

Examples & Exercises

6.4. Let $w(x) = Ax + t$ be an affine transformation in two dimensions. Make the change of coordinates $h(x) = x' = x - x_f$, under the assumption that $\det(I - A) \neq 0$, and show that $w'(x') = h \circ w \circ h^{-1}(x') = Ax'$, that $w(x) = (h^{-1} \circ w' \circ h)(x) = A(x - x_f) + x_f$, and hence that

$$w^{\circ n}(x) = A^n(x - x_f) + x_f \qquad \text{for } n = 0, 1, 2, 3, \ldots. \tag{2}$$

Give conditions on A such that it is contractive (a) in the Euclidean metric, and (b) in the Manhattan metric. Show that if $|A| < 1$, where $|A|$ denotes any appropriate norm of A viewed as a linear operator on a two-dimensional vector space, then $\{w^{\circ n}(x)\}$ is a Cauchy sequence that converges to x_f, for each $x \in \mathbb{R}^2$.

6.5. Let $f : \blacksquare \to \blacksquare$ be a contraction mapping on ($\blacksquare$, Euclidean). Show that Figure III.59 gives the right idea.

6.6. Let $f : \mathbb{R} \to \mathbb{R}$ be the affine transformation $f(x) = \frac{1}{2}x + \frac{1}{2}$. Verify f is a contraction mapping and deduce

$$\lim_{n \to \infty} f^{\circ n}(x) = x_f \qquad \text{for each } x \in \mathbb{R}.$$

Use this formula with $x = 1$ to obtain a geometrical series for the fixed point $x_f \in \mathbb{R}$. Observe, however, $f(\mathbb{R}) = \mathbb{R}$; indeed f is invertible.

6.7. Let (X, d) be a compact metric space that contains more than one point. Show that the situation in exercise 6.6 cannot occur for any contraction mapping $f : X \to X$. That is, show that $f(X) \subset X$ but $f(X) \neq X$. That is, show that a contraction mapping on a nontrivial compact metric space is not invertible. Hint: use the compactness of the space to show that there is a point in the space that is farthest away from the fixed point. Then show that there is a point that is not in $f(X)$.

6.8. Show that the set of contraction mappings on a metric space forms a semi-group.

6.9. Show that the affine transformation $w : \Delta \to \Delta$ defined by $w(x) = Ax + t$ is a contraction, where

$$A = \begin{pmatrix} \frac{1}{2}\cos 120° & -\frac{1}{2}\sin 120° \\ \frac{1}{2}\sin 120° & \frac{1}{2}\cos 120° \end{pmatrix} \quad \text{and } t = \begin{pmatrix} \frac{1}{2} \\ 0 \end{pmatrix}.$$

Here Δ is an equilateral Sierpinski triangle with a vertex at the origin and one at $(1, 0)$. You need to begin by verifying that w does indeed map Δ into itself! Locate the fixed point x_f. Make a picture of this contraction mapping "doing its work, mapping all of the compact metric space Δ toward the fixed point." Use different colors to denote the successive regions $f^{\circ(n)}(\Delta) \setminus f^{\circ(n+1)}(\Delta)$ for $n = 0, 1, 2, 3, \ldots$.

Figure III.60. The existence of a positive eigenvalue of an "angle-squeezing" linear transformation.

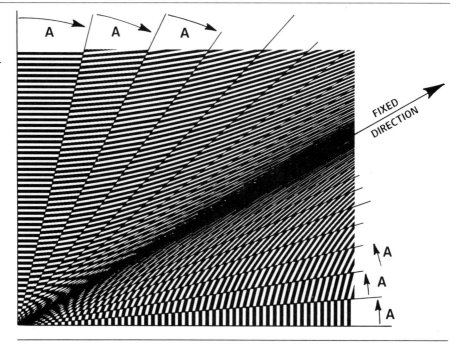

6.10. Define a mapping on the code space of two symbols $\{0, 1\}$ by $f(x_1 x_2 x_3 x_4 \ldots)$ $= 1 x_1 x_2 x_3 x_4 \ldots$. (Recall that the metric is $d(x, y) = \sum_{i=1}^{\infty} \frac{|x_i - y_i|}{3^i}$, or equivalent.) Show that f is a contraction mapping. Locate the fixed point of f.

6.11. Let $(\mathbf{X}, d)$ be a compact metric space, and let $f : \mathbf{X} \to \mathbf{X}$ be a contraction mapping. Show that $\{f^{\circ n}(\mathbf{X})\}_{n=0}^{\infty}$ is a Cauchy sequence of points in $(\mathcal{H}(\mathbf{X}), h)$ and $\lim_{n \to \infty} f^{\circ n}(\mathbf{X}) = \{x_f\}$, where x_f is the fixed point of f.

6.12. Let $(\mathbf{X}, d)$ be a compact metric space. Let $f : \mathbf{X} \to \mathbf{X}$ have the property $\lim_{n \to \infty} f^{\circ n}(\mathbf{X}) = x_f$. Find a metric $\tilde{d}$ on $\mathbf{X}$ such that f is a contraction mapping, and the identity is a homeomorphism from $(\overline{\mathbf{X}}, d) \to (\overline{\mathbf{X}}, \tilde{d})$.

6.13. Let $Ax = \begin{pmatrix} a & b \\ c & d \end{pmatrix} \begin{pmatrix} x_1 \\ x_2 \end{pmatrix}$ with $a, b, c, d \in \mathbb{R}$, all strictly positive, be a linear transformation on $\mathbb{R}^2$. Show that A maps the positive quadrant $\{(x_1, x_2) : x_1 \geq 0, x_2 \geq 0\}$ into itself. Let a mapping $f : [0, 90°] \to [0, 90°]$ be defined by

$$A \begin{pmatrix} \cos \theta \\ \sin \theta \end{pmatrix} = \text{(some positive number)} \begin{pmatrix} \cos f(\theta) \\ \sin f(\theta) \end{pmatrix}.$$

Show that $\{f^{\circ n}(\theta)\}$ converges to the unique fixed point of f. Deduce that there exists a unique positive number λ, and an angle $0 < \theta < 90°$ such that $A \begin{pmatrix} \cos \theta \\ \sin \theta \end{pmatrix} = \lambda \begin{pmatrix} \cos \theta \\ \sin \theta \end{pmatrix}$. See Figure III.60.

7   Contraction Mappings on the Space of Fractals

Let $(\mathbf{X}, d)$ be a metric space and let $(\mathcal{H}(\mathbf{X}), h(d))$ denote the corresponding space of nonempty compact subsets, with the Hausdorff metric $h(d)$. We introduce the notation $h(d)$ to show that d is the underlying metric for the Hausdorff metric h. For example, we may discuss $(\mathcal{H}(\hat{\mathbb{C}}), h(\text{spherical}))$ or $(\mathcal{H}(\mathbb{R}^2), h(\text{Manhattan}))$. We will drop this additional notation when we evaluate Hausdorff distances.

We have repeatedly refused to define fractals: we have agreed that they are subsets of simple geometrical spaces, such as $(\mathbb{R}^2, \text{Euclidean})$ and $(\hat{\mathbb{C}}, \text{Spherical})$. If we were to define a *deterministic fractal*, we *might* say that it is a fixed point of a contractive transformation on $(\mathcal{H}(\mathbf{X}), h(d))$. We would require that the underlying metric space $(\mathbf{X}, d)$ be "geometrically simple." We would require also that the contraction mapping be constructed from simple, easily specified, contraction mappings on $(\mathbf{X}, d)$, as described below.

Lemma 7.1 *Let $w : \mathbf{X} \to \mathbf{X}$ be a contraction mapping on the metric space $(\mathbf{X}, d)$. Then w is continuous.*

Proof Let $\epsilon > 0$ be given. Let $s > 0$ be a contractivity factor for w. Then

$$d(w(x), w(y)) \leq s d(x, y) < \epsilon$$

whenever $d(x, y) < \delta$, where $\delta = \epsilon/s$. This completes the proof.

Lemma 7.2 *Let $w : \mathbf{X} \to \mathbf{X}$ be a continuous mapping on the metric space $(\mathbf{X}, d)$. Then w maps $\mathcal{H}(\mathbf{X})$ into itself.*

Proof Let S be a nonempty compact subset of $\mathbf{X}$. Then clearly $w(S) = \{w(x) : x \in S\}$ is nonempty. We want to show that $w(S)$ is compact. Let $\{y_n = w(x_n)\}$ be an infinite sequence of points in S. Then $\{x_n\}$ is an infinite sequence of points in S. Since S is compact there is a subsequence $\{x_{N_n}\}$ that converges to a point $\hat{x} \in S$. But then the continuity of w implies that $\{y_{N_n} = f(x_{N_n})\}$ is a subsequence of $\{y_n\}$ that converges to $\hat{y} = f(\hat{x}) \in w(S)$. This completes the proof.

The following lemma tells us how to make a contraction mapping on $(\mathcal{H}(\mathbf{X}), h)$ out of a contraction mapping on $(\mathbf{X}, d)$.

Lemma 7.3 *Let $w : \mathbf{X} \to \mathbf{X}$ be a contraction mapping on the metric space $(\mathbf{X}, d)$ with contractivity factor s. Then $w : \mathcal{H}(\mathbf{X}) \to \mathcal{H}(\mathbf{X})$ defined by*

$$w(B) = \{w(x) : x \in B\} \forall B \in \mathcal{H}(\mathbf{X})$$

is a contraction mapping on $(\mathcal{H}(\mathbf{X}), h(d))$ with contractivity factor s.

Proof From Lemma 7.1 it follows that $w : \mathbf{X} \to \mathbf{X}$ is continuous. Hence by Lemma 7.2 w maps $\mathcal{H}(\mathbf{X})$ into itself. Now let $B, C \in \mathcal{H}(\mathbf{X})$. Then

$$d(w(B), w(C)) = \max\{\min\{d(w(x, y), w(y)) : y \in C\} : x \in B\}$$
$$\leq \max\{\min\{s \cdot d(x, y) : y \in C\} : x \in B\} = s \cdot d(B, C).$$

Similarly, $d(w(C), w(B)) \leq s \cdot d(C, B)$. Hence

$$h(w(B), w(C)) = d(w(B), w(C)) \vee d(w(C), w(B)) \leq s \cdot d(B, C) \vee d(C, B)$$
$$\leq s \cdot d(B, C).$$

This completes the proof.

The following lemma gives a characteristic property of the Hausdorff metric which we will shortly need. The proof follows at once from exercise 6.13 of Chapter II.

Lemma 7.4 *For all B, C, D, and E, in $\mathcal{H}(\mathbf{X})$*

$$h(B \cup C, D \cup E) \leq h(B, D) \vee h(C, E),$$

where as usual h is the Hausdorff metric.

The next lemma provides an important method for combining contraction mappings on $(\mathcal{H}(\mathbf{X}), h)$ to produce new contraction mappings on $(\mathcal{H}(\mathbf{X}), h)$. This method is distinct from the obvious one of composition.

Lemma 7.5 *Let $(\mathbf{X}, d)$ be a metric space. Let $\{w_n : n = 1, 2, \ldots, N\}$ be contraction mappings on $(\mathcal{H}(\mathbf{X}), h)$. Let the contractivity factor for w_n be denoted by s_n for each n. Define $W : \mathcal{H}(\mathbf{X}) \to \mathcal{H}(\mathbf{X})$ by*

$$W(B) = w_1(B) \cup w_2(B) \cup \ldots \cup w_n(B)$$
$$= \cup_{n=1}^{n} w_n(B), \qquad \text{for each } B \in \mathcal{H}(\mathbf{X}).$$

Then W is a contraction mapping with contractivity factor $s = \max\{s_n : n = 1, 2, \ldots, N\}$.

Proof We demonstrate the claim for $N = 2$. An inductive argument then completes the proof. Let $B, C \in \mathcal{H}(\mathbf{X})$. We have

$$h(W(B), W(C)) = h(w_1(B) \cup w_2(B), w_1(C) \cup w_2(C))$$
$$\leq h(w_1(B), w_1(C)) \vee h(w_2(B), w_2(C)) \text{ (by Lemma 7.2)}$$
$$\leq s_1 h(B, C) \vee s_2 h(B, C) \leq s h(B, C).$$

This completes the proof.

Definition 7.1 *A (hyperbolic) iterated function system consists of a complete metric space $(\mathbf{X}, d)$ together with a finite set of contraction mappings $w_n : \mathbf{X} \to \mathbf{X}$, with respective contractivity factors s_n, for $n = 1, 2, \ldots, N$. The abbreviation "IFS" is used for "iterated function system." The notation for the IFS just announced is $\{\mathbf{X}; w_n, n = 1, 2, \ldots, N\}$ and its contractivity factor is $s = \max\{s_n : n = 1, 2, \ldots, N\}$.*

We put the word "hyperbolic" in parentheses in this definition because it is sometimes dropped in practice. Moreover, we will sometimes use the nomenclature "IFS" to mean simply a finite set of maps acting on a metric space, with no particular conditions imposed upon the maps.

The following theorem summarizes the main facts so far about a hyperbolic IFS.

Theorem 7.1 *Let* $\{X; w_n, n = 1, 2, \ldots, N\}$ *be a hyperbolic iterated function system with contractivity factor* s. *Then the transformation* $W : \mathcal{H}(X) \to \mathcal{H}(X)$ *defined by*

$$W(B) = \cup_{n=1}^{n} w_n(B)$$

for all $B \in \mathcal{H}(X)$, *is a contraction mapping on the complete metric space* $(\mathcal{H}(X), h(d))$ *with contractivity factor* s. *That is*

$$h(W(B), W(C)) \leq s \cdot h(B, C)$$

for all $B, C \in \mathcal{H}(X)$. *Its unique fixed point,* $A \in \mathcal{H}(X)$, *obeys*

$$A = W(A) = \cup_{n=1}^{n} w_n(A)$$

and is given by $A = \lim_{n \to \infty} W^{\circ n}(B)$ *for any* $B \in \mathcal{H}(X)$.

Definition 7.2 *The fixed point* $A \in \mathcal{H}(X)$ *described in the theorem is called the* attractor *of the IFS.*

Sometimes we will use the name "attractor" in connection with an IFS that is simply a finite set of maps acting on a complete metric space X. By this we mean that one can make an assertion analagous to the last sentence of Theorem 7.1.

We wanted to use the words "deterministic fractal" in place of "attractor" in Definition 7.2. We were tempted, but resisted. The nomenclature "iterated function system" is meant to remind one of the name "dynamical system." We will introduce dynamical systems in Chapter 4. Dynamical systems often possess attractors, and when these are interesting to look at they are called *strange attractors*.

Examples & Exercises

7.1. This exercise takes place in the metric spaces $(\mathbb{R}, \text{Euclidean})$ and $(\mathcal{H}(R), h(\text{Euclidean}))$. Consider the IFS $\{\mathbb{R}; w_1, w_2\}$, where $w_1(x) = \frac{1}{3}x$ and $w_2(x) = \frac{1}{3}x + \frac{2}{3}$. Show that this is indeed an IFS with contractivity factor $s = \frac{1}{3}$. Let $B_0 = [0, 1]$. Calculate $B_n = W^{\circ n}(B_0)$, $n = 1, 2, 3, \ldots$. Deduce that $A = \lim_{n \to \infty} B_n$ is the classical Cantor set. Verify directly that $A = \frac{1}{3}A \cup \{\frac{1}{3}A + \frac{2}{3}\}$. Here we use the following notation: for a subset A of $\mathbb{R}$, $xA = \{xy : y \in A\}$ and $A + x = \{y + x : y \in A\}$.

7.2. With reference to example 7.1, show that if $w_1(x) = s_1 x$ and $w_2(x) = (1 - s_1)x + s_1$, where s_1 is a number such that $0 < s_1 < 1$, then $B_1 = B_2 = B_3 = \ldots$. Find the attractor.

7.3. Repeat example 7.1 with $w_1(x) = \frac{1}{3}x$ and $w_2(x) = \frac{1}{2}x + \frac{1}{2}$. In this case $A = \lim_{n \to \infty} B_n$ will not be the classical Cantor set, but it will be something like it. Describe A. Show that A contains no intervals. How many points does A contain?

7.4. Consider the IFS $\{\mathbb{R}; \frac{1}{4}x + \frac{3}{4}, \frac{1}{2}x, \frac{1}{4}x + \frac{1}{4}\}$. Verify that the attractor looks like the image in Figure III.61. Show, precisely, how the set in Figure III.61 is a union of three "shrunken copies of itself." This attractor is interesting: it contains countably many holes and countably many intervals.

Figure III.61. Attractor for three affine maps on the real line. Can you find the maps?

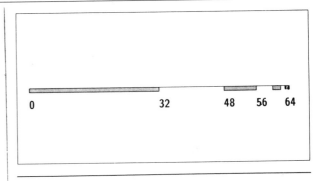

Figure III.62. A sequence of sets converging to a line segment.

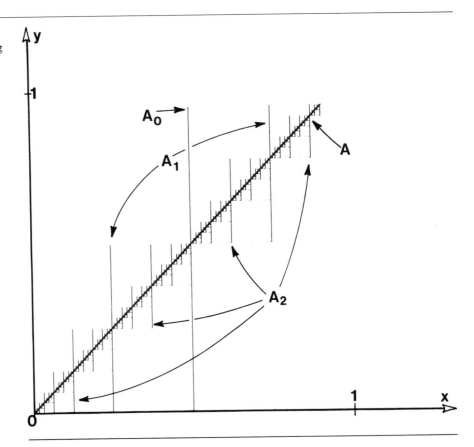

7.5. Show that the attractor of an IFS having the form $\{\mathbb{R}; w_1(x) = ax + b, w_2(x) = cx + d\}$, where $a, b, c,$ and $d \in \mathbb{R}$, is either connected or totally disconnected.

7.6. Does there exist an IFS of three affine maps in $\mathbb{R}^2$ whose attractor is the union of two disjoint closed intervals?

7.7. Consider the IFS

$$\left\{ \mathbb{R}^2; \begin{pmatrix} \frac{1}{2} & 0 \\ 0 & \frac{1}{2} \end{pmatrix} \begin{pmatrix} x \\ y \end{pmatrix} + \begin{pmatrix} \frac{1}{2} \\ \frac{1}{2} \end{pmatrix}, \begin{pmatrix} \frac{1}{2} & 0 \\ 0 & \frac{1}{2} \end{pmatrix} \begin{pmatrix} x \\ y \end{pmatrix} \right\}.$$

Let $A_0 = \{(\frac{1}{2}, y) : 0 \le y \le 1\}$, and let $W^{\circ n}(A_0) = A_n$, where W is defined on $\mathcal{H}(\mathbb{R}^2)$ in the usual way. Show that the attractor is $A = \{(x, y) : x = y, 0 \le x \le 1\}$ and that Figure III.62 is correct. Draw a sequence of pictures to show what happens if $A_0 = \{(x, y) \in \mathbb{R}^2 : 0 \le x \le 1, 0 \le y \le 1\}$.

7.8. Consider the attractor for the IFS $\{\mathbb{R}; w_1(x) = 0, w_2(x) = \frac{2}{3}x + \frac{1}{3}\}$. Show that it consists of a countable increasing sequence of real points $\{x_n : n = 0, 1, 2, \ldots\}$ together with $\{1\}$. Show that x_n can be expressed as the nth partial sum of an infinite geometric series. Give a succinct formula for x_n.

7.9. Describe the attractor A for the IFS $\{[0, 2]; w_1(x) = \frac{1}{9}x^2, w_2(x) = \frac{3}{4}x + \frac{1}{2}\}$ by describing a sequence of sets which converges to it. Show that A is totally disconnected. Show that A is perfect. Find the contractivity factor for the IFS.

7.10. Let $(r, \theta), 0 \le r \le \infty, 0 \le \theta < 2\pi$ denote the polar coordinates of a point in the plane, $\mathbb{R}^2$. Define $w_1(r, \theta) = (\frac{1}{2}r + \frac{1}{2}, \frac{1}{2}\theta)$, and $w_2(r, \theta) = (\frac{2}{3}r + \frac{1}{3}, \frac{2}{3}\theta + \frac{2\pi}{3})$. Show that $\{\mathbb{R}^2; w_1, w_2\}$ is not a hyperbolic IFS because both maps w_1 and w_2 are discontinuous on the whole plane. Show that $\{\mathbb{R}^2; w_1, w_2\}$ nevertheless has an attractor; find it (just consider r and θ separately).

7.11. Show that the sequence of sets illustrated in Figure III.63 can be written in the form $A_n = W^{\circ n}(A_0)$ for $n = 1, 2, \ldots$, and find $W : \mathcal{H}(\mathbb{R}^2) \to \mathcal{H}(\mathbb{R}^2)$.

7.12. Describe the collection of functions that constitutes the attractor A for the IFS

$$\{C[0, 1]; w_1(f(x)) = \frac{1}{2}f(x), w_2(f(x)) = \frac{1}{2}f(x) + 2x(1 - x)\}.$$

Find the contractivity factor for the IFS.

7.13. Let $C^0[0, 1] = \{f \in C[0, 1] : f(0) = f(1) = 0\}$, and define $d(f, g) = \max\{|f(x) - g(x)| : x \in [0, 1]\}$. Define $w_1 : C^0[0, 1] \to C^0[0, 1]$ by $(w_1(f))(x) = \frac{1}{2}f(2x \bmod 1) + 2x(1 - x)$ and $(w_2(f))(x) = \frac{1}{2}f(x)$. Show that $\{C^0[0, 1]; w_1, w_2\}$ is an IFS, find its contractivity factor, and find its attractor. Draw a picture of the attractor.

7.14. Find conditions such that the Möbius transformation $w(x) = (ax + b)/(cx + d), a, b, c, d \in \mathbb{C}, ad - bc \ne 0$, provides a contraction mapping on the unit disk

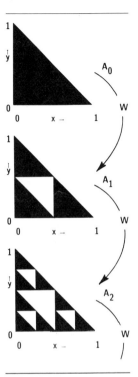

Figure III.63. The first three sets $A_0, A_1,$ and A_2 in a convergent sequence of sets in $\mathcal{H}(\mathbb{R}^2)$. Can you find a transformation $W : \mathcal{H}(\mathbb{R}^2) \to \mathcal{H}(\mathbb{R}^2)$ such that $A_{n+1} = W(A_n)$?

$X = \{z \in \mathbb{C} : |z| \leq 1\}$. Find an upper bound for the contractivity factor. Construct an IFS using two Möbius transformations on X, and describe its attractor.

7.15. Show that a Möbius transformation on $\hat{\mathbb{C}}$ is never a contraction in the spherical metric.

7.16. Let (Σ, d) be the code space of three symbols $\{0, 1, 2\}$, with metric

$$d(x, y) = \sum_{n=1}^{\infty} \frac{|x_n - y_n|}{4^n}.$$

Define $w_1 : \Sigma \to \Sigma$ by $w_1(x) = 0x_1x_2x_3 \ldots$ and $w_2(x) = 2x_1x_2x_3 \ldots$. Show that w_1 and w_2 are both contraction mappings and find their contractivity factors. Describe the attractor of the IFS $\{\Sigma; w_1, w_2\}$. What happens if we include in the IFS a third transformation defined by $w_3x = 1x_1x_2x_3 \ldots$?

7.17. Let $\triangle \subset \mathbb{R}^2$ denote the compact metric space constisting of an equilateral Sierpinski triangle with vertices at $(0, 0)$, $(1, 0)$, and $(\frac{1}{2}, \frac{\sqrt{3}}{2})$, and consider the IFS $\{\triangle, \frac{1}{2}z + \frac{1}{2}, \frac{1}{2}e^{2\pi i/3}z + \frac{1}{2}\}$ where we use complex number notation. Let $A_0 = \triangle$, and $A_n = W^{\circ n}(A_0)$ for $n = 1, 2, 3, \ldots$. Describe A_1, A_2, and the attractor A. What happens if the third transformation $w_3(z) = \frac{1}{2}z + \frac{1}{4} + (\sqrt{3}/4)i$ is included in the IFS?

---

8   Two Algorithms for Computing Fractals from Iterated Function Systems

In this section we take time out from the mathematical development to provide two algorithms for rendering pictures of attractors of an IFS on the graphics display device of a microcomputer or workstation. The reader should establish a computer-graphical environment that includes one or both of the software tools suggested in this section.

The algorithms presented are (1) the Deterministic Algorithm and (2) the Random Iteration Algorithm. The Deterministic Algorithm is based on the idea of directly computing a sequence of sets $\{A_n = W^{\circ n}(A)\}$ starting from an initial set A_0. The Random Iteration Algorithm is founded in ergodic theory; its mathematical basis will be presented in Chapter IX. An intuitive explanation of why it works is presented in Chapter IV. We defer important questions concerning discretization and accuracy. Such questions are considered to some extent in later chapters.

For simplicity we restrict attention to hyperbolic IFS of the form $\{\mathbb{R}^2; w_n : n = 1, 2, \ldots, N\}$, where each mapping is an affine transformation. We illustrate the algorithms for an IFS whose attractor is a Sierpinski triangle. Here's an example of such an IFS:

$$w_1 \begin{bmatrix} x_1 \\ x_2 \end{bmatrix} = \begin{bmatrix} 0.5 & 0 \\ 0 & 0.5 \end{bmatrix} \begin{bmatrix} x_1 \\ x_2 \end{bmatrix} + \begin{bmatrix} 1 \\ 1 \end{bmatrix},$$

$$w_2 \begin{bmatrix} x_1 \\ x_2 \end{bmatrix} = \begin{bmatrix} 0.5 & 0 \\ 0 & 0.5 \end{bmatrix} \begin{bmatrix} x_1 \\ x_2 \end{bmatrix} + \begin{bmatrix} 1 \\ 50 \end{bmatrix},$$

$$w_3 \begin{bmatrix} x_1 \\ x_2 \end{bmatrix} = \begin{bmatrix} 0.5 & 0 \\ 0 & 0.5 \end{bmatrix} \begin{bmatrix} x_1 \\ x_2 \end{bmatrix} + \begin{bmatrix} 25 \\ 50 \end{bmatrix}.$$

This notation for an IFS of affine maps is cumbersome. Let us agree to write

$$w_i(x) = w_i \begin{bmatrix} x_1 \\ x_2 \end{bmatrix} = \begin{bmatrix} a_i & b_i \\ c_i & d_i \end{bmatrix} \begin{bmatrix} x_1 \\ x_2 \end{bmatrix} + \begin{bmatrix} e_i \\ f_i \end{bmatrix} = A_i x + t_i.$$

Then Table III.1 is a tidier way of conveying the same iterated function system.

Table III.1 also provides a number p_i associated with w_i for $i = 1, 2, 3$. These numbers are in fact probabilities. In the more general case of the IFS $\{X; w_n : n = 1, 2, \ldots, N\}$, there would be N such numbers $\{p_i : i = 1, 2, \ldots, N\}$ that obey

$$p_1 + p_2 + p_3 + \cdots + p_n = 1 \text{ and } p_i > 0 \qquad \text{for } i = 1, 2, \ldots, N.$$

These probabilities play an important role in the computation of images of the attractor of an IFS using the Random Iteration Algorithm. They play no role in the Deterministic Algorithm. Their mathematical significance is discussed in later chapters. For the moment we will use them only as a computational aid, in connection with the Random Iteration Algorithm. To this end we take their values to be given approximately by

$$p_i \approx \frac{|\det A_i|}{\sum_{i=1}^{N} |A_i|} = \frac{|a_i d_i - b_i c_i|}{\sum_{i=1}^{N} |a_i d_i - b_i c_i|} \qquad \text{for } i = 1, 2, \ldots, N.$$

Here the symbol $\approx$ means "approximately equal to." If, for some i, $\det A_i = 0$, then p_i should be assigned a small positive number, such as 0.001. Other situations should be treated empirically. We refer to the data in Table III.1 as an IFS *code*. Other IFS codes are given in Tables III.2, III.3, and III.4.

Algorithm 8.1 The Deterministic Algorithm. *Let* $\{X; w_1, w_2, \ldots, w_N\}$ *be a hyperbolic IFS. Choose a compact set* $A_0 \subset \mathbb{R}^2$. *Then compute successively* $A_n = W^{\circ n}(A)$ *according to*

$$A_{n+1} = \cup_{j=1}^{n} w_j(A_n) \qquad \text{for } n = 1, 2, \ldots.$$

Thus construct a sequence $\{A_n : n = 0, 1, 2, 3, \ldots\} \subset \mathcal{H}(X)$. Then by Theorem 7.1 the sequence $\{A_n\}$ converges to the attractor of the IFS in the Hausdorff metric.

Table III.1. IFS code for a Sierpinski triangle.

w	a	b	c	d	e	f	p
1	0.5	0	0	0.5	1	1	0.33
2	0.5	0	0	0.5	1	50	0.33
3	0.5	0	0	0.5	50	50	0.34

Table III.2.   IFS code for a square.

w	a	b	c	d	e	f	p
1	0.5	0	0	0.5	1	1	0.25
2	0.5	0	0	0.5	50	1	0.25
3	0.5	0	0	0.5	1	50	0.25
4	0.5	0	0	0.5	50	50	0.25

Table III.3.   IFS code for a fern.

w	a	b	c	d	e	f	p
1	0	0	0	0.16	0	0	0.01
2	0.85	0.04	-0.04	0.85	0	1.6	0.85
3	0.2	-0.26	0.23	0.22	0	1.6	0.07
4	-0.15	0.28	0.26	0.24	0	0.44	0.07

Table III.4.   IFS code for a fractal tree.

w	a	b	c	d	e	f	p
1	0	0	0	0.5	0	0	0.05
2	0.42	-0.42	0.42	0.42	0	0.2	0.4
3	0.42	0.42	-0.42	0.42	0	0.2	0.4
4	0.1	0	0	0.1	0	0.2	0.15

We illustrate the implementation of the algorithm. The following program computes and plots successive sets A_{n+1} starting from an initial set A_0, in this case a square, using the IFS code in Table III.1. The program is written in BASIC. It should run without modification on an IBM PC with Color Graphics Adaptor or Enhanced Graphics Adaptor, and Turbobasic. It can be modified to run on any personal computer with graphics display capability. On any line the words preceded by a ' are comments and not part of the program.

Program 1.   (Example of the Deterministic Algorithm)

```
screen 1 : cls 'initialize graphics
dim s(100,100) : dim t(100,100) 'allocate two arrays of pixels
a(1)=0.5:b(1)=0:c(1)=0:d(1)=0.5:e(1)=1:f(1)=1 'input the IFS code
a(2)=0.5:b(2)=0:c(2)=0:d(2)=0.5:e(2)=50:f(2)=1
a(3)=0.5:b(3)=0:c(3)=0:d(3)=0.5:e(3)=25:f(3)=50
for i=1 to 100 'input the initial set A(0), in this case
      a square, into the array t(i,j)
```

```
t(i,1)=1: pset(i,1) 'A(0) can be used as a condensation set
t(1,i)=1:pset(1,i)  'A(0) is plotted on the screen
t(100,i)=1:pset(100,i)
t(i,100)=1:pset(i,100)
next: do
for i=1 to 100 'apply W to set A(n) to make A(n+1) in the
       array s(i,j)
for j=1 to 100 : if t(i,j)=1 then
s(a(1)*i+b(1)*j+e(1),c(1)*i+d(1)*j+f(1))=1 'and apply W to A(n)
s(a(2)*i+b(2)*j+e(2),c(2)*i+d(2)*j+f(2))=1
s(a(3)*i+b(3)*j+e(3),c(3)*i+d(3)*j+f(3))=1
end if: next j: next i
cls 'clears the screen--omit to obtain sequence with a A(0) as
       condensation set (see section 9 in Chapter II)
for i=1 to 100 : for j=1 to 100
t(i,j)=s(i,j) 'put A(n+1) into the array t(i,j)
s(i,j)=0 'reset the array s(i,j) to zero
if t(i,j)=1 then
pset(i,j) 'plot A(n+1)
end if : next : next
loop until instat 'if a key has been pressed then stop,
       otherwise compute A(n+1)=W(A(n+1))
```

The result of running a higher-resolution version of this program on a Masscomp 5600 workstation and then printing the contents of the graphics screen is presented in Figure III.64. In this case we have kept each successive image produced by the program.

Notice that the program begins by drawing a box in the array $t(i, j)$. This box has no influence on the finally computed image of a Sierpinski triangle. One could just as well have started from any other (nonempty) set of points in the array $t(i, j)$, as illustrated in Figure III.65.

To adapt Program 1 so that it runs with other IFS codes will usually require changing coordinates to ensure that each of the transformations of the IFS maps the pixel array $s(i, j)$ into itself. Change of coordinates in an IFS is discussed in exercise 10.14. As it stands in Program 1, the array $s(i, j)$ is a discretized representation of the square in $\mathbb{R}^2$ with lower left corner at $(1, 1)$ and upper right corner at $(100, 100)$. Failure to adjust coordinates correctly will lead to unpredictable and exciting results!

Algorithm 8.2 The Random Iteration Algorithm. *Let* $\{\mathbf{X}; w_1, w_2, \ldots, w_N\}$ *be a hyperbolic IFS, where probability* $p_i > 0$ *has been assigned to to* w_i *for* $i = 1, 2, \ldots, N$, *where* $\sum_{i=1}^{n} p_i = 1$. *Choose* $x_0 \in \mathbf{X}$ *and then choose recursively, independently,*

$$x_n \in \{w_1(x_{n-1}), w_2(x_{n-1}), \ldots, w_N(x_{n-1})\} \qquad \text{for } n = 1, 2, 3, \ldots,$$

where the probability of the event $x_n = w_i(x_{n-1})$ is p_i. Thus, construct a sequence $\{x_n : n = 0, 1, 2, 3, \ldots\} \subset X$.

★ The reader should skip the rest of this paragraph and come back to it after reading Section 9. If $\{\mathbf{X}, w_0, w_1, w_2, \ldots, w_N\}$ is an IFS with condensation map w_0 and associated condensation set $C \subset \mathcal{H}(\mathbf{X})$, then the algorithm is modified by (a) attaching a probability $p_0 > 0$ to w_0, so now $\sum_{i=0}^{n} p_i = 1$; (b) whenever $w_0(x_{n-1})$ is selected for some n, choose x_n "at random" from C. Thus, in this case too, we construct a sequence $\{x_n : n = 0, 1, 2, \ldots\}$ of points in $\mathbf{X}$.

The sequence $\{x_n\}_{n=0}^{\infty}$ "converges to" the attractor of the IFS, under various conditions, in a manner that will be made precise in Chapter IX.

We illustrate the implementation of the algorithm. The following program computes and plots a thousand points on the attractor corresponding to the IFS code in Table III.1. The program is written in BASIC. It runs without modification on an

Figure III.64. The result of running the Deterministic Algorithm (Program 1) with various values of N, for the IFS code in Table III.1.

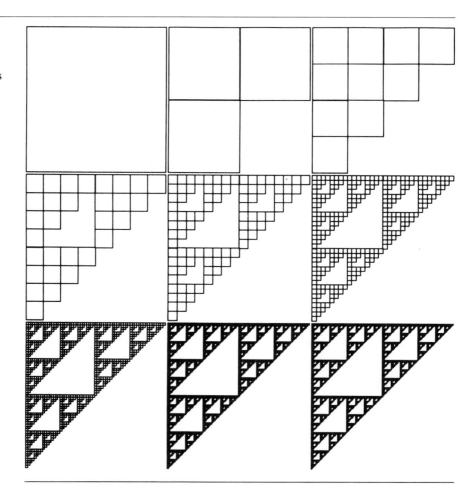

IBM PC with Enhanced Graphics Adaptor and Turbobasic. On any line the words preceded by a ' are comments: they are not part of the program.

Program 2.   (Example of the Random Iteration Algorithm)

```
'Iterated Function System Data
a[1] = 0.5 : b[1] =0 : c[1] =0 : d[1] =.5 : e[1] =1 : f[1] =1
a[2] = 0.5 : b[2] =0 : c[2] =0 : d[2] =.5 : e[2] =50 : f[2] =1
a[3] = 0.5 : b[3] =0 : c[3] =0 : d[3] =.5 : e[3] =50 : f[3] =50

screen 1 : cls 'initialize computer graphics
window (0,0)-(100,100) 'set plotting window to 0<x<100, 0<y<100
x =0   : y  =  0: numits =1000 'initialize (x,y) and define
      the number of iterations, numits
```

Figure III.65. The result of running the Deterministic Algorithm (Program 1), again for the IFS code in Table III.1, but starting from a different initial array. The final result is always the same!

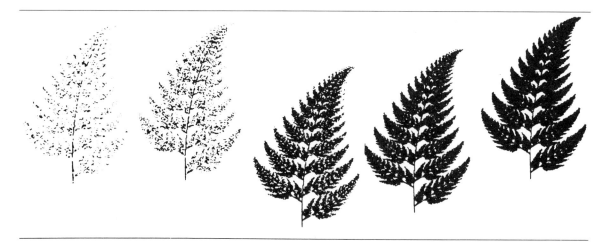

Figure III.66.   The result of running the Random Iteration Algorithm for increasing numbers of iterations. The randomly dancing point starts to suggest the structure of the attractor of the IFS given in Table III.3.

```
for n =1  to numits 'Random Iteration begins!
k  =  int(3*rnd-0.00001)  +1  'choose one of the numbers 1, 2,
     and 3 with equal probability

'apply affine transformation number k to (x,y)
newx =a[k]*x+b[k]*y+e[k]  : newy =c[k]*x+d[k]*y+f[k]
x =newx  : y =newy  'set (x,y) to the point thus obtained
if n  >  10 then pset (x,y) 'plot (x,y) after the first 10
     iterations
next :    end
```

The result of running an adaptation of this program on a Masscomp workstation and then printing the contents of the graphics screen is presented in Figure III.66. Notice that if the size of the plotting window is decreased, for example by replacing the window call by WINDOW (0,0)–(50,50), then only a portion of the image is plotted, but at a higher resolution. Thus we have a simple means for "zooming in" on images of IFS attractors. The number of iterations may be increased to improve the quality of the computed image.

Examples & Exercises

8.1. Rewrite Programs 1 and 2 in a form suitable for your own computer environment, then run them and obtain hardcopy of the output. Compare their performance.

8.2. Modify Programs 1 and 2 so that they will compute images associated with the IFS code given in Table III.2.

8.3. Modify Program 2 so that it will compute images associated with the IFS codes given in Tables III.3 and III.4.

8.4. By changing the window size in Program 2, obtain images of "zooms" on the Sierpinski triangle. For example, use the following windows: $(1, 1) - (50, 50)$; $(1, 1) - (25, 25)$; $(1, 1) - (12, 12)$; ...; $(1, 1) - (N, N)$. How must the total number of iterations be adjusted as a function of N in order that (approximately) the number of points that land within the window remains constant? Make a graph of the total number of iterations against the window size.

8.5. What should happen, theoretically, to the sequence of images computed by Program 1 if the set A_0 is changed? What happens in practice? Make a computational experiment to see if there is any difference in say A_{10} corresponding to two different choices for A_0.

8.6. Rewrite Program 2 so that it applies the transformation w_i with probability p_i, where the probabilities are input by the user. Compare the number of iterations needed to produce a "good" rendering of the Sierpinski triangle, for the cases (a) $p_1 = 0.33$, $p_2 = 0.33$, $p_3 = 0.34$; (b) $p_1 = 0.2$, $p_2 = 0.46$, $p_3 = 0.34$; (c) $p_1 = 0.1$, $p_2 = 0.56$, $p_3 = 0.34$.

9   Condensation Sets

There is another important way of making contraction mappings on $\mathcal{H}(\mathbf{X})$.

Definition 9.1 *Let* $(\mathbf{X}, d)$ *be a metric space and let* $C \in \mathcal{H}(\mathbf{X})$. *Define a transformation* $w_0 : \mathcal{H}(\mathbf{X}) \to \mathcal{H}(\mathbf{X})$ *by* $w_0(B) = C$ *for all* $B \in \mathcal{H}(\mathbf{X})$. *Then* w_0 *is called a* condensation transformation *and* C *is called the associated* condensation set.

Observe that a condensation transformation $w_0 : \mathcal{H}(\mathbf{X}) \to \mathcal{H}(\mathbf{X})$ is a contraction mapping on the metric space $(\mathcal{H}(\mathbf{X}), h(d))$, with contractivity factor equal to zero, and that it possesses a unique fixed point, namely the condensation set.

Definition 9.2 *Let* $\{\mathbf{X}; w_1, w_2, \ldots, w_n\}$ *be a hyperbolic IFS with contractivity factor* $0 \leq s < 1$. *Let* $w_0 : \mathcal{H}(\mathbf{X}) \to \mathcal{H}(\mathbf{X})$ *be a condensation transformation. Then* $\{\mathbf{X}; w_0, w_1, \ldots, w_n\}$ *is called a* hyperbolic IFS with condensation, *with contractivity factor* s.

Theorem 7.1 can be modified to cover the case of an IFS with condensation.

Theorem 9.1 *Let* $\{\mathbf{X}; w_n : n = 0, 1, 2, \ldots, N\}$ *be a hyperbolic iterated function system with condensation, with contractivity factor* s. *Then the transformation* $W : \mathcal{H}(\mathbf{X}) \to \mathcal{H}(\mathbf{X})$ *defined by*

$$W(B) = \cup_{n=0}^{n} w_n(B) \forall B \in \mathcal{H}(\mathbf{X})$$

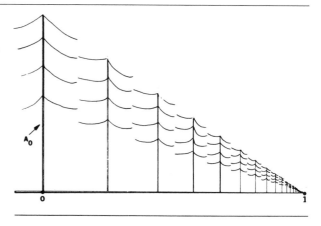

Figure III.67. A geometric series of pine trees, the attractor of an IFS with condensation.

is a contraction mapping on the complete metric space $(\mathcal{H}(\mathbf{X}), h(d))$ with contractivity factor s. That is

$$h(W(B), W(C)) \leq s \cdot h(B, C) \forall B, C \in \mathcal{H}(\mathbf{X}).$$

Its unique fixed point, $A \in \mathcal{H}(\mathbf{X})$, obeys

$$A = W(A) = \cup_{n=0}^{n} w_n(A)$$

and is given by $A = \lim_{n \to \infty} W^{\circ n}(B)$ for any $B \in \mathcal{H}(\mathbf{X})$.

Examples & Exercises

9.1. A sequence of sets $\{A_n \subset \mathbf{X}\}_{n=0}^{\infty}$, where $(\mathbf{X}, d)$ is a metric space, is said to be *increasing* if $A_0 \subset A_1 \subset A_2 \subset \cdots$ and *decreasing* if $A_0 \supset A_1 \supset A_2 \supset \cdots$. The inclusions are not necessarily strict. A decreasing sequence of sets $\{A_n \subset \mathcal{H}(\mathbf{X})\}_{n=0}^{\infty}$ is a Cauchy sequence (prove it!). If $\mathbf{X}$ is compact then an increasing sequence of sets $\{A_n \subset \mathcal{H}(\mathbf{X})\}_{n=0}^{\infty}$ is a Cauchy sequence (prove it!). Let $\{\mathbf{X}; w_0, w_1, \ldots, w_n\}$ be a hyperbolic IFS with condensation set C, and let $\mathbf{X}$ be compact. Let $W_0(B) = \cup_{n=0}^{n} w_n(B) \forall B \in \mathcal{H}(\mathbf{X})$ and let $W(B) = \cup_{n=1}^{n} w_n(B)$. Define $\{C_n = W_0^{\circ n}(C)\}_{n=0}^{\infty}$. Then Theorem 9.1 tells us $\{C_n\}$ is a Cauchy sequence in $\mathcal{H}(\mathbf{X})$ that converges to the attractor of the IFS. Independently of the theorem observe that

$$C_n = C \cup W(C) \cup W^{\circ 2}(C) \cup \ldots \cup W^{\circ n}(C)$$

provides an increasing sequence of compact sets. It follows immediately that the limit set A obeys $W_0(A) = A$.

9.2. This example takes place in $(\mathbb{R}^2$, Euclidean). Let $C = \qquad = A_0 \subset \mathbb{R}^2$

denote a set that looks like a scorched pine tree standing at the origin, with its trunk perpendicular to the x-axis. Let

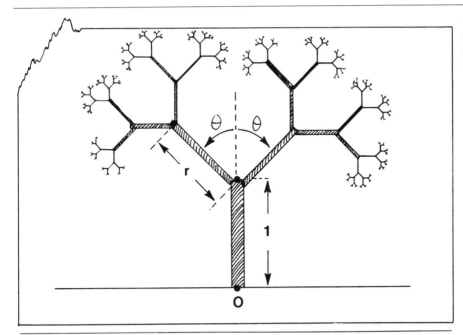

Figure III.68. Sketch of a fractal tree, the attractor of an IFS with condensation.

$$w_1 \begin{pmatrix} x \\ y \end{pmatrix} = \begin{pmatrix} 0.75 & 0 \\ 0 & 0.75 \end{pmatrix} \begin{pmatrix} x \\ y \end{pmatrix} + \begin{pmatrix} 0.25 \\ 0 \end{pmatrix}.$$

Show that $\{\mathbb{R}^2; w_0, w_1\}$ is an IFS with condensation and find its contractivity factor. Let $A_n = W^{\circ n}(A_0)$ for $n = 1, 2, 3, \ldots$, where $W(B) = \cup_{n=0}^n w_n(B)$ for $B \in \mathcal{H}(\mathbb{R}^2)$. Show that A_n consists of the first $(n + 1)$ pine trees reading from left to right in Figure III.67. If the first tree required 0.1% of the ink in the artist's pen to draw, and if the artist had been very meticulous in drawing the whole attractor correctly, find the total amount of ink used to draw the whole attractor.

9.3. What happens to the trees in Figure III.67 if $w_1 \begin{pmatrix} x \\ y \end{pmatrix}$ is replaced by

$$w_1 \begin{pmatrix} x \\ y \end{pmatrix} = \begin{pmatrix} 0.5 & 0 \\ 0 & 0.75 \end{pmatrix} \begin{pmatrix} x \\ y \end{pmatrix} + \begin{pmatrix} 0.5 \\ 0 \end{pmatrix}$$

in exercise 9.2?

9.4. Find the attractor for the IFS with condensation $\{\mathbb{R}^2; w_0, w_1\}$, where the condensation set is the interval $[0, 1]$ and $w_1(x) = \frac{1}{2}x + 2$. What happens if $w_1(x) = \frac{1}{2}x$?

9.5. Find an IFS with condensation that generates the treelike set in Figure III.68. Give conditions on r and θ such that the tree is simply connected. Show that the tree is either simply connected or infinitely connected.

9.6. Find an IFS with condensation that generates Figure III.69.

9.7. You are given a condensation map $w_0(x)$ in $\mathbb{R}^2$ that provides the largest tree

Figure III.69. An endless spiral of little men.

in Figure III.46. Find a hyperbolic IFS with condensation, of the form $\{\mathbb{R}^2; w_0, w_1, w_2\}$, which produces the whole orchard. What is the contractivity factor for this IFS? Find the attractor of the IFS $\{\mathbb{R}^2; w_1, w_2\}$.

9.8. Explain why removing the command that clears the screen ("cls") from Program 1 will result in the computation of an image associated with an IFS with condensation. Identify the condensation set. Run your version of Program 1 with the "cls" command removed.

10   How to Make Fractal Models with the Help of the Collage Theorem

The following theorem is central to the design of IFS's whose attractors are close to given sets.

Theorem 10.1 (The Collage Theorem, (Barnsley 1985b)). *Let $(\mathbf{X}, d)$ be a complete metric space. Let $L \in \mathcal{H}(\mathbf{X})$ be given, and let $\epsilon \geq 0$ be given. Choose an IFS (or IFS with condensation) $\{\mathbf{X}; (w_0), w_1, w_2, \ldots, w_n\}$ with contractivity factor*

$0 \leq s < 1$, *so that*

$$h(L, \cup^n_{\substack{n=1 \\ (n=0)}} w_n(L)) \leq \epsilon,$$

where $h(d)$ is the Hausdorff metric. Then

$$h(L, A) \leq \epsilon/(1 - s),$$

where A is the attractor of the IFS. Equivalently,

$$h(L, A) \leq (1 - s)^{-1} h(L, \cup^n_{\substack{n=1 \\ (n=0)}} w_n(L)) \qquad \text{for all } L \in \mathcal{H}(\mathbf{X}).$$

The proof of the Collage Theorem is given in the next section. The theorem tells us that to find an IFS whose attractor is "close to" or "looks like" a given set, one must endeavor to find a set of transformations—contraction mappings on a suitable space within which the given set lies—such that the union, or collage, of the images of the given set under the transformations is near to the given set. Nearness is measured using the Hausdorff metric.

Examples & Exercises

10.1. This example takes place in $(\mathbb{R}, \text{Euclidean})$. Observe that $[0, 1] = [0, \frac{1}{2}] \cup [\frac{1}{2}, 1]$. Hence the attractor is $[0, 1]$ for any pair of contraction mappings $w_1 : \mathbb{R} \to \mathbb{R}$ and $w_2 : \mathbb{R} \to \mathbb{R}$ such that $w_1([0, 1]) = [0, \frac{1}{2}]$ and $w_2([0, 1]) = [\frac{1}{2}, 1]$. For example, $w_1(x) = \frac{1}{2}x$ and $w_2(x) = \frac{1}{2}x + \frac{1}{2}$ does the trick. The unit interval is a collage of two smaller "copies" of itself.

10.2. Suppose we are using a trial-and-error procedure to adjust the coefficients in two affine transformations $w_1(x) = ax + b$, $w_2(x) = cx + d$, where $a, b, c, d \in \mathbb{R}$, to look for an IFS $\{\mathbb{R}; w_1, w_2\}$ whose attractor is $[0, 1]$. We might come up with $w_1(x) = 0.51x - 0.01$ and $w_2(x) = 0.47x + 0.53$. How far from $[0, 1]$ will the attractor for the IFS be? To find out compute

$$h\left([0, 1], \cup^2_{i=1} w_i([0, 1])\right) = h([0, 1], [-0.0l, 0.5] \cup [0.53, 1]) = 0.015$$

and observe that the contractivity factor of the IFS is $s = 0.51$. So by the Collage Theorem, if A is the attractor,

$$h([0, 1], A) \leq 0.015/0.49 < 0.04.$$

10.3. Figure III.70 shows a target set $L \subset \mathbb{R}^2$, a leaf, represented by the polygonalized boundary of the leaf. Four affine transformations, contractive, have been applied to the boundary at lower left, producing the four smaller deformed leaf boundaries. The Hausdorff distance between the union of the four copies and the original is approximately 1.0 units, where the width of the whole frame is taken to be 10 units. The contractivity of the associated IFS $\{\mathbb{R}^2; w_1, w_2, w_3, w_4\}$ is approximately 0.6. Hence the Hausdorff distance $h(\text{Euclidean})$ between the original target leaf L and the attractor A of the IFS will be less than 2.5 units. (This is not promising much!) The actual attractor, translated to the right, is shown at lower right. Not surprisingly,

Figure III.70. The Collage Theorem applied to a region bounded by a polygonalized leaf boundary.

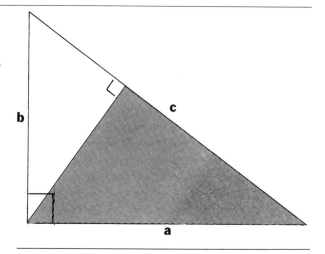

Figure III.71. The region bounded by a right-angle triangle is the union of the results of two similitudes applied to it.

it does not look much like the original leaf! An improved collage is shown at the upper left. The distance $h(L, \cup_{n=1}^4 w_n(L))$ is now less than 0.02 units, while the contractivity of the IFS is still approximately 0.6. Hence $h(L, A)$ should now be less than 0.05 units, and we expect that the attractor should look quite like L at the resolution of the figure. A, translated to the right, is shown at the upper right.

10.4. To find an IFS whose attractor is a region bounded by a right-angle triangle, observe the collage in Figure III.71.

Figure III.72. Use the Collage Theorem to help you find an IFS consisting of two affine maps in $\mathbb{R}^2$ whose attractor is close to this set.

10.5. A nice proof of Pythagoras' Theorem is obtained from the collage in Figure III.71. Clearly both transformations involved are similitudes. The contractivity factors of these similitudes involved are (b/c) and (a/c). Hence the area $\mathcal{A}$ obeys $\mathcal{A} = (b/c)^2\mathcal{A} + (a/c)^2\mathcal{A}$. This implies $c^2 = a^2 + b^2$ since $\mathcal{A} > 0$.

10.6. Figures III.72–III.76 provide exercises in the application of the Collage Theorem. Condensation sets are not allowed in working these examples!

10.7. It is straightforward to see how the Collage Theorem gives us sets of maps for IFS's that generate $\mathbb{A}$. A Menger Sponge looks like this: . Find an IFS for which it is the attractor.

10.8. The IFS that generates the *Black Spleenwort* fern, shown in Figure III.77, consists of four affine maps in the form

$$w_i \begin{pmatrix} x \\ y \end{pmatrix} = \begin{pmatrix} r\cos\theta & -s\sin\theta \\ r\sin\theta & s\cos\theta \end{pmatrix}\begin{pmatrix} x \\ y \end{pmatrix} + \begin{pmatrix} h \\ k \end{pmatrix} (i = 1, 2, 3, 4);$$

see Table III.5.

10.9. Find a collage of affine transformations in $\mathbb{R}^2$, corresponding to Figure III.78.

10.10. A collage of a leaf is shown in Figure III.79 (a). This collage implies the IFS $\{\mathbb{C}; w_1, w_2, w_3, w_4\}$ where, in complex notation,

$$w_i(z) = s_i z + (1 - s_i)a_i \qquad \text{for } i = 1, 2, 3, 4.$$

Verify that in this formula a_i is the fixed point of the transformation. The values found for s_i and a_i are listed in Table III.6. Check that these make sense in relation to the collage. The attractor for the IFS is shown in Figure III.79 (b).

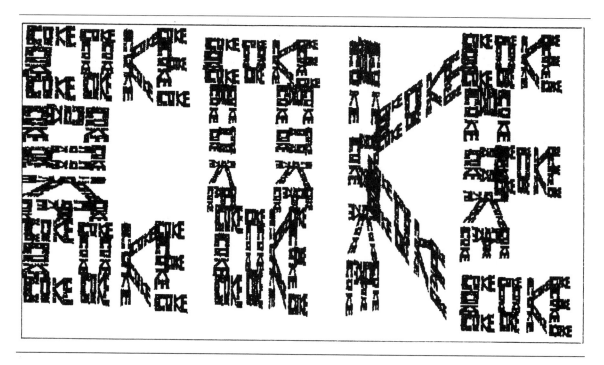

Figure III.73. This image represents the attractor of 14 affine transformations in $\mathbb{R}^2$. Use the Collage Theorem to help you find them.

Figure III.74. Use the Collage Theorem to help find a hyperbolic IFS of the form $\{\mathbb{R}^2; w_1, w_2, w_3\}$, where w_1, w_2, and w_3 are similitudes in $\mathbb{R}^2$, whose attractor is represented here. You choose the coordinate system.

10.11. The attractor in Figure III.80 is determined by two affine maps. Locate the fixed points of two such affine transformations on $\mathbb{R}^2$.

10.12. Figure III.81 shows the attractor for an IFS $\{\mathbb{R}^2; w_i, i = 1, 2, 3, 4\}$ where each w_i is a three-dimensional affine transformation. See also Color Plate 3. The attractor is contained in the region $\{(x_1, x_2, x_3) \in \mathbb{R}^3 : -10 \le x_1 \le 10, 0 \le x_2 \le 10, -10 \le x_3 \le 10\}$.

$$w_1 \begin{bmatrix} x_1 \\ x_2 \\ x_3 \end{bmatrix} = \begin{bmatrix} 0 & 0 & 0 \\ 0 & 0.18 & 0 \\ 0 & 0 & 0 \end{bmatrix} \begin{bmatrix} x_1 \\ x_2 \\ x_3 \end{bmatrix} + \begin{bmatrix} 0 \\ 0 \\ 0 \end{bmatrix}$$

$$w_2 \begin{bmatrix} x_1 \\ x_2 \\ x_3 \end{bmatrix} = \begin{bmatrix} 0.85 & 0 & 0 \\ 0 & 0.85 & 0.1 \\ 0 & -0.1 & 0.85 \end{bmatrix} \begin{bmatrix} x_1 \\ x_2 \\ x_3 \end{bmatrix} + \begin{bmatrix} 0 \\ 1.6 \\ 0 \end{bmatrix}$$

$$w_3 \begin{bmatrix} x_1 \\ x_2 \\ x_3 \end{bmatrix} = \begin{bmatrix} 0.2 & -0.2 & 0 \\ 0.2 & 0.2 & 0 \\ 0 & 0 & 0.3 \end{bmatrix} \begin{bmatrix} x_1 \\ x_2 \\ x_3 \end{bmatrix} + \begin{bmatrix} 0 \\ 0.8 \\ 0 \end{bmatrix}$$

$$w_4 \begin{bmatrix} x_1 \\ x_2 \\ x_3 \end{bmatrix} = \begin{bmatrix} -0.2 & 0.2 & 0 \\ 0.2 & 0.2 & 0 \\ 0 & 0 & 0.3 \end{bmatrix} \begin{bmatrix} x_1 \\ x_2 \\ x_3 \end{bmatrix} + \begin{bmatrix} 0 \\ 0.8 \\ 0 \end{bmatrix}$$

10.13. Find an IFS of similitudes in $\mathbb{R}^2$ such that the attractor is represented by the shaded region in Figure III.82. The collage should be "just-touching," by which we mean that the transforms of the region provide a tiling of the region: they should fit together like the pieces of a jigsaw puzzle.

10.14. This exercise suggests how to change the coordinates of an IFS. Let $\{X_1, d_1\}$ and $\{X_2, d_2\}$ be metric spaces. Let $\{X_1; w_1, w_2, \ldots, w_N\}$ be a hyperbolic IFS with attractor A_1. Let $\theta : X_1 \to X_2$ be an invertible continuous transformation. Consider the IFS $\{X_2; \theta \circ w_1 \circ \theta^{-1}, \theta \circ w_2 \circ \theta^{-1}, \ldots, \theta \circ w_N \circ \theta^{-1}\}$. Use θ to define a metric on X_2 such that the new IFS is indeed a hyperbolic IFS. Prove that if $A_2 \in \mathcal{H}(X_2)$ is

Table III.5.   The IFS code for the Black Spleenwort, expressed in scale and angle formats.

	Translations		Rotations		Scalings	
Map	h	k	θ	ϕ	r	s
1	0.0	0.0	0	0	0.0	0.16
2	0.0	1.6	-2.5	-2.5	0.85	0.85
3	0.0	1.6	49	49	0.3	0.34
4	0.0	0.44	120	-50	0.3	0.37

Figure III.75.   Find an IFS of the form $\{\mathbb{R}^2; w_1, w_2, w_3, w_4\}$, where the w_i's are affine transformations on $\mathbb{R}^2$, whose attractor when rendered contains this image. Check your conclusion using Program 2.

the attractor of the new IFS, then $A_2 = \theta(A_1)$. Thus we can readily construct an IFS whose attractor is a transform of the attractor of another IFS.

10.15. Find some of the affine transformations used in the design of the fractal scene in Figure III.83.

10.16. Use the Collage Theorem to find an IFS whose attractor approximates the set in Figure III.84.

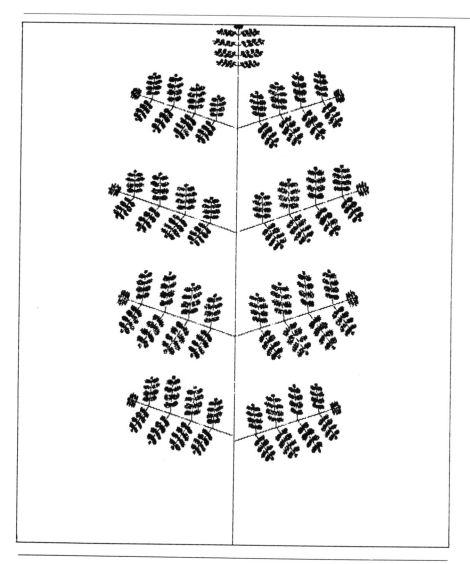

Figure III.76. How many affine transformations in $\mathbb{R}^2$ are needed to generate this attractor? You do not need to use a condensation set.

10.17. Solve the problems proposed in the captions of (a) Figure III.85, (b) Figure III.86, (c) Figure III.87.

11   Blowing in the Wind: The Continuous Dependence of Fractals on Parameters

The Collage Theorem provides a way of approaching the inverse problem: given a set L, find an IFS for which L is the attractor. The underlying mathematical principle

Figure III.77. The Black Spleenwort fern. The top image illustrates one of the four affine transformations in the IFS whose attractor was used to render the fern. The transformation takes the triangle ABC to triangle abc. The Collage Theorem provides the other three transformations. The IFS coded for this image is given in Table III.3. Observe that the stem is the image of the whole set under one of the transformations. Determine to which map number in Table III.3 the stem corresponds.The bottom image shows the Black Spleenwort fern and a close-up.

is very easy: the proof of the Collage Theorem is just the proof of the following lemma.

Lemma 11.1 *Let $(\mathbf{X}, d)$ be a complete metric space. Let $f : \mathbf{X} \to \mathbf{X}$ be a contraction mapping with contractivity factor $0 \leq s < 1$, and let the fixed point of f be $x_f \in \mathbf{X}$. Then*

$$d(x, x_f) \leq (1 - s)^{-1} \cdot d(x, f(x)) \text{ for all } x \in \mathbf{X}.$$

Figure III.78. Use the Collage Theorem to find the four affine transformations corresponding to this image. Can you find a transformation which will put in the "missing corner"?

(a) Collage

(b) Attractor

Proof The distance function $d(a, b)$, for fixed $a \in \mathbf{X}$, is continuous in $b \in \mathbf{X}$. Hence

$$d(x, x_f) = d\left(x, \lim_{n \to \infty} f^{\circ n}(x)\right) = \lim_{n \to \infty} d(x, f^{\circ n}(x))$$

$$\leq \lim_{n \to \infty} \sum_{m=1}^{n} d(f^{\circ(m-1)}(x), f^{\circ(m)}(x))$$

$$\leq \lim_{n \to \infty} d(x, f(x))(1 + s + \cdots + s^{n-1}) \leq (1 - s)^{-1} d(x, f(x)).$$

This completes the proof.

The following results are important and closely related to the above material. They establish the continuous dependence of the attractor of a hyperbolic IFS on parameters in the maps that constitute the IFS.

Table III.6. Scaling factors and fixed points for the collage in Figure III.79.

s	a
0.6	0.45 + 0.9i
0.6	0.45 + 0.3i
0.4 - 0.3i	0.60 + 0.3i
0.4 + 0.3i	0.30 + 0.3i

Figure III.79. A collage of a leaf is obtained using four similitudes, as illustrated in (a). The corresponding IFS is presented in complex notation in Table III.6. The attractor of the IFS is rendered in (b).

Lemma 11.2 *Let (P, d_p) and $(\mathbf{X}, d)$ be metric spaces, the latter being complete. Let $w : P \times \mathbf{X} \to \mathbf{X}$ be a family of contraction mappings on $\mathbf{X}$ with contractivity factor $0 \le s < 1$. That is, for each $p \in P$, $w(p, \cdot)$ is a contraction mapping on $\mathbf{X}$. For each fixed $x \in \mathbf{X}$ let w be continuous on P. Then the fixed point of w depends continuously on p. That is, $x_f : P \to \mathbf{X}$ is continuous.*

Proof Let $x_f(p)$ denote the fixed point of w for fixed $p \in P$. Let $p \in P$ and $\epsilon > 0$ be given. Then for all $q \in P$,

$$
\begin{aligned}
d(x_f(p), x_f(q)) &= d(w(p, x_f(p)), w(q, x_f(q))) \\
&\le d(w(p, x_f(p)), w(q, x_f(p))) \\
&\quad + d(w(q, x_f(p)), w(q, x_f(q))) \\
&\le d(w(p, x_f(p)), w(q, x_f(p))) + s d(x_f(p), x_f(q)),
\end{aligned}
$$

which implies

$$
d(x_f(p), x_f(q)) \le (1 - s)^{-1} d(w(p, x_f(p)), w(q, x_f(p))).
$$

The right-hand side here can be made arbitrarily small by restricting q to be sufficiently close to p. (Notice that if there is a real constant C such that

$$
d(w(p, x), w(q, x)) \le C d(p, q) \qquad \text{for all } p, q \in P, \qquad \text{for all } x \in \mathbf{X},
$$

then $d(x_f(p), x_f(q)) \le (1 - s)^{-1} \cdot C \cdot d(p, q)$, which is a useful estimate.) This completes the proof.

Examples & Exercises

11.1. The fixed point of the contraction mapping $w : \mathbb{R} \to \mathbb{R}$ defined by $w(x) = \frac{1}{2} x + p$ depends continuously on the real parameter p. Indeed, $x_f = 2p$.

Figure III.80. Locate the fixed points of a pair of affine transformations in $\mathbb{R}^2$ whose attractor is rendered here.

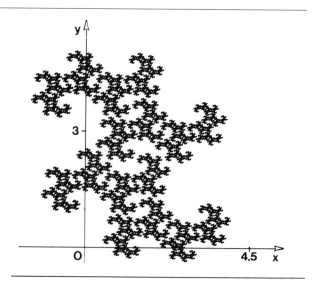

Figure III.81. Single three-dimensional fern. The attractor of an IFS of affine maps in $\mathbb{R}^3$.

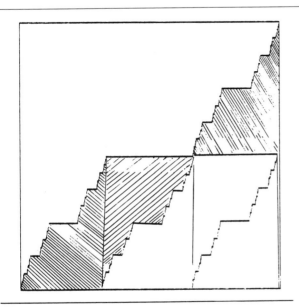

Figure III.82. Find a "just-touching" collage of the area under this Devil's Staircase.

Figure III.83. Determine some of the affine transformations used in the design of this fractal scene. For example, where do the dark sides of the largest mountain come from?

11.2. Show that the fixed function for the transformation $w : C^0[0, 1] \to C^0[0, 1]$ defined by $w(f(x)) = pf(2x \bmod 1) + x(1 - x)$ is continuous in p for $p \in (-1, 1)$. Here, $C^0[0, 1] = \{f \in C[0, 1] : f(0) = f(1) = 0\}$ and the distance is $d(f, g) = \max\{|f(x) - g(x)| : x \in [0, 1]\}$.

In order for this to be of use to us, we need some method of moving the continuous dependence on the parameter p to $\mathcal{H}(X)$. We cannot do this just because the image of a point in some set B depends continuously on p, since, although this gives

Figure III.84. "Typical" fractals are not pretty: use the Collage Theorem to find an IFS whose attractor approximates this set.

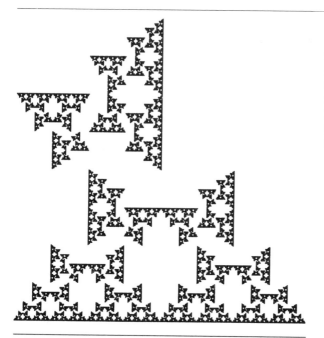

Figure III.85. Determine the affine transformations for an IFS corresponding to this fractal. Can you see, just by looking at the picture, if the linear part of any of the transformations has a negative determinant?

Figure III.86. Use the Collage Theorem to analyze this fractal. On how many different scales is the whole image apparently repeated here? How many times is the smallest clearly discernible copy repeated?

us a δ to constrain p with in order that $w(p, x)$ moves by less than ϵ, this relation is still dependent on the point (p, x). A set $B \in \mathcal{H}(X)$, which is interesting, contains an infinite number of such points, giving us no δ greater than 0 to constrain p with to limit the change in the whole set. We can get such a condition by further restricting $w(p, x)$. Many constraints will do this; we pick one that is simple to understand. For our IFS, parametrized by $p \in P$, that is $\{X: w_{1_p}, \ldots, w_{N_p}\}$, we want the conditions under which given $\epsilon > 0$, we can find a $\delta > 0$ such that

$$d_p(p, q) < \delta \Rightarrow h(w_p(B), w_q(B)) < \epsilon.$$

Figure III.87. Consider the white areas in this figure to represent a set S in $\mathbb{R}^2$. Locate the boundary of the largest pathwise-connected subset of S. It is recommended that you work with a photocopy of the image, a magnifying glass, and a fine red felt-tip pen.

Suppose that for every $p \in P$, $w_{i_p}(x)$ is a continuous function on X. Furthermore, we ask that there is a $k > 0$, independent of x and p such that for each fixed $x \in X$ and for each w_{i_p}, the condition

$$d(w_{i_p}(x), w_{i_q}(x)) \leq k \cdot d_p(p, q)$$

holds. This condition is called *Lipshitz* continuity. It is not the most general condition to prove what we need; we really only need some continuous function of $d(p, q)$ which is independent of x on the right-hand side. We choose Lipshitz continuity here because for the maps we are interested in, it is the easiest condition to check. If we can show that for any set $B \in \mathcal{H}(X)$ we have

$$h(w_{i_p}(B), w_{i_q}(B)) \leq k \cdot d_p(p, q),$$

then we can easily get the condition we want from the Collage Theorem. Proving this is simply a matter of writing down the definitions for the metric h.

$$h(w_p(B), w_q(B)) = d(w_p(B), w_q(B)) \vee d(w_q(B), w_p(B)),$$

where

$$d(w_p(B), w_q(B)) = \max_{x \in w_p(B)} (d(x, w_q(B)))$$

$$d(x, w_q(B)) = \min_{y \in w_q(B)} (d(x, y)).$$

Now, $x \in w_p(B)$ implies that there is an $\tilde{x} \in B$ such that $x = w_p(\tilde{x})$. Then there is a point $w_q(\tilde{x}) \in w_q(B)$, which is the image of $\tilde{x}$ under w_q. For this point, our condition holds, and

$$d(x, w_q(\tilde{x})) \leq k \cdot d_p(p, q) \Rightarrow \min_{y \in w_q(B)} (d(x, y)) \leq d(x, w_q(\tilde{x})) \leq k \cdot d_p(p, q)$$

Since this condition holds, for every $x \in w_p(B)$ the maximum over these points is at most $k \cdot d_p(p, q)$, and we have

$$d(w_p(B), w_q(B)) \leq k \cdot d_p(p, q).$$

The argument is nearly identical for $d(w_q(B), w_p(B))$, so we have

$$h(w_p(B), w_q(B)) \leq k \cdot d_p(p, q),$$

and a small change in the parameter on a particular map produces a small change in the image of any set $B \in \mathcal{H}(X)$. For a finite set of maps, $w_{1_p}, \ldots, w_{N_p}$, and their corresponding constants $k_1, \ldots, k_N$, it is then certainly the case that if $k = max_{i=1,\ldots,N}(k_i)$, we have

$$h(w_{i_p}(B), w_{i_q}(B)) \leq k \cdot d_p(p, q).$$

Now the union of such image sets cannot vary from parameter to parameter by more than the maximum Hausdorff distance above, consequently,

$$h(W_p(B), W_q(B)) \le k \cdot d_p(p, q).$$

We now apply the results of Lemma 11.2 to the complete metric space $\mathcal{H}(X)$, yielding

$$h(A_p, A_q) \le (1 - s)^{-1} h(A_p, W_q(A_p)) \le (1 - s)^{-1} k \cdot d_p(p, q).$$

Theorem 11.1 *Let (X, d) be a complete metric space. Let $\{X; w_1, \ldots, w_N\}$ be a hyperbolic IFS with contractivity s. For $n = 1, 2, \ldots, N$, let w_n depend on the parameter $p \in (P, d_p)$ subject to the condition $d(w_{n_p}(x), w_{n_q}(x)) \le k \cdot d_p(p, q)$ for all $x \in X$ with k independent of n, p, or x. Then the attractor $A(p) \in \mathcal{H}(X)$ depends continuously on the parameter $p \in P$ with respect to the Hausdorff metric $h(d)$.*

In other words, small changes in the parameters will lead to small changes in the attractor, provided that the system remains hyperbolic. This is very important because it tells us that we can continuously control the attractor of an IFS by adjusting parameters in the transformations, as is done in image compression applications. It also means we can smoothly interpolate between attractors: this is useful for image animation, for example.

Examples & Exercises

11.3. Construct a one-parameter family of IFS, of the form $\{\mathbb{R}^2; w_1, w_2, w_3\}$, where each w_i is affine and the parameter p lies in the interval $[0, 24]$. The attractor should tell the time, as illustrated in Figure III.88. $A(p)$ denotes the attractor at time p.

11.4. Imagine a slightly more complicated clockface, generated by using a one-parameter family of IFS of the form $\{\mathbb{R}^2; w_0, w_1, w_2, w_3\}$, $p \in [0, 24]$. w_0 creates the clockface, w_1 and w_2 are as in Exercise 11.3, and w_3 is a similitude that places a copy of the clockface at the end of the hour hand, as illustrated in Figure III.89. Then as p goes from 0 to 12 the hour hand sweeps through 360°, the hour hand on the smaller clockface sweeps through 720°, and the hour hand on the yet smaller clockface sweeps through 1080°, and so on. Thus as p advances, there exist lines on the attractor which are rotating at arbitrarily great speeds. Nonetheless we have continuous dependence of the image on p in the Hausdorff metric! At what times do all of the hour hands point in the same direction?

11.5. Find a one-parameter family of IFS in $\mathbb{R}^2$, whose attractors include the three trees in Figure III.90.

11.6. Run your version of Program 1 or Program 2, making small changes in the IFS code. Convince yourself that resulting rendered images "vary continuously" with respect to these changes.

11.7. Solve the following problems with regard to the images (a)–(f) in Figure III.91. Recall that a "just-touching" collage in $\mathbb{R}^2$ is one where the transforms of the target set do not overlap. They fit together like the pieces of a jigsaw puzzle.

Figure III.88. A one-parameter family of IFS that tells the time!

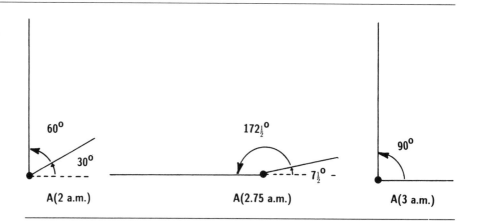

Figure III.89. This fractal clockface depends continuously on time in the Hausdorff metric.

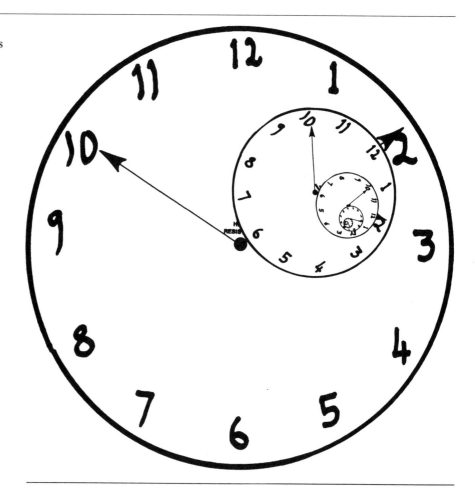

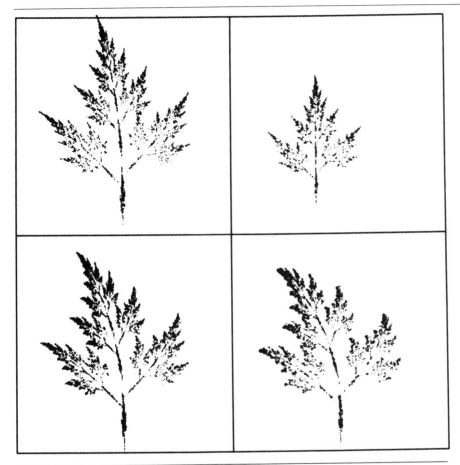

Figure III.90. Blowing in the wind. Find a one-parameter family of IFS whose attractors include the trees shown here. The Random Iteration Algorithm was used to compute these images.

(a) Find a one-parameter family collage of affine transformations.

(b) Find a "just-touching" collage of affine transformations.

(c) Find a collage using similitudes only. What is the smallest number of affine transformations in $\mathbb{R}^2$, such that the boundary is the attractor?

(d) Find a one-parameter family collage of affine transformations.

(e) Find a "just-touching" collage, using similitudes only, parameterized by the real number p.

(f) Find a collage for circles and disks.

Figure III.91. Classical collages. Can you find an IFS corresponding to each of these classical geometrical objects?

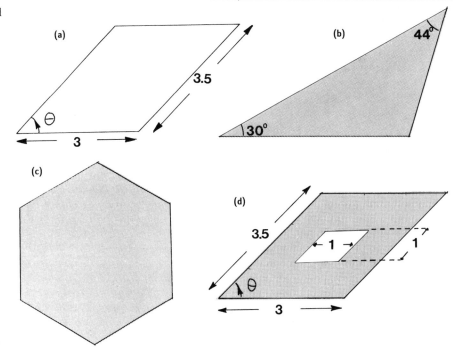

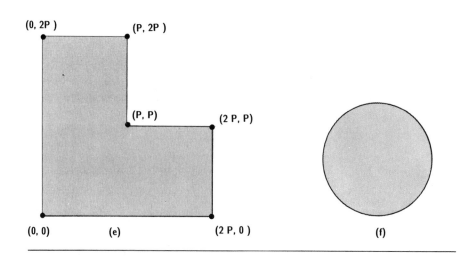

Plate 1

ce in trees exhibits natural affine redundancy.

Plate 2

The same IFS with two different coloring schemes: color is determined by the Hausdorff distance from the fern attractor.

Plate 3
Three-dimensional ferns.

Plate 4
A truly four-dimensional fern. As we rotate this weird object in the three dimensions in which we live, what will we see emerging from the fourth dimension? Computed using 4-D Fernmaker™.

Plate 5
This sequence of images illustrates the continuous dependence of the attractor on the parameters in the IFS code.

Plate 6
The result of running a version of Program 1 in Chapter VII on a Masscomp 5600 workstation with Aurora graphics.

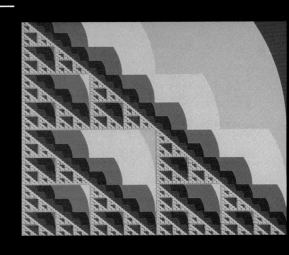

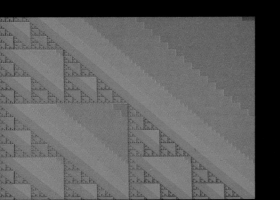

Plate 7
The result of running a version of Program 1 in Chapter VII, with $(a, b) = (0, 0)$, $(c, d) = (5 \times 10^{-18}, 5 \times 10^{-18})$, and numits = 65. This viewing window is minute, the colors arranged in a wonderful pattern: we are stunned by the geometrical complexity of this image at such high magnification.

Plate 8
A seed beginning to germinate?

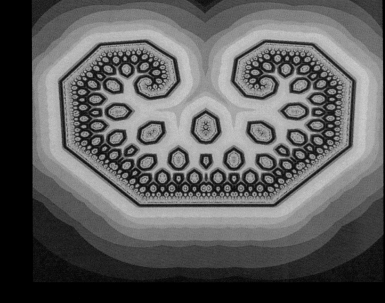

Plate 9
The Escape Time Algorithm applied
to an IFS of affine transformations.
Such images can be magnified
enormously, without loss of mystery.

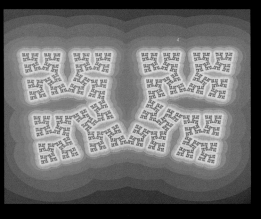

Plate 10
Primitive art?

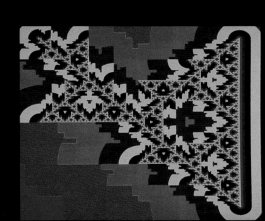

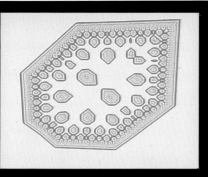

Plate 11

The Escape Time Algorithm is applied to an overlapping IFS of two affine transformations. Can such sets be used to model sections of plants, as can be seen under a microscope?

Plate 12

The Escape Time Algorithm is applied to an IFS of two maps. Here the magnification is extreme. Do not race by. Look closely and think.

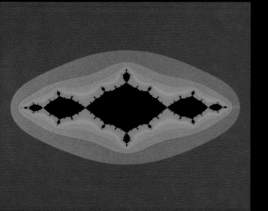

Plate 13

The Escape Time Algorithm is applied to produce an image of the Julia set for $z^2 - 1$.

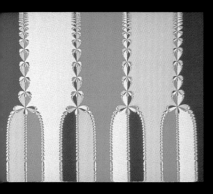

Plate 14

The Escape Time Algorithm is used to analyze the Newton transformation associated with the polynomial $z^4 - 1$. Here the Riemann sphere is mapped onto a rectangle using Mercator's projection. The North Pole at the top corresponds to the Point at Infinity, while the South Pole, at the bottom, corresponds to the Origin. The roots are $a_1 = 1$, $a_2 = -1$, $a_3 = i$, $a_4 = -i$. Points whose orbits converge to a_1 are plotted in dark green; points whose orbits converge to a_2 are plotted in very light green; points whose orbits converge to a_3 are plotted in light green; and points whose orbits converge to a_4 are plotted in white.

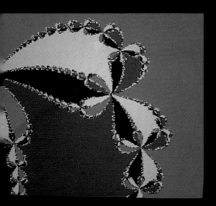

Plate 15

Zoom in on a piece of Plate 14 and change the color scheme. Every point on a boundary between two colors is actually on the boundary of all four colors!

Plate 16

Part of the Mandelbrot set for $z^2 - \lambda$, painted in colors according to "escape times." The green land represents values of the complex parameter λ for which the associated Julia set is connected. The blue sea corresponds to totally disconnected Julia sets. The exciting place is the coastline, where the white foam of the breaking waves symbolizes the transition between connection and disconnec-
tion

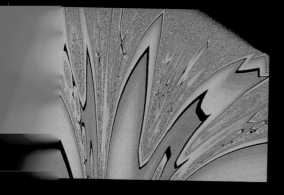

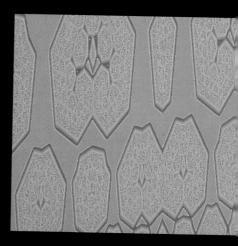

Plate 17

Part of the parameter space $P = \{\lambda \in \mathbb{C}: |\lambda| < 1\}$
for the family of dynamical systems
$$f(z) = \begin{cases} (z-1)/\lambda & \text{if } \operatorname{Re} z \geq 0; \\ (z+1)/\lambda^* & \text{if } \operatorname{Re} z < 0. \end{cases}$$
ordered according to the "escape time" of a point
$0 \in \mathbb{R}^2$.

Plate 18

An extreme close-up
a piece of the param
space in Color Plate

Plate 19

A map of a small piece of the parameter space for
the family of dynamical systems in Example 4.2
in Chapter VIII. The map was computed using
Algorithm 4.1 in Chapter VIII. The grainy
multicolored areas resemble the repelling sets
of the dynamical system for the corresponding
values of λ.

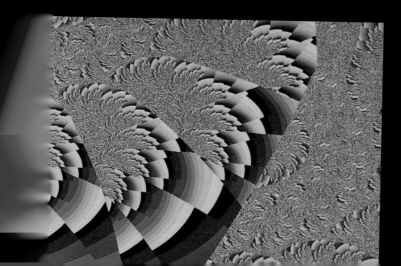

Plate 20

An extreme close-up on
a grainy region in Color
Plate 19.

Plate 22
Zoom on the smoke coming out of a chimney.

Chapter IV

Chaotic Dynamics on Fractals

1 The Addresses of Points on Fractals

We begin by considering informally the concept of the *addresses* of points on the attractor of a hyperbolic IFS. Figure IV.92 shows the attractor of the IFS:

$$\{\mathbb{C}; w_1(z) = (0.13 + 0.64i)z, w_2(z) = (0.13 + 0.64i)z + 1\}.$$

This attractor, A, is the union of two disjoint sets, $w_1(A)$ and $w_2(A)$, lying to the left and right, respectively, of the dotted line ab. In turn, each of these two sets is made of two disjoint sets:

$$w_1(A) = w_1(w_1(A)) \cup w_2(w_1(A)), w_2(A) = w_2(w_2(A)) \cup w_2(w_2(A)).$$

This leads to the idea of addressing points in terms of the sequences of transformations, applied to A, which lead to them. All points belonging to A, in the subset $w_1(w_1(A))$, are situated on the piece of the attractor that lies below dc and to the left of ab, and their addresses all begin $11\ldots$.. Clearly, the more precisely we specify geometrically where a point in A lies, the more bits to the address we can provide. For example, every point to the right of ab, below ef, to the left of gh, has an address that begins $212\ldots$.. In Theorem 2.1 we prove that, in examples such as this one, it is possible to assign a unique address to every point of A. In such cases we say that the IFS is "totally disconnected."

Here is a different type of example. Consider the IFS

$$\{\mathbb{C}; w_1(z) = \frac{1}{2}z, w_2(z) = \frac{1}{2}z + \frac{1}{2}, w_3(z) = \frac{1}{2}z + \frac{1}{2}i\}.$$

The attractor, A, of this IFS is a Sierpinski triangle with vertices at $(0, 0)$, $(1, 0)$, and $(0, 1)$. Again we can address points on A according to the sequences of transformations that lead to them. This time there are at least three points in A that

Figure IV.92. Addresses of points on an attractor. The lines ab, cd, ef, and gh are not part of the attractor.

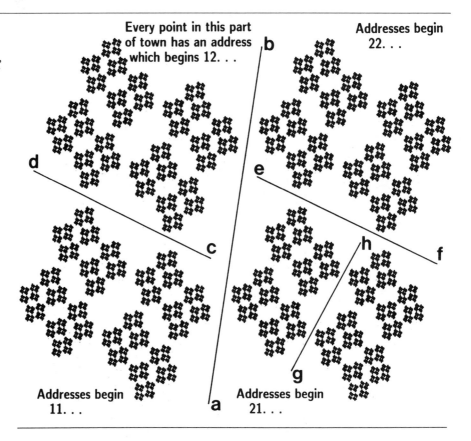

Every point in this part of town has an address which begins 12. . .

Addresses begin 22. . .

Addresses begin 11. . .

Addresses begin 21. . .

have two addresses, because there is a point in each of the sets $w_1(A) \cap w_2(A)$, $w_2(A) \cap w_3(A)$, and $w_3(A) \cap w_1(A)$, as illustrated in Figure IV.93.

On the other hand, some points on the Sierpinski triangle have only one address, such as the three vertices $(0, 0)$, $(1, 0)$, and $(0, 1)$. Although the attractor is connected, the proportion of points with multiple addresses is "small," in a sense we do not yet make precise. In such cases as this we say that the IFS is "just-touching." Notice that this terminology refers to the IFS itself rather than to its attractor.

Let us look at a third, fundamentally different example. Consider the hyperbolic IFS

$$\{[0, 1]; \frac{1}{2}x, \frac{3}{4}x + \frac{1}{4}\}.$$

The attractor is $A = [0, 1]$, but now

$$w_1(A) \cap w_2(A) = [0, \frac{1}{2}] \cap [\frac{1}{4}, 1] = [\frac{1}{4}, \frac{1}{2}];$$

so $w_1(A) \cap w_2(A)$ is a significant piece of the attractor. The attractor would look very different if the overlapping piece $[\frac{1}{4}, \frac{1}{2}]$ were missing. Now observe that every

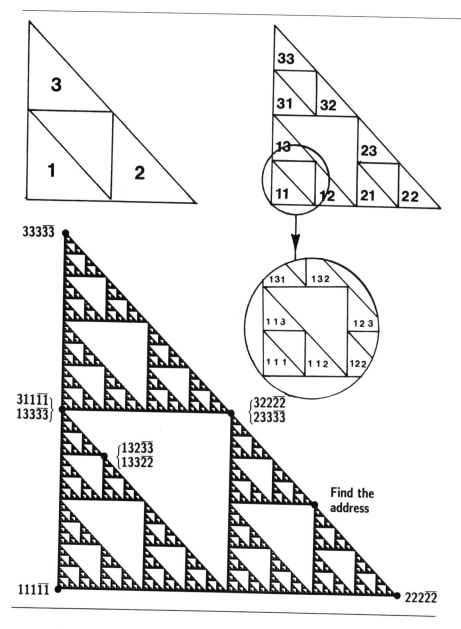

Figure IV.93. Some points on this Sierpinski triangle have two addresses, while others have only one address. Overlining on the last symbols, in an expression such as $311\overline{11}$, means that the overlined symbols are repeated endlessly. For example, $311\overline{11} =$ 31111111111111111 1111..., and $31\overline{123} =$ 31123123123123...

point in $[\frac{1}{4}, \frac{1}{2}]$ has at least two addresses. On the other hand, the points 0 and 1 have only one address each. Nonetheless, it appears that the proportion of points with multiple addresses is large. In such cases we say that the IFS is "overlapping."

The terminologies "totally disconnected," "just-touching," and "overlapping" refer to the IFS itself rather than to the attractor. The reason for this is that the same set may be the attractor of several different hyperbolic IFS's. Consider, for example,

Figure IV.94. Different IFS's with the same attractor provide different addressing schemes. Here the symbols {0, 1} are used in place of {1, 2} for obvious reasons.

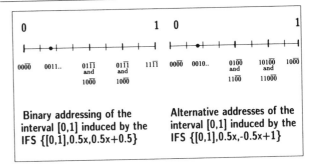

Binary addressing of the interval [0,1] induced by the IFS {[0,1],0.5x,0.5x+0.5}

Alternative addresses of the interval [0,1] induced by the IFS {[0,1],0.5x,-0.5x+1}

the two IFS's

$$\{[0, 1]; w_1(x) = \frac{1}{2}x, w_2(x) = \frac{1}{2}x + \frac{1}{2}\}$$

and

$$\{[0, 1]; w_1(x) = \frac{1}{2}x, w_2(x) = -\frac{1}{2}x + 1\}.$$

The attractor of each one is the real interval $[0, 1]$. We can obtain two different addressing schemes for the points in $[0, 1]$, as illustrated in Figure IV.94.

These two IFS are just-touching. However, the IFS

$$\{[0, 1]; w_1(x) = \frac{1}{2}x, w_2(x) = \frac{3}{4}x + \frac{1}{4}\}$$

is overlapping, while its attractor is also $[0, 1]$.

Examples & Exercises

1.1. Figure IV.95 shows the attractor of an IFS of the form $\{\mathbb{R}^2; w_n, n = 1, 2, 3\}$, where each of the transformations $w_n : \mathbb{R}^2 \to \mathbb{R}^2$ is affine. The addresses of several points are given. Find the addresses of $a, b,$ and c.

1.2. In Figure IV.95 locate the point whose address is $111\overline{11}$.

1.3. A *quadtree* is an addressing scheme used in computer science for addressing small squares in the unit square $\blacksquare = \{(x_1, x_2) \in \mathbb{R}^2 : 0 \le x_1 \le 1, 0 \le x_2 \le 1\}$ as follows. The square is broken into four quarters. Points in the first quarter have addresses that begin 0, points in the second quarter have addresses that begin 1, and so on, as illustrated in Figure IV.96. Find an IFS that gives rise to the addressing scheme suggested in Figure IV.96. Is this a totally disconnected, just-touching, or overlapping IFS?

1.4. Addresses are assigned to the Sierpinski triangle, as in Figure IV.93. Characterize the addresses of the set of points that lie on the outermost boundary, the triangle with vertices $\overline{11}, \overline{22},$ and $\overline{33}$.

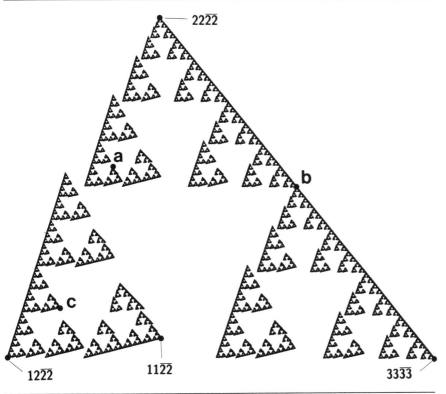

The figure shows addresses: $22\overline{2}$, $12\overline{2}$, $11\overline{2}2$, $33\overline{3}$, with points labeled a, b, c.

Figure IV.95. Can you find the addresses of a, b, and c?

1.5. Characterize the addresses of points belonging to the boundary of the largest hole in Figure IV.97.

1.6. Consider a hyperbolic IFS with condensation set C. Suppose the condensation set is itself the attractor of another hyperbolic IFS. Design an addressing scheme for the attractor of the IFS with condensation. Can all possible addresses occur?

1.7. Figure IV.98 shows an "overlapping" IFS attractor, for two affine transformations in $\mathbb{R}^2$. Choose one point in each of the marked regions on the attractor. Find the first four numbers in two different addresses for each of these points. The first few numbers in the addresses of some points on the attractor are included in the figure to remove possible ambiguities.

1.8. ⋆ Identify the set of addresses of points on the attractor, A, of a hyperbolic IFS with code space. Argue that nearby codes correspond to points on A which are nearby.

1.9. Address the real number 0.7513 in each of the two coding schemes given in Figure IV.94.

In thinking about the addresses of points on fractals, already we have been led to

33	32	23	22
30	31	20	21
03	02	13	12
00	01	10	11

Figure IV.96. Addresses at depth two in a quadtree.

try to compare "how many" points have a certain property to how many have another property. For example, in the case of the addressing scheme on the Sierpinski triangle described above, we wanted to compare the number of points with multiple addresses to the number of points with single addresses. It turns out that both numbers are infinite. Yet still we want to compare their numbers. One way in which this may be done is through the concept of countability.

Definition 1.1 *Let S be a set. S is* countable *if it is empty or if there is an onto transformation $c : I \rightarrow S$, where I is either one of the sets*

$$\{1\}, \{1, 2\}, \{1, 2, 3\}, \ldots, \{1, 2, 3, \ldots, n\}, \ldots,$$

or the positive integers $\{1, 2, 3, 4, \ldots\}$. S is uncountable *if it is not countable.*

We think of an uncountable set as being larger than a countable set.

We are going to make fundamental use of code space to formalize the concept of addresses. How many points does code space contain?

Theorem 1.1 *Code space on two or more symbols is uncountable.*

Proof We prove it here for the code space on the two symbols $\{1, 2\}$. Denote an element of code space Σ by $\omega = \omega_1 \omega_2 \omega_3 \ldots$, where each $\omega_i \in \{1, 2\}$. Define

---

Figure IV.97. Can you describe the addresses of the points on the boundary of the central white region?

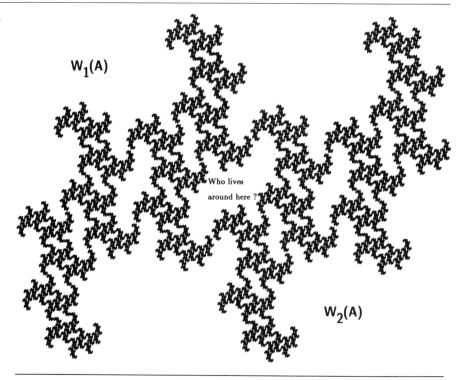

$W_1(A)$

Who lives around here ?

$W_2(A)$

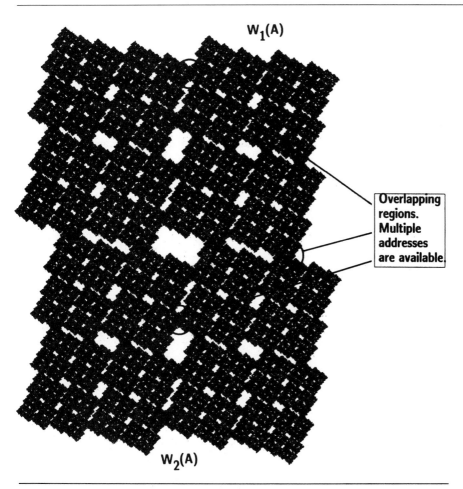

$W_1(A)$

$W_2(A)$

Overlapping
regions.
Multiple
addresses
are available.

Figure IV.98. Attractor of a hyperbolic IFS in the overlapping case. In the overlapping regions multiple addresses are available.

$\rho : \{1, 2\} \to \{1, 2\}$ by $\rho(1) = 2$ and $\rho(2) = 1$. Suppose code space is countable. Let the counting function be $c : \{1, 2, 3, \ldots\} \to \Sigma$. Consider the point $\sigma \in \Sigma$ defined by

$$\sigma = \sigma_1 \sigma_2 \sigma_3 \ldots,$$

where $\sigma_n = \rho((c(n))_n)$, and $(c(n))_n$ means the nth symbol of $c(n)$. When does the counting function reach σ? Never! For example, $c(3) \neq \sigma$ because their third symbols are different! This completes the proof.

Examples & Exercises

1.10. The set of integers $\mathbb{N} = \{0, \pm 1, \pm 2, \ldots, \}$ is countable. Define $c : N \to \mathbb{N}$ by $c(z) = (z - 1)/2$ if z is odd, $c(z) = -z/2$ if z is even.

1.11. Prove that a countable set of countable sets is countable. Show that an uncountable set, take away a countable set, is uncountable.

1.12. The *rational* numbers are countable. A rational number can be written in the form p/q, where p and q are integers with $q \neq 0$. Figure IV.99 shows how to count the positive ones, some numbers being counted more than once. Make a rule that gets rid of the redundant countings. Also, show how to include the negative rationals in the scheme.

1.13. Show that a Sierpinski triangle contains countably many triangles.

1.14. Let S be a perfect subset of a metric space. Suppose that S contains more than one point. Prove that S is uncountable.

1.15. Characterize the addresses of the missing pieces in Figure IV.100.

---

2 Continuous Transformations from Code Space to Fractals

Definition 2.1 *Let* $\{X; w_1, w_2, \ldots, w_N\}$ *be a hyperbolic IFS. The* code space *associated with the IFS,* (Σ, d_C), *is defined to be the code space on* N *symbols* $\{1, 2, \ldots, N\}$, *with the metric* d_C *given by*

$$d_C(\omega, \sigma) = \sum_{n=1}^{\infty} \frac{|\omega_n - \sigma_n|}{(N+1)^n} \qquad \text{for all } \omega, \sigma \in \Sigma.$$

Our goal is to construct a continuous transformation ϕ from the code space associated with an IFS onto the attractor of the IFS. This will allow us to formalize our notion of addresses. In order to make this construction we will need two lemmas. The first lemma tells us that if we have a hyperbolic IFS acting on a complete metric space, but we are only interested in studying how the IFS acts in relation to a fixed compact subset of X, then we can treat the IFS as though it were defined on a compact metric space.

Lemma 2.1 *Let* $\{X; w_n : n = 1, 2, \ldots, N\}$ *be a hyperbolic IFS, where* (X, d) *is a complete metric space. Let* $K \in \mathcal{H}(X)$. *Then there exists* $\tilde{K} \in \mathcal{H}(X)$ *such that* $K \subset \tilde{K}$ *and* $w_n : \tilde{K} \to \tilde{K}$ *for* $n = 1, 2, \ldots, N$. *In other words,* $\{\tilde{K}; w_n : n = 1, 2, 3, \ldots, N\}$ *is a hyperbolic IFS where the underlying space is compact.*

Proof Define $W : \mathcal{H}(X) \to \mathcal{H}(X)$ by

$$W(B) = \cup_{n=1}^{N} w_n(B) \qquad \text{for all } B \in \mathcal{H}(X).$$

To construct $\tilde{K}$ consider the IFS with condensation $\{X; w_n; n = 0, 1, 2, \ldots, N\}$, where the condensation map w_0 is associated with the condensation set K. By Theorem 7.1 in Chapter III the attractor of this IFS belongs to $\mathcal{H}(X)$. By exercise 9.1 in Chapter III it can be written

$$\tilde{K} = \text{Closure of } (K \cup W^{\circ 1}(K) \cup W^{\circ 2}(K) \cup W^{\circ 3}(K) \cup \ldots \cup W^{\circ n}(K) \cup \ldots \ldots).$$

It is readily seen that $K \subset \tilde{K}$ and that $W(\tilde{K}) \subset \tilde{K}$. This completes the proof.

The next lemma makes the first step in linking code space to IFS attractors, by introducing a certain transformation ϕ, which maps the Cartesian product space $\Sigma \times N \times \mathbf{X}$ into $\mathbf{X}$. By taking appropriate limits, in Theorem 2.1 below, we will eliminate the dependence on N and $\mathbf{X}$ to provide the desired connection between Σ and $\mathbf{X}$.

Lemma 2.2 *Let* $\{\mathbf{X}; w_n : n = 1, 2, \ldots, N\}$ *be a hyperbolic IFS of contractivity* s, *where* $(\mathbf{X}, d)$ *is a complete metric space. Let* (Σ, d_C) *denote the code space associated with the IFS. For each* $\sigma \in \Sigma, n \in N$, *and* $x \in \mathbf{X}$, *define*

$$\phi(\sigma, n, x) = w_{\sigma_1} \circ w_{\sigma_2} \circ \ldots \circ w_{\sigma_n}(x).$$

Let K *denote a compact nonempty subset of* $\mathbf{X}$. *Then there is a real constant* D *such that*

$$d(\phi(\sigma, m, x_1), \phi(\sigma, n, x_2)) \leq Ds^{m \wedge n}$$

for all $\sigma \in \Sigma$, *all* $m, n \in N$, *and all* $x_1, x_2 \in K$.

Proof Let σ, m, n, x_1, and x_2 be as stated in the lemma. Construct $\tilde{K}$ from K as in Lemma 2.1. Without any loss of generality we can suppose that $m < n$. Then observe that

$$\phi(\sigma, n, x_2) = \phi(\sigma, m, \phi(\omega, n - m, x_2)),$$

where

$$\omega = \sigma_{n-m+1} \sigma_{n-m+2} \ldots \sigma_n \ldots \in \Sigma.$$

Let $x_3 = \phi(\omega, n - m, x_2)$. Then x_3 belongs to $\tilde{K}$. Hence we can write

$$d(\phi(\sigma, m, x_1), \phi(\sigma, n, x_2)) = d(\phi(\sigma, m, x_1), \phi(\sigma, m, x_3))$$
$$\leq sd(w_{\sigma_2} \circ \ldots \circ w_{\sigma_m}(x_1), w_{\sigma_2} \circ \ldots \circ w_{\sigma_m}(x_3))$$
$$\leq s^2 d(w_{\sigma_3} \circ \ldots \circ w_{\sigma_m}(x_1), w_{\sigma_3} \circ \ldots \circ w_{\sigma_m}(x_3))$$
$$\leq s^m d(x_1, x_3) \leq s^m D,$$

where $D = \max\{d(x_1, x_3) : x_1, x_3 \in \tilde{K}\}$. D is finite because $\tilde{K}$ is compact. This completes the proof.

Theorem 2.1 *Let* $(\mathbf{X}, d)$ *be a complete metric space. Let* $\{\mathbf{X}; w_n : n = 1, 2, \ldots, N\}$ *be a hyperbolic IFS. Let* A *denote the attractor of the IFS. Let* (Σ, d_C) *denote the code space associated with the IFS. For each* $\sigma \in \Sigma, n \in N$, *and* $x \in \mathbf{X}$, *let*

$$\phi(\sigma, n, x) = w_{\sigma_1} \circ w_{\sigma_2} \circ \ldots \circ w_{\sigma_n}(x).$$

Then

$$\phi(\sigma) = \lim_{n \to \infty} \phi(\sigma, n, x)$$

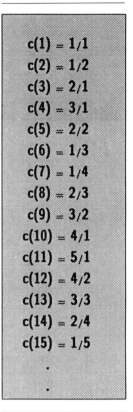

$$c(1) = 1/1$$
$$c(2) = 1/2$$
$$c(3) = 2/1$$
$$c(4) = 3/1$$
$$c(5) = 2/2$$
$$c(6) = 1/3$$
$$c(7) = 1/4$$
$$c(8) = 2/3$$
$$c(9) = 3/2$$
$$c(10) = 4/1$$
$$c(11) = 5/1$$
$$c(12) = 4/2$$
$$c(13) = 3/3$$
$$c(14) = 2/4$$
$$c(15) = 1/5$$
.
.

Figure IV.99. How to count the positive rational numbers. What is $c(24)$?

Figure IV.100. Characterize the addresses of the missing pieces.

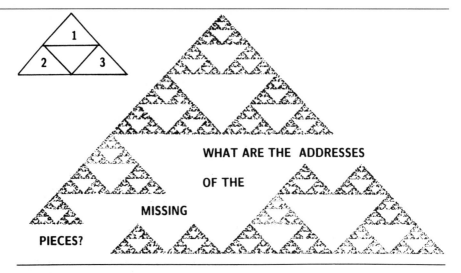

WHAT ARE THE ADDRESSES

OF THE

MISSING

PIECES?

exists, belongs to A, and is independent of $x \in \mathbf{X}$. If K is a compact subset of $\mathbf{X}$, then the convergence is uniform over $x \in K$. The function $\phi : \Sigma \to A$ thus provided is continuous and onto.

Proof Let $x \in \mathbf{X}$. Let $K \in \mathcal{H}(\mathbf{X})$ be such that $x \in K$. Construct $\tilde{K}$ as in Lemma 2.1. Define $W : \mathcal{H}(\mathbf{X}) \to \mathcal{H}(\mathbf{X})$ in the usual way. By Theorem 7.1 in Chapter III, W is a contraction mapping on the metric space $(\mathcal{H}(\mathbf{X}), h(d))$; and we have

$$A = \lim_{n \to \infty} \{W^{\circ n}(K)\}.$$

In particular $\{W^{\circ n}(K)\}$ is a Cauchy sequence in $(\mathcal{H}, h)$. Notice that $\phi(\sigma, n, x) \in W^{\circ n}(K)$. It follows from Theorem 7.1 in Chapter II that if $\lim_{n \to \infty} \phi(\sigma, n, x)$ exists, then it belongs to A.

That the latter limit does exist follows from the fact that, for fixed $\sigma \in \Sigma$, $\{\phi(\sigma, n, x)\}_{n=1}^{\infty}$ is a Cauchy sequence: by Lemma 2.2

$$d(\phi(\sigma, m, x), \phi(\sigma, n, x)) \le Ds^{m \wedge n} \qquad \text{for all } x \in K,$$

and the right-hand side here tends to zero as m and n tend to infinity. The uniformity of the convergence follows from the fact that the constant D is independent of $x \in K$.

Next we prove that $\phi : \Sigma \to A$ is continuous. Let $\epsilon > 0$ be given. Choose n so that $s^n D < \epsilon$, and let $\sigma, \omega \in \Sigma$ obey

$$d_C(\sigma, \omega) < \sum_{m=n+2}^{\infty} \frac{N}{(N+1)^m} = \frac{1}{(N+1)^{n+1}}.$$

Then one can verify that σ must agree with ω through n terms; that is, $\sigma_1 = \omega_1, \sigma_2 = \omega_2, \ldots, \sigma_n = \omega_n$. It follows that for each $m \ge n$ we can write

$$d(\phi(\sigma, m, x), \phi(\omega, m, x)) = d(\phi(\sigma, n, x_1), \phi(\sigma, n, x_2)),$$

for some pair $x_1, x_2 \in \tilde{K}$. By Lemma 2.2 the right-hand side here is smaller than $s^n D$ which is smaller than ϵ. Taking the limit as $m \to \infty$ we find

$$d(\phi(\sigma), \phi(\omega)) < \epsilon.$$

Finally, we prove that ϕ is onto. Let $a \in A$. Then, since

$$A = \lim_{n \to \infty} W^{\circ n}(\{x\}),$$

it follows from Theorem 7.1 in Chapter II that there is a sequence $\{\omega^{(n)} \in \Sigma : n = 1, 2, 3, \ldots\}$ such that

$$\lim_{n \to \infty} \phi(\omega^{(n)}, n, x) = a.$$

Since (Σ, d_C) is compact, it follows that $\{\omega^{(n)} : n = 1, 2, 3, \ldots\}$ possesses a convergent subsequence with limit $\omega \in \Sigma$. Without loss of generality assume $\lim_{n \to \infty} \omega^{(n)} = \omega$. Then the number of successive initial agreements between the components of $\omega^{(n)}$ and ω increases without limit. That is, if

$$\alpha(n) = \text{ number of elements in } \{j \in N : \omega_k^{(n)} = \omega_k \text{ for } 1 \le k \le j\},$$

where $N = \{1, 2, 3, \ldots\}$, then $\alpha(n) \to \infty$ as $n \to \infty$. It follows that

$$d(\phi(\omega, n, x), \phi(\omega^{(n)}, n, x)) \le s^{\alpha(n)} D.$$

By taking the limit on both sides as $n \to \infty$ we find $d(\phi(\omega), a) = 0$, which implies $\phi(\omega) = a$. Hence $\phi : \Sigma \to A$ is onto. This completes the proof.

Definition 2.2 *Let* $\{X; w_n, n = 1, 2, 3, \ldots, N\}$ *be a hyperbolic IFS with associated code space* Σ. *Let* $\phi : \Sigma \to A$ *be the continuous function from code space onto the attractor of the IFS constructed in Theorem 1. An* address *of a point* $a \in A$ *is any member of the set*

$$\phi^{-1}(A) = \{\omega \in \Sigma : \phi(\omega) = a\}.$$

This set is called the set of addresses *of* $a \in A$. *The IFS is said to be* totally disconnected *if each point of its attractor possesses a unique address. The IFS is said to be* just-touching *if it is not totally disconnected yet its attractor contains an open set* $\mathcal{O}$ *such that*

(1) $w_i(\mathcal{O}) \cap w_j(\mathcal{O}) = \emptyset \forall i, j \in \{1, 2, \ldots, N\}$ *with* $i \ne j$;
(2) $\cup^N \lim_{i=1} w_i(\mathcal{O}) \subset \mathcal{O}$.

An IFS whose attractor obeys (i) and (ii) is said to obey the open set *condition. The IFS is said to be* overlapping *if it is neither just-touching nor disconnected.*

Theorem 2.2 *Let* $\{X; w_n, n = 1, 2, \ldots, N\}$ *be a hyperbolic IFS with attractor A. The IFS is* totally disconnected *if and only if*

$$w_i(A) \cap w_j(A) = \emptyset \forall i, \qquad j \in \{1, 2, \ldots, N\} \qquad \text{with } i \ne j. \tag{1}$$

Proof If the IFS is totally disconnected, then each point on its attractor possesses a unique address. This implies Equation 3. If the IFS is not totally disconnected, then some point on its attractor possesses two different addresses. These must disagree at some first place: choose inverse images to get this place out front, to produce a contradiction to Equation 3. This completes the proof.

Examples & Exercises

2.1. Show that the attractor of the IFS $\{\mathbb{R}; \frac{1}{2}x, \frac{1}{2}x + \frac{1}{2}\}$ is just-touching. Classify the attractor for the IFS $\{\mathbb{R}; \frac{1}{2}x, 1\}$.

2.2. Prove that the attractor of the IFS $\{\mathbb{R}; \frac{1}{2}x, \frac{3}{4}x + \frac{1}{4}\}$ is overlapping.

2.3. Consider the IFS $\{[0, 1], w_n(x) = \frac{n-1}{10} + \frac{1}{10}x, n = 1, 2, 3, \ldots, 10\}$ and for the associated code space use the symbols $\{0, 1, 2, \ldots, 9\}$. Show that the attractor of the IFS is $[0, 1]$ and that it is just-touching. Identify the addresses of points with multiple addresses. Show that the address of a point is just its decimal representation. Comment on the fact that some numbers have two decimal representations.

2.4. Prove that the attractor to the IFS $\{[0, 1]; w_1(x) = \frac{1}{3}x, w_2(x) = \frac{1}{3}x + \frac{2}{3}\}$ is totally disconnected.

2.5. Prove that the IFS that generates the *Black Spleenwort* fern, given in Chapter 2, is just-touching.

2.6. Show that the IFS $\{[0, 1]; w_1(x) = \frac{1}{2}, w_2(x) = \frac{1}{2}\}$ is overlapping.

We need to understand the structure of code space. Theorem 2.1 told us that the code space on N symbols is the mother of all hyperbolic IFS consisting of N maps. We will use the following theorem to show that the mother is metrically equivalent to a classical Cantor set.

Theorem 2.3 *Let Σ denote the code space of the N symbols, $\{1, 2, \ldots, N\}$, and define two different metrics on Σ by*

$$d_1(x, y) = \sum_{i=1}^{\infty} \frac{|x_i - y_i|}{(N + 1)^i}, \qquad d_2(x, y) = \left| \sum_{i=1}^{\infty} \frac{x_i - y_i}{(N + 1)^i} \right|.$$

Then (Σ, d_1) and (Σ, d_2) are equivalent metric spaces.

Proof We give the proof for the case $N = 10$. Let $x, y \in \Sigma$ be given. Clearly we have $d_2(x, y) \leq d_1(x, y)$. We must show that there is a constant C so that $Cd_1(x, y) \leq d_2(x, y)$, where C is independent of x and y. Here we pick $C = \frac{1}{19}$ and show that it works.

We can suppose that for some $k \in \{1, 2, 3, \ldots\}$, $x_1 = y_1, x_2 = y_2, \ldots, x_{k-1} = y_{k-1}, x_k \neq y_k$. Then

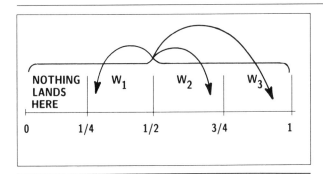

Figure IV.101. Nothing lands here.

$$d_2(x, y) = |\sum_{i=k}^{\infty} \frac{x_i - y_i}{11^i}| \geq \frac{|x_k - y_k|}{11^k} - \sum_{i=k+1}^{\infty} \frac{|x_i - y_i|}{11^i}$$

$$\geq \frac{|x_k - y_k|}{11^k} - \sum_{i=k+1}^{\infty} \frac{9}{11^i} = (|x_k - y_k| - \frac{9}{10}) \frac{1}{11^k}$$

$$\geq \frac{1}{19}(|x_k - y_k| + \frac{9}{10}) \frac{1}{11^k},$$

(verify this by checking it for $|x_k - y_k| \in \{1, 2, \ldots, 9\}$),

$$\geq \frac{1}{19}(\frac{|x_k - y_k|}{11^k} + \sum_{i=k+1}^{\infty} \frac{9}{11^i}) \geq \frac{1}{19}(\frac{|x_k - y_k|}{11^k} + \sum_{i=k+1}^{\infty} \frac{|x_i - y_i|}{11^i})$$

$$\geq \frac{1}{19} \sum_{n=1}^{\infty} \frac{|x_i - y_i|}{11^i} = \frac{1}{19} d_1(x, y).$$

This completes the proof.

We now show that code space is metrically equivalent to a totally disconnected Cantor subset of $[0, 1]$. Define a hyperbolic IFS by $\{[0, 1]; w_n(x) = \frac{1}{(N+1)}x + \frac{n}{N+1} : n = 1, 2, \ldots, N\}$. Thus

$$w_n([0, 1]) = [\frac{n}{N + 1}, \frac{n + 1}{N + 1}] \qquad \text{for } n = 1, 2, \ldots, N,$$

as illustrated for $N = 3$ in Figure IV.101.

The attractor for this IFS is totally disconnected, as illustrated in Figure IV.102 for $N = 3$.

In the case $N = 3$, the attractor is contained in $[\frac{1}{3}, 1]$. The fixed points of the three transformations $w_1(x) = \frac{1}{4}x + \frac{1}{4}$, $w_2(x) = \frac{1}{4}x + \frac{1}{2}$, $w_2(x) = \frac{1}{4}x + \frac{3}{4}$ are $\frac{1}{3}$, $\frac{2}{3}$, and 1, respectively. Moreover, the address of any point on the attractor is exactly the same as the string of digits that represents it in base $N + 1$. What is happening here is this. At level zero we begin with all numbers in $[0, 1]$ represented in base $(N + 1)$. We remove all those points whose first digit is 0. For example, in the case $N = 3$ this eliminates the interval $[0, \frac{1}{4}]$. At the second level we remove from the remaining points all those that have digit 0 in the second place. And so on. We end up with

Figure IV.102. A special ternary Cantor set in the making.

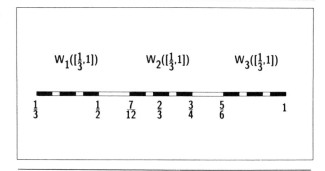

those numbers whose expansion in base $(N + 1)$ does not contain the digit 0. Now consider the continuous transformation $\phi : (\Sigma, d_C) \to (A,$ Euclidean$)$. It follows from Theorem 2.3 that the two metric spaces are equivalent. ϕ is the transformation that provides the equivalence. Thus, we have a realization, a way of picturing code space.

Examples & Exercises

2.7. Find the figure analogous to Figure IV.102, corresponding to the case $N = 9$.

2.8. What is the smallest number in $[0, 1]$ whose decimal expansion contains no zeros?

We continue to discuss the relationship between the attractor A of a hyperbolic IFS $\{\mathbf{X}; w_1, w_2, \ldots, w_N\}$ and its associated code space Σ. Let $\phi : \Sigma \to \mathbf{X}$ be the code space map constructed in Theorem 2.1. Let $\omega = \omega_1\omega_2\omega_3\omega_4 \ldots$ be an address of a point $x \in A$. Then

$$\tilde{\omega} = j\omega_1\omega_2\omega_3\omega_4 \ldots$$

is an address of $w_j(x)$, for each $j \in \{1, 2, \ldots, N\}$.

Definition 2.3 *Let* $\{\mathbf{X}, w_1, w_2, \ldots, w_N\}$ *be a hyperbolic IFS with attractor A. A point* $a \in A$ *is called a* periodic point *of the IFS if there is a finite sequence of numbers* $\{\sigma(n) \in \{1, 2, \ldots, N\}\}_{n=1}^{P}$ *such that*

$$a = w_{\sigma(P)} \circ w_{\sigma(P-1)} \circ \ldots \circ w_{\sigma(1)}(a). \tag{2}$$

If $a \in A$ *is periodic, then the smallest integer P such that the latter statement is true is called the* period *of a.*

Thus, a point on an attractor is periodic if we can apply a sequence of w_n's to it, in such a way as to get back to exactly the same point after finitely many steps. Let $a \in A$ be a periodic point that obeys (2). Let σ be the point in the associated code space, defined by

$$\sigma = \sigma(P)\sigma(P-1)\ldots\sigma(1)\sigma(P)\sigma(P-1)\ldots\sigma(1)\sigma(P)\sigma(P-1)\ldots$$
$$= \overline{\sigma(P)\sigma(P-1)\ldots\sigma(1)}. \tag{3}$$

Then, by considering $\lim_{n\to\infty} \phi(\sigma, n, a)$, we see that $\phi(\sigma) = a$.

Definition 2.4 *A point in code space whose symbols are periodic, as in (3), is called a* periodic address. *A point in code space whose symbols are periodic after a finite initial set is omitted is called* eventually periodic.

Examples & Exercises

2.9. An example of a periodic address is

121212121212121212121212121212121212121212121212121212...,

where 12 is repeated endlessly. An example of an eventually periodic address is:

112111111211111211111211111212212112121212121212121212121212121...,

where 21 is repeated endlessly.

2.10. Prove the following theorem: "Let $\{\mathbf{X}; w_1, w_2, \ldots, w_N\}$ be a hyperbolic IFS with attractor A. Then the following statements are equivalent:

(1) $x \in A$ is a periodic point;
(2) $x \in A$ possesses a periodic address;
(3) $x \in A$ is a fixed point of an element of the semigroup of transformations generated by $\{w_1, w_2, \ldots, w_N\}$."

2.11. Show that a point $x \in [0, 1]$ is a periodic point of the IFS

$$\{[0, 1]; \frac{1}{2}x, \frac{1}{2}x + \frac{1}{2}\}$$

if and only if it can be written $x = p/(2^N - 1)$ for some integer $0 \le p \le 2^N - 1$ and some integer $N \in \{1, 2, 3, \ldots\}$.

2.12. Let $\{\mathbf{X}; w_1, w_2, \ldots, w_N\}$ denote a hyperbolic IFS with attractor A. Define $W(S) = \cup_{n=1}^{N} w_n(S)$ when S is a subset of $\mathbf{X}$. Let P denote the set of periodic points of the IFS. Show that $W(P) = P$.

2.13. Locate all the periodic points of period 3 for the IFS $\{\mathbb{R}^2; \frac{1}{2}z, \frac{1}{2}z + \frac{1}{2}, \frac{1}{2}z + \frac{1}{2}\}$. Mark the positions of these points on $\mathbb{A}$.

2.14. Locate all periodic points of the IFS $\{\mathbb{R}; w_1(x) = 0, w_2(x) = \frac{1}{2}x + \frac{1}{2}\}$.

Theorem 2.4 *The attractor of an IFS is the closure of its periodic points.*

Proof Code space is the closure of the set of periodic codes. Lift this statement to A using the code space map $\phi : \Sigma \to A$. (ϕ is a continuous mapping from a metric space Σ onto a metric space A. If $S \subset \Sigma$ is such that its closure equals Σ, then the closure of $f(S)$ equals A.)

Examples & Exercises

2.15. Prove that the attractor of a totally disconnected hyperbolic IFS of two or more maps is uncountable.

2.16. Under what conditions does the attractor of a hyperbolic IFS contain uncountably many points with multiple addresses? Do not try to give a complete answer; just some conditions: think about the problem.

2.17. Under what conditions do there exist points in the attractor of a hyperbolic IFS with uncountably many addresses? As in 2.16, do not try to give a full answer.

2.18. In the standard construction of the classical Cantor set C, described in exercise 1.5 in Chapter III, a succession of open subintervals of $[0, 1]$ is removed. The endpoints of each of these intervals belong to C. Show that the set of such interval endpoints is countable. Show that C itself is uncountable. C is the attractor of the IFS $\{[0, 1]; \frac{1}{3}x, \frac{1}{3}x + \frac{2}{3}\}$. Characterize the addresses of the set of interval endpoints in C.

3   Introduction to Dynamical Systems

We introduce the idea of a dynamical system and some of the associated terminology.

Definition 3.1 *A dynamical system is a transformation* $f : \mathbf{X} \to \mathbf{X}$ *on a metric space* $(\mathbf{X}, d)$. *It is denoted by* $\{\mathbf{X}; f\}$. *The* orbit *of a point* $x \in \mathbf{X}$ *is the sequence* $\{f^{\circ n}(x)\}_{n=0}^{\infty}$.

As we will discover, dynamical systems are sources of deterministic fractals. The reasons for this are deeply intertwined with IFS theory, as we will see. Later we will introduce a special type of dynamical system, called a shift dynamical system, which can be associated with an IFS. By studying the orbits of these systems we will learn more about fractals. One of our goals is to learn why the Random Iteration Algorithm, used in Program 2 in Chapter III, successfully calculates the images of attractors of IFS. More information about the deep structure of attractors of IFS will be discovered.

Examples & Exercises

3.1. Define a function on code space, $f : \Sigma \to \Sigma$, by

$$f(x_1 x_2 x_3 x_4 \ldots) = x_2 x_3 x_4 x_5 \ldots.$$

Then $\{\Sigma; f\}$ is a dynamical system.

3.2. $\{[0, 1]; f(x) = \lambda x(1 - x)\}$ is a dynamical system for each $\lambda \in [0, 4]$. We say that we have a *one-parameter family* of dynamical systems.

3.3. Let $w(x) = Ax + t$ be an affine transformation in $\mathbb{R}^2$. Then $\{\mathbb{R}^2; w\}$ is a dynamical system.

3.4. Define $T : C[0, 1] \to C[0, 1]$ by

$$(Tf)(x) = \frac{1}{2}f(\frac{1}{2}x) + \frac{1}{2}f(\frac{1}{2}x + \frac{1}{2}).$$

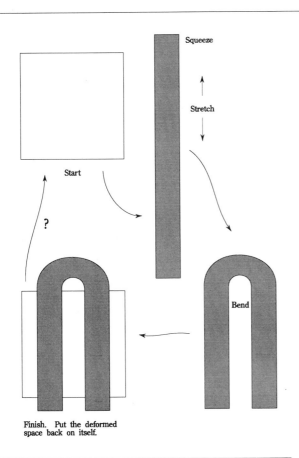

Figure IV.103. An example of a "stretch, squeeze, and bend" dynamical system (Smale horseshoe function).

Squeeze

Stretch

Start

?

Bend

Finish. Put the deformed space back on itself.

Then $\{C[0, 1]; T\}$ is a dynamical system.

3.5. Let $w: \hat{\mathbb{C}} \to \hat{\mathbb{C}}$ be a Möbius transformation. That is $w(z) = (az + b)/(cz + d)$, where $a, b, c, d \in \mathbb{C}$, and $(ad - bc) \neq 0$. Then $\{\hat{\mathbb{C}}; w(z)\}$ is a dynamical system.

3.6. $\{[0, 1]; 2x \bmod 1\}$ is a dynamical system. Here $2x \bmod 1 = 2x - [2x]$, where $[2x]$ denotes the greatest integer less than or equal to $2x$.

3.7. Define a transformation $f : \blacksquare \to \blacksquare$ as illustrated in Figure IV.103. $\{\blacksquare; f\}$ is a dynamical system.

In dynamical systems theory one is interested in what happens when one follows a typical orbit: is there some kind of attractor that usually occurs? Dynamical systems become interesting when the transformations involved are *not* contraction mappings, so that a single transformation suffices to produce interesting behavior. The orbit of a single point may be a geometrically complex set. Some thought about horizontal slices through Figure IV.104 will quickly suggest to the inquisitive student that there

Figure IV.104. One million iterations of a small black square in a "stretch, squeeze, and bend" dynamical system. Can you find a relationship to IFS theory?

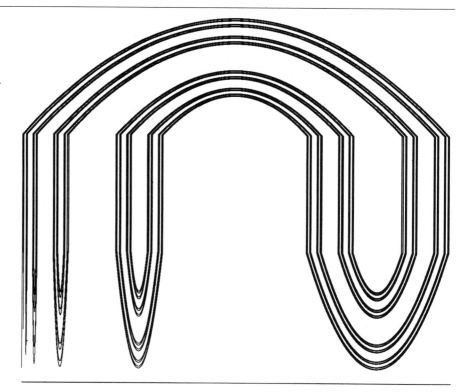

is a close relationship between this noncontractive dynamical system and a hyperbolic IFS.

Definition 3.2 *Let* $\{\mathbf{X}; f\}$ *be a dynamical system. A* periodic point *of* f *is a point* $x \in \mathbf{X}$ *such that* $f^{\circ n}(x) = x$ *for some* $n \in \{1, 2, 3, \ldots\}$. *If* x *is a periodic point of* f, *then an integer* n *such that* $f^{\circ n}(x) = x, n \in \{1, 2, 3, \ldots\}$ *is called a period of* x. *The least such integer is called the* minimal period *of the periodic point* x. *The orbit of a periodic point of* f *is called a* cycle *of* f. *The minimal period of a cycle is the number of distinct points it contains. A period of a cycle of* f *is a period of a point in the cycle.*

Definition 3.3 *Let* $\{\mathbf{X}; f\}$ *be a dynamical system and let* $x_f \in \mathbf{X}$ *be a fixed point of* f. *The point* x_f *is called an* attractive fixed point *of* f *if there is a number* $\epsilon > 0$ *so that* f *maps the ball* $B(x_f, \epsilon)$ *into itself, and moreover* f *is a contraction mapping on* $B(x_f, \epsilon)$. *Here* $B(x_f, \epsilon) = \{y \in \mathbf{X} : d(x_f, y) \le \epsilon\}$. *The point* x_f *is called a* repulsive fixed point *of* f *if there are numbers* $\epsilon > 0$ *and* $C > 1$ *such that*

$$d(f(x_f), f(y)) \ge Cd(x_f, y) \qquad \text{for all } y \in B(x_f, \epsilon).$$

A periodic point of f of period n is *attractive* if it is an attractive fixed point of $f^{\circ n}$. A cycle of period n is an *attractive cycle* of f if the cycle contains an attractive periodic point of f of period n. A periodic point of f of period n is *repulsive* if it

ON THE

SPHERE

REPULSIVE FIXED POINT

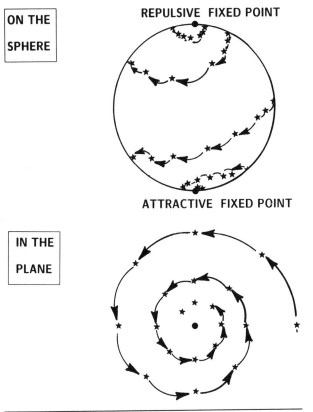

ATTRACTIVE FIXED POINT

Figure IV.105. The dynamics of a simple Möbius transformation. Points spiral away from one fixed point and spiral in toward the other. What happens if the fixed points coincide?

IN THE

PLANE

is a repulsive fixed point of $f^{\circ n}$. A cycle of period n is a *repulsive cycle* of f if the cycle contains a repulsive periodic point of f of period n.

Definition 3.4 *Let* $\{\mathbf{X}, f\}$ *be a dynamical system. A point* $x \in \mathbf{X}$ *is called an* eventually periodic *point of* f *if* $f^{\circ m}(x)$ *is periodic for some positve integer* m.

Remark: The definitions given here for attractive and repulsive points are consistent with the definitions we use for metric equivalence and will be used throughout the text. The definitions used in dynamical systems theory are usually more topological in nature. These are given later in exercises 5.4 and 5.5.

Examples & Exercises

3.8. The point $x_f = 0$ is an attractive fixed point for the dynamical system $\{\mathbb{R}; \frac{1}{2}x\}$, and a repulsive fixed point for the dynamical system $\{\mathbb{R}; 2x\}$.

3.9. The point $z = 0$ is an attractive fixed point, and $z = \infty$ is a repulsive fixed point, for the dynamical system

$$\{\hat{\mathbb{C}}; (\cos 10° + i \sin 10°)(0.9)z\}.$$

Figure IV.106. Points belonging to an orbit of a Möbius transformation on a sphere.

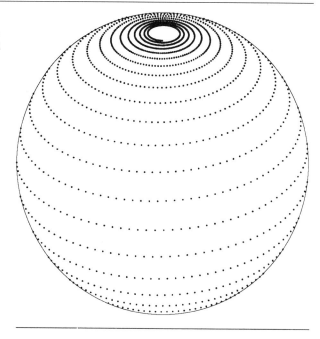

Figure IV.107. This shows an example of a web diagram. A web diagram is a means for displaying and analyzing the orbit of a point $x_0 \in \mathbb{R}$ for a dynamical system $(\mathbb{R}, f)$. The geometrical construction of a web diagram makes use of the graph of $f(x)$.

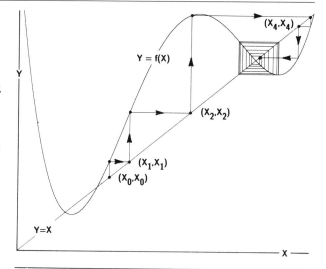

A typical orbit, starting from near the point of infinity on the sphere, is shown in Figures IV.105 and IV.106.

3.10. The point $x_f = 111\overline{111}$ is a repulsive fixed point for the dynamical system $\{\Sigma; f\}$ where $f : \Sigma \to \Sigma$ is defined by

$$f(x_1 x_2 x_3 x_4 x_5 \ldots) = x_2 x_3 x_4 x_5 \ldots .$$

Show that $x = 12121\overline{2}$ is a repulsive fixed point of period 2, and that $\{12\overline{12}, 21\overline{21}\}$ is a repulsive cycle of period 2.

3.11. The dynamical system $\{[0, 1]; \frac{1}{2}x(1 - x)\}$ possesses the attractive fixed point $x_f = 0$. Can you find a repulsive fixed point for this system?

There is a delightful construction for representing orbits of a dynamical system of the special form $\{\mathbb{R}; f(x)\}$. It utilizes the graph of the function $f : \mathbb{R} \to \mathbb{R}$. We describe here how it is used to represent the orbit $\{x_n = f^{\circ n}(x_0)\}_{n=1}^{\infty}$ of a point $x_0 \in \mathbb{R}$.

For simplicity we suppose that $f : [0, 1] \to [0, 1]$. Draw the square $\{(x, y) : 0 \le x \le 1, 0 \le y \le 1\}$ and sketch the graphs of $y = f(x)$ and $y = x$ for $x \in [0, 1]$. Start at the point (x_0, x_0) and connect it by a straight-line segment to the point $(x_0, x_1 = f(x_0))$. Connect this point by a straight-line segment to the point (x_1, x_1). Connect this point by a straight-line segment to the point $(x_1, x_2 = f(x_1))$; and continue. The orbit itself shows up on the 45° line $y = x$, as the sequence of points (x_0, x_0), (x_1, x_1), (x_2, x_2), ... We call the result of this geometrical construction a *web diagram*.

It is straightforward to write computergraphical routines that plot web diagrams on the graphics display device of a microcomputer. The following program is written in BASIC. It runs without modification on an IBM PC with Color Graphics Adaptor and Turbobasic. On any line the words preceded by a ' are comments: they are not part of the program.

Program 1.

```
l=3.79 : xn=0.95            'parameter value 3.79, orbit starts
                             at 0.95
def fnf(xn)=l*xn*(1-xn)      'change this function f(x) for other
                             dynamical systems
screen 1 : cls               'initialize computer graphics
window (0,0)-(1,1)           'set plotting window to 0 < x < 1 ,
                             0 < y < 1
for k=1 to 400               'plot the graph of the f(x)
pset(k/400, fnf(k/400))
next k
do                           'the main computational loop
n=n+1                        'increment the counter, $n$
y=fnf(xn)                    'compute the next point on the orbit
line (xn,xn)-(xn,y), n       'draw a line from (xn,xn) to (xn,y)
                             in color n
line (xn,y)-(y,y), n         'draw a line segment from (xn,y) to
                             (y,y) in color n
xn=y                         'set xn to be the most recently computed
                             point on the orbit

loop until instat : end      'stop running if a key is pressed.
```

Two examples of some web diagrams computed using this program are shown in Figure IV.108. The dynamical system used in this case is $\{[0, 1]; f(x) = 3.79x(1 - x)\}$.

Examples & Exercises

3.12. Rewrite Program 1 in a form suitable for your own computer environment. Use the resulting system to study the dynamical systems $\{[0, 1]; \lambda x(1 - x)\}$ for $\lambda = 0.55, 1.3, 2.225, 3.014, 3.794$. Try to classify the various species of web diagrams that occur for this one-parameter family of dynamical systems.

3.13. Divide $[0, 1]$ into 16 subintervals $[0, \frac{1}{16}), [\frac{1}{16}, \frac{1}{16}), \ldots, [\frac{14}{16}, \frac{15}{16}), [\frac{15}{16}, 1]$. Let $f : [0, 1] \to [0, 1]$ be defined by $f(x) = \lambda x(1 - x)$, where $\lambda \in [0, 4]$ is a parameter. Compute $\{f^{\circ n}(\frac{1}{2}) : n = 0, 1, 2, \ldots, 5000\}$ and keep track of the *frequency* with which $f^{\circ n}(\frac{1}{2})$ falls in the kth interval for $k = 1, 2, 4, 8, 16$, and $\lambda = 0.55, 1.3, 2.225, 3.014, 3.794$. Make histograms of your results.

3.14. Describe the behavior for the one-parameter family of dynamical system $s\{\mathbb{R} \cup \{\infty\}; \lambda x\}$, where λ is a real parameter, in the cases (i) $\lambda = 0$; (ii) $0 < |\lambda| < 1$; (iii) $\lambda = -1$; (iv) $\lambda = 1$; (v) $1 < \lambda < \infty$.

3.15. Analyze possible behaviors of $\{\mathbb{R}^2; Ax + t\}$, where $Ax + t$ is an affine transformation.

3.16. Study possible behaviors of orbits for the dynamical system $\{\hat{\mathbb{C}}; \text{Möbius transformation}\}$. You should make appropriate changes of coordinates to simplify the discussion.

3.17. Show that all points are eventually periodic for the slide-and-fold dynamical system $\{\mathbb{R}; f\}$, where

$$f(x) = \begin{cases} x + 1 & \text{if } x \leq 0, \\ -x + 1 & \text{if } x \geq 0. \end{cases}$$

This system is illustrated in Figure IV.109.

3.18. Let $\{X; w_1, w_2, \ldots, w_N\}$ be a hyperbolic IFS. Then $\{\mathcal{H}(X); W\}$ is a dynamical system, where

$$W(B) = \cup_{n=1}^{N} w_n(B) \text{ for all } B \in \mathcal{H}(X).$$

Dynamical systems that act on sets in place of points are sometimes called *set dynamical systems*. Show that the attractor of the IFS is an attractive fixed point of the dynamical system $\{\mathcal{H}(X); W\}$. You should quote appropriate results from earlier theorems.

3.19. We consider again our two-dimensional code space, having both past and future, called the space of shifts (see exercise 1.12 in Chapter II). In this space, the operation of the shift transformation is a homeomorphism of the space to itself (it is frequently called the *shift automorphism*). There is a very geometrical interpretation

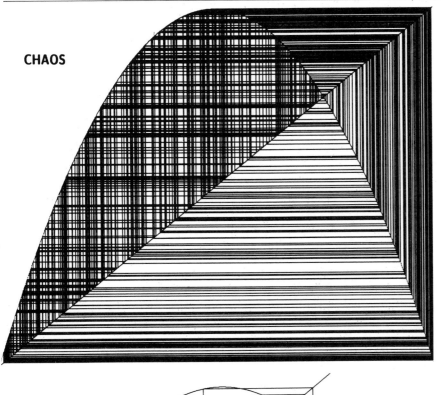

CHAOS

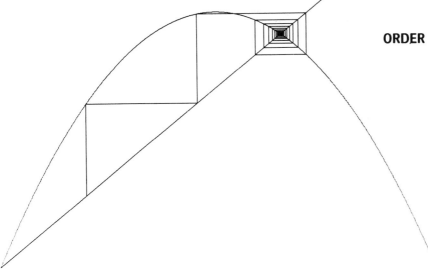

ORDER

Figure IV.108. Two examples of web diagrams computed using Program 1. The dynamical system in this case is $\{[0, 1]; f(x) = \lambda x(1 - x)\}$, for two different values of $\lambda \in (0, 4)$. The system corresponding to the lower value of λ is orderly; the other is close to being chaotic.

Figure IV.109. An orbit of the "slide-and-fold" dynamical system described in example 3.17. Can you prove that all orbits are eventually periodic?

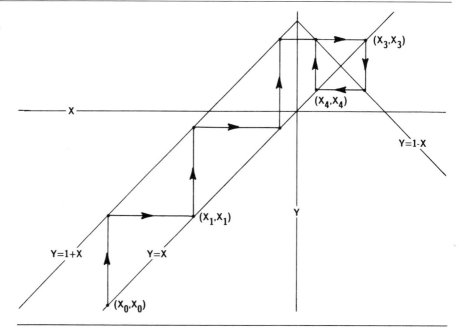

of the shift automorphism here. We arrive at this by looking at the action of the shift transformation with the metric d_k with $k = N$, as in exercise 2.19, Chapter II. To simplify the discussion, assume $N = 2$. The space of shifts is a two dimensional code space with points

$$(x, y) = (x_1 x_2 x_3 \ldots, y_1 y_2 y_3 \ldots)$$

on which we put the "Euclidean" metric (see exercise 2.6 in Chapter II),

$$d((x_1, y_1), (x_2, y_2)) = \sqrt{\left(\sum_{i=1}^{\infty} \frac{x_{1_i} - x_{2_i}}{2^i}\right)^2 + \left(\sum_{i=1}^{\infty} \frac{y_{1_i} - y_{2_i}}{2^i}\right)^2}.$$

The shift transformation here is best described by writing

$$(x, y) = \ldots y_3 y_2 y_1.x_1 x_2 x_3 \ldots.$$

We now shift by moving the dot one place to the right, to get

$$T(x, y) = \ldots y_2 y_1 x_1.x_2 x_3 x_4 \ldots = (x_2 x_3 \ldots, x_1 y_1 y_2 \ldots).$$

With the metric just mentioned, we can relate it to the square $[0, 1] \times [0, 1]$. Each point in this square has a binary expansion in terms of ones and zeros, so that a point (x, y) can be written $(.x_1 x_2 \ldots, .y_1 y_2 \ldots)$, with precisely the same symbols and metric (Euclidean). The shift operation can now be seen as doing the following:

stretch x: double x. This shifts the first digit up so that it is in the ones place.

Figure IV.110. A sign of things to come.

squeeze y: halve y. This shifts a zero into the first digit and shifts all other digits down.

raise half the interval: If the digit now in the ones place for x is a 1, replace the new 0 first digit of y with a 1. This adds a half to y.

put it on top: If the digit in the ones place for x is a 1, discard it. This brings the x values to those between 0 and 1, so this half of the points is put above the other half.

What we have done is stretch the square to twice its width (double x) and half its height (halve y), cut the rectangle into two pieces at $x = 1$, and put the right half on top of the bottom half (add $1/2$ to y if the new x is greater than 1). This operation of stretching out the square, cutting it, and stacking the pieces is called a *baker's transformation*, because it resembles a baker rolling, cutting, and stacking dough (to make pastry, for example). It is identical to the shift transformation on the space of shifts so long as the dough remains in distinct layers (unlike $\mathbb{R}^2$, $\overline{01} \neq \overline{10}$).

Definition 3.5 *This transformation is famous because it is the heart of any invertible "mixing" function (one has to allow any number of cuts and for uneven rolling). A mixing function is a function f such that given any set A (from some class of sets; here let's say sets with an interior), and any other set B from the same class, there is an N such that $f^n(A) \cap B \neq \emptyset$ for any $n > N$.*

The term *mixing* is appropriate: if A is red and B is blue, then eventually they are both somewhat purple (have both red and blue in them). A nice property of this mixing business is that there is at least one point in the space such that $\{f^n(x) : n = 1, 2, \ldots\}$ is dense, that is, given an open set $\mathcal{O}$, there is an n such that $f^n(x) \in \mathcal{O}$. When f has this property, we say that it has a *dense orbit*. T is mixing on the space of shifts and on code space, and it has a dense orbit as a result.

Examples & Exercises

3.20. Prove that for any code space Σ on N symbols, there is a point $\sigma \in \Sigma$, such that σ has a dense orbit under the shift transformation, that is $\{T^n(\sigma) : n = 1, 2, 3, \ldots\}$ is dense in Σ.

3.21. Show that T is mixing on code space for the class of open sets.

4   Dynamics on Fractals: Or How to Compute Orbits by Looking at Pictures

We continue with the main theme for this chapter, namely dynamical systems on fractals. We will need the following result.

Lemma 4.1 *Let $\{X; w_n, n = 1, 2, \ldots, N\}$ be a hyperbolic IFS with attractor A. If the IFS is totally disconnected, then for each $n \in \{1, 2, \ldots, N\}$, the transformation $w_n : A \rightarrow A$ is one-to-one.*

Proof We use a code space argument. Suppose that there is an integer $n \in \{1, 2, \ldots, N\}$ and distinct points $a_1, a_2 \in A$ so that $w_n(a_1) = w_n(a_2) = a \in A$. If a_1 has address ω and a_2 has address σ, then a has the two addresses $n\omega$ and $n\sigma$. This is impossible because A is totally disconnected. This completes the proof.

Lemma 4.1 shows that the following definition is good.

Definition 4.1 *Let $\{X; w_n, n = 1, 2, \ldots, N\}$ be a totally disconnected hyperbolic IFS with attractor A. The associated shift transformation on A is the transformation $S : A \rightarrow A$ defined by*

$$S(a) = w_n^{-1}(a) \qquad \text{for } a \in w_n(A),$$

where w_n is viewed as a transformation on A. The dynamical system $\{A; S\}$ is called the shift dynamical system associated with the IFS.

Examples & Exercises

4.1. Figure IV.111 shows the attractor of the IFS

$$\left\{ \mathbb{R}^2; 0.47 \begin{pmatrix} x_1 \\ x_2 \end{pmatrix}, 0.47 \begin{pmatrix} x_1 \\ x_2 \end{pmatrix} + \begin{pmatrix} 1 \\ 0 \end{pmatrix}, 0.47 \begin{pmatrix} x_1 \\ x_2 \end{pmatrix} + \begin{pmatrix} 1 \\ 0 \end{pmatrix} \right\}.$$

Figure IV.111 also shows an eventually periodic orbit $\{a_n = S^{\circ n}(a_0)\}_{n=0}^{\infty}$ for the associated shift dynamical system. This orbit actually ends up at the fixed point $\phi(2\overline{2}2\overline{2})$. The orbit reads $a_0 = \phi(13132\overline{22})$, $a_1 = \phi(313122\overline{22})$, $a_2 = \phi(13222\overline{22})$, $a_3 = \phi(3222\overline{22})$, $a_4 = \phi(\overline{22}22)$, where $\phi : \Sigma \rightarrow A$ is the associated code space map. $a_4 \in A$ is clearly a repulsive fixed point of the dynamical system. Notice how one can read off the orbit of the point a_0 from its address. Start from another point very close to a_0 and see what happens. Notice how the dynamics depend not only on A itself, but also on the IFS. A different IFS with the same attractor will in general lead to different shift dynamics.

4.2. Both Figures IV.112 and IV.113 show attractors of IFS's. In each case the implied IFS is the obvious one. Give the addresses of the points $\{a_n = S^{\circ n}(a_0)\}_{n=0}^{\infty}$ of the eventually periodic orbit in Figure IV.112. Show that the cycle to which the

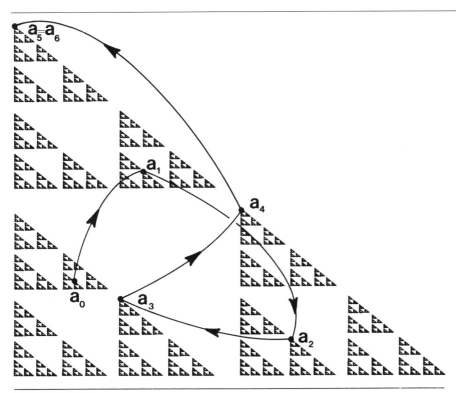

Figure IV.111. An orbit of a shift dynamical system on a fractal.

Figure IV.112. This orbit ends up in a cycle of period 3.

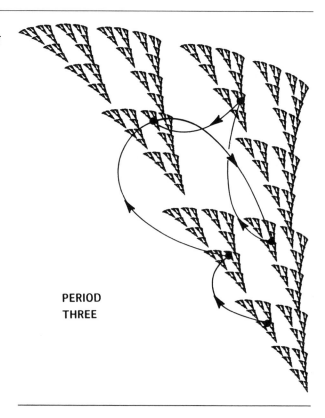

PERIOD
THREE

orbit converges is a repulsive cycle of period 3. The orbit in Figure IV.113 is either very long or infinitely long: why is it hard for us to know which?

4.3. Figure IV.114 shows an orbit of a point under the shift dynamical system associated with a certain IFS $\{\mathbb{R}^2; w_1, w_2, w_3\}$, where w_1, w_2, and w_3 are affine transformations. Deduce the orbits of the points marked b and c in the figure.

4.4. Figure IV.115 shows the start of an orbit of a point under the shift dynamical system associated with a certain hyperbolic IFS. The IFS is of the form $\{\mathbb{R}; w_1, w_2, w_3\}$, where the transformations $w_n : \mathbb{R} \to \mathbb{R}$ are affine and the attractor is [0, 1]. Sketch part of the orbit of the point labelled b in the figure. (Notice that this IFS is actually just-touching: nonetheless it is straightforward to define uniquely the associated shift dynamics on $\mathcal{O} \cap A$ where $\mathcal{O}$ is the open set referred to in Definition 2.2.)

We can sharpen up the definition of the overlapping IFS with the aid of the mixing properties discussed in section 3. Let $\{\mathbf{X}; w_1, \ldots, w_N\}$ be a hyperbolic IFS, and define the set

$$M = \bigcup_{i \neq j} (w_i(A) \cap w_j(A))$$

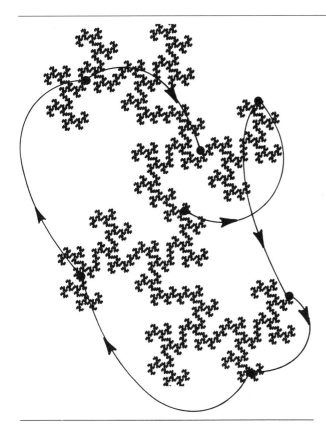

Figure IV.113. A chaotic orbit getting started. The shift dynamics are often wild. Why?

of points in various intersections of the maps of the IFS. Then the following properties hold:

open interior: If there is a set $\mathcal{O}$, open with respect to A, such that $\mathcal{O} \subset M$, then the IFS is overlapping. This allows the IFS to be declared overlapping easily in some cases. The proof is not too difficult: Suppose this to be the case, namely that M contains an open set $\mathcal{O}$. Suppose that $\mathcal{O}_1$ were an open set that we thought might satisfy the open set condition for just-touching IFS. Then $W^n(\mathcal{O}_1) \cap \mathcal{O} = \emptyset$ for all n, since $\mathcal{O}_1$ can't contain points in the overlap, and maps inside itself. Using the continuous map $\phi : \Sigma \to A$, we know that we would then have $\phi^{-1}(\mathcal{O}_1)$ and $\phi^{-1}(\mathcal{O})$ both open sets in code space. But $T^n(\phi^{-1}(\mathcal{O}))$ must intersect $\phi^{-1}(\mathcal{O}_1)$ in code space for some n, due to mixing, and an address in the intersection corresponds to a point a on the attractor such that $W^n(\{a\})$ thus intersects $\mathcal{O}$. Hence $\mathcal{O}_1$ cannot exist, and the IFS is overlapping.

dense address: Notice that in order to prevent the IFS from being just-touching in the proof just given, the orbit of $\phi^{-1}(M)$ only needs to be dense in Σ to end up with points in the image of any open set in A. Consequently,

Figure IV.114. The orbit of the point a is shown. Can you plot the first few points of the orbits of b and c? Warning! The IFS here is not the usual one. See how the knowledge of some dynamics can imply some more!

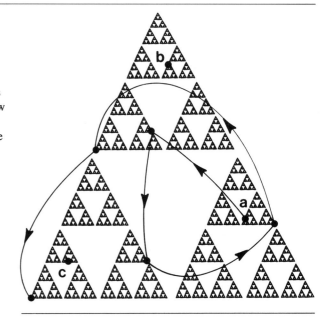

Figure IV.115. This figure shows a sketch of part of an orbit of an IFS $\{[0, 1]; w_1, w_2, w_3\}$ on its attractor $[0, 1]$. The transformation $w_1 : [0, 1] \to [0, 1]$ is affine for $i = 1, 2, 3$. Sketch part of the orbit of b.

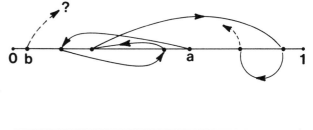

an IFS is overlapping if the orbit of $\phi^{-1}(M)$ in code space under the shift transformation is dense.

empty interiors: If the IFS is made up of affine maps, and it "looks" just-touching, that is, M does not have an interior, then it is just-touching. The important property of affine maps used here is that they map boundary points to boundary points, and boundary points come from boundary points.

4.5. The empty interiors property is not general, but is useful when it applies. To see why it is restricted, consider the following IFS, made of six translated copies of the following map:

On the interval $[-1, 1]$, let $\theta(x) = \text{Arccos}(x)$, that is, map x to the point directly above it on the unit circle. Then take $\theta(x)$ to the point $\theta(x) - \alpha \sin \theta(x)$, where $\alpha \in [0, 1/2]$. Then map the new point on the circle back to $[-1, 1]$ by taking $\theta(x)$ to $x' = \cos \theta(x)$. Now fold the interval over at 0 with the map x'^2, and ensure that it is contractive by dividing by 3. The interval has now been mapped to [0, 1/3]. Call this map $v(x)$. Explicitly, we have

$$v(x) = \frac{1}{3}\cos^2(\text{Arccos}(x) - \sin(\text{Arccos}(x))).$$

We now form 6 maps $w_1, \ldots, w_6$ by translations and inversions of $v(x)$:

$$w_1(x) = v(x) - 1 \qquad w_2(x) = -v(x) - \frac{1}{3}$$

$$w_3(x) = v(x) - \frac{1}{3} \qquad w_4(x) = -v(x) + \frac{1}{3}$$

$$w_5(x) = v(x) + \frac{1}{3} \qquad w_6(x) = -v(x) + 1.$$

The reader should be able to verify that the attractor of the IFS

$$\{[-1, 1]; w_1, w_2, w_3, w_4, w_5, w_6\}$$

is the interval $[-1, 1]$, and that each of these maps touches any neighbor at a single point. There is a point $x_0(\alpha)$ such that

$$\text{Arccos}(x_0) - \alpha \sin \text{Arccos}(x_0) = \pi/2,$$

whose image is the points

$$\{-1, -1/3, 1/3, 1\},$$

and whose endpoints map to $\{-2/3, 0, 2/3\}$. By choosing α at different values we can move x_0 around the interval $[-1/3, 0]$. If we pick $\sigma \in \Sigma$ to be a dense orbit in Σ under the shift transformation, we can successively approximate this address for x_0 such that the address of x_0 is 3σ, and we can do this for each such $\sigma \in \Sigma$. We can also do this for a variety of periodic orbits, which are not dense.

It turns out that for *most* values of $x_0 \in [-1/3, 0]$ (the probability of a value in this interval being one of these is 100%), this IFS is overlapping, although between every two values for which it is overlapping, there is a value for which it is just-touching. The attractors of this family of IFS are identical, as are the intersection points. This is thus both an example of an IFS that has a finite set of intersection points and is (sometimes) overlapping, and an example of one that does not go smoothly through the succession from totally disconnected to just-touching to overlapping. Small wonder these properties are defined for the IFS and not the attractor; they are really properties governing the behavior of addresses in code space.

5   Equivalent Dynamical Systems

Definition 5.1 *Two metric spaces* $(\mathbf{X}_1, d_1)$ *and* $(\mathbf{X}_2, d_2)$ *are said to be* topologically equivalent *if there is a homeomorphism* $f: \mathbf{X}_1 \rightarrow \mathbf{X}_2$. *Two subsets* $S_1 \subset \mathbf{X}_1$ *and* $S_2 \subset$

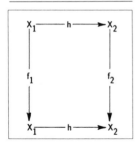

Figure IV.116. A commutative diagram that establishes the equivalence between the two dynamical systems $\{X_1; f_1\}$ and $\{X_2; f_2\}$. The function $h : X_1 \to X_2$ is a homeomorphism.

X_2 *are topologically equivalent, or* homeomorphic, *if the metric spaces* (S_1, d_1) *and* (S_2, d_2) *are topologically equivalent.* S_1 *and* S_2 *are* metrically equivalent *if* (S_1, d_1) *and* (S_2, d_2) *are equivalent metric spaces.*

The Cantor set and code space, discussed following Theorem 2.3 in Chapter IV, are metrically equivalent. Theorem 8.5 in Chapter II tells us that if $f : X_1 \to X_2$ is a continuous one-to-one mapping from a compact metric (X_1, d_1) onto a compact metric space (X_2, d_2) then f is a homeomorphism. So by means of the code space mapping $\phi : \Sigma \to A$ (Theorem 2.1) one readily establishes that the attractor of a totally disconnected hyperbolic IFS is topologically equivalent to a classical Cantor set.

Topological equivalence permits a great deal more "stretching" and "compression" to take place than is permitted by metric equivalence. Later we will define a quantity called the fractal dimension. The fractal dimension of a subset of a metric space such as $(\mathbb{R}^2,$ Euclidean) provides a measure of the geometrical complexity of the set; it measures the wildness of the set, and it may be used to predict your excitement and wonder when you look at a picture of the set. We will show that two metrically equivalent sets have the same fractal dimension. If they are merely topologically equivalent, their fractal dimensions may be different.

With the naturally implied metrics, $[0, 1]$ is homeomorphic to $[0, 2]$. ■ is homeomorphic to ●. What is more, ⋏ is even homeomorphic to ⋏ , and ——— is homeomorphic to ⋏⋏⋏ .

In fractal geometry we are especially interested in the *geometry* of sets, and in the way they *look,* when they are represented by pictures. Thus we use the restrictive condition of metric equivalence to start to define mathematically what we mean when we say that two sets are alike. However, in dynamical systems theory we are interested in *motion* itself, in the dynamics, in the way points move, in the existence of periodic orbits, in the asymptotic behavior of orbits, and so on. These structures are not damaged by homeomorphisms, as we will see, and hence we say that two dynamical systems are alike if they are related via a homeomorphism.

Definition 5.2 *Two dynamical systems* $\{X_1; f_1\}$ *and* $\{X_2; f_2\}$ *are said to be* equivalent, *or* topologically conjugate, *if there is a homeomorphism* $\theta : X_1 \to X_2$ *such that*

$$f_1(x_1) = \theta^{-1} \circ f_2 \circ \theta(x_1) \text{ for all } x_1 \in X_1,$$

$$f_2(x_2) = \theta \circ f_1 \circ \theta^{-1}(x_2) \text{ for all } x_2 \in X_2.$$

In other words, the two dynamical systems are related by the commutative diagram shown in Figure IV.116.

The following theorem expresses formally what should already be clear intuitively from our experience with shift dynamics on fractals.

Theorem 5.1 *Let* $\{X; w_1, w_2, \ldots, w_N\}$ *be a totally disconnected hyperbolic IFS and let* $\{A; S\}$ *be the associated shift dynamical system. Let* Σ *be the associated code space of N symbols and let* $T : \Sigma \to \Sigma$ *be defined by*

$$T(\sigma_1\sigma_2\sigma_3 \ldots) = \sigma_2\sigma_3\sigma_4 \ldots \qquad \text{for all } \sigma = \sigma_1\sigma_2\sigma_3 \ldots \in \Sigma.$$

Then the two dynamical systems $\{A; S\}$ *and* $\{\Sigma; T\}$ *are equivalent. The homeomorphism that provides this equivalence is* $\phi : \Sigma \to A$, *as defined in Theorem 4.2.1. Moreover,* $\{a_1, a_2, \ldots, a_p\}$ *is a repulsive cycle of period p for S if, and only if,* $\{\phi(a_1), \phi(a_2), \ldots, \phi(a_p)\}$ *is a repulsive cycle of period p for T.*

Examples & Exercises

5.1. Let $\{X_1; f_1\}$ and $\{X_2; f_2\}$ be equivalent dynamical systems. Let a homeomorphism that provides this equivalence be denoted by $\theta : X_1 \to X_2$. Show that

$$\{x_1, x_2, \ldots, x_p\}$$

is a cycle of period p for $\{X_1; f_1\}$ if and only if

$$\{\theta(x_1), \theta(x_2), \ldots, \theta(x_p)\}$$

is a cycle of period p for $\{X_2; f_2\}$. Suppose that $\{x_1, x_2, \ldots, x_p\}$ is an attractive cycle for f_1. Show that this does not imply that $\{\theta(x_1), \ldots, \theta(x_p)\}$ is an attractive cycle for f_2.

5.2. Let $\{X_1; f_1\}$ and $\{X_2; f_2\}$ be equivalent dynamical systems. Let a homeomorphism that provides this equivalence be denoted by $\theta : X_1 \to X_2$. Let $\{f_1^{\circ n}(x)\}_{n=0}^{\infty}$ be an eventually periodic orbit of f_1. Show that $\{f_2^{\circ n}(\theta(x))\}_{n=0}^{\infty}$ is an eventually periodic orbit of f_2.

5.3. Let $\{X_1; f_1\}$ and $\{X_2; f_2\}$ be equivalent dynamical systems. Let a homeomorphism that provides this equivalence be denoted by $\theta : X_1 \to X_2$. Let this homeomorphism be such as to make the two spaces (X_1, d_1) and (X_2, d_2) metrically equivalent. Construct an example where $x_f \in X_1$ is a repulsive fixed point of the dynamical system $\{X_1, f_1\}$ yet $\theta(x_f)$ is not a repulsive fixed point of $\{X_2, d_2\}$.

5.4. Let $\{X_1; f_1\}$ and $\{X_2; f_2\}$ be equivalent metric spaces. Let a homeomorphism that provides their equivalence be denoted by $\theta : X_1 \to X_2$. Let $x_f \in X_1$ be a fixed point of f_1. Suppose there is an open set $\mathcal{O}$ that contains x_f and is such that $x \in \mathcal{O}$ implies $\lim_{n \to \infty} f_1^{\circ n}(x) = x_f$. Show that there is an open neighborhood of $\theta(x_f)$ in X_2 with a similar property.

5.5. Our definition of *attractive* and *repulsive* fixed points and cycles, Definition 3.4, has the feature that it depends heavily on the metric. It is motivated by the situation of analytic dynamics where small disks are almost mapped into disks. Show how one can use exercise 5.4 to make a definition of an attractive cycle in such a way that attractiveness of cycles is preserved under topological conjugacy.

Figure IV.117. Attractive and repulsive fixed points in a web diagram for a differentiable dynamical system. Analyze the ? points.

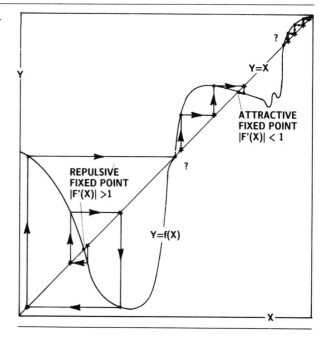

5.6. Let $A \subset \mathbb{R}$. Then a function $f : A \to A$ is *differentiable* at a point $x_0 \in A$ if

$$\lim_{\substack{x \to x_0 \\ x \in A}} \left\{ \frac{f(x) - f(x_0)}{x - x_0} \right\}$$

exists. If this limit exists it is denoted by $f'(x_0)$. Let $\{\mathbb{R}; w_1, w_2, \ldots, w_N\}$ be a totally disconnected hyperbolic IFS acting on the metric space $(\mathbb{R}, \text{Euclidean})$. Suppose that, for each $n = 1, 2, \ldots, N$, $w_n(x)$ is differentiable, with $|w_n'(x)| > 0$ for all $x \in \mathbb{R}$. Show that the associated shift dynamical system $\{A; S\}$ is such that S is differentiable at each point $x_0 \in A$ and, moreover, $|S'(x_0)| > 1$ for all $x \in A$.

5.7. Let $\{\mathbb{R}; f\}$ and $\{\mathbb{R}; g\}$ be equivalent dynamical systems. Let a homeomorphism that provides their equivalence be denoted by $\theta : \mathbb{R} \to \mathbb{R}$. If $\theta(x)$ is differentiable for all $x \in \mathbb{R}$, then the dynamical systems are said to be *diffeomorphic*. Prove that a_1 is an attractive fixed point of f if and only if $\theta(a_1)$ is an attractive fixed point of g.

5.8. Let $\{\mathbb{R}; f\}$ be a dynamical system such that f is differentiable for all $x \in \mathbb{R}$. Consider the web diagrams associated with this system. Show that the fixed points of f are exactly the intersections of the line $y = x$ with the graph $y = f(x)$. Let a be a fixed point of f. Show that a is an attractive fixed point of f if and only if $|f'(a)| < 1$. Generalize this result to cycles. Note that if $\{a_1, a_2, \ldots, a_p\}$ is a cycle of period p, then $\frac{d}{dx}(f^{\circ p}(x)|_{x=a_1} = f'(a_1)f'(a_2) \ldots f'(a_p)$. Assure yourself that the situation is correctly summarized in the web diagram shown in Figure IV.117.

5.9. Consider the dynamical system $\{[0, 1]; f(x)\}$ where

$$f(x) = \begin{cases} 1 - 2x & \text{when } x \in [0, \tfrac{1}{2}], \\ 2x - 1 & \text{when } x \in [\tfrac{1}{2}, 1]. \end{cases}$$

Consider also the just-touching IFS $\{[0, 1], \tfrac{1}{2}x + \tfrac{1}{2}, -\tfrac{1}{2}x + \tfrac{1}{2}\}$. Show that it is possible to define a "shift transformation," S, on the attractor, A, of this IFS in such a way that $\{[0, 1]; S\}$ and $\{[0, 1]; f(x)\}$ are equivalent dynamical systems. To do this you should define $S : A \to A$ in the obvious manner for points with unique addresses; and you should make a suitable definition for the action of S on points with multiple addresses.

5.10. Let $\{\mathbb{R}^2; w_1, w_2, w_3\}$ denote a one-parameter family of IFS, where

$$w_1 \begin{pmatrix} x \\ y \end{pmatrix} = \begin{pmatrix} (\tfrac{1+p}{4}) & 0 \\ 0 & (\tfrac{1+p}{4}) \end{pmatrix} \begin{pmatrix} x \\ y \end{pmatrix};$$

$$w_2 \begin{pmatrix} x \\ y \end{pmatrix} = \begin{pmatrix} (\tfrac{1+p}{4}) & 0 \\ 0 & (\tfrac{1+p}{4}) \end{pmatrix} \begin{pmatrix} x \\ y \end{pmatrix} + \begin{pmatrix} \tfrac{3+p}{8} \\ \tfrac{p}{2} \end{pmatrix},$$

$$w_3 \begin{pmatrix} x \\ y \end{pmatrix} = \begin{pmatrix} \tfrac{1+p}{4} & 0 \\ 0 & (\tfrac{1+p}{4}) \end{pmatrix} \begin{pmatrix} x \\ y \end{pmatrix} + \begin{pmatrix} \tfrac{3-p}{4} \\ 0 \end{pmatrix} \quad \text{for } p \in [0, 1].$$

Let the attractor of this IFS be denoted by $A(p)$. Show that $A(0)$ is a Cantor set and $A(1)$ is a Sierpinski triangle. Consider the associated family of code space maps $\phi(p) : \Sigma \to A(p)$. Show that $\phi(p)(\sigma)$ is continuous in p for fixed $\sigma \in \Sigma$; that is $\phi(p)(\sigma) : [0, 1] \to \mathbb{R}^2$ is a continuous path. Draw some of these paths, including ones that meet at $p = 1$. Interpret these observations in terms of the Cantor set becoming "joined to itself" at various points to make a Sierpinski triangle, as suggested in Figure IV.118.

Since the IFS is totally disconnected when $p = 0$, $\phi(p = 0) : \Sigma \to A(0)$ is invertible. Hence we can define a continuous transformation $\theta : A(0) \to A(1)$ by $\theta(x) = \phi(p = 1)(\phi^{-1}(p = 0)(x))$. Show that if we define a set $J(x) = \{y \in A(0) : \theta(y) = x\}$ for each $x \in A(1)$, then $J(x)$ is the set of points in $A(0)$ whose associated paths meet at $x \in A(1)$ when $p = 1$. Invent shift dynamics on paths.

6 The Shadow of Deterministic Dynamics

Our goal in this section is to extend the definition of the shift dynamical system associated with a totally disconnected hyperbolic IFS to cover the just-touching and overlapping cases. This will lead us to the idea of a random shift dynamical system and to the discovery of a beautiful theorem. This theorem will be called the Shadow Theorem.

Let $\{X; w_1, w_2, \ldots, w_N\}$ denote a hyperbolic IFS, and let A denote its attractor. Assume that $w_n : A \to A$ is invertible for each $n = 1, 2, \ldots, N$, but that the IFS is

Figure IV.118. Continuous transformation of a Cantor set into a Sierpinski triangle. The inverse transformation would involve some ripping.

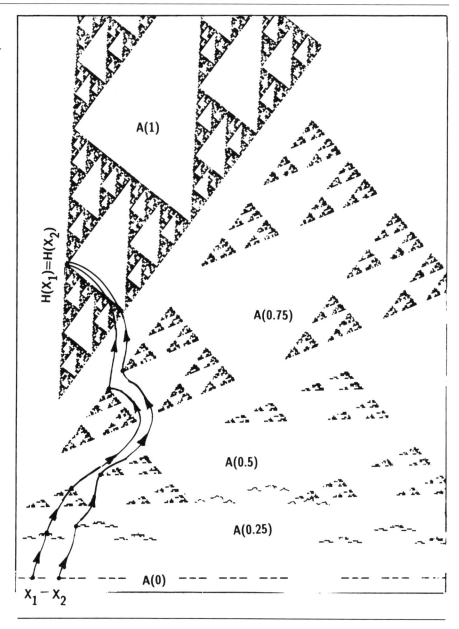

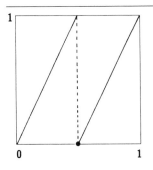 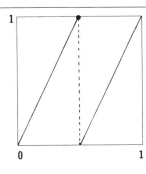

Figure IV.119. The two possible shift dynamical systems associated with the just-touching IFS $\{[0, 1]; \frac{1}{2}x, \frac{1}{2}x + \frac{1}{2}\}$ are represented by the two possible graphs of $S(x)$. "Most" orbits are unaffected by the difference between the two systems.

not totally disconnected. We want to define a dynamical system $\{A; S\}$ analogous to the shift dynamical system defined earlier. Clearly, we should define

$$S(x) = w_n^{-1}(x) \qquad \text{when } x \in w_n(A), \text{ but } x \notin w_m(A) \qquad \text{for } m \neq n,$$

for each $n = 1, 2, \ldots, N$.

However, at least one of the intersections $w_m(A) \cap w_n(A)$ is nonempty for some $m \neq n$. One idea is simply to make an assignment of which inverse map is to be applied in the overlapping region. For the case $N = 2$ we might define, for example,

$$S(x) = \begin{cases} w_1^{-1}(x) & \text{when } x \in w_1(A), \\ w_2^{-1}(x) & \text{when } x \in A \setminus w_1(A). \end{cases}$$

In the just-touching case the assignment of where S takes points that lie in the overlapping regions does not play a very important role: only a relatively small proportion of points will have somewhat arbitrarily specified orbits. We look at some examples, just to get the flavor.

Examples & Exercises

6.1. Consider the shift dynamical systems associated with the IFS

$$\{[0, 1]; \frac{1}{2}x, \frac{1}{2}x + \frac{1}{2}\}.$$

We have $S(x) = 2x$ for $x \in [0, \frac{1}{2})$ and $S(x) = 2x - 1$ for $x \in (\frac{1}{2}, 1]$. We can define the value of $S(\frac{1}{2})$ to be either 1 or 0. The two possible graphs for $S(x)$ are shown in Figure IV.119. The only points $x \in [0, 1] = A$ whose orbits are affected by the definition are those rational numbers whose binary expansions end $\ldots 01\overline{1}$ or $\ldots 10\overline{0}$, the dyadic rationals.

6.2. Show that if we follow the ideas introduced above, there is only one dynamical system $\{A; S\}$ that can be associated with the just-touching IFS $\{[0, 1]; -\frac{1}{2}x + \frac{1}{2}, \frac{1}{2}x\}$. The key here is that $w_1^{-1}(x) = w_2^{-1}(x)$ for all $x \in w_1(A) \cap w_2(A)$.

6.3. Consider some possible "shift" dynamical systems $\{A; S\}$ that can be associated with the IFS

Figure IV.120. Two possible shift dynamical systems that can be associated with the overlapping IFS $\{[0, 1]; \frac{1}{2}x, \frac{3}{4}x + \frac{1}{4}\}$. In what ways are they alike?

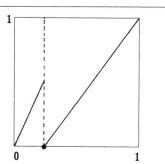 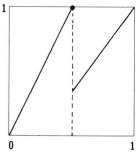

$$\{C; \frac{1}{2}z, \frac{1}{2}z + \frac{1}{2}, \frac{1}{2}z + \frac{i}{2}\}.$$

The attractor, $\mathbb{A}$, is overlapping at the three points $a = w_1(\mathbb{A}) \cap w_2(\mathbb{A})$, $b = w_2(\mathbb{A}) \cap w_3(\mathbb{A})$, and $c = w_3(\mathbb{A}) \cap w_1(\mathbb{A})$. We might define $S(a) = w_1^{-1}(a)$ or $w_2^{-1}(a)$, $S(b) = w_2^{-1}(b)$ or $w_3^{-1}(b)$, and $S(c) = w_3^{-1}(c)$ or $w_1^{-1}(c)$. Show that regardless of which definition is made, the orbits of a, b, and c are eventually periodic.

6.4. Consider a just-touching IFS of the form $\{\mathbb{R}^2; w_1, w_2, w_3\}$ whose attractor is an equilateral Sierpinski triangle $\mathbb{A}$. Assume that each of the maps is a similitude of scaling factor 0.5. Consider the possibility that each map involves a rotation through $0°$, $120°$, or $240°$. The attractor, $\mathbb{A}$, is overlapping at the three points $a = w_1(\mathbb{A}) \cap w_2(\mathbb{A})$, $b = w_2(\mathbb{A}) \cap w_3(\mathbb{A})$, and $c = w_3(\mathbb{A}) \cap w_1(\mathbb{A})$. Show that it is possible to choose the maps so that $w_1^{-1}(a) = w_2^{-1}(a)$, $w_2^{-1}(b) = w_3^{-1}(b)$, and $w_3^{-1}(c) = w_1^{-1}(c)$.

6.5. Is code space on two symbols topologically equivalent to code space on three symbols? Yes! Construct a homeomorphism that establishes this equivalence.

6.6. Consider the hyperbolic IFS $\{\Sigma; t_1, t_2, \ldots, t_N\}$, where Σ is code space on N symbols $\{1, 2, \ldots, N\}$ and

$$t_n\sigma = n\sigma \qquad \text{for all } \sigma \in \Sigma.$$

Show that the associated shift dynamical system is exactly $\{\Sigma; T\}$ defined in Theorem 4.5.1. Can two such shift dynamical systems be equivalent for different values of N? To answer this question consider how many fixed points the dynamical system $\{\Sigma; T\}$ possesses for different values of N.

6.7. Consider the overlapping hyperbolic IFS $\{[0, 1]; \frac{1}{2}x, \frac{3}{4}x + \frac{1}{4}\}$. Compare the two associated shift dynamical systems whose graphs are shown in Figure IV.120. What features do they share in common?

6.8. Demonstrate that code space on two symbols is not metrically equivalent to code space on three symbols.

In considering exercises such as 6.7, where two different dynamical systems are

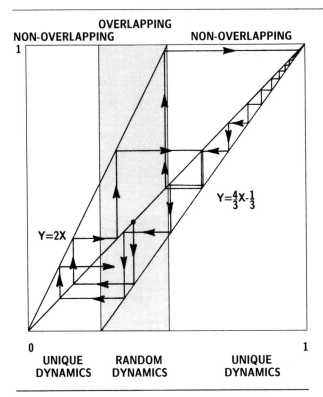

Figure IV.121. A partially random and partially deterministic shift dynamical system associated with the IFS $\{[0, 1]; \frac{1}{2}x, \frac{3}{4}x + \frac{1}{4}\}$.

associated with an IFS in the overlapping case, we are tempted to entertain the idea that no particular definition of the shift dynamics in the overlapping regions is to be preferred. This suggests that we define the dynamics in overlapping regions in a somewhat random manner. Whenever a point on an orbit lands in an overlapping region we should allow the possibility that the next point on the orbit is obtained by applying any one of the available inverse transformations. This idea is illustrated in Figure IV.121, which should be compared with Figure IV.120.

Definition 6.1 *Let* $\{X; w_1, w_2\}$ *be a hyperbolic IFS. Let* A *denote the attractor of the IFS. Assume that both* $w_1 : A \to A$ *and* $w_2 : A \to A$ *are invertible. A sequence of points* $\{x_n\}_{n=0}^{\infty}$ *in* A *is called an orbit of the* random *shift dynamical system associated with the IFS if*

$$x_{n+1} = \begin{cases} w_1^{-1}(x_n) & \text{when } x_n \in w_1(A) \text{ and } x_n \notin w_1(A) \cap w_2(A), \\ w_2^{-1}(x_n) & \text{when } x_n \in w_2(A) \text{ and } x_n \notin w_1(A) \cap w_2(A), \\ \text{one of } \{w_1^{-1}(x_n), w_2^{-1}(x_n)\} & \text{when } x_n \in w_1(A) \cap w_2(A), \end{cases}$$

for each $n \in \{0, 1, 2, \ldots\}$. *We will use the notation* $x_{n+1} = S(x_n)$ *although there may be no well-defined transformation* $S : A \to A$ *that makes this true. Also we will write* $\{A; S\}$ *to denote the collection of possible orbits defined here, and we will call* $\{A; S\}$ *the* random *shift dynamical system associated with the IFS.*

Notice that if $w_1(A) \cap w_2(A) = \emptyset$ then the IFS is totally disconnected and the orbits defined here are simply those of the shift dynamical system $\{A; S\}$ defined earlier.

We now show that there is a completely deterministic dynamical system acting on a higher-dimensional space, whose projection into the original space X yields the "random dynamics" we have just described. Our random dynamics are seen as the shadow of deterministic dynamics. To achieve this we turn the IFS into a totally disconnected system by introducing an additional variable. To keep the notation succinct we restrict the following discussion to IFS's of two maps.

Definition 6.2 *The* lifted *IFS associated with a hyperbolic IFS $\{X; w_1, w_2\}$ is the hyperbolic IFS $\{X \times \Sigma; \tilde{w}_1, \tilde{w}_2\}$, where Σ is the code space on two symbols $\{1, 2\}$, and*

$$\tilde{w}_1(x, \sigma) = (w_1(x), 1\sigma) \qquad \text{for all } (x, \sigma) \in X \times \Sigma;$$
$$\tilde{w}_2(x, \sigma) = (w_2(x), 2\sigma) \qquad \text{for all } (x, \sigma) \in X \times \Sigma.$$

What is the nature of the attractor $\tilde{A} \subset X \times \Sigma$ of the lifted IFS? It should be clear that

$$A = \{x \in A : (x, \sigma) \in \tilde{A}\} \text{ and } \Sigma = \{\sigma \in \Sigma : (x, \sigma) \in \tilde{A}\}.$$

In other words, the projection of the attractor of the lifted IFS into the original space X is simply the attractor A of the original IFS. The projection of $\tilde{A}$ into Σ is Σ. Recall that Σ is equivalent to a classical Cantor set. This tells us that the attractor of the lifted IFS is totally disconnected.

Lemma 6.1 *Let $\{X; w_1, w_2\}$ be a hyperbolic IFS with attractor A. Let the two transformations $w_1 : A \to A$ and $w_2 : A \to A$ be invertible. Then the associated lifted IFS is hyperbolic and totally disconnected.*

Definition 6.3 *Let $\{X; w_1, w_2\}$ be a hyperbolic IFS. Let the two transformations $w_1 : A \to A$ and $w_2 : A \to A$ be invertible. Let $\tilde{A}$ denote the attractor of the associated lifted IFS. Then the shift dynamical system $\{\tilde{A}; \tilde{S}\}$ associated with the lifted IFS is called the* lifted shift dynamical system *associated with the IFS.*

Notice that

$$\tilde{S}(x, \sigma) = (w_{\sigma_1}^{-1}(x), T(\sigma)) \qquad \text{for all}(X, \sigma) \in \tilde{A},$$

where

$$T(\sigma_1 \sigma_2 \sigma_3 \sigma_4 \ldots) = \sigma_2 \sigma_3 \sigma_4 \sigma_5 \ldots \qquad \text{for all } \sigma = \sigma_1 \sigma_2 \sigma_3 \sigma_4 \ldots . \in \Sigma.$$

Theorem 6.1 *[(The Shadow Theorem).] Let $\{X; w_1, w_2\}$ be a hyperbolic IFS of invertible transformations w_1 and w_2 and attractor A. Let $\{x_n\}_{n=0}^{\infty}$ be any orbit of the associated random shift dynamical system $\{A; S\}$. Then there is an orbit $\{\tilde{x}_n\}_{n=0}^{\infty}$ of the lifted dynamical system $\{\tilde{A}; \tilde{S}\}$ such that the first component of $\tilde{x}_n$ is x_n for all n.*

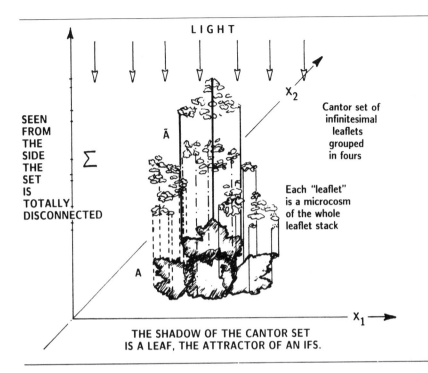

LIGHT

SEEN
FROM
THE
SIDE
THE
SET
IS
TOTALLY
DISCONNECTED

Σ

$\tilde{A}$

x_2

Cantor set of
infinitesimal
leaflets
grouped
in fours

Each "leaflet"
is a microcosm
of the whole
leaflet stack

A

$x_1 \longrightarrow$

THE SHADOW OF THE CANTOR SET
IS A LEAF, THE ATTRACTOR OF AN IFS.

Figure IV.122. The lift of the overlapping leaf attractor is totally disconnected. Deterministic shift dynamics become possible. See also Figure IV.123.

We leave the proofs of Lemma 6.1 and Theorem 6.1 as exercises. It is fun, however, and instructive to look in a couple of different geometrical ways at what is going on here.

Examples & Exercises

6.9. Consider the IFS $\{C; w_1(z), w_2(z), w_3(z), w_4(z)\}$ where, in complex notation,

$$w_1(z) = (0.5)(\cos 45° - \sqrt{-1}\sin 45°)z + (0.4 - 0.2\sqrt{-1}),$$
$$w_2(z) = (0.5)(\cos 45° + \sqrt{-1}\sin 45°)z - (0.4 + 0.2\sqrt{-1}),$$
$$w_3(z) = (0.5)z + \sqrt{-1}(0.3),$$
$$w_4(z) = (0.5)z - \sqrt{-1}(0.3).$$

A sketch of its attractor is included in Figure IV.122. It looks like a maple leaf. The leaf is made of four overlapping leaflets, which we think of as separate entities, at different heights "above" the attractor. In turn, we think of each leaflet as consisting of four smaller leaflets, again at different heights. One quickly gets the idea: one ends up with a set of heights distributed on a Cantor set in such a way that the shadow of the whole collection of infinitesimal leaflets is the leaf attractor in the C plane. The Cantor set is essentially Σ. The lifted attractor is totally disconnected; it supports deterministic shift dynamics, as illustrated in Figure IV.123.

Figure IV.123. A picture of the Shadow Theorem. Deterministic dynamics on a totally disconnected dust has a shadow that is dancing random shift dynamics on a leaf attractor.

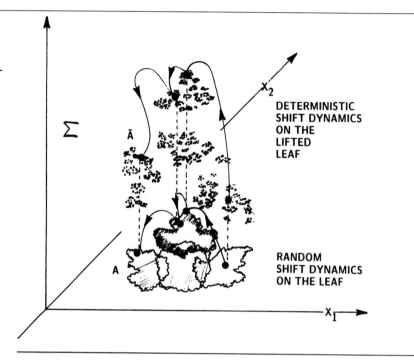

6.10. Consider the overlapping hyperbolic IFS $\{\mathbb{R}; \frac{1}{2}x, \frac{3}{4}x + \frac{1}{4}\}$. We can lift this to the hyperbolic IFS $\{\mathbb{R}^2; w_1(x), w_2(x)\}$, where

$$w_1 \begin{pmatrix} x_1 \\ x_2 \end{pmatrix} = \begin{pmatrix} \frac{1}{2} & 0 \\ 0 & \frac{1}{3} \end{pmatrix} \begin{pmatrix} x_1 \\ x_2 \end{pmatrix};$$

$$w_2 \begin{pmatrix} x_1 \\ x_2 \end{pmatrix} = \begin{pmatrix} \frac{3}{4} & 0 \\ 0 & \frac{1}{3} \end{pmatrix} \begin{pmatrix} x_1 \\ x_2 \end{pmatrix} + \begin{pmatrix} \frac{1}{4} \\ \frac{2}{3} \end{pmatrix}.$$

The attractor $\tilde{A}$ of this lifted system is shown in Figure IV.124, which also shows an orbit of the associated shift dynamical system. The shadow of this orbit is an apparently random orbit of the original system. The Shadow Theorem asserts that *any* orbit $\{x_n\}_{n=0}^{\infty}$ of a random shift dynamical system associated with the IFS $\{\mathbb{R}; \frac{1}{2}x, \frac{3}{4}x + \frac{1}{4}\}$ is the projection, or shadow, of some orbit for the shift dynamical system associated with the lifted IFS.

6.11. As a compelling illustration of the Shadow Theorem, consider the IFS

$$\{\mathbb{R}; \frac{1}{2}x, \frac{3}{4}x + \frac{1}{4}\}.$$

Let us look at the orbits $\{x_n\}_{n=0}^{\infty}$ of the shift dynamical system specified in the left-hand graph of Figure IV.120. In this case we *always* choose $S(x) = w_2^{-1}(x)$ in the overlapping region. What orbits $\{\tilde{x}_n\}_{n=0}^{\infty}$ of the lifted system, described in exercise 6.7, are these orbits the shadows of? Look again at Figure IV.124! Define

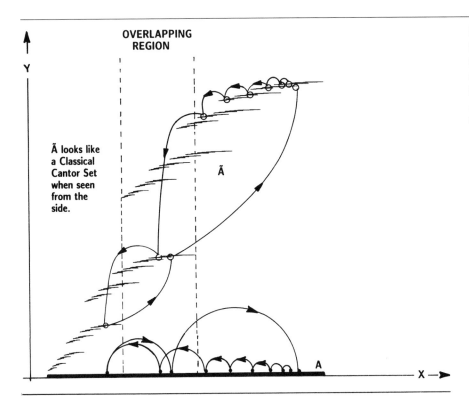

Figure IV.124. The Shadow Theorem asserts that the random shift dynamical system orbit on the overlapping attractor A is the shadow of a deterministic orbit on $\tilde{A}$.

the top of $\tilde{A}$ as

$$\tilde{A}_{top} = \{(x, y) \in \tilde{A} : (z, y) \in \tilde{A} \Rightarrow z \le x, \quad \text{and } y \in [0, 1]\}.$$

Notice that $\tilde{S} : \tilde{A}_{top} \to \tilde{A}_{top}$. It is easy to see that there is a one-to-one correspondence between orbits of the lifted system $\{\tilde{A}_{top}; \tilde{S}\}$ and orbits of the original system specified through the left-hand graph of Figure IV.120. Indeed,

$$\{(x_n, y_n)\}_{n=0}^{\infty} \text{ is an orbit of the lifted system and } (x_0, y_0) \in \tilde{A}_{top}$$

$$\Updownarrow$$

$$\{x_n\}_{n=0}^{\infty} \text{ is an orbit of the left-hand graph of Figure IV.120}$$

6.12. Draw some pictures to illustrate the Shadow Theorem in the case of the just-touching IFS $\{[0, 1]; \frac{1}{2}x, \frac{1}{2}x + \frac{1}{2}\}$.

6.13. Illustrate the Shadow Theorem using the overlapping IFS $\{[0, 1]; -\frac{3}{4}x + \frac{3}{4}, \frac{3}{4}x + \frac{1}{4}\}$. Find an orbit of period 2 whose lift has minimal period 4. Do there exist periodic orbits whose lifts are not periodic?

6.14. Prove Lemma 6.1.

6.15. Prove Theorem 6.1.

6.16. The IFS $\{\Sigma; w_1(\sigma), \ldots, w_N(\sigma)\}$ given by

$$w_n(\sigma) = n\sigma$$

for each $n = 1, 2, \ldots, N$, has an interesting lift. Show that the lift of this IFS, with a suitably defined inverse, is the shift automorphism on the space of shifts and therefore equivalent to the baker's transformation.

6.17. In section 5 it was shown that the associated shift dynamical system of any totally disconnected IFS is equivalent to the shift transformation on code space. Then we may replace the second map in the lift for the Shadow Theorem with such a totally disconnected IFS. That is, we could take a map like the leaf shown in Figures IV.122 and IV.123, and define the map

$$\{\mathbb{R}^2 \times \Delta; \tilde{w}_1(x, y), \ldots \tilde{w}_4(x, y)\},$$

where $\tilde{w}_i = (w_i^{-1}(x, y), v_i(x, y))$, where v_i are the maps of the totally disconnected IFS

$$v_1 \begin{pmatrix} x \\ y \end{pmatrix} = \begin{pmatrix} 1/3 & 0 \\ 0 & 1/3 \end{pmatrix} \begin{pmatrix} x \\ y \end{pmatrix},$$

$$v_2 \begin{pmatrix} x \\ y \end{pmatrix} = \begin{pmatrix} 1/3 & 0 \\ 0 & 1/3 \end{pmatrix} \begin{pmatrix} x \\ y \end{pmatrix} + \begin{pmatrix} 1 \\ 0 \end{pmatrix},$$

$$v_3 \begin{pmatrix} x \\ y \end{pmatrix} = \begin{pmatrix} 1/3 & 0 \\ 0 & 1/3 \end{pmatrix} \begin{pmatrix} x \\ y \end{pmatrix} + \begin{pmatrix} 1 \\ 1 \end{pmatrix},$$

$$v_2 \begin{pmatrix} x \\ y \end{pmatrix} = \begin{pmatrix} 1/3 & 0 \\ 0 & 1/3 \end{pmatrix} \begin{pmatrix} x \\ y \end{pmatrix} + \begin{pmatrix} 0 \\ 1 \end{pmatrix}.$$

Since this IFS produces an attractor that is totally disconnected, and therefore a copy of code space, the resulting lift is totally disconnected. What would a rendition of the lifted system look like if the maple leaf were lifted using a totally disconnected tree?

---

7   The Meaningfulness of Inaccurately Computed Orbits Is Established by Means of a Shadowing Theorem

Let $\{X; w_1, w_2, \ldots, w_N\}$ be a hyperbolic IFS of contractivity $0 < s < 1$. Let A denote the attractor of the IFS, and assume that $w_n : A \to A$ is invertible for each $n = 1, 2, \ldots, N$. If the IFS is totally disconnected, let $\{A; S\}$ denote the associated shift dynamical system; otherwise let $\{A; S\}$ denote the associated random shift dynamical system. Consider the following model for the inaccurate calculation of an orbit of a point $x_0 \in A$. This model will surely describe the reader's experiences in computing shift dynamics directly on pictures of fractals. Moreover, it is a reasonable model for the occurrence of numerical errors when machine computation is used to compute an orbit.

Let an exact orbit of the point $x_0 \in A$ be denoted by $\{x_n\}_{n=0}^{\infty}$, where $x_n = S^{\circ n}(x_0)$ for each n. Let an approximate orbit of the point $x_0 \in A$ be denoted by $\{\tilde{x}_n\}_{n=0}^{\infty}$ where $\tilde{x}_0 = x_0$. Then we suppose that at each step there is made an error of at most θ for some $0 \le \theta < \infty$; that is,

$$d(\tilde{x}_{n+1}, S(\tilde{x}_n)) \le \theta \qquad \text{for } n = 0, 1, 2, \ldots .$$

We proceed to analyze this model. It is clear that the inaccurate orbit $\{\tilde{x}_n\}_{n=0}^{\infty}$ will usually start out by diverging from the exact orbit $\{x_n\}_{n=0}^{\infty}$ at an exponential rate. It may well occur "accidentally" that $d(x_n, \tilde{x}_n)$ is small for various large values of n, due to the compactness of A. But typically, if $d(x_n, \tilde{x}_n)$ is small enough, then $d(x_{n+j}, \tilde{x}_{n+j})$ will again grow exponentially with increasing j. To be precise, suppose $d(\tilde{x}_1, S(\tilde{x}_0)) = \theta$ and that we make no further errors. Suppose also that for some integer M, and some integers $\sigma_1, \sigma_2, \ldots, \sigma_M \in \{1, 2, \ldots, N\}$, we have

$$\tilde{x}_n \text{ and } x_n \in w_{\sigma_n}(A), \text{ for } n = 0, 1, 2, \ldots, M.$$

Moreover, suppose that

$$x_{n+1} = w_{\sigma_n}^{-1}(x_n) \text{ and } \tilde{x}_{n+1} = w_{\sigma_n}^{-1}(\tilde{x}_n), \text{ for } n = 0, 1, 2, \ldots, M.$$

Then we have

$$d(x_{n+1}, \tilde{x}_{n+1}) \ge s^{-n}\theta, \text{ for } n = 0, 1, 2, \ldots, M.$$

For some integer $J > M$ it is likely to be the case that

$$x_{J+1} = w_{\sigma_J}^{-1}(x_n) \text{ and } \tilde{x}_{n+1} = w_{\tilde{\sigma}_J}^{-1}(\tilde{x}_n), \text{ for some } \sigma_J \neq \tilde{\sigma}_J.$$

Then, without further assumptions, we cannot say anything more about the correlation between the exact orbit and the approximate orbit. Of course, we always have the error bound

$$d(x_n, \tilde{x}_n) \le \text{diam}(A) = \max\{d(x, y) : x \in A, y \in A\}, \text{ for all } n = 1, 2, 3, \ldots .$$

Do the above comments make the situation hopeless? Are all of the calculations of shift dynamics we have done in this chapter without point because they are riddled with errors? No! The following wonderful theorem tells us that however many errors we make, there is an exact orbit that lies at every step within a small distance of our errorful one. This orbit shadows the errorful orbit. This type of theorem is extremely important in dynamics, and in any class of dynamical systems that has one (such as IFS) behavior that can be accurately analyzed using graphics on computers. Here we are use the word "shadows" in the sense of a secret agent who shadows a spy. The agent is always just out of sight, not too far away, usually not too close, but forever he follows the spy.

Theorem 7.1 The Shadowing Theorem. *Let* $\{\mathbf{X}; w_1, w_2, \ldots, w_N\}$ *be a hyperbolic IFS of contractivity* s, *where* $0 < s < 1$. *Let* A *denote the attractor of the IFS and suppose that each of the transformations* $w_n : A \to A$ *is invertible. Let* $\{A; S\}$ *denote the associated shift dynamical system in the case that the IFS is totally* *disconnected; otherwise let $\{A; S\}$ denote the associated random shift dynamical system. Let $\{\tilde{x}_n\}_{n=0}^{\infty} \subset A$ be an approximate orbit of S, such that*

$$d(\tilde{x}_{n+1}, S(\tilde{x}_n)) \le \theta \qquad \text{for all } n = 0, 1, 2, 3, \ldots,$$

for some fixed constant θ with $0 \le \theta \le \text{diam}(A)$. Then there is an exact orbit $\{x_n = S^{\circ n}(x_0)\}_{n=0}^{\infty}$ for some $x_0 \in A$, such that

$$d(\tilde{x}_{n+1}, x_{n+1}) \le \frac{s\theta}{(1-s)} \qquad \text{for all } n = 0, 1, 2, \ldots.$$

Proof As usual we exploit code space! For $n = 1, 2, 3, \ldots$, let $\sigma_n \in \{1, 2, \ldots, N\}$ be chosen so that $w_{\sigma_1}^{-1}, w_{\sigma_2}^{-1}, w_{\sigma_3}^{-1}, \ldots$, is the actual sequence of inverse maps used to compute $S(\tilde{x}_0), S(\tilde{x}_1), S(\tilde{x}_2), \ldots$. Let $\phi : \Sigma \to A$ denote the code space map associated with the IFS. Then define

$$x_0 = \phi(\sigma_1 \sigma_2 \sigma_3 \ldots).$$

Then we compare the exact orbit of the point x_0,

$$\{x_n = S^{\circ n}(x_0) = \phi(\sigma_{n+1}\sigma_{n+2}\ldots)\}_{n=0}^{\infty}$$

with the errorful orbit $\{\tilde{x}_n\}_{n=0}^{\infty}$.

Let M be a large positive integer. Then, since x_M and $S(\tilde{x}_{M-1})$ both belong to A, we have

$$d(S(x_{M-1}), S(\tilde{x}_{M-1}) \le \text{diam}(A) < \infty.$$

Since $S(x_{M-1})$ and $S(\tilde{x}_{M-1})$ are both computed with the same inverse map $w_{\sigma_M}^{-1}$ it follows that

$$d(x_{M-1}, \tilde{x}_{M-1}) \le s \, \text{diam}(A).$$

Hence

$$\begin{aligned} d(S(x_{M-2}), S(\tilde{x}_{M-2})) &= d(x_{M-1}, S(\tilde{x}_{M-2})) \\ &\le d(x_{M-1}, \tilde{x}_{M-1}) + d(\tilde{x}_{M-1}, S(\tilde{x}_{M-2})) \\ &\le \theta + s \, \text{diam}(A); \end{aligned}$$

and repeating the argument used above we now find

$$d(x_{M-2}, \tilde{x}_{M-2})) \le s(\theta + s \, \text{diam}(A)).$$

Repeating the same argument k times we arrive at

$$d(x_{M-k}, \tilde{x}_{M-k}) \le s\theta + s^2\theta + \cdots + s^{k-1}\theta + s^k \, \text{diam}(A).$$

Hence for any positive integer M and any integer n such that $0 < n < M$, we have

$$d(x_n, \tilde{x}_n) \le s\theta + s^2\theta + \cdots + s^{M-n-1}\theta + s^{M-n} \, \text{diam}(A).$$

Now take the limit of both sides of this equation as $M \to \infty$ to obtain

$$d(x_n, \tilde{x}_n) \le s\theta(1 + s + s^2 + \cdots) = \frac{s\theta}{(1-s)}, \qquad \text{for all } n = 1, 2, \ldots.$$

This completes the proof.

Examples & Exercises

7.1. Let us apply the Shadowing Theorem to an orbit on the Sierpinski triangle, using the random shift dynamical system associated with the IFS

$$\{C; \frac{1}{2}z, \frac{1}{2}z + \frac{1}{2}, \frac{1}{2}z + \frac{i}{2}\}.$$

Since the system is just-touching we must assign values to the shift transformation applied to the just-touching points. We do this by defining

$$S(x_1 + ix_2) = 2x_1 \bmod 1 + i(2x_2 \bmod 1).$$

We consider the orbit of the point $\tilde{x}_0 = (0.2147, 0.0353)$. We compute the first 11 points on the exact orbit of this point, and compare it to the results obtained when a deliberate error $\theta = 0.0001$ is introduced at each step. We obtain:

Errorful	*Exact*
$\tilde{x}_0 = (0.2147, 0.0353)$	$S^{\circ 0}(\tilde{x}_0) = (0.2147, 0.0353)$
$\tilde{x}_1 = (0.4295, 0.0705)$	$S^{\circ 1}(\tilde{x}_0) = (0.4294, 0.0706)$
$\tilde{x}_2 = (0.8591, 0.1409)$	$S^{\circ 2}(\tilde{x}_0) = (0.8588, 0.1412)$
$\tilde{x}_3 = (0.7183, 0.2817)$	$S^{\circ 3}(\tilde{x}_0) = (0.7176, 0.2824)$
$\tilde{x}_4 = (0.4365, 0.5635)$	$S^{\circ 4}(\tilde{x}_0) = (0.4352, 0.5648)$
$\tilde{x}_5 = (0.8731, 0.1269)$	$S^{\circ 5}(\tilde{x}_0) = (0.8704, 0.1296)$
$\tilde{x}_6 = (0.7463, 0.2537)$	$S^{\circ 6}(\tilde{x}_0) = (0.7408, 0.2592)$
$\tilde{x}_7 = (0.4927, 0.5073)$	$S^{\circ 7}(\tilde{x}_0) = (0.4816, 0.5184)$
$\tilde{x}_8 = (0.9855, 0.0145)$	$S^{\circ 8}(\tilde{x}_0) = (0.9632, 0.0368)$
$\tilde{x}_9 = (0.9711, 0.0289)$	$S^{\circ 9}(\tilde{x}_0) = (0.9264, 0.0736)$
$\tilde{x}_{10} = (0.9423, 0.0577)$	$S^{\circ 10}(\tilde{x}_0) = (0.8528, 0.1472)$

Notice how the orbit with errors diverges from the exact orbit of $\tilde{x}_0$. Nonetheless, the shadowing theorem asserts that there is an *exact* orbit $\{x_n\}$ such that

$$d(x_n, \tilde{x}_n) \le \frac{\frac{1}{2}}{1 - \frac{1}{2}}(0.0001) = 0.0001,$$

where $d(\cdot, \cdot)$ denotes the Manhattan metric. This really *seems* unlikely; but it must be true! Here's an example of such a shadowing orbit, also computed exactly.

Exact Shadowing Orbit $x_n = S^{\circ n}(x_0)$	$d(x_n, \tilde{x}_n) \le 0.0001$
$x_0 = (0.21478740234375, 0.03521259765625)$	0.00009
$x_1 = (0.4295748046875, 0.0704251953125)$	0.00008
$x_2 = (0.8591496093750, 0.1408503906250)$	0.00005
$x_3 = (0.7182992187500, 0.2817007812500)$	0.000001
$x_4 = (0.4365984375000, 0.5634015625000)$	0.0001
$x_5 = (0.8731968750000, 0.1268031250000)$	0.0001
$x_6 = (0.7463937500000, 0.2536062500000)$	0.0001
$x_7 = (0.4927875000000, 0.5072125000000)$	0.00009
$x_8 = (0.9855750000000, 0.0144250000000)$	0.00008

Figure IV.125. The Shadowing Theorem tells us there is an exact orbit closer to $\{\tilde{x}_n\}$ than 0.03 for all n.

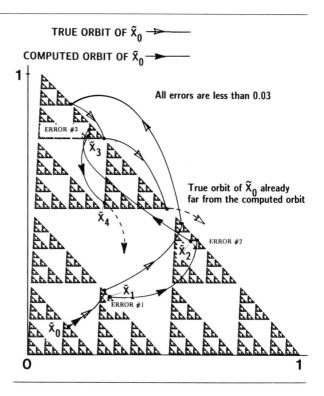

$x_9 = (0.9711500000000, 0.0288500000000)$ 0.00005
$x_{10} = (0.9423000000000, 0.0577000000000)$ 0.000000

Figure IV.125 illustrates the idea.

7.2. Consider the shift dynamical system $\{\Sigma; T\}$ on the code space of two symbols $\{1, 2\}$. Show that the sequence of points $\{\tilde{x}_n\}$ given by

$$\tilde{x}_0 = 21\overline{2}, \text{ and } \tilde{x}_n = 1\overline{2} \qquad \text{for all } n = 1, 2, 3, \ldots$$

is an errorful orbit for the system. Illustrate the divergence of $T^{\circ n}\tilde{x}_0$ from $\tilde{x}_n$. Find a shadowing orbit $\{x_n\}_{n=0}^{\infty}$ and verify the error estimate provided by the Shadowing Theorem.

7.3. Illustrate the Shadowing Theorem by constructing an erroneous orbit, and an orbit that shadows it, for the shift dynamical system $\{[0, 1]; \frac{1}{3}x, \frac{1}{2}x + \frac{1}{2}\}$.

7.4. Compute an orbit for a random shift dynamical system associated with the overlapping IFS $\{[0, 1]; \frac{3}{4}x, \frac{1}{2}x + \frac{1}{2}\}$.

7.5. An orbit of the shift dynamical system associated with the IFS

$$\left\{ \mathbb{R}^2; \frac{1}{2}\begin{pmatrix} x \\ y \end{pmatrix}, \frac{3}{4}\begin{pmatrix} x \\ y \end{pmatrix} + \frac{1}{4}\begin{pmatrix} 1 \\ 1 \end{pmatrix}, \frac{1}{2}\begin{pmatrix} x \\ y \end{pmatrix} + \begin{pmatrix} 2 \\ 0 \end{pmatrix}, \frac{1}{8}\begin{pmatrix} x \\ y \end{pmatrix} + \begin{pmatrix} 0 \\ 7 \end{pmatrix} \right\},$$

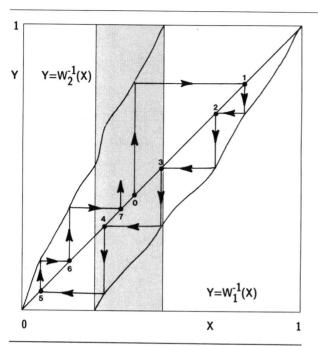

Figure IV.126. An exact orbit shadows the orbit "computed" by "drawing" in this web diagram for a random shift dynamical system.

is computed to accuracy 0.0005. How close a shadowing orbit does there exist? Use the Manhattan metric.

7.6. In Figure IV.126 an orbit of the random shift dynamical system associated with the overlapping IFS $\{[0, 1], w_1(x), w_2(x)\}$ is computed by drawing a web diagram. The computer in this case consists of a pencil and a drafting table. Estimate the errors in the drawing and then deduce how closely an exact orbit shadows the plotted one. You will need to estimate the contractivity of the IFS. Also draw a tube around the plotted orbit, within which an exact orbit lies.

7.7. Figure IV.127 shows an orbit $\{x_n\}$ of the random shift dynamical system associated with the IFS $\{[0, 1]; w_1(x), w_2(x)\}$. It was obtained by defining $S(x) = w_2^{-1}(x)$ for $x \in w_1(A) \cap w_2(A)$. A contractivity factor for the IFS is readily estimated from the drawing to be $\frac{3}{5}$. Hence if the web diagram is accurate to within 1 mm at each iteration, that is

$$d(\tilde{x}_{n+1}, S(\tilde{x}_n)) \leq 1 \,\text{mm},$$

then there is an exact orbit $\{x_n = S^{\circ n}(x_0)\}_{n=0}^{\infty}$ such that

$$d(x_n, \tilde{x}_n) \leq \frac{(\frac{3}{5})}{(\frac{2}{5})} = 1.5 \,\text{mm}.$$

Thus there is an actual orbit that remains within the "orbit tube" shown in Figure IV.127.

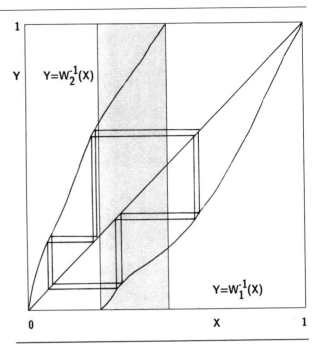

Figure IV.127. Only the *Shadow* knows. Inside the "orbit tube" there is an *exact* orbit $\{x_n\}_{n=0}^{\infty}$ of the random shift dynamical system associated with the IFS.

8   Chaotic Dynamics on Fractals

The shift dynamical system $\{A; S\}$ associated with a totally disconnected hyperbolic IFS is equivalent to the shift dynamical system $\{\Sigma, T\}$, where Σ is the code space associated with the IFS. As we have seen, this equivalence means that the two systems have a number of properties in common; for example, the two systems have the same number of cycles of minimal period 7. A particularly important property that they share is that they are both "chaotic" dynamical systems, a concept that we explain in this section. First, however, we want to underline that the two systems are deeply different from the point of view of the interplay of their dynamics with the geometry of the underlying spaces.

Consider the case of an IFS of three transformations. Let Σ denote the code space of the three symbols $\{1, 2, 3\}$, and look at the orbit of the point $\sigma \in \Sigma$ given by

$$\sigma =$$
12311121321222331323311111
12113121122123131132133211212213
22122222323123223331131231332132
23233313323331111111211131121112
21123113111321133121112121213122
11222122312311232123313
1113121212 . . . . **FOREVER.**

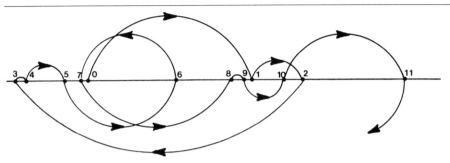

Figure IV.128. The start of a chaotic orbit on a Ternary Cantor set.

This orbit $\{T^{\circ n}\sigma\}_{n=0}^{\infty}$ may be plotted on a Cantor set of three symbols, as sketched in Figure IV.128. This can be compared with the orbit $\{S^{\circ n}(\phi(\sigma))\}_{n=0}^{\infty}$ of the shift dynamical system $\{A, S\}$ associated with an IFS of three maps, as plotted in Figure IV.129. Figure IV.130 shows an equivalent orbit, but this time for the just-touching IFS $\{[0, 1]; \frac{1}{3}x, \frac{1}{3}x + \frac{1}{3}, \frac{1}{3}x + \frac{2}{3}\}$, displayed using a web diagram.

In each case the "same" dynamics look entirely different. The qualities of beauty and harmony present in the observed orbits are different. This is not suprising: the equivalence of the dynamical systems is a topological equivalence. It does not provide much information about the interplay of the dynamics with the geometries of the spaces on which they act. This interplay is an open area for research. For example, what are the special conserved properties of two metrically equivalent dynamical systems? Can you quantify the grace and delicacy of a dancing orbit on a fractal?

This said, we turn our attention back to an important collection of properties shared by all shift dynamical systems. For simplicity we formalize the discussion for the case of the shift dynamical system $\{A, S\}$ associated with a totally disconnected hyperbolic IFS.

Definition 8.1 *Let $(\mathbf{X}, d)$ be a metric space. A subset $B \subset \mathbf{X}$ is said to be* dense *in $\mathbf{X}$ if the closure of B equals $\mathbf{X}$. A sequence $\{x_n\}_{n=0}^{\infty}$ of points in $\mathbf{X}$ is said to be dense in $\mathbf{X}$ if, for each point $a \in \mathbf{X}$, there is an subsequence $\{x_{\sigma_n}\}_{n=0}^{\infty}$ that converges to a. In particular an orbit $\{x_n\}_{n=0}^{\infty}$ of a dynamical system $\{\mathbf{X}, f\}$ is said to be dense in $\mathbf{X}$ if the sequence $\{x_n\}_{n=0}^{\infty}$ is dense in $\mathbf{X}$.*

By now you will have had some experience with using the random iteration algorithm, Program 2 of Chapter III, for computing images of the attractor A of IFS in $\mathbb{R}^2$. If you run the algorithm starting from a point $x_0 \in A$, then all of the computed points lie on A. Apparently, the sequences of points we plot are examples of sequences that are dense in the metric space (A, d).

The property of being dense is invariant under homeomorphism : if B is dense in a metric space $(\mathbf{X}, d)$ and if $\theta : \mathbf{X} \to \mathbf{Y}$ is a homeomorphism, then $\theta(B)$ is dense in $\mathbf{Y}$. If $\{\mathbf{X}; f\}$ and $\{\mathbf{Y}, g\}$ are equivalent dynamical systems under θ; and if $\{x_n\}$ is an orbit of f dense in $\mathbf{X}$, then $\{\theta(x_n)\}$ is an orbit of g dense in $\mathbf{Y}$.

Figure IV.129. The start of an orbit of a deterministic shift dynamical system. This orbit is chaotic. It will visit the part of the attractor inside each of these little circles infinitely many times.

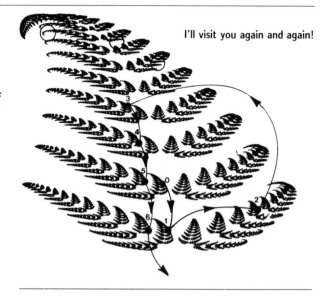

I'll visit you again and again!

Figure IV.130. Equivalent orbit to the one in Figures IV.128 and IV.129, this time ploted using a web diagram. The starting point has address 12311121321222331…. This manifestation of an orbit, which goes arbitrarily close to any point, takes place on a just-touching attractor.

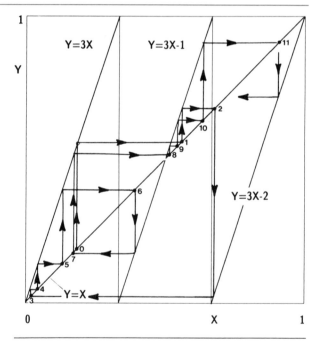

Definition 8.2 *A dynamical system* {**X**, *f*} *is* transitive *if, whenever* $\mathcal{U}$ *and* $\mathcal{V}$ *are open subsets of the metric space* (**X**, *d*)*, there exists a finite integer n such that*

$$\mathcal{U} \cap f^{\circ n}(\mathcal{V}) \neq \emptyset.$$

The dynamical system {[0, 1]; $f(x) = \min\{2x, 2 - 2x\}$} is topologically transitive. To verify this just let $\mathcal{U}$ and $\mathcal{V}$ be any pair of open intervals in the metric space ([0, 1], Euclidean). Clearly, each application of the transformation increases the length of the interval $\mathcal{U}$ in such a way that it eventually overlaps $\mathcal{V}$.

Definition 8.3 *The dynamical system* {**X**; *f*} *is* sensitive to initial conditions *if there exists* $\delta > 0$ *such that, for any* $x \in$ **X** *and any ball* $B(x, \epsilon)$ *with radius* $\epsilon > 0$, *there is* $y \in B(x, \epsilon)$ *and an integer* $n \geq 0$ *such that* $d(f^{\circ n}(x), f^{\circ n}(y)) > \delta$.

Roughly, orbits that begin close together get pushed apart by the action of the dynamical system. For example, the dynamical system {[0, 1]; $2x \bmod 1$} is sensitive to initial conditions.

Examples & Exercises

8.1. Show that the rational numbers are dense in the metric space ($\mathbb{R}$, Euclidean).

8.2. Let $C(n)$ be a counting function that counts all of the rational numbers that lie in the interval [0, 1]. Let $r_{C(n)}$ denote the nth rational number in [0, 1]. Prove that the sequence of real numbers {$r_{C(n)} \in [0, 1] : n = 1, 2, 3, ...$} is dense in the metric space ([0, 1], Euclidean).

8.3. Consider the dynamical system {[0, 1]; $f(x) = 2x \bmod 1$}. Find a point $x_0 \in$ [0, 1] whose orbit is dense in [0, 1].

8.4. Show that the dynamical system {[0, ∞) : $f(x) = 2x$} is sensitive to initial conditions, but that the dynamical system {[0, ∞) : $f(x) = (0.5)x$} is not.

8.5. Show that the shift dynamical system {Σ; T}, where Σ is the code space of two symbols, is transitive and sensitive to initial conditions.

8.6. Let {**X**, *f*} and {**Y**, *g*} be equivalent dynamical systems. Show that {**X**, *f*} is transitive if and only if {**Y**, *g*} is transitive. In other words, the property of being transitive is preserved between equivalent dynamical systems.

Definition 8.4 *A dynamical system* {**X**, *f*} *is* chaotic *if*

(1) it is transitive;
(2) it is sensitive to initial conditions;
(3) the set of periodic orbits of f is dense in **X**.

Theorem 8.1 *The shift dynamical system associated with a totally disconnected hyperbolic IFS of two or more transformations is chaotic.*

Sketch of Proof: First one establishes that the shift dynamical system {Σ; T} is chaotic where Σ is the code space of N symbols, with $N \geq 2$. One then uses the code space map $\phi : \Sigma \to A$ to carry the results over to the equivalent dynamical system $\{A; S\}$.

Theorem 1 applies to the lifted IFS associated with a hyperbolic IFS. Hence the lifted shift dynamical system associated with an IFS of two or more transformations is chaotic. In turn this implies certain characteristics to the behavior of the projection of a lifted shift dynamical system, namely a random shift dynamical system.

Let us consider now why the random iteration algorithm works, from an intuitive point of view. Consider the hyperbolic IFS $\{\mathbb{R}^2; w_1, w_2\}$. Let $a \in A$; suppose that the address of a is $\sigma \in \Sigma$, the associated code space. That is

$$a = \phi(\sigma).$$

With the aid of a random-number generator, a sequence of one million ones and twos is selected. For example, suppose that the the actual sequence produced is the following one, *which has been written from right to left*,

$$21\ldots121211212111211121111121121112111121121212221\,1$$

By this we mean that the first number chosen is a 1, then a 1, then three 2's, and so on. Then the following sequence of points on the attractor is computed:

$$a = \phi(\sigma)$$

$$w_1(a) = \phi(1\sigma)$$

$$w_1 \circ w_1(a) = \phi(11\sigma)$$

$$w_2 \circ w_1 \circ w_1(a) = \phi(211\sigma)$$

$$w_2 \circ w_2 \circ w_1 \circ w_1(a) = \phi(2211\sigma)$$

$$w_2 \circ w_2 \circ w_2 \circ w_1 \circ w_1(a) = \phi(22211\sigma)$$

$$w_1 \circ w_2 \circ w_2 \circ w_2 \circ w_1 \circ w_1(a) = \phi(122211\sigma)$$

$$w_2 \circ w_1 \circ w_2 \circ w_2 \circ w_2 \circ w_1 \circ w_1(a) = \phi(2122211\sigma)$$

$$w_1 \circ w_2 \circ w_1 \circ w_2 \circ w_2 \circ w_2 \circ w_1 \circ w_1(a) = \phi(12122211\sigma)$$

$$w_2 \circ w_1 \circ w_2 \circ w_1 \circ w_2 \circ w_2 \circ w_2 \circ w_1 \circ w_1(a) = \phi(212122211\sigma)$$

$$w_1 \circ w_2 \circ w_1 \circ w_2 \circ w_1 \circ w_2 \circ w_2 \circ w_2 \circ w_1 \circ w_1(a) = \phi(1212122211\sigma)$$

$$w_1 \circ w_1 \circ w_2 \circ w_1 \circ w_2 \circ w_1 \circ w_2 \circ w_2 \circ w_2 \circ w_1 \circ w_1(a) = \phi(11212122211\sigma)$$

$$\vdots$$

$$w_2 \circ w_1 \circ \ldots w_1 \circ w_1 \circ w_2 \circ w_1 \circ w_2 \circ w_1 \circ w_2 \circ w_2 \circ w_2 \circ w_1 \circ w_1(a) = \phi(21\ldots11212122211\sigma)$$

We imagine that instead of plotting the points as they are computed, we keep a list of the one million computed points. This done, we plot the points in the reverse order from the order in which they were computed. That is, we begin by plotting the point $\phi(21\ldots11212122211\sigma)$ and we finish by plotting the point $\phi(\sigma)$. What will we see? We will see a million points on the orbit of the shift dynamical system $\{A; S\}$; namely, $\{S^{\circ n}(\phi(21\ldots11212122211\sigma))\}_{n=0}^{1,000,000}$.

Now from our experience with shift dynamics and from our theoretical knowledge and intuitions what do we expect of such an orbit? We expect it to be *chaotic* and to visit a widely distributed collection of points on the attractor. We are looking at part of a "randomly chosen" orbit of the shift dynamical system; we expect it to be dense in the attractor.

For example, suppose that you are doing shift dynamics on a picture of a totally disconnected fractal, or a fern. You should be convinced that by making sly adjustments in the orbit at each step, as in the Shadowing Theorem, you can most easily coerce an orbit into visiting, to within a distance $\epsilon > 0$, each point in the image. But then the Shadowing Theorem ensures that there is an actual orbit close to our artificial one, and it too goes close to every point on the fractal, say to within a distance of 2ϵ of each point on the image. This suggests that "most" orbits of the shift dynamical system are dense in the attractor.

Examples & Exercises

8.7. Make experiments on a picture of the attractor of a totally disconnected hyperbolic IFS to verify the assertion in the last paragraph that "by making sly adjustments in an orbit ... you can most easily coerce the orbit into visiting to within a distance $\epsilon > 0$ of each point in the image." Can you make some experimental estimates of how many orbits go to within a distance $\epsilon > 0$, for several values of ϵ, of every point in the picture? One way to do this might be to work with a discretized image and to try to count the number of available orbits.

8.8. Run the Random Iteration Algorithm, Program 2 in Chapter III, to produce an image of a fractal, for example a fern without a stem as used in Figure IV.129. As the points are calculated and plotted, keep a list of them. Then plot the points over again in reverse order, this time making them flash on and off on the picture of the attractor on the screen, so that you can see where they land. This way you will see the interplay of the geometry with the shift dynamics on the attractor. See if the orbit is beautiful. If you think that it is, try to make your impression objective.

We want to begin to formulate the idea that "most" orbits of the shift dynamical system associated with a totally disconnected IFS are dense in the attractor. The following lemma counts the number of cycles of minimal period p.

Lemma 8.1 *Let $\{A; S\}$ be the shift dynamical system associated with a totally disconnected hyperbolic IFS $\{\mathbf{X}; w_1, w_2, \ldots, w_N\}$. Let $\mathcal{N}(p)$ denote the number of distinct cycles of minimal period p, for $p \in \{1, 2, 3, \ldots\}$. Then*

$$\mathcal{N}(p) = \left(N^p - \sum_{\substack{k=1 \\ k \text{ divides } p}}^{p-1} k\mathcal{N}(k) \right) / p \qquad \text{for } p = 1, 2, 3, \ldots.$$

Proof It suffices to restrict attention to code space, and to give the main idea, consider only the case $N = 2$. For $p = 1$, the cycles of period 1 are the fixed points of T. The equation

$$T\sigma = \sigma\sigma \in \Sigma$$

implies $\sigma = \overline{1111}$ or $\sigma = \overline{2222}$. Thus $\mathcal{N}(1) = 2$. For $p = 2$, any point that lies on a cycle of period 2 must be a fixed point of $T^{\circ 2}$, namely

$$T^{\circ 2}\sigma = \sigma,$$

where $\sigma = \overline{11}, \overline{12}, \overline{21}$, or $\overline{22}$. The only cycles here that are not of minimal period 2 must have minimal period 1. Furthermore, there are two distinct points on a cycle of minimal period 2, so

$$\mathcal{N}(2) = (2^2 - \mathcal{N}(1))/2 = 2/2 = 1.$$

One quickly gets the idea. Mathematical induction on p completes the proof for $N = 2$.

For $N = 2$, we find, for example, $\mathcal{N}(2) = 1$, $\mathcal{N}(3) = 2$, $\mathcal{N}(4) = 3$, $\mathcal{N}(5) = 6$, $\mathcal{N}(6) = 9$, $\mathcal{N}(7) = 18$, $\mathcal{N}(8) = 30$, $\mathcal{N}(9) = 56$, $\mathcal{N}(10) = 99$, $\mathcal{N}(11) = 186$, $\mathcal{N}(12) = 335$, $\mathcal{N}(13) = 630$, $\mathcal{N}(14) = 1161$, $\mathcal{N}(15) = 2182$, $\mathcal{N}(16) = 4080$, $\mathcal{N}(17) = 7710$, $\mathcal{N}(18) = 14532$, $\mathcal{N}(19) = 27594$, $\mathcal{N}(20) = 52377$. In particular, 99.9% of all points lying on cycles of period 20 lie on cycles of minimal period 20.

Here is the idea we are getting at. We know that the set of periodic cycles are dense in the attractor of a hyperbolic IFS. It follows that we may approximate the attractor by the set of all cycles of some finite period, say period 12 billion. Thus we replace the attractor A by such an approximation $\tilde{A}$, which consists of $2^{12,000,000,000}$ points. Suppose we pick one of these points at random. Then this point is extremely likely to lie on a cycle of *minimal* period 12 billion. Hence the orbit of a point chosen "at random" on the approximate attractor $\tilde{A}$ is extremely likely to consist of 12 billion *distinct* points on A.

In fact one can show that a statistically random sequence of symbols contains every possible finite subsequence. So we expect that the set of 12 billion distinct points on A is likely to contain at least one representative from each part of the attractor!

Chapter V

Fractal Dimension

---

1 Fractal Dimension

How big is a fractal? When are two fractals similar to one another in some sense? What experimental measurements might we make to tell if two different fractals may be metrically equivalent? What is the same about the two fractals in Figure V.131?

There are various numbers, associated with fractals, which can be used to compare them. They are generally referred to as fractal dimensions. They are attempts to quantify a subjective feeling we have about how densely the fractal occupies the metric space in which it lies. Fractal dimensions provide an objective means for comparing fractals.

Fractal dimensions are important because they can be defined in connection with real-world data, and they can be measured approximately by means of experiments. For example, one can measure the "fractal dimension" of the coastline of Great Britain; its value is about 1.2. Fractal dimensions can be attached to clouds, trees, coastlines, feathers, networks of neurons in the body, dust in the air at an instant in time, the clothes you are wearing, the distribution of frequencies of light reflected by a flower, the colors emitted by the sun, and the wrinkled surface of the sea during a storm. These numbers allow us to compare sets in the real world with the laboratory fractals, such as attractors of IFS.

We restrict attention to compact subsets of metric spaces. This fits well with the idea of modelling the real physical world by subsets of metric spaces. Suppose that an experimentalist is studying a physical entity, and he wishes to model this entity by means of a subset of $\mathbb{R}^3$. Then he can use a compact set for his model. For example, he can assume that the distances he measures are Euclidean distances, and he can assume that the universe is bounded. He can assume that any Cauchy sequence of points in his model set converges to a point in his model set, because he cannot experimentally invalidate this assumption. Although mathematically we can distinguish between a set and its closure, we cannot make the same distinction

Figure V.131. Do the two implied fractals have the same dimension?

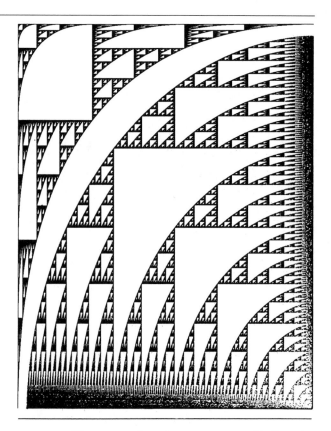

between their physical counterparts. The assumption of compactness will allow the model to be handled theoretically with relative ease.

Let (X, d) denote a complete metric space. Let $A \in \mathcal{H}(X)$ be a nonempty compact subset of X. Let $\epsilon > 0$. Let $B(x, \epsilon)$ denote the closed ball of radius ϵ and center at a point $x \in X$. We wish to define an integer, $\mathcal{N}(A, \epsilon)$, to be the least number of closed balls of radius ϵ needed to cover the set A. That is

$$\mathcal{N}(A, \epsilon) = \text{ smallest positive integer } M \text{ such that } A \subset \cup_{n=1}^{M} B(x_n, \epsilon),$$

for some set of distinct points $\{x_n : n = 1, 2, \ldots, M\} \subset X$. How do we know that there is such a number $\mathcal{N}(A, \epsilon)$? Easy! The logic is this: surround every point $x \in A$ by an *open* ball of radius $\epsilon > 0$ to provide a cover of A by open sets. Because A is compact this cover possesses a finite subcover, consisting of an integer number, say $\hat{M}$, of open balls. By taking the closure of each ball, we obtain a cover consisting of $\hat{M}$ *closed* balls. Let C denote the set of covers of A by at most $\hat{M}$ closed balls of radius ϵ. Then C contains at least one element. Let $f : C \to \{1, 2, 3, \ldots, \hat{M}\}$ be defined by $f(c) = $ number of balls in the cover $c \in C$. Then $\{f(c) : c \in C\}$ is a finite set of positive integers. It follows that it contains a least integer, $\mathcal{N}(A, \epsilon)$.

The intuitive idea behind fractal dimension is that a set A has fractal dimension D if:

$$\mathcal{N}(A, \epsilon) \approx C\epsilon^{-D} \text{ for some positive constant C.}$$

Here we use the notation "≈" as follows. Let $f(\epsilon)$ and $g(\epsilon)$ be real valued functions of the positive real variable ϵ. Then $f(\epsilon) \approx g(\epsilon)$ means that $\lim_{\epsilon \to 0}\{\ln(f(\epsilon))/\ln (g(\epsilon))\} = 1$. If we "solve" for D we find that

$$D \approx \frac{\ln \mathcal{N}(A, \epsilon) - \ln C}{\ln(1/\epsilon)}.$$

We use the notation $\ln(x)$ to denote the logarithm to the base e of the positive real number x. Now notice that the term $\ln C / \ln(1/\epsilon)$ approaches 0 as $\epsilon \to 0$. This leads us to the following definition.

Definition 1.1 *Let $A \in \mathcal{H}(X)$ where (X, d) is a metric space. For each $\epsilon > 0$ let $\mathcal{N}(A, \epsilon)$ denote the smallest number of closed balls of radius $\epsilon > 0$ needed to cover A. If*

$$D = \lim_{\epsilon \to 0} \left\{ \frac{\ln(\mathcal{N}(A, \epsilon))}{\ln(1/\epsilon)} \right\}$$

exists, then D is called the fractal dimension *of A. We will also use the notation $D = D(A)$ and will say "A has fractal dimension D."*

Examples & Exercises

1.1. This example takes place in the metric space ($\mathbb{R}^2$, Euclidean). Let $a \in X$ and let $A = \{a\}$. A consists of a single point in the space. For each $\epsilon > 0, \mathcal{N}(A, \epsilon) = 1$. It follows that $D(A) = 0$.

1.2. This example takes place in the metric space ($\mathbb{R}^2$, Manhattan). Let A denote the line segment [0, 1]. Let $\epsilon > 0$. Then it is quite easy to see that $\mathcal{N}(A, \epsilon) = -[-1/\epsilon]$, where $[x]$ denotes the integer part of the real number x. In Figure V.132 we have plotted the graph of $\ln(\mathcal{N}(A, \epsilon))$ as a function of $\ln(1/\epsilon)$. Despite a rough start, it appears clear that

$$\lim_{\epsilon \to 0} \left\{ \frac{\ln(\mathcal{N}(A, \epsilon))}{\ln(1/\epsilon)} \right\} = 1.$$

In fact for $0 < \epsilon < 1$

$$\frac{\ln(1/\epsilon)}{\ln(1/\epsilon)} \le \frac{\ln(-[-1/\epsilon])}{\ln(1/\epsilon)} = \frac{\ln(\mathcal{N}(A, \epsilon))}{\ln(1/\epsilon)} \le \frac{\ln(1/\epsilon + 1)}{\ln(1/\epsilon)} = \frac{\ln(1 + \epsilon) + \ln(1/\epsilon)}{\ln(1/\epsilon)},$$

Both sides here converge to 1 as $\epsilon \to 0$. Hence the quantity in the middle also converges to 1. We conclude that the fractal dimension of a closed line segment is one. We would have obtained the same result if we had used the Euclidean metric.

1.3. Let (X, d) be a metric space. Let $a, b, c \in X$, and let $A = \{a, b, c\}$. Prove that $D(A) = 0$.

The following two theorems simplify the process of calculating the fractal dimension. They allow one to replace the continuous variable ϵ by a discrete variable.

Figure V.132. Plot of $\ln([1/x])$ as a function of $\ln(1/x)$. This illustrates that in the computation of the fractal dimension one usually evaluates the limiting "slope" of a discontinuous function. In the present example this slope is 1.

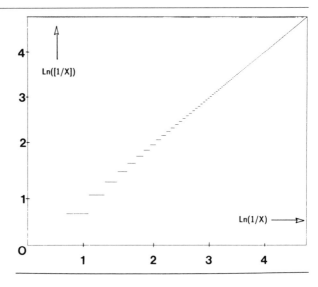

Theorem 1.1 *Let $A \in \mathcal{H}(X)$, where (X, d) is a metric space. Let $\epsilon_n = Cr^n$ for real numbers $0 < r < 1$ and $C > 0$, and integers $n = 1, 2, 3, \ldots$. If*

$$D = \lim_{n \to \infty} \left\{ \frac{\ln(\mathcal{N}(A, \epsilon_n))}{\ln(1/\epsilon_n)} \right\},$$

then A has fractal dimension D.

Proof Let the real numbers r and C, and the sequence of numbers $E = \{\epsilon_n : n = 1, 2, 3, \ldots\}$ be as defined in the statement of the theorem. Define $f(\epsilon) = \max\{\epsilon_n \in E : \epsilon_n \le \epsilon\}$. Assume that $\epsilon \le r$. Then

$$f(\epsilon) \le \epsilon \le f(\epsilon)/r \text{ and } \mathcal{N}(A, f(\epsilon)) \ge \mathcal{N}(A, \epsilon) \ge \mathcal{N}(A, f(\epsilon)/r).$$

Since $\ln(x)$ is an increasing positive function of x for $x \ge 1$, it follows that

$$\left\{ \frac{\ln(\mathcal{N}(A, f(\epsilon)/r))}{\ln(1/f(\epsilon))} \right\} \le \left\{ \frac{\ln(\mathcal{N}(A, \epsilon))}{\ln(1/\epsilon)} \right\} \tag{1}$$

$$\le \left\{ \frac{\ln(\mathcal{N}(A, f(\epsilon)))}{\ln(r/f(\epsilon))} \right\}. \tag{2}$$

Assume that $\mathcal{N}(A; \epsilon) \to \infty$ as $\epsilon \to 0$; if not then the theorem is true. The right-hand side of equation 2 obeys

$$\lim_{\epsilon \to 0} \left\{ \frac{\ln(\mathcal{N}(A, f(\epsilon)))}{\ln(r/f(\epsilon))} \right\} = \lim_{n \to \infty} \left\{ \frac{\ln(\mathcal{N}(A, \epsilon_n))}{\ln(r/\epsilon_n)} \right\}$$

$$= \lim_{n \to \infty} \left\{ \frac{\ln(\mathcal{N}(A, \epsilon_n))}{\ln(r) + \ln(1/\epsilon_n)} \right\}$$

$$= \lim_{n \to \infty} \left\{ \frac{\ln(\mathcal{N}(A, \epsilon_n))}{\ln(1/\epsilon_n)} \right\}.$$

The left-hand side of equation 2 obeys

$$\lim_{\epsilon \to 0} \left\{ \frac{\ln(\mathcal{N}(A, f(\epsilon)/r))}{\ln(1/f(\epsilon))} \right\} = \lim_{n \to \infty} \left\{ \frac{\ln(\mathcal{N}(A, \epsilon_{n-1}))}{\ln(1/\epsilon_n)} \right\}$$

$$= \lim_{n \to \infty} \left\{ \frac{\ln(\mathcal{N}(A, \epsilon_{n-1}))}{\ln(1/r) + \ln(1/\epsilon_{n-1}))} \right\}$$

$$= \lim_{n \to \infty} \left\{ \frac{\ln(\mathcal{N}(A, \epsilon_n))}{\ln(1/\epsilon_n)} \right\}.$$

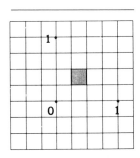

Figure V.133. Closed boxes of side $(1/2^n)$ cover $\mathbb{R}^2$. Here $n = 2$. See Theorem 1.2.

So as $\epsilon \to 0$ both the left-hand side and the right-hand side of equation 2 approach the same value, claimed in the theorem. By the Sandwich Theorem of calculus, the limit as $\epsilon \to 0$ of the quantity in the middle of equation 2 also exists, and it equals the same value. This completes the proof of the theorem.

Theorem 1.2 The Box Counting Theorem. *Let $A \in \mathcal{H}(\mathbb{R}^m)$, where the Euclidean metric is used. Cover $\mathbb{R}^m$ by closed square boxes of side length $(1/2^n)$, as exemplified in Figure V.133 for $n = 2$ and $m = 2$. Let $\mathcal{N}_n(A)$ denote the number of boxes of side length $(1/2^n)$ which intersect the attractor. If*

$$D = \lim_{n \to \infty} \left\{ \frac{\ln(\mathcal{N}_n(A))}{\ln(2^n)} \right\},$$

then A has fractal dimension D.

Proof We observe that for $m = 1, 2, 3, \ldots,$

$$2^{-m} \mathcal{N}_{n-1} \le \mathcal{N}(A, 1/2^n) \le \mathcal{N}_{k(n)} \text{ for all } n = 1, 2, 3, \ldots,$$

where $k(n)$ is the smallest integer k satisfying $k \ge n - 1 + \frac{1}{2} \log_2 m$. The first inequality holds because a ball of radius $1/2^n$ can intersect at most 2^m "on-grid" boxes of side $1/2^{n-1}$. The second follows from the fact that a box of side s can fit inside a ball of radius r provided $r^2 \ge (\frac{s}{2})^2 + (\frac{s}{2})^2 + cldots + (\frac{s}{2})^2 = m(\frac{s}{2})^2$ by the theorem of Pythagoras. Now

$$\lim_{n \to \infty} \left\{ \frac{\ln(\mathcal{N}_{k(n)})}{\ln(2^n)} \right\} = \lim_{n \to \infty} \left\{ \frac{\ln(2^{k(n)})}{\ln(2^n)} \frac{\ln(\mathcal{N}_{k(n)})}{\ln(2^{k(n)})} \right\} = D,$$

since $\frac{k(n)}{n} \to 1$. Since also

$$\lim_{n \to \infty} \left\{ \frac{\ln 2^{-m} \mathcal{N}_{n-1}}{\ln(2^n)} \right\} = \lim_{n \to \infty} \left\{ \frac{\ln \mathcal{N}_{n-1}}{\ln(2^{n-1})} \right\} = D,$$

Theorem 1.1 with $r = 1/2$ completes the proof.

There is nothing magical about using boxes of side $(1/2)^n$ in Theorem 1.2. One can equally well use boxes of side Cr^n, where $C > 0$ and $0 < r < 1$ are fixed real numbers.

Figure V.134. It re-quires $(1/2^n)^{-2}$ boxes of side $(1/2^n)$ to cover $\blacksquare \subset \mathbb{R}^2$. We deduce, with a feeling of relief, that the fractal dimension of $\blacksquare$ is 2. To which collage is this related?

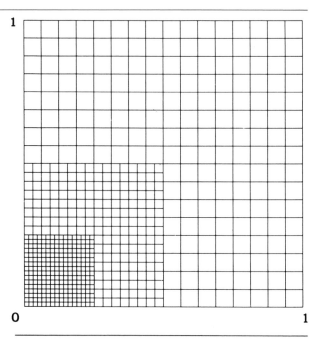

Examples & Exercises

1.4. Consider the $\blacksquare \subset \mathbb{R}^2$. It is easy to see that $\mathcal{N}_1(\blacksquare) = 4, \mathcal{N}_2(\blacksquare) = 16, \mathcal{N}_3(\blacksquare) = 64$, $\mathcal{N}_4(\blacksquare) = 256$, and in general that $\mathcal{N}_n(\blacksquare) = 4^n$ for $n = 1, 2, 3, \ldots$; see Figure V.134.

Theorem 1.2 implies that

$$D(\blacksquare) = \lim_{n \to \infty} \left\{ \frac{\ln(\mathcal{N}_n(\blacksquare))}{\ln(2^n)} \right\} = \lim_{n \to \infty} \left\{ \frac{\ln(4^n)}{\ln(2^n)} \right\} = 2.$$

1.5. Consider the Sierpinski triangle $\triangle$, in Figure V.135, as a subset of $(\mathbb{R}^2,$ Euclidean).

We see that $\mathcal{N}_1(\triangle) = 3, \mathcal{N}_2(\triangle) = 9, \mathcal{N}_3(\triangle) = 27, \mathcal{N}_4(\triangle) = 81$, and in general $\mathcal{N}_n(\triangle) = 3^n$ for $n = 1, 2, 3, \ldots$.

Theorem 1.2 implies that

$$D(\triangle) = \lim_{n \to \infty} \left\{ \frac{\ln(\mathcal{N}_n(\triangle))}{\ln(2^n)} \right\} = \lim_{n \to \infty} \left\{ \frac{\ln(3^n)}{\ln(2^n)} \right\} = \frac{\ln(3)}{\ln(2)}.$$

1.6. Use the Box Counting Theorem, but with boxes of side length $(1/3)^n$, to calculate the fractal dimension of the classical Cantor set $\mathcal{C}$ described in exercise 1.5 in Chapter III.

1.7. Use the Box Counting Theorem to estimate the fractal dimension of the fractal subset of $\mathbb{R}^2$ shown in Figure V.136. You will need to take as your first box the

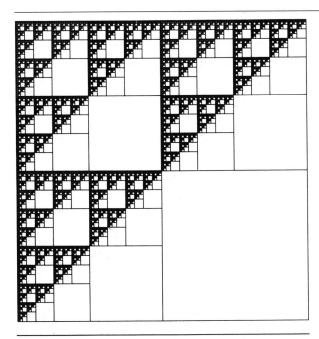

Figure V.135. It requires 3^n closed boxes of side $(1/2)^n$ to cover the Sierpinski triangle $\mathbb{A} \subset \mathbb{R}^2$. We deduce that its fractal dimension is $\ln(3)/\ln(2)$.

obvious one suggested by the figure. You should then find that there appears to be a pattern to the sequence of numbers $\mathcal{N}_1, \mathcal{N}_2, \mathcal{N}_3, \ldots$.

1.8. The same problem as 1.7, this time applied to Figure V.137. By making the right choice of Cartesian coordinate system, you will make this problem easy.

What happens to the fractal dimension of a set if we deform it "with bounded distortion"? The following theorem tells us that metrically equivalent sets have the same fractal dimension. For example, the two fractals in Figure V.131 have the same fractal dimension!

Theorem 1.3 *Let the metric spaces (X_1, d_1) and (X_2, d_2) be metrically equivalent. Let $\theta : X_1 \to X_2$ be a transformation that provides the equivalence of the spaces. Let $A_1 \in \mathcal{H}(X_1)$ have fractal dimension D. Then $A_2 = \theta(A_1)$ has fractal dimension D. That is*

$$D(A_1) = D(\theta(A_2)).$$

Proof This proof makes use of the concepts of the lim sup and lim inf of a function. (The lim sup is discussed briefly following Definition 2.1 in the next section.)

Since the two spaces (X_1, d_1) and (X_2, d_2) are equivalent under θ, there exist positive constants e_1 and e_2 such that

$$e_1 d_2(\theta(x), \theta(y)) < d_1(x, y) < e_2 d_2(\theta(x), \theta(y)) \qquad \text{for all } x, y \in X_1. \quad (3)$$

Without loss of generality we assume that $e_1 < 1 < e_2$. Equation 3 implies

$$d_2(\theta(x), \theta(y)) < \frac{d_1(x, y)}{e_1} \text{ for all } x, y \in X_1.$$

Figure V.136. What other well-known fractal has the same fractal dimension?

This implies

$$\theta(B(x, \epsilon)) \subset B(\theta(x), \epsilon/e_1)) \quad \text{for all } x \in X_1. \tag{4}$$

Now, from the definition of $\mathcal{N}(A_1, \epsilon)$, we know that there is a set of points $\{x_1, x_2, \ldots, x_\mathcal{N}\} \subset X_1$, where $\mathcal{N} = \mathcal{N}(A_1, \epsilon)$, such that the set of closed balls $\{B(x_n, \epsilon) : n = 1, 2, \ldots, \mathcal{N}(A_1, \epsilon)\}$ provides a cover of A_1. It follows that $\{\theta(B(x_n, \epsilon)) : n = 1, 2, \ldots, \mathcal{N}(A_1, \epsilon)\}$ provides a cover of A_2. Equation 4 now implies that $\{B(\theta(x_n), \epsilon/e_1)) : n = 1, 2, \ldots, \mathcal{N}(A_1, \epsilon)\}$ provides a cover of A_2. Hence

$$\mathcal{N}(A_2, \epsilon/e_1) \leq \mathcal{N}(A_1, \epsilon).$$

Hence, when $\epsilon < 1$,

$$\frac{\ln(\mathcal{N}(A_2, \epsilon/e_1))}{\ln(1/\epsilon)} \leq \frac{\ln(\mathcal{N}(A_1, \epsilon))}{\ln(1/\epsilon)}.$$

It follows that

$$\limsup_{\epsilon \to 0} \left\{ \frac{\ln(\mathcal{N}(A_2, \epsilon))}{\ln(1/\epsilon)} \right\} \tag{5}$$

$$= \limsup_{\epsilon \to 0} \left\{ \frac{\ln(\mathcal{N}(A_2, \epsilon/e_1))}{\ln(1/\epsilon)} \right\} \leq \lim_{\epsilon \to 0} \left\{ \frac{\ln(\mathcal{N}(A_1, \epsilon))}{\ln(1/\epsilon)} \right\} = D(A_1). \tag{6}$$

We now seek an inequality in the opposite direction. Equation 3 implies that

$$d_1(\theta^{-1}(x), \theta^{-1}(y)) < e_2 d_2(x, y) \quad \text{for all } x, y \in X_2.$$

This tells us that

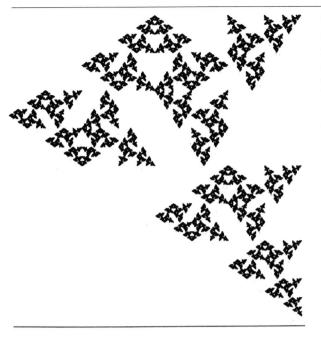

Figure V.137. If you choose the "first" box just right, the fractal dimension of this fractal is easily estimated. Count the number $\mathcal{N}_n$ of boxes of side $1/2^n$ which intersect the set, for $n = 1, 2, 3, \ldots,$ and apply the Box Counting Theorem.

$$\theta^{-1}(B(x, \epsilon)) \subset B(\theta^{-1}(x), e_2\epsilon)) \text{ for all } x \in X_2,$$

and this in turn implies

$$\mathcal{N}(A_1, e_2\epsilon) \leq \mathcal{N}(A_2, \epsilon).$$

Hence, when $\epsilon < 1$,

$$\frac{\ln(\mathcal{N}(A_1, e_2\epsilon))}{\ln(1/\epsilon)} \leq \frac{\ln(\mathcal{N}(A_2, \epsilon))}{\ln(1/\epsilon)}.$$

It follows that

$$\begin{aligned}
D(A_1) &= \lim_{\epsilon \to 0} \left\{ \frac{\ln(\mathcal{N}(A_1, \epsilon))}{\ln(1/\epsilon)} \right\} \\
&= \lim_{\epsilon \to 0} \left\{ \frac{\ln(\mathcal{N}(A_1, e_2\epsilon))}{\ln(1/\epsilon)} \right\} \\
&\leq \liminf_{\epsilon \to 0} \left\{ \frac{\ln(\mathcal{N}(A_2, \epsilon))}{\ln(1/\epsilon)} \right\}.
\end{aligned} \tag{7}$$

By combining equations 6 and 7 we obtain

$$\liminf_{\epsilon \to 0} \left\{ \frac{\ln(\mathcal{N}(A_2, \epsilon))}{\ln(1/\epsilon)} \right\} = D(A_1) = \limsup_{\epsilon \to 0} \left\{ \frac{\ln(\mathcal{N}(A_2, \epsilon))}{\ln(1/\epsilon)} \right\}.$$

From this it follows that

$$D(A_2) = \lim_{\epsilon \to 0} \left\{ \frac{\ln(\mathcal{N}(A_2, \epsilon))}{\ln(1/\epsilon)} \right\} = D(A_1).$$

This completes the proof.

Examples & Exercises

1.9. Let $\mathcal{C}$ denote the classical Cantor set, living in $[0, 1]$ and obtained by omitting "middle thirds." Let $\tilde{\mathcal{C}}$ denote the Cantor set obtained by starting from the closed interval $[0, 3]$ and omitting "middle thirds." Use Theorem 1.3 to show that they have the same fractal dimension. Verify the conclusion by means of a box-counting argument.

1.10. Let A be a compact nonempty subset of $\mathbb{R}^2$. Suppose that A has fractal dimension D_1 when evaluated using the Euclidean metric and fractal dimension D_2 when evaluated using the Manhattan metric. Show that $D_1 = D_2$.

1.11. This example takes place in the metric space $(\mathbb{R}^2, \text{Manhattan})$. Let A_1 and A_2 denote the attractors of the following two hyperbolic IFS

$$\{\mathbb{R}^2; w_1(x, y), w_2(x, y), w_3(x, y)\} \quad \text{and} \quad \{\mathbb{R}^2; w_4(x, y), w_5(x, y), w_6(x, y)\},$$

where

$$w_1 \begin{pmatrix} x \\ y \end{pmatrix} = \begin{pmatrix} \frac{1}{2} & 0 \\ 0 & \frac{1}{2} \end{pmatrix} \begin{pmatrix} x \\ y \end{pmatrix} + \begin{pmatrix} 1 \\ 0 \end{pmatrix},$$

$$w_2 \begin{pmatrix} x \\ y \end{pmatrix} = \begin{pmatrix} \frac{1}{2} & 0 \\ 0 & \frac{1}{2} \end{pmatrix} \begin{pmatrix} x \\ y \end{pmatrix},$$

$$w_3 \begin{pmatrix} x \\ y \end{pmatrix} = \begin{pmatrix} \frac{1}{2} & 0 \\ 0 & \frac{1}{2} \end{pmatrix} \begin{pmatrix} x \\ y \end{pmatrix} + \begin{pmatrix} 0 \\ 1 \end{pmatrix},$$

and

$$w_4 \begin{pmatrix} x \\ y \end{pmatrix} = \begin{pmatrix} \frac{1}{2} & 0 \\ 0 & \frac{1}{2} \end{pmatrix} \begin{pmatrix} x \\ y \end{pmatrix} + \begin{pmatrix} 2 \\ 0 \end{pmatrix},$$

$$w_5 \begin{pmatrix} x \\ y \end{pmatrix} = \begin{pmatrix} \frac{1}{2} & 0 \\ 0 & \frac{1}{2} \end{pmatrix} \begin{pmatrix} x \\ y \end{pmatrix},$$

$$w_6 \begin{pmatrix} x \\ y \end{pmatrix} = \begin{pmatrix} \frac{1}{2} & 0 \\ 0 & \frac{1}{2} \end{pmatrix} \begin{pmatrix} x \\ y \end{pmatrix} + \begin{pmatrix} 1 \\ 1 \end{pmatrix}.$$

By finding a suitable change of coordinates, show A_1 and A_2 have the same fractal dimensions.

2   The Theoretical Determination of the Fractal Dimension

The following definition extends Definition 1.1. It provides a value for the fractal dimension for a wider collection of sets.

Definition 2.1 *Let (X, d) be a complete metric space. Let $A \in \mathcal{H}(X)$. Let $\mathcal{N}(\epsilon)$ denote the minimum number of balls of radius ϵ needed to cover A. If*

$$D = \lim_{\epsilon \to 0} \left\{ \sup \left\{ \frac{\ln \mathcal{N}(\tilde{\epsilon})}{\ln(1/\tilde{\epsilon})} : \tilde{\epsilon} \in (0, \epsilon) \right\} \right\}$$

exists, then D is called the fractal dimension *of A. We will also use the notation* $D = D(A)$, *and will say "A has fractal dimension D."*

In stating this definition we have "spelled out" the lim sup. For any function $f(\epsilon)$, defined for $0 < \epsilon < 1$, for example, we have

$$\limsup_{\epsilon \to 0} f(\epsilon) = \lim_{\epsilon \to 0} \{\sup\{f(\epsilon) : \tilde{\epsilon} \in (0, \epsilon)\}\}.$$

It can be proved that Definition 2.1 is consistent with Definition 1.1: if a set has fractal dimension D according to Definition 1.1 then it has the same dimension according to Definition 2.1. Also, all of the theorems in this book apply with either definition. The broader definition provides a fractal dimension in some cases where the previous definition makes no assertion.

Theorem 2.1 *Let m be a positive integer; and consider the metric space* $(\mathbb{R}^m,$ Euclidean). *The fractal dimension* $D(A)$ *exists for all* $A \in \mathcal{H}(\mathbb{R}^m)$. *Let* $B \in \mathcal{H}(\mathbb{R}^m)$ *be such that* $A \subset B$; *and let* $D(B)$ *denote the fractal dimension of B. Then* $D(A) \le D(B)$. *In particular,*

$$0 \le D(A) \le m.$$

Proof We prove the theorem for the case $m = 2$. Without loss of generality we can suppose that $A \subset \blacksquare$. It follows that $\mathcal{N}(A, \epsilon) \le \mathcal{N}(\blacksquare, \epsilon)$ for all $\epsilon > 0$. Hence for all ϵ such that $0 < \epsilon < 1$ we have

$$0 \le \frac{\ln(\mathcal{N}(A, \epsilon))}{\ln(1/\epsilon)} \le \frac{\ln(\mathcal{N}\blacksquare, \epsilon))}{\ln(1/\epsilon)}.$$

It follows that

$$\limsup_{\epsilon \to 0} \left\{ \frac{\ln(\mathcal{N}(A, \epsilon))}{\ln(1/\epsilon)} \right\} \le \limsup_{\epsilon \to 0} \left\{ \frac{\ln(\mathcal{N}(\blacksquare, \epsilon))}{\ln(1/\epsilon)} \right\}.$$

The lim sup on the right-hand side exists and has value 2. It follows that the lim sup on the left-hand side exists and is bounded above by 2. Hence the fractal dimension $D(A)$ is defined and bounded above by 2. Also $D(A)$ is nonnegative.

If $A, B \in \mathcal{H}(\mathbb{R}^2)$ with $A \subset B$, then the fractal dimensions of A and B are defined. The above argument wherein $\blacksquare$ is replaced by B shows that $D(A) \le D(B)$. This completes the proof.

The following theorem helps us to calculate the fractal dimension of the union of two sets.

Theorem 2.2 *Let m be a positive integer; and consider the metric space* $(\mathbb{R}^m,$ Euclidean). *Let A and B belong to* $\mathcal{H}(\mathbb{R}^m)$. *Let A be such that its fractal dimension is given by*

$$D(A) = \lim_{\epsilon \to 0} \left\{ \frac{\ln(\mathcal{N}(A, \epsilon))}{\ln(1/\epsilon)} \right\}.$$

Let $D(B)$ and $D(A \cup B)$ denote the fractal dimensions of B and $A \cup B$, respectively. Suppose that $D(B) < D(A)$. Then

$$D(A \cup B) = D(A).$$

Proof From Theorem 2.1 it follows that $D(A \cup B) \geq D(A)$. We want to show that $D(A \cup B) \leq D(A)$. We begin by observing that, for all $\epsilon > 0$,

$$\mathcal{N}(A \cup B, \epsilon) \leq \mathcal{N}(A, \epsilon) + \mathcal{N}(B, \epsilon).$$

It follows that

$$
\begin{aligned}
D(A \cup B) &= \limsup_{\epsilon \to 0} \left\{ \frac{\ln(\mathcal{N}(A \cup B, \epsilon))}{\ln(1/\epsilon)} \right\} \\
&\leq \limsup_{\epsilon \to 0} \left\{ \frac{\ln(\mathcal{N}(A, \epsilon) + \mathcal{N}(B, \epsilon))}{\ln(1/\epsilon)} \right\} \\
&\leq \limsup_{\epsilon \to 0} \left\{ \frac{\ln(\mathcal{N}(A, \epsilon))}{\ln(1/\epsilon)} \right\} + \limsup_{\epsilon \to 0} \left\{ \frac{\ln(1 + \mathcal{N}(B, \epsilon)/\mathcal{N}(A, \epsilon))}{\ln(1/\epsilon)} \right\}.
\end{aligned}
$$

The proof is completed by showing that $\mathcal{N}(B, \epsilon)/\mathcal{N}(A, \epsilon)$ is less than 1 when ϵ is sufficiently small. This would imply that the second limit on the right here is equal to zero. The first limit on the right converges to $D(A)$.

Notice that

$$\sup \left\{ \frac{\ln(\mathcal{N}(B, \tilde{\epsilon}))}{\ln(1/\tilde{\epsilon})} : \tilde{\epsilon} < \epsilon \right\}$$

is a decreasing function of the positive variable ϵ. It follows that

$$\frac{\ln(\mathcal{N}(B, \epsilon))}{\ln(1/\epsilon)} < D(A) \text{ for all sufficiently small } \epsilon > 0.$$

Because the limit explicitly stated in the theorem exists, it follows that

$$\frac{\ln(\mathcal{N}(B, \epsilon))}{\ln(1/\epsilon)} < \frac{\ln(\mathcal{N}(A, \epsilon))}{\ln(1/\epsilon)} \text{ for all sufficiently small } \epsilon > 0.$$

This allows us to conclude that

$$\frac{\mathcal{N}(B, \epsilon)}{\mathcal{N}(A, \epsilon)} < 1 \text{ for all sufficiently small } \epsilon > 0.$$

This completes the proof.

Examples & Exercises

2.1. The fractal dimension of the hairy set $A \subset \mathbb{R}^2$, suggested in Figure V.138, is 2. The contribution from the hairs to $\mathcal{N}(A, \epsilon)$ becomes exponentially small compared to the contribution from ■, as $\epsilon \to 0$.

We now give you a wonderful theorem that provides the fractal dimension of the attractor of an important class of IFS. It will allow you to estimate fractal dimensions "on the fly," simply from inspection of pictures of fractals, once you get used to it.

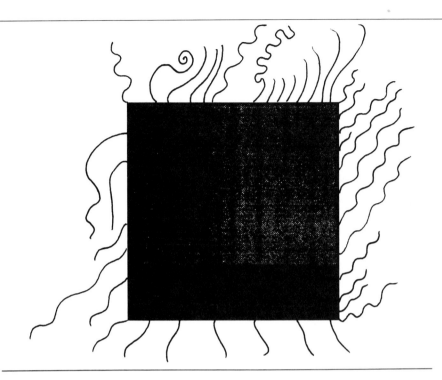

Figure V.138. Picture of a hairy box. The fractal dimension of the subset of $\mathbb{R}^2$ suggested here is the same as the fractal dimension of the box. The hairs are overpowered.

Theorem 2.3 *Let $\{\mathbb{R}^m; w_1, w_2, \ldots, w_N\}$ be a hyperbolic IFS, and let A denote its attractor. Suppose w_n is a similitude of scaling factor s_n for each $n \in \{1, 2, 3, \ldots, N\}$. If the IFS is totally disconnected or just-touching then the attractor has fractal dimension $D(A)$, which is given by the unique solution of*

$$\sum_{n=1}^{N} |s_n|^{D(A)} = 1, \qquad D(A) \in [0, m].$$

If the IFS is overlapping, then $\overline{D} \geq D(A)$, where $\overline{D}$ is the solution of

$$\sum_{n=1}^{N} |s_n|^{\overline{D}} = 1, \qquad \overline{D} \in [0, \infty).$$

Sketch of proof The full proof can be found in [Bedford 1986], [Hardin 1985], [Hutchinson 1981], and [Reuter 1987]. The following argument gives a valuable insight into the fractal dimension. We restrict attention to the case where the IFS $\{\mathbb{R}^m; w_1, w_2, \ldots, w_N\}$ is totally disconnected. We suppose that the scaling factor s_i associated with the similitude w_i is nonzero for each $i \in \{1, 2, \ldots, N\}$. Let $\epsilon > 0$. We begin by making two observations.

Observation (i): Let $i \in \{1, 2, \ldots, N\}$. Since w_i is a similitude of scaling factor s_i, it maps closed balls onto closed balls, according to

$$w_i(B(x, \epsilon)) = B(w_i(x), |s_i|\epsilon).$$

Assume that $s_i \neq 0$. Then w_i is invertible, and obtain

$$w_i^{-1}(B(x, \epsilon)) = B(w_i^{-1}(x), |s_i|^{-1}\epsilon).$$

The latter two relations allow us to establish that for all $\epsilon > 0$,

$$\mathcal{N}(A, \epsilon) = \mathcal{N}(w_i(A), |s_i|\epsilon);$$

which is equivalent to

$$\mathcal{N}(w_i(A), \epsilon) = \mathcal{N}(A, |s_i|^{-1}\epsilon). \tag{1}$$

This applies for each $i \in \{1, 2, 3, \ldots, N\}$.

Observation (ii): The attractor A of the IFS is the disjoint union

$$A = w_1(A) \cup w_2(A) \cup \ldots \cup w_N(A),$$

where each of the sets $w_n(A)$ is compact. Hence we can choose the positive number ϵ so small that if, for some point $x \in \mathbb{R}^2$ and some integer $i \in \{1, 2, \ldots, N\}$, we have $B(x, \epsilon) \cap w_i(A) \neq \emptyset$, then $B(x, \epsilon) \cap w_j(A) = \emptyset$ for all $j \in \{1, 2, \ldots, N\}$ with $j \neq i$. It follows that if the number ϵ is sufficiently small we have

$$\mathcal{N}(A, \epsilon) = \mathcal{N}(w_1(A), \epsilon) + \mathcal{N}(w_2(A), \epsilon) + \mathcal{N}(w_3(A), \epsilon) + \cdots + \mathcal{N}(w_N(A), \epsilon)$$

We put our two observations together. Substitute from equation 1 into the last equation to obtain

$$\begin{aligned}
\mathcal{N}(A, \epsilon) = {} & \mathcal{N}(A, |s_1|^{-1}\epsilon) + \mathcal{N}(A, |s_2|^{-1}\epsilon) \\
& + \mathcal{N}(A, |s_3|^{-1}\epsilon) + \cdots + \mathcal{N}(A, |s_N|^{-1}\epsilon).
\end{aligned} \tag{2}$$

This functional equation is true for all positive numbers ϵ that are sufficiently small. The proof is completed by showing formally that this implies the assertion in the theorem.

Here we demonstrate the reasonableness of the last step. Let us make the assumption $\mathcal{N}(A, \epsilon) \sim C\epsilon^{-D}$. Then substituting into equation 2 we obtain the equation:

$$C\epsilon^{-D} \approx C|s_1|^D\epsilon^{-D} + C|s_2|^D\epsilon^{-D} + C|s_3|^D\epsilon^{-D} + \cdots + C|s_N|^D\epsilon^{-D}.$$

From this we deduce that

$$1 = |s_1|^D + |s_2|^D + |s_3|^D + \cdots + |s_N|^D.$$

This completes our sketch of the proof of Theorem 2.3.

Examples & Exercises

2.2. This example takes place in the metric space ($\mathbb{R}^2$, Euclidean). A Sierpinski triangle is the attractor of a just-touching IFS of three similitudes, each with scaling factor 0.5. Hence the fractal dimension is the solution D of the equation

$$(0.5)^D + (0.5)^D + (0.5)^D = 1$$

Figure V.139. The Castle fractal. This is an example of a self-similar fractal, and its fractal dimension may be calculated with the aid of Theorem 2.3. The associated IFS code is given in Table V.1.

from which we find

$$D = \frac{\ln(1/3)}{\ln(0.5)} = \frac{\ln(3)}{\ln(2)}.$$

2.3. Find a just-touching IFS of similitudes in $\mathbb{R}^2$ whose attractor is ■. Verify that Theorem 2.3 yields the correct value for the fractal dimension of ■.

2.4. The classical Cantor set is the attractor of the hyperbolic IFS

$$\{[0, 1]; w_1(x) = \frac{1}{3}x; w_1(x) = \frac{1}{3}x + \frac{2}{3}\}.$$

Use Theorem 2.3 to calculate its fractal dimension.

2.5. The attractor of a just-touching hyperbolic IFS $\{\mathbb{R}^2; w_i(x), i = 1, 2, 3, 4\}$ is represented in Figure V.139. The affine transformations $w_i : \mathbb{R}^2 \to \mathbb{R}^2$ are similitudes and are given in tabular form in Table V.1. Use Theorem 2.3 to calculate the fractal dimension of the attractor.

2.6. The attractor of a just-touching hyperbolic IFS $\{\mathbb{R}^2; w_i(x), i = 1, 2, 3\}$ is represented in Figure V.140. The affine transformations $w_i : \mathbb{R}^2 \to \mathbb{R}^2$ are similitudes.

Table V.1.   IFS code for a Castle.

w	a	b	c	d	e	f	p
1	0.5	0	0	0.5	0	0	0.25
2	0.5	0	0	0.5	2	0	0.25
3	0.4	0	0	0.4	0	1	0.25
4	0.5	0	0	0.5	2	1	0.25

Figure V.140. To calculate the fractal dimension of the subset of $\mathbb{R}^2$ represented here, first apply the Collage Theorem to find a corresponding set of similitudes. Then use Theorem 2.3.

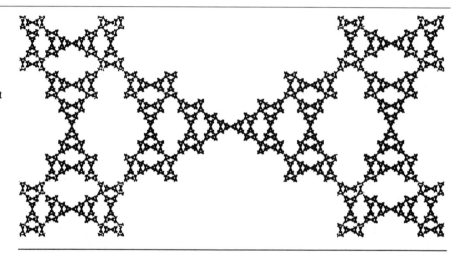

Use the Collage Theorem to find the similitudes, and then use Theorem 2.3 to calculate the fractal dimension of the attractor.

2.7. Figure V.141 represents the attractor of an overlapping hyperbolic IFS

$$\{\mathbb{R}^2; w_i(x), i = 1, 2, 3, 4\}.$$

Use the Collage Theorem and Theorem 2.3 to obtain an upper bound to the fractal dimension of the attractor.

2.8. Calculate the fractal dimension of the subset of $\mathbb{R}^2$ represented by Figure V.142.

2.9. Consider the attractor A of a totally disconnected hyperbolic IFS

$$\{\mathbb{R}^7; w_i(x), i = 1, 2\}$$

where the two maps

$$w_1 : \mathbb{R}^7 \to \mathbb{R}^7 \quad \text{and} \quad w_2 : \mathbb{R}^7 \to \mathbb{R}^7$$

are similitudes, of scaling factors s_1 and s_2, respectively. Show that A is also the attractor of the totally disconnected hyperbolic IFS $\{\mathbb{R}^7; v_i(x), i = 1, 2, 3, 4\}$ where $v_1 = w_1 \circ w_1$, $v_2 = w_1 \circ w_2$, $v_3 = w_2 \circ w_1$, and $v_4 = w_2 \circ w_2$. Show that $v_i(x)$ is a similitude, and find its scaling factor, for $i = 1, 2, 3, 4$. Now apply Theorem 2.3 to

Figure V.141. An upper bound to the fractal dimension of the attractor of an overlapping IFS, corresponding to this picture, can be computed with the aid of Theorem 2.3.

Figure V.142. Calculate the fractal dimension of the subset of $\mathbb{R}^2$ represented by this image.

yield two apparently different equations for the fractal dimension of A. Prove that these two equations have the same solution.

3   The Experimental Determination of the Fractal Dimension

In this section we consider the experimental determination of the fractal dimension of sets in the physical world. We model them, as best we can, as subsets of ($\mathbb{R}^2$, Euclidean) or ($\mathbb{R}^3$, Euclidean). Then, based on the definition of the fractal dimension, and sometimes in addition on one or another of the preceding theorems, such as the Box Counting Theorem, we analyze the model to provide a fractal dimension for the real-world set.

In the following examples we emphasize that when the fractal dimension of a physical set is quoted, some indication of how it was calculated must also be provided. There is not yet a broadly accepted unique way of associating a fractal dimension with a set of experimental data.

Example

3.1. There is a curious cloud of dots in the woodcut in Figure V.143. Let us try to estimate its fractal dimension by direct appeal to Definition 1.1.

We begin by covering the cloud of points by disks of radius ϵ for a range of ϵ-values from $\epsilon = 3$ cm down to $\epsilon = 0.3$ cm; and in each case we count the number of

Figure V.143. Covering a cloud of dots in a woodcut by balls of radius $\epsilon > 0$.

Table V.2. Minimal numbers of balls, of various radii, needed to cover a "dust" in a woodcut.

ϵ	$\mathcal{N}(A, \epsilon)$
3 cm	2
2 cm	3
1.5 cm	4
1.2 cm	6
1 cm	7
0.75 cm	10
0.5 cm	16
0.4 cm	23
0.3 cm	31
0.015 cm	267

Table V.3. The data in Table III.1 is tabulated in log–log form. These values are used to obtain the fractal dimension.

$\ln(1/\epsilon)$	$\ln(\mathcal{N}(A, \epsilon))$
-1.1	0.69
-0.69	1.09
-0.405	1.39
-0.182	1.79
0	1.95
0.29	2.30
0.693	2.77
0.916	3.13
1.204	3.43
4.2	5.59

disks needed. This provides the set of approximate values for $\mathcal{N}(A, \epsilon)$ given in Table III.1. The data is redisplayed in log–log format in Table V.3. The data in Table V.3 is plotted in Figure V.144. A straight line that approximately passes through the points is drawn. The slope of this straight line is our approximation to the fractal dimension of the cloud of points.

The experimental number $\mathcal{N}(A, 0.015\ \text{cm})$ is not very accurate. It is a very rough estimate based on the size of the dots themselves and is not included in the plot in Figure V.144. The slope of the straight line in Figure V.144 gives

$$D(A) \simeq 1.2, \ \text{over the range } 0.3\,\text{cm to 3\,cm,} \tag{1}$$

where A denotes the set of points whose dimension we are approximating.

The straight line in Figure V.144 was drawn "by eye." Thus if one was to repeat

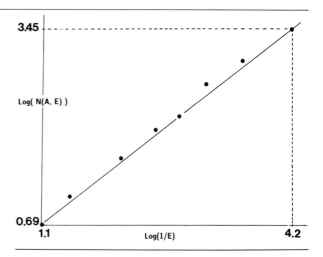

Figure V.144. Log–log plot to estimate the fractal dimension D for the cloud of dots in the woodcut in Figure V.143. The data is in Tables III.1 and V.3.

the experiment, a different value for $D(A)$ may be obtained. In order to make the results consistent from experiment to experiment, the straight line should be estimated by a least squares method.

In proceeding by direct appeal to Definition 1.1, the estimates of $\mathcal{N}(A, \epsilon)$ need to be made very carefully. One needs to be quite sure that $\mathcal{N}(A, \epsilon)$ is indeed the *least* number of balls of radius ϵ needed. For large sets of data this could be very time-consuming.

It is clearly important to state the range of scales used: we have no idea or definition concerning the structure of the dots in Figure V.143 at higher resolutions than say 0.015 cm. Moreover, regardless of how much experimental data we have, and regardless of how many scales of observation are available to us, we will always end up estimating the slope of a straight line corresponding to a finite range of scales. If we include the data point (0.015 cm, 267) in the above estimation, we obtain

$$D(A) \simeq 0.9, \text{ over the range of scales } 0.015 \text{ to } 5 \text{ cm.} \qquad (2)$$

We comment on the difference between the estimates in equations 1 and 2. If we restrict ourselves to the range of scales in equation 1, there little information present in the data to distinguish the cloud of points from a very irregular curve. However, the data used to obtain equation 2 contains values for $\mathcal{N}(A, \epsilon)$ for several values of ϵ such that the corresponding coverings of A are disconnected. The data is "aware" that A is disconnected. This lowers the experimentally determined value of D.

3.2. In this example we consider the physical set labelled A in Figure V.145. A is actually an approximation to a classical Cantor set. In this case we make an experimental estimate of the fractal dimension, based on the Box Counting Theorem. A Cartesian coordinate system is set up as shown and we attempt to count the number of square boxes $\mathcal{N}_n(A)$ of side $(1/2^n)$ which intersect A. We are able to obtain fairly accurate values of $\mathcal{N}_n(A)$ for $n = 0, 1, 2, 3, 4, 5,$ and 6. These values are pre-

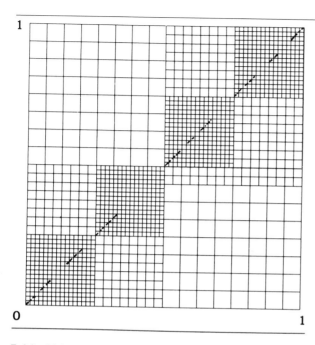

Figure V.145. Successive subdivision of overlaying grid to obtain the box counts needed for the application of Theorem 1.2 to estimate the fractal dimension of the Cantor set A. The counts are presented in Table V.4.

Table V.4. The data determined from Figure 145, in the experimental calculation of the fractal dimension of the physical set A.

n	$\mathcal{N}_n(A)$	$\ln \mathcal{N}_n(A)$	$n \ln 2$
0	1	0	0
1	3	1.10	0.69
2	7	1.95	1.38
3	10	2.30	2.08
4	19	2.94	2.77
5	33	3.50	3.46
6	58	4.06	4.16

sented in Table V.4. We note that these values depend on the choice of coordinate system. Nonetheless the values of $\mathcal{N}_n(A)$ are much easier to measure than the values of $\mathcal{N}(A, \epsilon)$ used in example 3.1.

The analysis of the data proceeds just as in example 3.1. It is represented in Table V.4 and Figure V.146. We obtain

$$D(A) \simeq 0.8, \text{ over the range } \frac{1}{8} \text{ inch to 8 inches.}$$

3.3. In this example we show how a good experimentalist [Strahle 1987] overcomes the inherent difficulties with the experimental determination of fractal dimensions. In so doing he obtains a major scientific result. The idea is to compare two sets of experimental data, obtained by different means, on the same physical system. The

Figure V.146. Slope of the plot of the data in Table V.4 gives an approximation to the fractal dimension of the set *A* in Figure V.145.

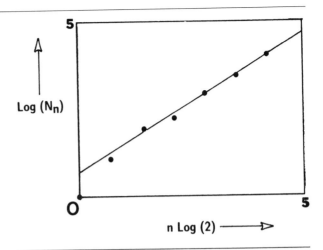

physical system is a laboratory jet flame. The data are time series for the temperature and velocity at two different points in the jet. The idea is to apply the same procedure to the analysis of the two sets of data to obtain a value for the fractal dimension. The two values are the same. Instead of drawing the conclusion that the two sets of data "have the same fractal dimension," he deduces that the two sets of data have a common source. That common source is physical, real-world chaos.

The experimental setup is as follows. A flame is probed by (a) a laser beam and (b) a very thin wire. These two probes, coupled with appropriate measuring devices, allow measurements to be made of the temperature and velocity in the jet, at two different points, as a function of time. In (a) the light bounces off the fast-moving molecules in the exhaust, and a receiver measures the characteristics of the bounced light. The output from the receiver is a voltage. This voltage, suitably rescaled, gives the temperature of the jet as a function of time. In (b) a constant temperature is maintained through a wire in the flame. The voltage required to hold the temperature constant is recorded. This voltage, suitably rescaled, gives the velocity of the jet as a function of time. In this way we obtain two independent readings of two different, but related, quantities.

Of course the experimental apparatus is much more sophisticated than it sounds from the above description. What is important is that the measuring devices are of very high resolution, accuracy, and sensitivity. A reading of the velocity can be made once every microsecond. In this example the temperature was read every 0.5×10^{-4} sec. Vast amounts of data can be obtained. A sample of the experimental output from (a) is shown in Figure V.147, where it is represented as the graph of voltage against time. It is a very complex curve. If one "magnifies up" the curve, one finds that its geometrical complexity in the curve continues to be present. It is just the sort of thing we fractal geometers like to analyze.

A sample of the experimental output from (b) is shown in Figure V.148, again represented as a graph of voltage against time. You should compare Figures V.147

RAYLEIGH SCATTERING VOLTAGE

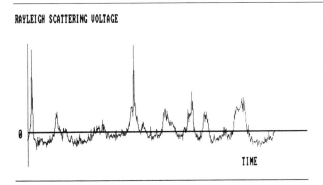

TIME

Figure V.147. Graph of voltage as a function of time from an experimental probe of a turbulent jet. In this case the probe measures scattering of a laser beam by the flame.

HOT FILM VOLTAGE

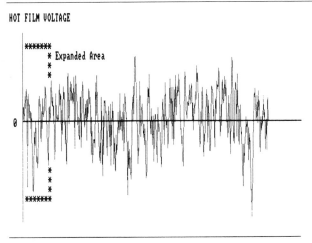

Figure V.148. Graph of voltage as a function of time from an experimental probe of a turbulent jet. In this case the probe measures the voltage across a wire in the flame. This data has a definite fractal character, as demonstrated by the expanded piece shown in Figure V.149.

and V.148. They look different. Is there a relationship between them? There should be: they both probe the same burning gas and they are in the same units.

In order to bring out the fractal character in the data, an expanded piece of the data in Figure V.148 is shown in Figure V.149.

The fractal dimensions of the graphs of the two time series, obtained from (a) and (b), is calculated using a method based on the Box Counting Theorem. Exactly the same method is applied to both sets of data, over the same range of scales. Figure V.150 shows the graphical analysis of the resulting box counts. Both experiments yield the same value

$$D \approx 1.5, \text{ over the range of scales } 2^6 \times 10^{-5} \text{ sec to } 2^{13} \times 10^{-5} \text{ sec.}$$

This suggests that, despite the different appearances of their graphs, there is a common source for the data.

We believe that this common source is chaotic dynamics of a certain special flavor and character, present in the jet flame. If so then fractal dimension provides an experimentally measurable parameter that can be used to characterize the brand of choas.

3.4. Use a method based on the direct application of Definition 1.1 to make an experimental determination of the fractal dimension of the physical set defined by the black ink in Figure V.151. Give the range of scales to which your result applies.

3.5. Use a method based on the Box Counting Theorem, as in Example 3.2, to estimate the fractal dimension of the "random dendrite" given in Figure V.152. State the range of scales over which your estimate applies. Make several complete experiments to obtain some idea of the accuracy of your result.

3.6. Make an experimental estimate of the fractal dimension of the dendrite shown in Figure V.153. Note that a grid of boxes of size (1/12)th inch by (1/12)th inch has been printed on top of the dendrite. Compare the result you obtain with the result of exercise 3.5. It is important that you follow exactly the same procedure in both experiments.

3.7. Make an experimental determination of the fractal dimension of the set in Figure V.142. Compare your result with a theoretical estimate based on Theorem 2.3, as in exercise 2.7.

3.8. Obtain maps of Great Britain of various sizes. Make an experimental determination of the fractal dimension of the coastline, over as wide a range of scales as possible.

3.9. Obtain data showing the variations of a Stock Market index, at several different time scales, for example, hourly, daily, monthly, and yearly. Make an experimental determination of the fractal dimension. Find a second economic indicator for the same system and analyze its fractal dimension. Compare the results.

Figure V.149. A blowup of a piece of the graph in Figure V.148.

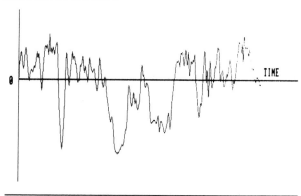

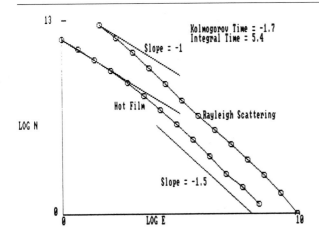

Figure V.150. Graphical analysis of box counts associated with experiments (a) and (b). The data analyzed is illustrated in Figures V.145 and V.148. The two data streams, which come from probes of a single turbulent system, when analyzed in exactly the same way, yield the same value $D = 1.5$ for the fractal dimension. This suggests that, despite the different appearances of their graphs, there is a common source for the data. This source is chaotic dynamics of a certain special flavor and character. The fractal dimension provides a measurable symptom of the brand of chaos.

4 The Hausdorff-Besicovitch Fractal Dimension

The Hausdorff-Besicovitch dimension of bounded subset of $\mathbb{R}^m$ is another real number that can be used to characterize the geometrical complexity of bounded subsets of $\mathbb{R}^m$. Its definition is more complex and subtle than that of the fractal dimension. One of the reasons for its importance is that it is associated with a method for comparing the "sizes" of sets whose fractal dimensions are the same. It is harder to work with than the fractal dimension, and its definition is not usually used as the basis of experimental procedures for the determination of fractal dimensions of physical sets.

Throughout we work in the metric space $(\mathbb{R}^m, d)$ where m is a positive integer and d denotes the Euclidean metric. Let $A \subset \mathbb{R}^m$ be bounded. Then we use the notation

$$\mathrm{diam}(A) = \sup\{d(x, y) : x, y \in A\}.$$

Let $0 < \epsilon < \infty$, and $0 \le p < \infty$. Let $\mathcal{A}$ denote the set of sequences of subsets $\{A_i \subset A\}$, such that $A = \cup_{i=1}^{\infty} A_i$. Then we define

Figure V.151. Make an experimental estimate of the fractal dimension of the set A of black ink, above, over the range of scales 5 inches to 0.1 inches. Base your experimental method directly on Definition 1.1.

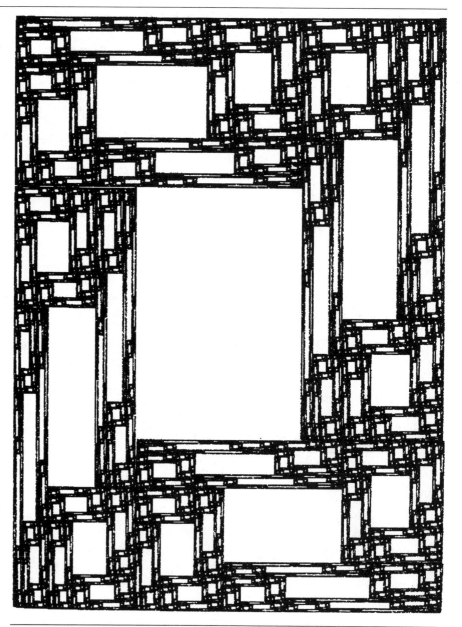

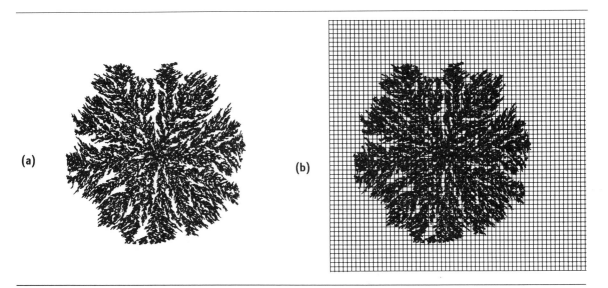

(a)

(b)

Figure V.152. Make an experimental estimate of the fractal dimension of this set in (a), over the range of scales 5 inches to (1/12)th inch, basing your method on the Box Counting Theorem and graphical analysis. In order to help you with your work, in (b) we have overlaid the set a grid of boxes (1/12)th inch by (1/12)th inch.

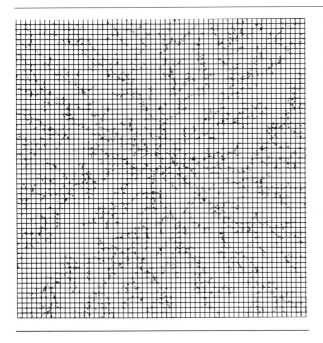

Figure V.153. Make an experimental estimate of the fractal dimension of the random dendrite shown here. Note that a grid of boxes of size (1/12)th inch by (1/12)th inch has been printed on top of the dendrite. Compare the experimental fractal dimension here with that of the dendrite in Figure V.152. In advance, which one do you expect will have lower fractal dimension?

$$\mathcal{M}(A, p, \epsilon) =$$

$$\inf \left\{ \sum_{i=1}^{\infty} (\text{diam}(A_i))^p : \{A_i\} \in \mathcal{A}, \text{ and } \text{diam}(A_i) < \epsilon \qquad \text{for } i = 1, 2, 3, \dots \right\}.$$

Here we use the convention that $(\text{diam}(A_i))^0 = 0$ when A_i is empty. $\mathcal{M}(A, p, \epsilon)$ is a number in the range $[0, \infty]$; its value may be zero, finite, or infinite. You should verify that it is a nonincreasing function of ϵ. We now define

$$\mathcal{M}(A, p) = \sup\{\mathcal{M}(A, p, \epsilon) : \epsilon > 0\}.$$

Then for each $p \in [0, \infty]$ we have $\mathcal{M}(A, p) \in [0, \infty]$.

Definition 4.1 *Let m be a positive integer and let A be a bounded subset of the metric space ($\mathbb{R}^m$, Euclidean). For each $p \in [0, \infty)$ the quantity $\mathcal{M}(A, p)$ described above is called the* Hausdorff p*-dimensional measure of A.*

Examples & Exercises

4.1. Show that $\mathcal{M}(A, p)$ is a nonincreasing function of $p \in [0, \infty]$.

4.2. Let A denote a set of seven distinct points in ($\mathbb{R}^2$, Euclidean). Show that $\mathcal{M}(A, 0) = 7$ and $\mathcal{M}(A, p) = 0$ for $p > 0$.

4.3. Let A denote a countable infinite set of distinct points in ($\mathbb{R}^2$, Euclidean). Show that $\mathcal{M}(A, 0) = \infty$ and $\mathcal{M}(A, p) = 0$ for $p > 0$

4.4. Let C denote the classical Cantor set in $[0,1]$. Show that $\mathcal{M}(C, 0) = \infty$ and $\mathcal{M}(C, 1) = 0$.

4.5. Let $\triangle$ denote a convenient Sierpinski triangle. Show that $\mathcal{M}(\triangle, 1) = \infty$ and $\mathcal{M}(\triangle, 2) = 0$. Can you evaluate $\mathcal{M}(\triangle, \ln(3)/\ln(2))$? At least try to argue why this might be an interesting number.

The Hausdorff p-dimensional measure $\mathcal{M}(A, p)$, as a function of $p \in [0, \infty]$, behaves in a remarkable manner. Its range consists of only one, two, or three values! The possible values are zero, a finite number, and infinity. In Figure V.154 we illustrate this behavior when A is a certain Sierpinski triangle.

Theorem 4.1 *Let m be a positive integer. Let A be a bounded subset of the metric space ($\mathbb{R}^m$, Euclidean). Let $\mathcal{M}(A, p)$ denote the function of $p \in [0, \infty)$ defined above. Then there is a unique real number $D_H \in [0, m]$ such that*

$$\mathcal{M}(A, p) = \begin{cases} \infty & \text{if } p < D_H \text{ and } p \in [0, \infty), \\ 0 & \text{if } p > D_H \text{ and } p \in [0, \infty). \end{cases}$$

Proof This can be found, for example, in [Federer 1969], section 2.10.3.

Definition 4.2 *Let m be a positive integer and let A be a bounded subset of the metric space ($\mathbb{R}^m$, Euclidean). The corresponding real number D_H, occurring in Theorem 4.1, is called the* Hausdorff-Besicovitch dimension *of the set A. This number will also be denoted by $D_H(A)$.*

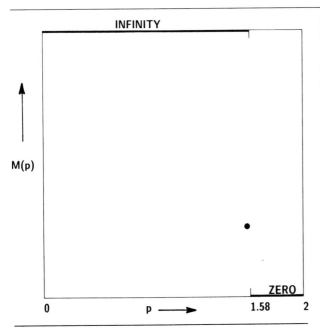

INFINITY

M(p)

ZERO

0   p ⟶   1.58   2

Figure V.154. Graph the function $\mathcal{M}(\mathbb{A}, p)$ when $\mathbb{A}$ is a certain Sierpinski triangle. It takes only three values.

Theorem 4.2 *Let m be a positive integer and let A be a subset of the metric space ($\mathbb{R}^m$, Euclidean). Let $D(A)$ denote the fractal dimension of A and let $D_H(A)$ denote the Hausdorff-Besicovitch dimension of A. Then*

$$0 \le D_H(A) \le D(A) \le m.$$

Examples & Exercises

4.6. Describe a situation where you would expect $D_H(A) < D(A)$.

4.7. Prove Theorem 4.2.

Theorem 4.3 *Let m be a positive integer. Let $\{\mathbb{R}^m; w_1, w_2, \ldots, w_N\}$ be a hyperbolic IFS, and let A denote its attractor. Let w_n be a similitude of scaling factor s_n for each $n \in \{1, 2, 3, \ldots, N\}$. If the IFS is totally disconnected or just-touching, then the Hausdorff-Besicovitch dimension $D_H(A)$ and the fractal dimension $D(A)$ are equal. In fact $D(A) = D_H(A) = D$, where D is the unique solution of*

$$\sum_{n=1}^{N} |s_n|^D = 1, \qquad D \in [0, m].$$

If D is positive, then the Hausdorff D-dimensional measure $\mathcal{M}(A, D_H(A))$ is a positive real number.

Proof This can be found in [Hutchinson 1981].

In the situation referred to in Theorem 4.3 the Hausdorff $D_H(A)$-dimensional measure can be used to compare the "sizes" of fractals that have the same fractional dimension. The larger the value of $\mathcal{M}(A, D_H(A))$, the "larger" the fractal. Of course, if two fractals have different fractal dimensions, then we say that the one with the higher fractal dimension is the "larger" one.

Examples & Exercises

4.8. Here we provide some intuition about the functions $\mathcal{M}(A, p, \epsilon)$ and $\mathcal{M}(A, p)$, and the "sizes" of fractals. We illustrate how these quantities can be estimated. The type of procedure we use can often be followed for attractors of just-touching and totally disconnected IFS whose maps are all similitudes and should lead to correct values. Formal justification is tedious and follows the lines suggested in [Hutchinson 1981].

Consider the Sierpinski triangle $\mathbb{A}$ with vertices at $(0, 0)$, $(0, 1)$, and $(1, 0)$. We work in $\mathbb{R}^2$ with the Euclidean metric. We begin by estimating the number $\mathcal{M}(\mathbb{A}, p, \epsilon)$ for $p \in [0, 1]$ for various values of ϵ. The values of ϵ we consider are $\epsilon = \sqrt{2}(1/2)^n$ for $n = 0, 1, 2, 3, \ldots$. Now notice that $\mathbb{A}$ can be covered very efficiently by 3^n closed disks of radius $\sqrt{2}(1/2)^n$. We guess that this covering is one for which the infinum in the definition of $\mathcal{M}(\mathbb{A}, p, \epsilon = \sqrt{2}(1/2)^n)$ is actually achieved. We obtain the estimate

$$\mathcal{M}(\mathbb{A}, p, \sqrt{2}(1/2)^n) = 3^n(\sqrt{2})^p(1/2)^{np} \text{ for } n = 1, 2, 3, \ldots.$$

The supremum in the definition of $\mathcal{M}(\mathbb{A}, p)$ can be replaced by a limit; so we obtain

$$\mathcal{M}(\mathbb{A}, p) = \lim_{n \to \infty} \{3^n(\sqrt{2})^p(1/2)^{np}\}$$

$$= \begin{cases} \infty & \text{if } p < \ln(3)/\ln(2), \\ (\sqrt{2})^{\ln(2)/\ln(3)} & \text{if } p = \ln(3)/\ln(2), \\ 0 & \text{if } p > \ln(3)/\ln(2). \end{cases}$$

This tells us that $D_H(\mathbb{A}) = \ln(3)/\ln(2)$, which we already know from Theorem 4.3. It also tells us that $\mathcal{M}(\mathbb{A}, D_H(\mathbb{A})) = (\sqrt{2})^{\ln(2)/\ln(3)}$. This is our estimate of the "size" of the particular Sierpinski triangle under consideration.

If one repeats the above steps for the Sierpinski triangle $\tilde{\mathbb{A}}$ with vertices at $(0, 0)$, $(0, 1/\sqrt{2})$, and $(1/\sqrt{2}, 0)$, one finds $\mathcal{M}(\tilde{\mathbb{A}}, D_H(\tilde{\mathbb{A}})) = 1$. Thus $\tilde{\mathbb{A}}$ is "smaller" than $\mathbb{A}$. Similar estimates can be made for pairs of attractors of totally disconnected or just-touching IFS whose maps are similitudes and whose fractal dimensions are equal. The comparison of "sizes" becomes exciting when the two attractors are not metrically equivalent.

4.9. Estimate the "sizes" of the two fractals represented in Figure V.155. Which one is "largest"? Does the computed estimate agree with your subjective feeling about which one is largest?

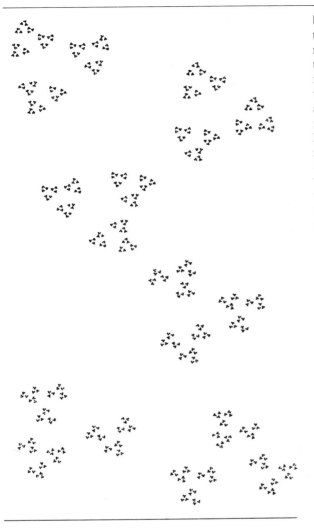

Figure V.155. The two images here represent the attractors of two different IFS of the form $\{\mathbb{R}^2; w_1, w_2, w_3\}$, where all of the maps are similitudes of scaling factor 0.4. Both sets have the same fractal dimension $\ln(3)/\ln(2.5)$. So which one is the "largest"? Compare their "sizes" by estimating their Hausdorff $\ln(3)/\ln(2.5)$-dimensional measures.

4.10. Prove that the Hausdorff-Besicovitch dimension of two metrically equivalent bounded subsets of $(\mathbb{R}^m, \text{Euclidean})$ is the same.

4.11. Let d denote a metric on $\mathbb{R}^2$ which is equivalent to the Euclidean metric. Let A denote a bounded subset of $\mathbb{R}^2$. Suppose that d is used in place of the Euclidean metric to calculate a "Hausdorff-Besicovitch" dimension of A, denoted by $\tilde{D}_H(A)$. Prove that $D_H(A) = \tilde{D}_H(A)$. Show, however, that the "size" of the set, $\mathcal{M}(A, D_H(A))$, may be different when computed using d in place of the Euclidean metric.

4.12. If distance in $\mathbb{R}^2$ is measured in inches, and a subset A of $\mathbb{R}^2$ has fractal dimension 1.391, what are the units of $\mathcal{M}(A, 1.391)$?

Figure V.156. Why are the fractal dimension and the Hausdorff-Besicovitch dimension of the attractor of the IFS represented by this image equal?

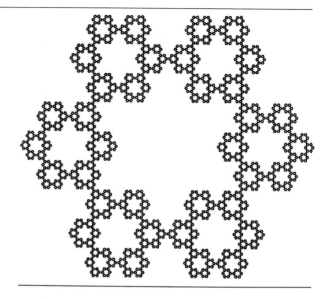

4.13. The image in Figure V.156 represents the attractor A of a certain hyperbolic IFS.

(1) Explain, with support from appropriate theorems, why the fractal dimension D and the Hausdorff-Besicovitch dimension D_H of the attractor of the IFS are equal.

(2) Evaluate D.

(3) Using inches as the unit, compare the Hausdorff-Besicovitch D-dimensional measures of A and $w(A)$, where $w(A)$ denotes one of the small "first-generation" copies of A.

4.14. By any means you like, estimate the Hausdorff-Besicovitch dimension of the coastline of Baron von Koch's Island, shown in Figure V.157. It is recommended that theoreticians try to make an experimental estimate, and that experimentalists try to make a theoretical estimate.

4.15. Does the work of some artists have a characteristic fractal dimension? Make a comparison of the empirical fractal dimensions of Romeo and Juliet, over an appropriate range of scales; see Figure V.158.

The
middle of
Baron von Koch's
Island is white to
save ink

Figure V.157. By any means you like, estimate the Hausdorff-Besicovitch dimension of the coastline of Baron von Koch's Island.

Figure V.158. Does the work of some artists have a characteristic fractal dimension? Make a comparison of the empirical fractal dimensions of Romeo and Juliet, over an appropriate range of scales.

Chapter VI

Fractal Interpolation

---

1   Introduction: Applications for Fractal Functions

Euclidean geometry, trigonometry, and calculus have taught us to think about modelling the shapes we see in the real world in terms of straight lines, circles, parabolas, and other simple curves. Consequences of this way of thinking are abundant in our everyday lives. They include the design of household objects; the common usage of drafting tables, straight-edges, and compasses; and the "applications" that accompany introductory calculus courses. We note in particular the provision of functions for drawing points, lines, polygons, and circles in computer graphics software such as MacPaint and Turbobasic. Most computer graphics hardware is designed specifically to provide rapid computation and display of classical geometrical shapes.

Euclidean geometry and elementary functions, such as sine, cosine, and polynomials, are the basis of the traditional method for analyzing experimental data. Consider an experiment that measures values of a real-valued function $F(x)$ as a function of a real variable x. For example, $F(x)$ may denote a voltage as a function of time, as in the experiments on the jet-engine exhaust described in Example 3.3 in Chapter V. The experiment may be a numerical experiment on a computer. In any case the result of the experiment will be a collection of data of the form:

$$\{(x_i, F_i) : i = 0, 1, 2, \ldots, N\}.$$

Here N is a positive integer, $F_i = F(x_i)$, and the x_i's are real numbers such that

$$x_0 < x_1 < x_2 < x_3 < \cdots < x_N.$$

The traditional method for analyzing this data begins by representing it graphically as a subset of $\mathbb{R}^2$. That is, the data points are plotted on graph paper. Next the graphical data is analyzed geometrically. For example, one may seek a straight line segment that is a good approximation to the graph of the data. Or else, one might construct a polynomial of as low degree as possible, whose graph is a good fit to the data over the interval $[x_0, x_N]$. In place of a polynomial, a linear combination of elementary functions might be used. The goal is always the same: to represent the

Figure VI.159. Illustration of the process whereby experimental data is represented graphically and modelled geometrically by means of a classical geometrical entity, such as a straight line or a polynomial fit to the data.

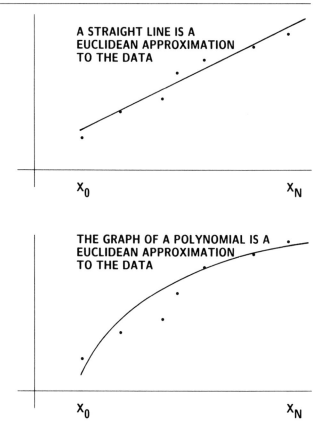

data, viewed as a subset of $\mathbb{R}^2$, by a classical geometrical entity. This entity is represented by a simple formula, one that can be communicated easily to someone else. The process is illustrated in Figure VI.159.

Elementary functions, such as trigonometric functions and rational functions, have their roots in Euclidean geometry. They share the feature that when their graphs are "magnified" sufficiently, locally they "look like" straight lines. That is, the tangent line approximation can be used effectively in the vicinity of most points. Moreover, the fractal dimension of the graphs of these functions is always 1. These elementary "Euclidean" functions are useful not only because of their geometrical content but because they can be expressed by simple formulas. We can use them to pass information easily from one person to another. They provide a common language for our scientific work. Moreover, elementary functions are used extensively in scientific computation, computer-aided design, and data analysis because they can be stored in small files and computed by fast algorithms.

Graphics systems founded on traditional geometry are effective for making pictures of man-made objects, such as bricks, wheels, roads, buildings, and cogs. This is not suprising, since these objects were designed in the first place using Euclidean geometry. However, it is desirable for graphics systems to be able to deal with a wider range of problems.

In this chapter we introduce fractal interpolation functions. The graphs of these functions can be used to approximate image components such as the profiles of mountain ranges, the tops of clouds, stalactite- hung roofs of caves, and horizons over forests, as illustrated in Figure VI.160. Rather than treating the image component as arising from a random assemblage of objects, such as individual mountains, cloudlets, stalactites, or tree tops, one models the image component as an interrelated single system. Such components are not well described by elementary functions or Euclidean graphics functions.

Fractal interpolation functions also provide a new means for fitting experimental data. Clearly it does not suffice to make a polynomial "least-squares" fit to the wild experimental data of Strahle for the temperature in a jet exhaust as a function of time, as illustrated in Figure V.147. Nor would classical geometry be a good tool for the analysis of voltages at a point in the human brain as read by an electorencephalograph. However, fractal interpolation functions can be used to "fit" such experimental data: that is, the graph of the fractal interpolation function can be made close, in the Hausdorff metric, to the data. Moreover, one can ensure that the fractal dimension of the graph of the fractal interpolation function agrees with that of the data, over an appropriate range of scales. This idea is illustrated in Figure VI.161.

Fractal interpolation functions share with elementary functions that they are of a geometrical character, that they can be represented succinctly by "formulas," and that they can be computed rapidly. The main difference is their fractal character. For example, they can have a noninteger fractal dimension. They are easy to work with— once one is accustomed to working with sets rather than points and with IFS theory using affine maps. If we start to pass them from one to another, fractal functions will become part of the common language of science. So read on!

Examples & Exercises

1.1. Write an essay on the influences of Euclidean geometry on the way in which we view the physical world. How does fractal geometry change that view?

1.2. Find the linear approximation $l(x)$ to the function $f(x) = \sin(x)$, about the point $x = 0$. Let $\epsilon > 0$. Find the linear change of coordinates $(x', y') = \theta(x, y)$ in $\mathbb{R}^2$, such that $\theta([0, \epsilon] \times [0, \epsilon]) = [0, 1] \times [0, 1]$. Let $l'(x')$ denote the function $l(x)$ represented in the new coordinate system. Let $f'(x')$ denote the function $f(x)$ in the new coordinate system. Let L denote the graph of $l'(x')$ for $x' \in [0, 1]$ and let G denote the graph of $f'(x')$ for $x' \in [0, 1]$. How small must ϵ be chosen to ensure that the Hausdorff distance from L to G is less than 0.01? The Hausdorff distance should be computed with respect to the Manhattan metric in $\mathbb{R}^2$.

Figure VI.160. The fractal interpolation functions introduced in this chapter may be used in computer graphics software packages to provide a simple means for rendering profiles of mountain ranges, the tops of clouds, and horizons over forests.

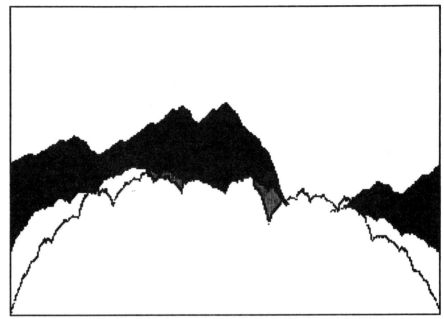

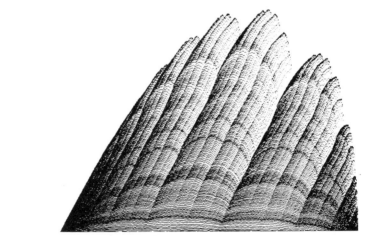

2   Fractal Interpolation Functions

Definition 2.1 *A set of* data *is a set of points of the form* $\{(x_i, F_i) \in \mathbb{R}^2 : i = 0, 1, 2, \ldots, N\}$, *where*

$$x_0 < x_1 < x_2 < x_3 < \ldots < x_N.$$

An interpolation function *corresponding to this set of data is a continuous function* $f : [x_0, x_N] \to \mathbb{R}$ *such that*

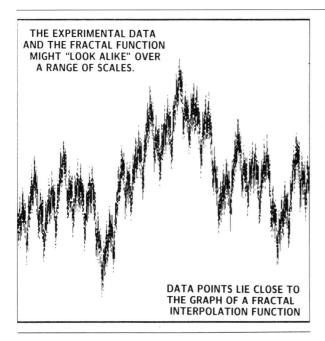

THE EXPERIMENTAL DATA
AND THE FRACTAL FUNCTION
MIGHT "LOOK ALIKE" OVER
A RANGE OF SCALES.

DATA POINTS LIE CLOSE TO
THE GRAPH OF A FRACTAL
INTERPOLATION FUNCTION

Figure VI.161. This figure illustrates the idea of using a fractal interpolation function to fit experimental data. The graph of the interpolation function may be close, in the Hausdorff metric, to the graph of the experimental data. The fractal dimension of the interpolation function may agree with that of the data over an appropriate range of scales.

$$f(x_i) = F_i \qquad \text{for } i = 1, 2, \ldots, N.$$

The points $(x_i, F_i) \in \mathbb{R}^2$ are called the interpolation points. *We say that the function f interpolates the data and that (the graph of) f passes through the interpolation points.*

Examples & Exercises

2.1. The function $f(x) = 1 + x$ is an interpolation function for the set of data

$$\{(0, 1), (1, 2)\}.$$

Consider the hyperbolic IFS $\{\mathbb{R}^2; w_1, w_2\}$, where

$$w_1\begin{pmatrix} x \\ y \end{pmatrix} = \begin{pmatrix} 0.5 & 0 \\ 0 & 0.5 \end{pmatrix}\begin{pmatrix} x \\ y \end{pmatrix} + \begin{pmatrix} 0 \\ 0.5 \end{pmatrix},$$

$$w_2\begin{pmatrix} x \\ y \end{pmatrix} = \begin{pmatrix} 0.5 & 0 \\ 0 & 0.5 \end{pmatrix}\begin{pmatrix} x \\ y \end{pmatrix} + \begin{pmatrix} 0.5 \\ 1 \end{pmatrix}.$$

Let G denote the attractor of the IFS. Then it is readily verified that G is the straight line segment that connects the pair of points $(0, 1)$ and $(1, 2)$. In other words, G is the graph of the interpolation function $f(x)$ over the interval $[0, 1]$.

2.2. Let $\{(x_i, F_i) : i = 0, 1, 2, \ldots, N\}$ denote a set of data. Let $f : [x_0, x_N] \to \mathbb{R}$ denote the unique continuous function that passes through the interpolation points and is linear on each of the subintervals $[x_{i-1}, x_i]$. That is,

Figure VI.162. Graph of the piecewise linear interpolation function $f(x)$ through the interpolation points $\{(F_i, x_i) : i = 0, 1, 2, 3, 4\}$. This graph is also the attractor of an IFS of the form $\{\mathbb{R}^2; w_n, n = 1, 2, 3, 4\}$, where the maps are affine.

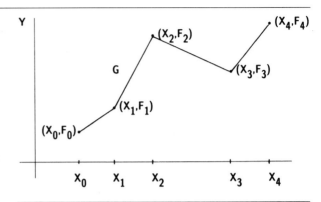

$$f(x) = F_{i-1} + \frac{(x - x_{i-1})}{(x_i - x_{i-1})}(F_i - F_{i-1}) \text{ for } x \in [x_{i-1}, x_i], i = 1, 2, \ldots, N.$$

The function $f(x)$ is called a piecewise linear interpolation function. The graph of $f(x)$ is illustrated in Figure VI.162. This graph, G, is also the attractor of an IFS of the form $\{\mathbb{R}^2; w_n, n = 1, 2, \ldots, N\}$, where the maps are affine. In fact,

$$w_n \begin{pmatrix} x \\ y \end{pmatrix} = \begin{pmatrix} a_n & 0 \\ c_n & 0 \end{pmatrix} \begin{pmatrix} x \\ y \end{pmatrix} + \begin{pmatrix} e_n \\ f_n \end{pmatrix},$$

where

$$a_n = \frac{(x_n - x_{n-1})}{(x_N - x_0)}, \quad e_n = \frac{(x_N x_{n-1} - x_0 x_n)}{(x_N - x_0)},$$

$$c_n = \frac{(F_n - F_{n-1})}{(x_N - x_0)}, \quad f_n = \frac{(x_N F_{n-1} - x_0 F_n)}{(x_N - x_0)}, \text{ for } n = 1, 2, \ldots, N.$$

Notice that the IFS may not be hyperbolic with respect to the Euclidean metric in $\mathbb{R}^2$. Can you prove that, nonetheless, G is the unique nonempty compact subset of $\mathbb{R}^2$ such that

$$G = \cup_{n=1}^{N} w_n(G)?$$

2.3. Verify the claims in exercise 2.2 in the case of the data set $\{(0, 0), (1, 3), (2, 0)\}$ by applying either the Deterministic Algorithm, Program 1, Chapter III, or the Random Iteration Algorithm, Program 2 of Chapter III. You will need to modify the programs slightly.

2.4. The parabola defined by $f(x) = 2x - x^2$ on the interval $[0, 2]$ is an interpolation function for the set of data $\{(0, 0), (1, 1), (2, 0)\}$. Let G denote the graph of $f(x)$. That is

$$G = \{(x, 2x - x^2) : x \in [0, 2]\}.$$

Then we claim that G is the attractor of the hyperbolic IFS $\{\mathbb{R}^2; w_1, w_2\}$, where

$$w_1\begin{pmatrix} x \\ y \end{pmatrix} = \begin{pmatrix} 0.5 & 0 \\ 0.5 & 0.25 \end{pmatrix}\begin{pmatrix} x \\ y \end{pmatrix},$$

$$w_2\begin{pmatrix} x \\ y \end{pmatrix} = \begin{pmatrix} 0.5 & 0 \\ -0.5 & 0.25 \end{pmatrix}\begin{pmatrix} x \\ y \end{pmatrix} + \begin{pmatrix} 1 \\ 1 \end{pmatrix}.$$

We verify this claim directly. We simply note that for all $x \in [0, 2]$,

$$w_1\begin{pmatrix} x \\ f(x) \end{pmatrix} = \begin{pmatrix} \frac{1}{2}x \\ 2(\frac{1}{2}x) - (\frac{1}{2}x)^2 \end{pmatrix} = \begin{pmatrix} \frac{1}{2}x \\ f(\frac{1}{2}x) \end{pmatrix},$$

$$w_2\begin{pmatrix} x \\ f(x) \end{pmatrix} = \begin{pmatrix} 1 + \frac{1}{2}x \\ 2(1 + \frac{1}{2}x) - (1 + \frac{1}{2}x)^2 \end{pmatrix} = \begin{pmatrix} 1 + \frac{1}{2}x \\ f(1 + \frac{1}{2}x) \end{pmatrix}.$$

As x varies over $[0, 2]$, the right-hand side of the first equation yields the part of the graph of $f(x)$ lying over the interval $[0, 1]$, while the right-hand side of the second equation yields the part of the graph of $f(x)$ lying over the interval $[1, 2]$. Hence $G = w_1(G) \cup w_2(G)$. Since $G \in \mathcal{H}(\mathbb{R}^2)$ we conclude that it is the attractor of the IFS. Notice that the IFS is just-touching.

2.5. Find a hyperbolic IFS of the form $\{\mathbb{R}^2; w_1, w_2\}$, where w_1 and w_2 are affine transformations in $\mathbb{R}^2$, whose attractor is the graph of the quadratic function that interpolates the data $\{(0, 0), (1, 1), (2, 4)\}$.

Let a set of data $\{(x_i, F_i) : i = 0, 1, 2, \ldots, N\}$ be given. We explain how one can construct an IFS in $\mathbb{R}^2$ such that its attractor, which we denote by G, is the graph of a continuous function $f : [x_0, x_N] \to \mathbb{R}$, which interpolates the data. Throughout we will restrict our attention to affine transformations. The usage of more general transformations is discussed in [Barnsley 1988f].

We consider an IFS of the form $\{\mathbb{R}^2; w_n, n = 1, 2, \ldots, N\}$, where the maps are affine transformations of the special structure

$$w_n\begin{pmatrix} x \\ y \end{pmatrix} = \begin{pmatrix} a_n & 0 \\ c_n & d_n \end{pmatrix}\begin{pmatrix} x \\ y \end{pmatrix} + \begin{pmatrix} e_n \\ f_n \end{pmatrix}.$$

The transformations are constrained by the data according to

$$w_n\begin{pmatrix} x_0 \\ F_0 \end{pmatrix} = \begin{pmatrix} x_{n-1} \\ F_{n-1} \end{pmatrix} \quad \text{and} \quad w_n\begin{pmatrix} x_N \\ F_N \end{pmatrix} = \begin{pmatrix} x_n \\ F_n \end{pmatrix} \quad \text{for } n = 1, 2, \ldots, N.$$

The situation is summarized in Figure VI.163.

Let $n \in \{1, 2, 3, \ldots, N\}$. The transformation w_n is specified by the five real numbers $a_n, b_n, c_n, d_n,$ and e_n, which must obey the four linear equations

$$a_n x_0 + e_n = x_{n-1},$$
$$a_n x_N + e_n = x_n,$$
$$c_n x_0 + d_n F_0 + f_n = F_{n-1},$$
$$c_n x_N + d_n F_N + f_n = F_n.$$

Figure VI.163. Two illustrations showing how an IFS of shear transformations is used to construct a fractal interpolation function. (Produced by Peter Massopust.)

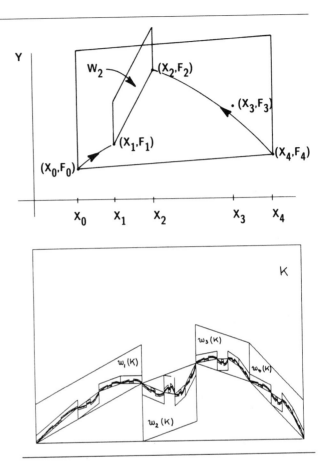

It follows that there is effectively one free parameter in each transformation. We choose this parameter to be d_n for the following reason. The transformation w_n is a *shear* transformation: it maps lines parallel to the y-axis into lines parallel to the y-axis. Let L denote a line segment parallel to the y-axis. Then $w_n(L)$ is also a line segment parallel to the y-axis. The ratio of the length of $w_n(L)$ to the length of L is $|d_n|$. We call d_n the *vertical scaling factor* in the transformation w_n. By choosing d_n to be the free parameter, we are able to specify the vertical scaling produced by the transformation. With $d_n = 0$, $n = 1, 2, \ldots, N$, one recovers the piecewise linear interpolation function. In section 6.2 we will show that these parameters determine the fractal dimension of the attractor of the IFS.

Let d_n be any real number. We demonstrate that we can always solve the above equations for a_n, c_n, e_n, and f_n in terms of the data and d_n. We find

$$a_n = \frac{(x_n - x_{n-1})}{(x_N - x_0)}, \tag{1}$$

$$e_n = \frac{(x_N x_{n-1} - x_0 x_n)}{(x_N - x_0)}, \tag{2}$$

$$c_n = \frac{(F_n - F_{n-1})}{(x_N - x_0)} - \frac{d_n(F_N - F_0)}{(x_N - x_0)}, \tag{3}$$

$$f_n = \frac{(x_N F_{n-1} - x_0 F_n)}{(x_N - x_0)} - d_n \frac{(x_N F_0 - x_0 F_N)}{(x_N - x_0)}. \tag{4}$$

Now let $\{\mathbb{R}^2; w_n, n = 1, 2, \ldots, N\}$ denote the IFS defined above. Let the vertical scaling factor d_n obey $0 \le d_n < 1$ for $n = 1, 2, \ldots, N$. Even with this condition, the IFS is not in general hyperbolic on the metric space ($\mathbb{R}^2$,Euclidean). Despite this, let us see what happens if we apply the Random Iteration Algorithm to the IFS.

Here we present Program 2 in Chapter III modified so that the input data consists of the interpolation points and vertical scaling factors. It is written for $N = 3$ and the data set

$$\{(0, 0), (30, 50), (60, 40), (100, 10)\}.$$

The vertical scaling factors are input by the user during execution of the code. The program calculates the coefficients in the shear transformations from the data, and then applies the Random Iteration Algorithm to the resulting IFS. The program is written in BASIC. It runs without modification on an IBM PC with enhanced graphics adaptor and Turbobasic. On any line the words preceded by a ' are comments: they are not part of the program.

Program 1.

```
x[0] =0 : x[1] =30 : x[2] =60 : x[3] =100 ' Data set

F[0] =0 : F[1] =50 : F[2] =40 : F[3] =10

'Vertical Scaling Factors
input "enter scaling factors d(1), d(2), and d(3)",
     d(1),d(2),d(3)

'Calculate the shear transformations from the Data
'and Vertical Scaling Factors
for n =1 to 3

b = x[3]-x[0] : a[n] = (x[n]-x[n-1])/b
```

```
e[n]  =  (x[3]*x[n-1]-x[0]*x[n])/b

c[n]  =  (F[n]-F[n-1]-d[n]*(F[3]-F[0]))/b

ff[n]  =  (x[3]*F[n-1]-x[0]*F[n]-d[n]*(x[3]*F[0]-x[0]*F[3]))/b

next

screen 2 : cls ' initalialize graphics

window(0,0)-(100,100) 'change this to zoom and/or pan

x =0: y =0 'initial point from which the random
          'iteration begins

for n =1 to 1000 'Random Iteration Algorithm

k =int(3*rnd-0.0001)+1

newx =   a[k]*x + e[k]

newy =   c[k]*x + d[k]*y + ff[k]

x =newx : y =newy

pset(x,y) 'plot the most recently computed point on the screen

next

end
```

The result of running an adaptation of this program on a Masscomp workstation and then printing the contents of the graphics screen is presented in Figure VI.164. In this case $d_1 = 0.5$, $d_2 = -0.5$, and $d_3 = 0.23$. Notice that if the size of the plotting window is decreased, for example by replacing the window call by WINDOW (0,0)– (50,50), then a portion of the image is plotted at a higher resolution. The number of iterations can be increased to impove the quality of the computed image.

Examples & Exercises

2.6. Rewrite Program 1 in a form suitable for your own computer environment, then run it and obtain hardcopy of the output.

2.7. Vary the data used by Program 1. Verify, by means of computergraphical experiments, that the corresponding IFS always seems to have a unique attractor, provided that the vertical scaling factors are less than 1 in norm. Verify that, provided suffi-

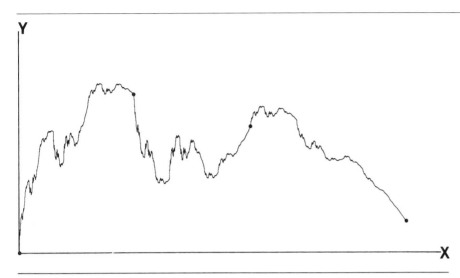

Figure VI.164. The result of running Program 1 with vertical scaling factors 0.5, −0.5, and 0.23. It appears that the corresponding IFS possesses a unique attractor that is the graph of a function that passes through the interpolation points {(0, 0), (30, 50), (60, 40), (100, 10)}. Is there a metric such that the IFS is hyperbolic?

ciently many points are plotted, the attractor always contains the data points, and that it looks like the graph of a function.

2.8. Show that the shear transformations w_n, descibed above, need not be contractions in the Euclidean metric, even though the magnitudes of the vertical scaling factors are less than 1. Once you have found such an example, use Program 1, suitably modified, to obtain graphical evidence concerning the possible existence of an attractor. You are supposed to discover that even though the IFS is not hyperbolic in the Euclidean metric, it appears to possess an attractor.

2.9. Use Program 1 to verify that the attractor of the IFS in exercise 2.4 is a parabola.

We now give the theoretical basis for our experimental observations.

Theorem 2.1 *Let N be a positive integer greater than 1. Let $\{\mathbb{R}^2; w_n, n = 1, 2, \ldots, N\}$ denote the IFS defined above, associated with the data set*

$$\{(x_n, F_n) : n = 1, 2, \ldots, N\}.$$

Let the vertical scaling factor d_n obey $0 \leq d_n < 1$ for $n = 1, 2, \ldots, N$. Then there is a metric d on $\mathbb{R}^2$, equivalent to the Euclidean metric, such that the IFS is hyperbolic with respect to d. In particular, there is a unique nonempty compact set $G \subset \mathbb{R}^2$ such that

$$G = \cup_{n=1}^{N} w_n(G).$$

Proof We define a metric d on $\mathbb{R}^2$ by

$$d((x_1, y_1), (x_2, y_2)) = |x_1 - x_2| + \theta|y_1 - y_2|,$$

where θ is a positive real number, which we specify below. We leave it as an exercise to the reader to prove that this metric is equivalent to the Euclidean metric on $\mathbb{R}^2$. Let $n \in \{1, 2, \ldots, N\}$. Let the numbers a_n, c_n, e_n, f_n, be defined by equations 1, 2, 3, and 4. Then we have

$$
\begin{aligned}
d(w_n&(x_1, y_1), w_n(x_2, y_2)) \\
&= d((a_n x_1 + e_n, c_n x_1 + d_n y_1 + f_n), \qquad (a_n x_2 + e_n, c_n x_2 + d_n y_2 + f_n)) \\
&= a_n |x_1 - x_2| + \theta |c_n(x_1 - x_2) + d_n(y_1 - y_2)| \\
&\leq (|a_n| + \theta |c_n|)|x_1 - x_2| + \theta |d_n||y_1 - y_2|.
\end{aligned}
$$

Now notice that $|a_n| = |x_n - x_{n-1}|/|x_N - x_0| < 1$ because $N \geq 2$. If $c_1 = c_2 \cdots = c_n = 0$, then we choose $\theta = 1$. Otherwise we choose

$$
\theta = \frac{\min\{(2 - |a_n|) : n = 1, 2, \ldots, N\}}{\max\{2|c_n| : n = 1, 2, \ldots, N\}}.
$$

Then it follows that

$$
\begin{aligned}
d(w_n(x_1, y_1), w_n(x_2, y_2)) &\leq (|a_n| + \theta |c_n|)|x_1 - x_2| + \theta |d_n||y_1 - y_2| \\
&\leq a|x_1 - x_2| + \theta \delta |y_1 - y_2| \\
&\leq \max\{a, \delta\} d((x_1, y_1), (x_2, y_2)),
\end{aligned}
$$

where

$$
a = (1 + a_n - \frac{\max\{|a_n| : n = 1, 2, \ldots, N\}}{2}) < 1,
$$

$$
\delta = \max\{|d_n| : n = 1, 2, \ldots, N\} < 1.
$$

This completes the proof.

Theorem 2.2 *Let N be a positive integer greater than 1. Let $\{\mathbb{R}^2; w_n, n = 1, 2, \ldots, N\}$ denote the IFS defined above, associated with the data set $\{(x_n, F_n) : n = 1, 2, \ldots, N\}$. Let the vertical scaling factor d_n obey $0 \leq d_n < 1$ for $n = 1, 2, \ldots, N$, so that the IFS is hyperbolic. Let G denote the attractor of the IFS. Then G is the graph of a continuous function $f : [x_0, x_N] \to \mathbb{R}$, which interpolates the data $\{(x_i, F_i) : i = 1, 2, \ldots, N\}$. That is,*

$$
G = \{(x, f(x)) : x \in [x_0, x_N]\},
$$

where

$$
f(x_i) = F_i \qquad \text{for } i = 0, 1, 2, 3, \ldots, N.
$$

Proof Let $\mathcal{F}$ denote the set of continuous functions $f : [x_0, x_1] \to \mathbb{R}$ such that $f(x_0) = F_0$ and $f(x_N) = F_N$. We define a metric d on $\mathcal{F}$ by

$$
d(f, g) = \max\{|f(x) - g(x)| : x \in [x_0, x_N]\} \qquad \text{for all } f, g \text{ in } \mathcal{F}.
$$

Then $(\mathcal{F}, d)$ is a complete metric space—see, for example [Rudin 1966], or prove it yourself.

Let the real numbers $a_n, c_n, e_n, f_n,$ be defined by equations 1, 2, 3, and 4. Define a mapping $T : \mathcal{F} \to \mathcal{F}$ by

$$(Tf)(x) = c_n l_n^{-1}(x) + d_n f(l_n^{-1}(x)) + f_n \text{ for } x \in [x_{n-1}, x_n], \quad \text{for } n = 1, 2, \ldots, N,$$

where $l_n : [x_0, x_N] \to [x_{n-1}, x_n]$ is the invertible transformation

$$l_n(x) = a_n x + e_n.$$

We verify that T does indeed take $\mathcal{F}$ into itself. Let $f \in \mathcal{F}$. Then the function $(Tf)(x)$ obeys the endpoint conditions because

$$\begin{aligned}
(Tf)(x_0) &= c_1 l_1^{-1}(x_0) + d_1 f(l_1^{-1}(x_0)) + f_n \\
&= c_1 x_0 + d_1 f(x_0) + f_n = c_1 x_0 + d_1 F_0 + f_n \\
&= F_0
\end{aligned}$$

and

$$\begin{aligned}
(Tf)(x_N) &= c_N l_N^{-1}(x_N) + d_N f(l_N^{-1}(x_N)) + f_N \\
&= c_N x_N + d_N f(x_N) + f_N \\
&= c_N x_N + d_N F_N.
\end{aligned}$$

The reader can prove that $(Tf)(x)$ is continuous on the interval $[x_{n-1}, x_n]$ for $n = 1, 2, \ldots, N$. Then it remains to demonstrate that $(Tf)(x)$ is continuous at each of the points $x_1, x_2, x_3, \ldots, x_{N-1}$. At each of these points the value of $(Tf)(x)$ is apparently defined in two different ways. For $n \in \{1, 2, \ldots, N - 1\}$ we have

$$\begin{aligned}
(Tf)(x_n) &= c_{n+1} l_{n+1}^{-1}(x_n) + d_{n+1} f(l_{n+1}^{-1}(x_n)) + f_{n+1} \\
&= c_{n+1} x_0 + d_{n+1} f(x_0) + f_{n+1} = F_n
\end{aligned}$$

and also

$$\begin{aligned}
(Tf)(x_n) &= c_n l_n^{-1}(x_n) + d_n f(l_n^{-1}(x_n)) + f_n \\
&= c_n x_N + d_n f(x_N) + f_n = F_n,
\end{aligned}$$

so both methods of evaluation lead to the same result. We conclude that T does indeed take $\mathcal{F}$ into $\mathcal{F}$.

Now we show that T is a contraction mapping on the metric space $(\mathcal{F}, d)$. Let $f, g \in \mathcal{F}$. Let $n \in \{1, 2, \ldots, N\}$ and $x \in [x_{n-1}, x_n]$. Then

$$|(Tf)(x) - (Tg)(x)| = |d_n| |f(l_n^{-1}(x)) - g(l_n^{-1}(x))| \le |d_n| d(f, g).$$

It follows that

$$d(Tf, Tg) \le \delta d(f, g) \quad \text{where } \delta = \max\{|d_n| : n = 1, 2, \ldots, N\} < 1.$$

We conclude that $T : \mathcal{F} \to \mathcal{F}$ is a contraction mapping. The Contraction Mapping Theorem implies that T possesses a unique fixed point in $\mathcal{F}$. That is, there exists a function $f \in \mathcal{F}$ such that

$$(Tf)(x) = f(x) \qquad \text{for all } x \in [x_0, x_N].$$

Figure VI.165. A sequence of functions $\{f_{n+1}(x) = (Tf_n)(x)\}$ converging to the fixed point of the mapping $T : \mathcal{F} \to \mathcal{F}$ used in the proof of Theorem 2.2. This is another example of a contraction mapping doing its work.

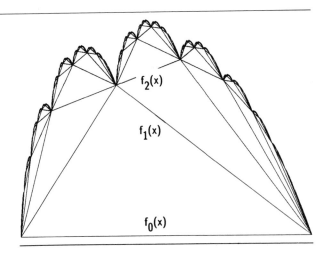

The reader should convince himself that f passes through the interpolation points.

Let $\tilde{G}$ denote the graph of f. Notice that the equations that define T can be rewritten

$$(Tf)(a_n x + e_n) = c_n x + d_n f(x) + f_n \text{ for } x \in [x_0, x_N], \text{ for } n = 1, 2, \ldots, N,$$

which implies that

$$\tilde{G} = \cup_{n=1}^N w_n(\tilde{G}).$$

But $\tilde{G}$ is a nonempty compact subset of $\mathbb{R}^2$. By Theorem 2.1 there is only one nonempty compact set G, the attractor of the IFS, which obeys the latter equation. It follows that $G = \tilde{G}$. This completes the proof.

Definition 2.2 *The function $f(x)$ whose graph is the attractor of an IFS as described in Theorems 2.1 and 2.2, above, is called a* fractal interpolation function *corresponding to the data $\{(x_i, F_i) : i = 1, 2, \ldots, N\}$.*

Figure VI.165 shows an example of a sequence of iterates $\{T^{\circ n} f_0 : n = 0, 1, 2, 3, \ldots\}$ obtained by repeated application of the contraction mapping T, introduced in the proof of Theorem 2.2. The initial function $f_0(x)$ is linear. The sequence converges to the fractal interpolation function f, which is the fixed point of T. Notice that the whole image can be interpreted as the attractor of an IFS with condensation, where the condensation set is the graph of the function $f_0(x)$.

The reader may wonder, in view of the proof of Theorem 2.2, why we go to the trouble of establishing that there is a metric such that the IFS is contractive. After all, we could simply use T to construct fractal interpolation functions. The answer has two parts, (a) and (b). (a) We can now apply the theory of hyperbolic IFS to fractal interpolation functions. Of special importance, this means that we can use IFS algorithms to compute fractal interpolation functions, that the Collage Theorem can be used as an aid to finding fractal interpolation functions that approximate given data, and that we can use the Hausdorff metric to discuss the accuracy of approximation of experimental data by a fractal interpolation function. (b) By treating fractal interpolation functions as attractors of IFS of affine transformations we provide a common language for the description of an important class of functions and sets: the same type of formula, namely an IFS code, can be used in all cases.

One consequence of the fact that the IFS $\{\mathbb{R}; w_n, n = 1, 2, \ldots, N\}$ associated with a set of data $\{(x_n, F_n) : n = 1, 2, \ldots, N\}$ is hyperbolic is that any set $A_0 \in \mathcal{H}(\mathbb{R}^2)$ leads to a Cauchy sequence of sets $\{A_n\}$ that converges to G in the Hausdorff metric. In the usual way we define $W : \mathcal{H}(\mathbb{R}^2) \to \mathcal{H}(\mathbb{R}^2)$ by

$$W(B) = \cup_{n=1}^{N} w_n(B) \qquad \text{for all } B \in \mathcal{H}(\mathbb{R}^2).$$

Then $\{A_n = W^{\circ n}(A_0)\}$ is a Cauchy sequence of sets which converges to G in the Hausdorff metric. This idea is illustrated in Figure VI.166. Notice that if A_0 is the graph of a function $f_0 \in \mathcal{F}$ then A_n is the graph of $T^{\circ n} f_0$.

Examples & Exercises

2.10. Prove that the metric on $\mathbb{R}^2$ introduced in the proof of Theorem 2.1 is equivalent to the Euclidean metric on $\mathbb{R}^2$.

2.11. Use the Collage Theorem to help you find a fractal interpolation function that approximates the function whose graph is shown in Figure VI.167.

2.12. Write a program that allows you to use the Deterministic Algorithm to compute fractal interpolation functions.

2.13. Explain why Theorems 2.1 and 2.2 have the restriction that N is greater than 1.

2.14. Let a set of data $\{(x_i, F_i) : i = 0, 1, 2, \ldots, N\}$ be given. Let the metric space $(\mathcal{F}, d)$ and the transformation $T : \mathcal{F} \to \mathcal{F}$ be defined as in the proof of Theorem 2.2. Prove that if $f \in \mathcal{F}$ then Tf is an interpolation function associated with the data. Deduce that if $f \in \mathcal{F}$ is a fixed point of T then f is an interpolation function associated with the data.

2.15. Make a nonlinear generalization of the theory of fractal interpolation functions. For example, consider what happens if one uses an IFS made up of nonlinear transformations $w_n : \mathbb{R}^2 \to \mathbb{R}^2$ of the form

$$w_n(x, y) = (a_n x + e_n, c_n x + d_n y + g_n y^2 + f_n),$$

where $a_n, c_n, d_n, g_n,$ and f_n are real constants. This example uses "quadratic scaling" in the vertical direction instead of linear scaling. Determine sufficient conditions for the IFS to be hyperbolic, with an attractor that is the graph of a function that interpolates the data $\{(x_i, F_i) : i = 0, 1, 2, \ldots, N\}$. Note that in certain circumstances the IFS generates the graph of a differentiable interpolation function.

2.16. Let $f(x)$ denote a fractal interpolation function associated with a set of data

Figure VI.166. Examples of the convergence of a sequence of sets $\{A_n\}$ in the Hausdorff metric, to the graph of a fractal interpolation function.

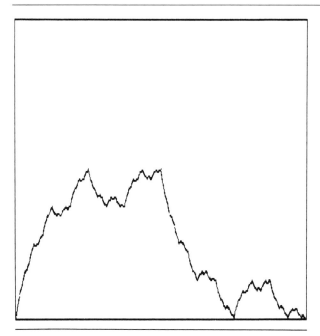

$\{(x_i, F_i) : i = 0, 1, 2, \ldots, N\}$, where $N > 1$. Let the metric space $(\mathcal{F}, d)$ and the transformation $T : \mathcal{F} \to \mathcal{F}$ be defined as in the proof of Theorem 2.2. The functional equation $Tf = f$ can be used to evaluate various integrals of f. As an example consider the problem of evaluating the integral

$$I = \int_{x_0}^{x_N} f(x)\, dx.$$

The integral is well defined because $f(x)$ is continuous. We have

$$I = \int_{x_0}^{x_N} (Tf)(x)\, dx = \sum_{n=1}^{N} \int_{x_{n-1}}^{x_n} (Tf)(x)\, dx$$

$$= \sum_{n=1}^{N} \int_{x_0}^{x_N} (c_n x + d_n f(x) + f_n) d(a_n x + e_n) = \alpha I + \beta,$$

where

$$\alpha = \left(\sum_{n=1}^{N} a_n d_n\right) \quad \text{and} \quad \beta = \sum_{n=1}^{N} a_n \int_{x_0}^{x_N} (c_n x + f_n)\, dx.$$

Show that, under the standard assumptions, $|\alpha| < 1$. Show also that

$$\beta = \int_{x_0}^{x_N} f_0(x)\, dx,$$

Figure VI.168. Illustration of the geometrical viewpoint concerning the integration of fractal interpolation functions.

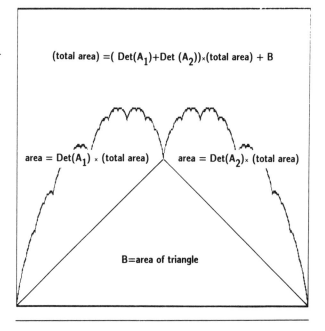

(total area) =(Det(A$_1$)+Det (A$_2$))×(total area) + B

area = Det(A$_1$) × (total area)   area = Det(A$_2$)× (total area)

B=area of triangle

where $f_0(x)$ is the piecewise linear interpolation function associated with the data. Conclude that

$$\int_{x_0}^{x_N} f(x)\,dx = \frac{\beta}{(1-\alpha)}.$$

Check this result for the case of the parabola, described in Exercise 2.4. In Figure VI.168 we illustrate a geometrical way of thinking about the integration of a fractal interpolation function.

2.17. Let $f(x)$ denote a fractal interpolation function associated with a set of data $\{(x_i, F_i) : i = 0, 1, 2, \ldots, N\}$, where $N > 1$. By following similar steps to those in exercise 2.16, find a formula for the integral

$$I_1 = \int_{x_0}^{x_N} xf(x)\,dx.$$

Check your formula by applying it to the parabola described in exercise 2.4.

2.18. Figure VI.169 shows a fractal interpolation function together with a zoom. Can you reproduce these images and then make a further zoom? What do you expect a very high-magnification zoom to look like?

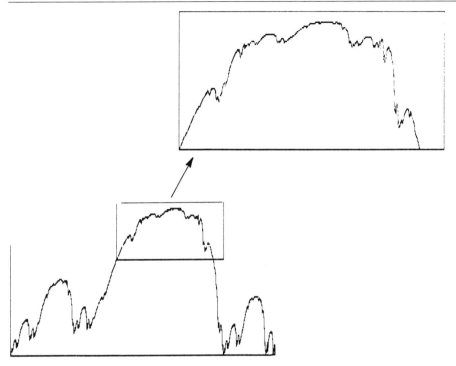

Figure VI.169. A fractal interpolation function together with a zoom. If the fractal dimension is equal to 1, what do you expect "most" very high magnification zooms to look like?

3   The Fractal Dimension of Fractal Interpolation Functions

The following excellent theorem tells us the fractal dimension of fractal interpolation functions.

Theorem 3.1   *Let N be a positive integer greater than 1. Let $\{(x_n, F_n) \in \mathbb{R}^2 : n = 1, 2, \ldots, N\}$ be a set of data. Let $\{\mathbb{R}^2; w_n, n = 1, 2, \ldots, N\}$ be an IFS associated with the data, where*

$$w_n \begin{pmatrix} x \\ y \end{pmatrix} = \begin{pmatrix} a_n & 0 \\ c_n & d_n \end{pmatrix} \begin{pmatrix} x \\ y \end{pmatrix} + \begin{pmatrix} e_n \\ f_n \end{pmatrix} \qquad \text{for } n = 1, 2, \ldots, N.$$

The vertical scaling factors d_n obey $0 \leq d_n < 1$; and the constants $a_n, c_n, e_n,$ and f_n are given by equations 1, 2, 3, and 4, for $n = 1, 2, \ldots, N$. Let G denote the attractor of the IFS, so that G is the graph of a fractal interpolation function associated with the data. If

$$\sum_{n=1}^{N} |d_n| > 1 \tag{1}$$

and the interpolation points do not all lie on a single straight line, then the fractal dimension of G is the unique real solution D of

Figure VI.170. The graph G of a fractal interpolation function is superimposed on a grid of closed square boxes of side length ϵ. $\mathcal{N}(\epsilon)$ is used to denote the number of boxes that intersect G. What is the value of $\mathcal{N}(\epsilon)$?

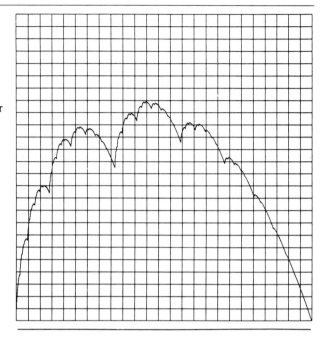

$$\sum_{n=1}^{N} |d_n| a_n^{D-1} = 1.$$

Otherwise the fractal dimension of G is 1.

Proof (Informal Demonstration). The formal proof of this theorem can be found in [Barnsley 1988f]. Here we give an informal argument for why it is true. We use the notation in the statement of the theorem.

Let $\epsilon > 0$. We consider G to be superimposed on a grid of closed square boxes of side length ϵ, as illustrated in Figure VI.170. Let $\mathcal{N}(\epsilon)$ denote the number of square boxes of side length ϵ which intersect G. These boxes are similar to the ones used in the Box Counting Theorem, Theorem 1.2 in Chapter V, except that their sizes are arbitrary. On the basis of the intuitive idea introduced in Chapter V, section 1, we suppose that G has fractal dimension D, where

$$\mathcal{N}(\epsilon) \approx \text{constant} \cdot \epsilon^{-D} \text{ as } \epsilon \to 0.$$

We want to estimate the value of D on the basis of this assumption.

Let $n \in \{1, 2, \ldots, N\}$. Let $\mathcal{N}_n(\epsilon)$ denote the number of boxes of side length ϵ which intersect $w_n(G)$ for $n = 1, 2, \ldots, N$. We suppose that ϵ is very small compared to $|x_N - x_0|$. Then because the IFS is just-touching it is reasonable to make the approximation

$$\mathcal{N}(\epsilon) \approx \mathcal{N}_1(\epsilon) + \mathcal{N}_2(\epsilon) + \mathcal{N}_3(\epsilon) + \cdots + \mathcal{N}_N(\epsilon). \tag{2}$$

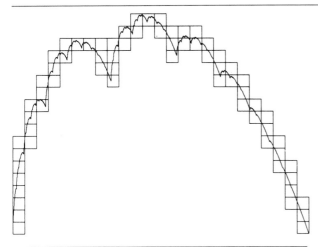

Figure VI.171. The boxes that intersect G can be thought of as organized in columns. The set of columns of boxes of side length ϵ which intersect G is denoted by $\{c_j(\epsilon) : j = 1, 2, \ldots, \mathcal{K}(\epsilon)\}$, where $\mathcal{K}(\epsilon)$ denotes the number of columns. What is the value of $\mathcal{K}(\epsilon)$ and how many boxes are there in $c_2(\epsilon)$, in this illustration?

We now look for a relationship between $\mathcal{N}(\epsilon)$ and $\mathcal{N}_n(\epsilon)$. The boxes that intersect G can be thought of as being organized into columns, as illustrated in Figure VI.171.

Let the set of columns of boxes of side length ϵ which intersect G be denoted by $\{c_j(\epsilon) : j = 1, 2, \ldots, \mathcal{K}(\epsilon)\}$, where $\mathcal{K}(\epsilon)$ denotes the number of columns. Under the conditions in equation 1, in the statement of the theorem, one can prove that the minimum number of boxes in a column increases without limit as ϵ approaches zero. To simplify the discussion we assume that

$$|d_n| > a_n \text{ for } n = 1, 2, \ldots, N.$$

(Notice that

$$\sum_{n=1}^{N} a_n = \sum_{n=1}^{N} \frac{(x_n - x_{n-1})}{(x_N - x_0)} = 1,$$

which tells us that this assumption is stronger than the assumption $\sum_{n=1}^{N} |d_n| > 1$.)

Then consider what happens to a column ofboxes $c_j(\epsilon)$ of side length ϵ when we apply the affine transformation w_n to it. It becomes a column of parallelograms. The width of the column is $a_n\epsilon$ and the height of the column is $|d_n|$ times the height of the column before transformation. Let $\mathcal{N}(c_j(\epsilon))$ denote the number of boxes in the column $c_j(\epsilon)$. Then the column $w_n(c_j(\epsilon))$ can be thought of as being made up of square boxes of side length $a_n\epsilon$, each of which intersects $w_n(G)$. How many boxes of side length $a_n\epsilon$ are there in this column? Approximately $|d_n|\mathcal{N}(c(\epsilon))/a_n$. Adding up the contribution to $\mathcal{N}_n(a_n\epsilon)$ from each column we obtain

$$\mathcal{N}_n(a_n\epsilon) \approx \sum_{j=1}^{\mathcal{K}(\epsilon)} \frac{|d_n|\mathcal{N}(c_j(\epsilon))}{a_n} = \frac{|d_n|}{a_n} \sum_{j=1}^{\mathcal{K}(\epsilon)} \mathcal{N}(c_j(\epsilon)) = \frac{|d_n|}{a_n}\mathcal{N}(\epsilon).$$

The situation is illustrated in Figure VI.172.

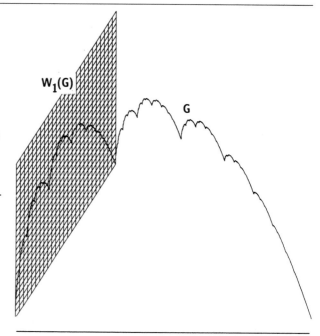

Figure VI.172. When the shear transformation w_1 is applied to the columns of boxes which cover the graph, G, the result is a set of thinner columns, of width $a_1\epsilon$, which cover $w_1(G)$. The new columns are made up of small parallelograms, but the number of square boxes of side length $a_1\epsilon$ which they contain is readily estimated.

From the last equation we deduce that when ϵ is very small compared to $[x_0, x_N]$,

$$\mathcal{N}_n(\epsilon) \approx \frac{|d_n|}{a_n}\mathcal{N}(\frac{\epsilon}{a_n}) \text{ for } n = 1, 2, \ldots, N. \tag{3}$$

We now substitute from equation 3 into 2 to obtain the functional equation

$$\mathcal{N}(\epsilon) \approx \frac{d_1}{a_1}\mathcal{N}(\frac{\epsilon}{a_1}) + \frac{d_2}{a_2}\mathcal{N}(\frac{\epsilon}{a_2}) + \frac{d_3}{a_3}\mathcal{N}(\frac{\epsilon}{a_3}) + \cdots + \frac{d_N}{a_N}\mathcal{N}(\frac{\epsilon}{a_N}).$$

Into this equation we substitute our assumption $\mathcal{N}(\epsilon) \approx$ constant $\cdot\epsilon^{-D}$ to obtain the equation

$$\epsilon^{-D} \approx |d_1|a_1{}^{D-1}\epsilon^{-D} + |d_2|a_2{}^{D-1}\epsilon^{-D} + |d_3|a_3{}^{D-1}\epsilon^{-D} + \cdots + |d_N|a_N{}^{D-1}\epsilon^{-D}.$$

The main formula in the statement of the theorem follows at once.

If the interpolation points are collinear, then the attractor of the IFS is the line segment that connects the point (x_0, F_0) to the point (x_N, F_N), and this has fractal dimension 1. If $\sum_{n=1}^{N} |d_n| \leq 1$, then one can show that $\mathcal{N}(\epsilon)$ behaves like a constant times ϵ^{-1}, whence the fractal dimension is 1. This completes our informal demonstration of the theorem.

Examples & Exercises

3.1. We consider the fractal dimension of a fractal interpolation function in the case where the interpolation points are equally spaced. Let $x_i = x_0 + \frac{i}{N}(x_N - x_0)$ for $i = 0, 1, 2, \ldots, N$. Then it follows that $a_n = \frac{1}{N}$ for $n = 1, 2, \ldots, N$. Hence if condition (1) in Theorem 3.1 holds then the fractal dimension D of the interpolation function obeys

$$\sum_{n=1}^{N} |d_n| \frac{1}{N})^{D-1} = (\frac{1}{N})^{D-1} \sum_{n=1}^{N} |d_n| = 1.$$

It follows that

$$D = 1 + \frac{\log \left(\sum_{n=1}^{N} |d_n| \right)}{\log(N)}.$$

This is a delightful formula for reasons of two types, (a) and (b). (a) This formula confirms our understanding of the fractal dimension of fractal interpolation functions. For example, notice that $\sum_{n=1}^{N} |d_n| < N$. Hence the dimension of a fractal interpolation function is less than 2: however, we can make it arbitrarily close to 2. Also, under the assumption that $\sum_{n=1}^{N} |d_n| > 1$, the fractal dimension is greater than 1: however, we can vary it smoothly down to 1. (b) It is remarkable that the fractal dimension does not depend on the values $\{F_i : i = 0, 1, 2, \ldots, N\}$, aside from the constraint that the interpolation points be noncollinear. Hence it is easy to explore a collection of fractal interpolation functions, all of which have the same fractal dimension, by imposing the following simple constraint on the vertical scaling factors:

$$\sum_{n=1}^{N} |d_n| = N^{D-1}.$$

Figure VI.173 illustrates some members of the family of fractal interpolation functions corresponding to the set of data $\{(0, 0), (1, 1), (2, 1), (3, 2)\}$, such that the fractal dimension ofeach member of the family is $D = 1.3$.

Figures 173 (a) and (c) illustrate members of a family of fractal interpolation functions parameterized by the fractal dimension D. Each function interpolates the same set of data.

3.2. Make an experimental estimate of the fractal dimension of the graphical data in Figure VI.175. Find a fractal interpolation function associated with the data $\{(0, 0), (50, 50), (100, 0)\}$, which has the same fractal dimension and two equal vertical scaling factors. Compare the graph of the fractal interpolation function with the graphical data.

3.3. Find a fractal interpolation function that approximates the experimental data shown in Figure V.147.

3.4. Figure VI.176 shows the graphs of functions belonging to various one-parameter families of fractal interpolation functions. Each graph is the attractor of an IFS consisting of two affine transformations. Find the IFS associated with one of the families.

Figure VI.173. Members of the family of fractal interpolation functions corresponding to the set of data $\{(0, 0), (1, 1), (2, 1), (3, 2)\}$, such that the fractal dimension of each member of the family is $D = 1.3$.

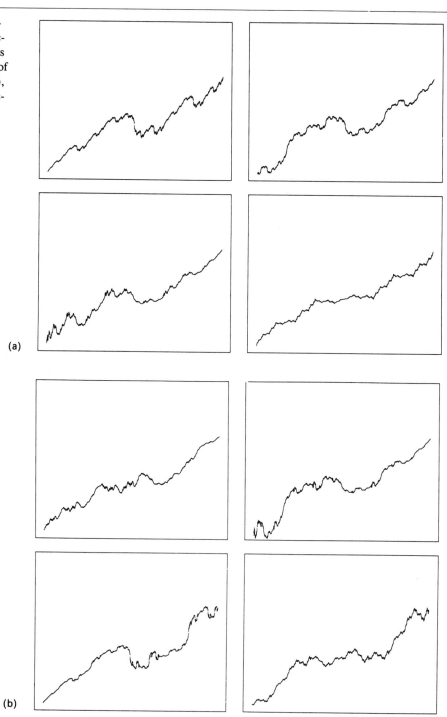

(a)

(b)

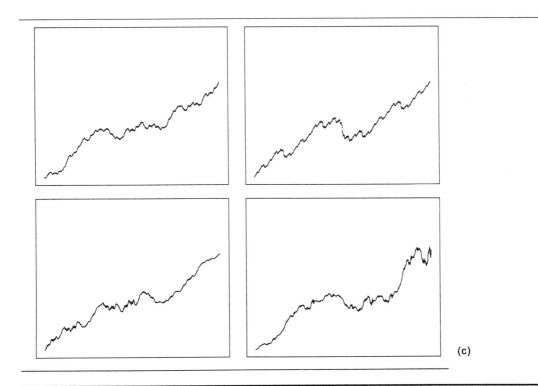

(c)

4 Hidden Variable Fractal Interpolation

We begin by generalizing the results of section 6.2. Throughout this section, let $(\mathbf{Y}, d_Y)$ denote a complete metric space.

Definition 4.1 *Let $I \subset \mathbb{R}$. Let $f : I \to \mathbf{Y}$ be a function. The* graph *of f is the set of points*

$$G = \{(x, f(x)) \in \mathbb{R} \times \mathbf{Y} : x \in I\}.$$

Definition 4.2 *A set of* generalized data *is a set of points of the form $\{(x_i, F_i) \in \mathbb{R} \times \mathbf{Y} : i = 0, 1, 2, \ldots, N\}$, where*

$$x_0 < x_1 < x_2 < x_3 < \cdots < x_N.$$

An interpolation function *corresponding to this set of data is a continuous function $f : [x_0, x_N] \to \mathbf{Y}$ such that*

$$f(x_i) = F_i \qquad \text{for } i = 1, 2, \ldots, N.$$

The points $(x_i, F_i) \in \mathbb{R} \times \mathbf{Y}$ are called the interpolation points. *We say that the function f interpolates the data and that (the graph of) f passes through the interpolation points.*

Let $\mathbf{X}$ denote the Cartesian product space $\mathbb{R} \times \mathbf{Y}$. Let θ denote a positive number.

Figure VI.174. Members of a one-parameter family of fractal interpolation functions. They correspond to the set of data $\{(0, 0), (1, 1), (2, 1), (3, 2)\}$ with vertical scaling factors $d_1 = -d_2 = d_3 = 3^{D-2}$ for $D = 1, 1.1, 1.2, 1.3, 1.4, 1.5, 1.6,$ and 1.7. D is the fractal dimension of the fractal interpolation function.

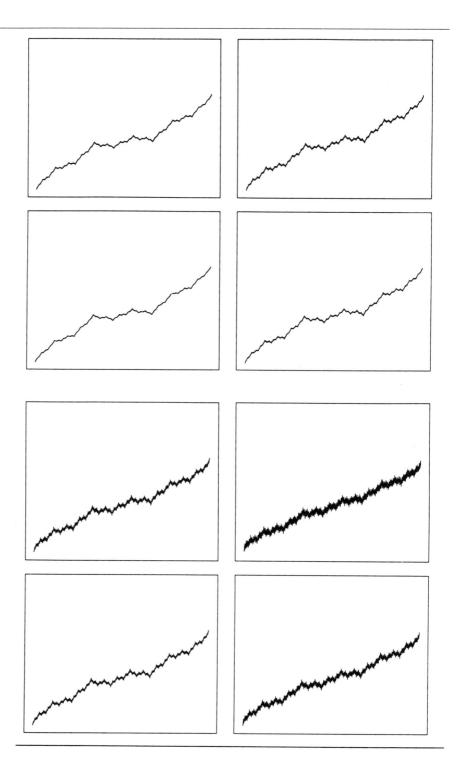

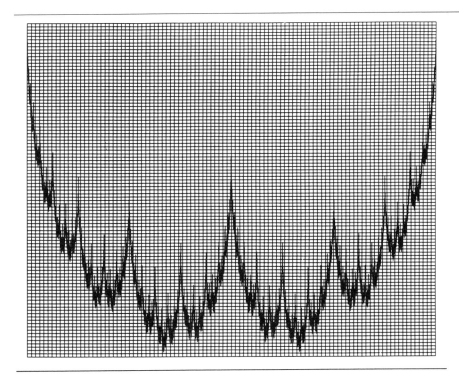

Figure VI.175. Make an experimental estimate of the fractal dimension of the graphical data shown here. Find a fractal interpolation function associated with the data $\{(0, 0), (50, -50), (100, 0)\}$, which has the same fractal dimension and two equal vertical scaling factors. Compare the graph of the fractal interpolation function with the graphical data.

Figure VI.176.   This figure shows the graphs of various one-parameter families of fractal interpolation functions. Each graph is the attractor of an IFS consisting of two affine transformations. Can you find the families?

Define a metric d on $\mathbf{X}$ by

$$d(X_1, X_2) = |x_1 - x_2| + \theta d_Y(y_1, y_2), \tag{1}$$

for all points $X_1 = (x_1, y_1)$ and $X_2 = (x_2, y_2)$ in $\mathbf{X}$. then $(\mathbf{X}, d)$ is a complete metric space.

Let N be an integer greater than 1. Let a set of generalized data $\{(x_i, F_i) \in \mathbf{X} : i = 0, 1, 2, \ldots, N\}$ be given. Let $n \in \{1, 2, \ldots, N\}$. Define $L_n : \mathbb{R} \to \mathbb{R}$ by

$$L_n(x) = a_n x + e_n \text{ where } a_n = \frac{(x_n - x_{n-1})}{(x_N - x_0)} \text{ and } e_n = \frac{(x_N x_{n-1} - x_0 x_n)}{(x_N - x_0)}. \tag{2}$$

so that $L_n([x_0, x_N]) = [x_{n-1}, x_n]$. Let c and s be real numbers, with $0 \leq s < 1$ and $c > 0$. For each $n \in \{1, 2, \ldots, N\}$ let $M_n : \mathbf{X} \to \mathbf{Y}$ be a function that obeys

$$d(M_n(a, y), M_n(b, y)) \leq c|a - b| \text{ for all } a, b \in \mathbb{R}, \tag{3}$$

and

$$d(M_n(x, a), M_n(x, b)) \leq s d_Y(a, b) \text{ for all } a, b \in \mathbf{Y}. \tag{4}$$

Define a transformation $w_n : \mathbf{X} \to \mathbf{X}$ by

$$w_n(x, y) = (L_n(x), M_n(x, y)) \text{ for all } (x, y) \in \mathbf{X}, n = 1, 2, \ldots, N.$$

Theorem 4.1 *Let the IFS $\{\mathbf{X}; w_n, n = 1, 2, \ldots, N\}$ be defined as above with $N > 1$. In particular, assume that there are real constants c and s such that $0 \leq s \leq 1, 0 < c$, and conditions 3 and 4 are obeyed. Let the constant θ in the definition of the metric d in equation 1 be defined by*

$$\theta = \frac{(1 - a)}{2c} \text{ where } a = \max\{a_i : i = 1, 2, \ldots, N\}.$$

Then the IFS $\{\mathbf{X}; w_n, n = 1, 2, \ldots, N\}$ is hyperbolic with respect to the metric d.

Proof This follows very similar lines to the proof of Theorem 2.1. We leave it as an exercise for enthusiastic readers. The proof can also be found in [Barnsley 1986].

We now constrain the hyperbolic IFS $\{\mathbf{X}; w_n, n = 1, 2, \ldots, N\}$, defined above, to ensure that its attractor includes the set of generalized data. We assume that

$$M_n(x_0, F_0) = F_{n-1} \text{ and } M_n(x_N, F_N) = F_n \text{ for } n = 1, 2, \ldots, N. \tag{5}$$

Then it follows that

$$w_n(x_0, F_0) = (x_{n-1}, F_{n-1}) \text{ and } w_n(x_N, F_N) = (x_n, F_n) \text{ for } n = 1, 2, \ldots, N.$$

Theorem 4.2 *Let N be a positive integer greater than 1. Let $\{\mathbf{X}; w_n, n = 1, 2, \ldots, N\}$ denote the IFS defined above, associated with the generalized data set $\{(x_i, F_i) \in \mathbb{R} \times \mathbf{Y} : i = 1, 2, \ldots, N\}$. In particular, assume that there are real constants c and s such that $0 \leq s \leq 1, 0 < c$, and conditions 3, 4, and 5 are obeyed. Let*

$G \in \mathcal{H}(\mathbf{X})$ *denote the attractor of the IFS. Then G is the graph of a continuous function* $f : [x_0, x_N] \to \mathbf{Y}$ *which interpolates the data* $\{(x_i, F_i) : i = 1, 2, \ldots, N\}$. *That is,*

$$G = \{(x, f(x)) : x \in [x_0, x_N]\},$$

where

$$f(x_i) = F_i \qquad \text{for } i = 0, 1, 2, 3, \ldots, N.$$

Proof Again we refer to [Barnsley 1988f]. The proof is analgous to the proof of Theorem 2.2.

Definition 4.3 *The function whose graph is the attractor of an IFS, as described in Theorems 4.1 and 4.2, above, is called a* generalized fractal interpolation function, *corresponding to the generalized data* $\{(x_i, F_i) : i = 1, 2, \ldots, N\}$.

We now show how to use the idea of generalized fractal interpolation functions to produce interpolation functions that are more flexible than heretofore. The idea is to construct a generalized fractal interpolation function, using affine transformations acting on $\mathbb{R}^3$, and to project its graph into $\mathbb{R}^2$. This can be done in such a way that the projection is the graph of a function that interpolates a set of data $\{(x_i, F_i) \in \mathbb{R}^2 : i = 1, 2, \ldots, N\}$. The extra degrees of freedom provided by working in $\mathbb{R}^3$ give us "hidden" variables. These variables can be used to adjust the shape and fractal dimension of the interpolation functions. The benefits of working with affine transformations are kept.

Let N be an integer greater than 1. Let a set of data $\{(x_i, F_i) \in \mathbb{R}^2 : i = 0, 1, 2, \ldots, N\}$ be given. Introduce a set of real parameters $\{H_i : i = 0, 1, 2, \ldots, N\}$. For the moment let us suppose that these parameters are fixed. Then we define a generalized set of data to be $\{(x_i, F_i, H_i) \in \mathbb{R} \times \mathbb{R}^2 : i = 0, 1, 2, \ldots, N\}$. In the present application of Theorem 4.2 we take $(\mathbf{Y}, d_Y)$ to be $(\mathbb{R}^2, \text{Euclidean})$. We consider an IFS $\{\mathbb{R}^3; w_n, n = 1, 2, \ldots, N\}$, where for $n \in \{1, 2, \ldots, N\}$ the map $w_n : \mathbb{R}^3 \to \mathbb{R}^3$ is an affine transformation of the special structure:

$$w_n \begin{bmatrix} x \\ y \\ z \end{bmatrix} = \begin{bmatrix} a_n & 0 & 0 \\ c_n & d_n & h_n \\ k_n & l_n & m_n \end{bmatrix} \begin{bmatrix} x \\ y \\ z \end{bmatrix} + \begin{bmatrix} e_n \\ f_n \\ g_n \end{bmatrix}.$$

Here $a_n, c_n, d_n, e_n, f_n, g_n, h_n, k_n, l_n,$ and m_n are real numbers. We assume that they obey the constraints

$$w_n \begin{bmatrix} x_0 \\ F_0 \\ G_0 \end{bmatrix} = w_n \begin{bmatrix} x_{n-1} \\ F_{n-1} \\ G_{n-1} \end{bmatrix},$$

and

$$w_n \begin{bmatrix} x_N \\ F_N \\ G_N \end{bmatrix} = w_n \begin{bmatrix} x_n \\ F_n \\ G_n \end{bmatrix}, \text{ for } n = 1, 2, \ldots, N.$$

Then we can write

$$w_n(x, y, z) = (L_n(x), M_n(x, y, z)) \text{ for all } (x, y, z) \in \mathbb{R}^3, n = 1, 2, \ldots, N,$$

where $L_n(x)$ is defined in Equation 2 and $M_n : \mathbb{R}^3 \to \mathbb{R}^2$ is defined by

$$M_n \begin{bmatrix} x \\ y \\ z \end{bmatrix} = A_n \begin{bmatrix} y \\ z \end{bmatrix} + \begin{bmatrix} f_n + c_n x \\ g_n + k_n x \end{bmatrix},$$

where

$$A_n = \begin{bmatrix} d_n & h_n \\ l_n & m_n \end{bmatrix} \text{ for } n = 1, 2, \ldots, N. \tag{6}$$

Let us replace F_n in condition 5 by (F_n, H_n). Then M_n obeys condition 5. Let us define

$$c = \max\{\max\{c_i, k_i\} : i = 1, 2, \ldots, N\}.$$

Then condition 3 is true. Lastly, assume that the linear transformation $A_n : \mathbb{R}^2 \to \mathbb{R}^2$ is contractive with contractivity factor s with $0 \le s < 1$. Then condition 4 is true. We conclude that, under the conditions given in this paragraph, the IFS $\{\mathbb{R}^3; w_n, n = 1, 2, \ldots, N\}$ satisfies the conditions of Theorem 4.2. It follows that the attractor of the IFS is the graph of a continuous function $f : [x_0, x_N] \to \mathbb{R}^2$ such that

$$f(x_i) = (F_i, H_i) \text{ for } i = 1, 2, \ldots, N.$$

Now write

$$f(x) = (f_1(x), f_2(x)).$$

Then $f_1 : [x_0, x_N] \to \mathbb{R}$ is a continuous function such that

$$f_1(x_i) = F_i \text{ for } i = 1, 2, \ldots, N.$$

Definition 4.4 *The function* $f_1 : [x_0, x_N] \to \mathbb{R}^2$ *constructed in the previous paragraph is called a* hidden variable *fractal interpolation function, associated with the set of data* $\{(x_i, F_i) \in \mathbb{R}^2 : i = 1, 2, \ldots, N\}$.

The easiest method for computing the graph of a hidden variable fractal interpolation function is with the aid of the Random Iteration Algorithm. Here we present an adaptation of Program 1. It computes points on the graph of a hidden variable fractal interpolation function and displays them on a graphics monitor. It is written for $N = 3$ and the data set

$$\{(0, 0), (30, 50), (60, 40), (100, 10)\}.$$

The "hidden" variables, namely the entries of the matrices A_n and the number H_n for $n = 1, 2, 3$, are input by the user during execution of the code. The program calculates the coefficients in the three-dimensional affine transformations from the data, and then applies the Random Iteration Algorithm to the resulting IFS. The first two coordinates of each successively computed point, which has three coordinates, are plotted on the screen of the graphics monitor. The program is written in BASIC. It runs without modification on an IBM PC with Enhanced Graphics Adaptor and Turbobasic. On any line the words preceded by a ' are comments: they are not part of the program.

Program 2.

```
x[0] =0 : x[1] =30 : x[2] =60 : x[3] =100   'Data set

F[0] =0 : F[1] =50 : F[2] =40 : F[3] =10

input "enter the hidden variables H[0], H[1], H[2] and H[3]",
     H[1],H[2],H[3],H[4]   'Hidden Variables

for n = 1 to 3 : print "for n = ", n

input "enter the hidden variables d, h, l, m",
     d[n],hh[n],l[n],m[n]   'More Hidden Variables

next

for n = 1 to 3  'Calculate the affine transformations from
     the Data and the Hidden Variables

p = F[n-1]-d[n]*F[0]-hh[n]*H[0] : q = H[n-1]-l[n]*F[0]-m[n]*H[0]

r = F[n]-d[n]*F[3]-hh[n]*H[3] : s = H[n]-l[n]*F[3]-m[n]*H[3]

   b = x[3]-x[0] : c[n] = (r-p)/b : k[n] = (s-q)/b

a[n] = (x[n]-x[n-1])/b : e[n] = (x[3]*x[n-1]-x[0]*x[n])/b

ff[n] = p-c[n]*x[0] : g[n] = q-k[n]*x[0]

next

screen 2 : cls  'initalialize graphics

window(0,0)-(100,100)  'change this to zoom and/or pan
```

```
x =0: y =0: z =hh[0]   'initial point from which the random
      iteration begins

for n = 1 to 1000 'Random Iteration Algorithm

   kk = int(3*rnd-0.0001)+1

   newx =  a[kk]*x + e[kk]

   newy =  c[kk]*x + d[kk]*y + hh[kk]*z + ff[kk]

   newz =  k[kk]*x + l[kk]*y + m[kk]*z + g[kk]

   x = newx : y = newy: z = newz

   pset(x,y),z  'plot the most recently computed point, in
      color z, on the screen

 next

 end
```

The result of running an adaptation of this program on a Masscomp workstation and then printing the contents of the graphics screen is presented in Figure VI.177. In this case $H[0] = 0$, $H[1] = 30$, $H[2] = 60$, $H[3] = 100$, $d(1) = d(2) = d(3) = 0.3$, $h(1) = h(2) = 0.2$, $h(3) = 0.1$, $l(1) = l(2) = l(3) = -0.1$, $m(1) = 0.3$, $m(2) = 0$, $m(3) = -0.1$. Remember that the linear transformation A_n must be contractive, so certainly do not enter values of magnitude larger than 1 for any of the numbers $d(n)$, $h(n)$, $l(n)$, and $m(n)$. The program renders each point in a color that depends on its z-coordinate. This helps the user to visualize the "hidden" three-dimensional character of the curve.

The important point about hidden variable fractal interpolation is this. Although the attractor of the IFS is a union of affine transformations applied to the attractor, this is not the case in general when we replace the word "attractor" by the phrase "projection of the attractor." The graph of the hidden variable fractal interpolation function $f_1(x)$ is not self-similar, or self-affine, or self-anything!

The idea of hidden variable fractal interpolation functions can be developed using any number of "hidden" dimensions. As the number of dimensions is increased, the process of specifying the function becomes more and more onerous, and the function itself, seen by us in flatland, becomes more and more random. One would never guess, from looking at pictures of them, that they are generated by *deterministic fractal geometry*.

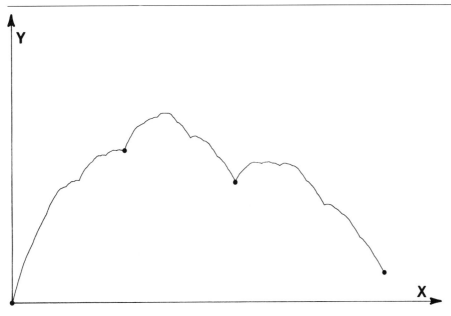

Figure VI.177. An example of a hidden variable fractal interpolation function. This graph was computed using Program 2 with the following "hidden" variables: $H[0] = 0$, $H[1] = 30$, $H[2] = 60$, $H[3] = 100$, $d(1) = d(2) = d(3) = 0.3$, $h(1) = h(2) = 0.2$, $h(3) = 0.1$, $l(1) = l(2) = l(3) = -0.1$, $m(1) = 0.3$, $m(2) = 0$, $m(3) = -0.1$.

Examples & Exercises

4.1. Generalize the proof of Theorem 2.1 to obtain a proof of Theorem 4.1.

4.2. Let $\mathcal{F}$ denote the set of continuous functions $f : [x_0, x_N] \to \mathbf{Y}$ such that

$$f(x_0) = F_0 \text{ and } f(x_N) = F_N.$$

Define a metric d on $\mathcal{F}$ by

$$d(f, g) = \max\{d_Y(f(x), g(x)) : x \in [x_0, x_N]\}.$$

Then $(\mathcal{F}, d)$ is a complete metric space; see, for example [Rudin 1966]. Use this fact to help you generalize the proof of Theorem 2.2 to provide a proof of Theorem 4.2.

4.3. Rewrite Program 2 in a form suitable for your own computer environment, then run it and obtain hardcopy of the output.

4.4. Modify your version of Program 2 so that you can adjust one of the "hidden" variables while it is running. In this way, make a picture that shows a one-parameter family of hidden variable fractal interpolation functions.

4.5. Modify your version of Program 2 so that you can see the projection of the attractor of the IFS into the (y, z) plane. To do this simply plot (y, z) in place of (x, y). Make hardcopy of the output.

4.6. Figure VI.178 shows three projections of the graph G of a generalized fractal interpolation function $f : [0, 1] \to \mathbb{R}^2$. The projections are (i) into the (x, y) plane, (ii) into the (x, z) plane, and (iii) into the (y, z) plane. G is the attractor of an IFS of the form $\{\mathbb{R}^3; w_1, w_2\}$, where w_1 and w_2 are affine transformations. Find w_1 and w_2. See also Figure VI.179.

4.7. Use a hidden-variable fractal interpolation function to fit the experimental data in Figure V.147. Here is one way to proceed. (a) Modify your version of Program 6.4.1 so that you can adjust the "hidden" variables from the keyboard. (b) Trace the data in the Figure V.147 onto a sheet of flexible transparent material, such as a viewgraph. (c) Attach the tracing to the screen of your graphics monitor using clear sticky tape. (d) Interactively adjust the "hidden" variables to provide a good visual fit to the data.

4.8. ⋆Show that, with hidden variables, one can use affine transformations to construct graphs of polynomials of any degree.

5 Space-Filling Curves

Here we make a delightful application of Theorem 4.2. Let A denote a nonempty pathwise-connected compact subset of $\mathbb{R}^2$. We show how to to construct a continuous function $f : [0, 1] \to \mathbb{R}^2$ such that $f([0, 1]) = A$.

Let $(\mathbf{Y}, d_Y)$ denote the metric space ($\mathbb{R}^2$, Euclidean). We represent points in $\mathbf{Y}$ using a Cartesian coordinate system defined by a y-axis and a z-axis. Thus, (y, z) may represent a point in $\mathbf{Y}$. To motivate the development we take $A = \blacksquare \subset \mathbf{Y}$. Consider the just-touching IFS $\{\mathbf{Y}; w_1, w_2, w_3, w_4\}$, where the maps are similitudes of scaling factor 0.5, corresponding to the collage in Figure VI.180.

Let
$$(F_0, H_0) = (0, 0), \qquad (F_1, H_1) = (0, 0.5), \qquad (F_2, H_2) = (0.5, 0.5),$$
$$(F_3, H_3) = (1, 0.5), \quad \text{and}$$
$$(F_4, H_4) = (1, 0).$$

The maps are chosen so that
$$w_n(F_0, H_0) = (F_{n-1}, H_{n-1}) \qquad \text{and} \quad w_n(F_4, H_4) = (F_n, H_n) \text{ for } n = 1, 2, 3, 4.$$

The IFS code for this IFS is given in Table VI.1.

Let $A_0 \in \mathcal{H}(\blacksquare)$ denote a simple curve that connects the point (F_0, H_0) to the point (F_4, H_4), such that $A_0 \cap \partial\blacksquare = \{(F_0, H_0), (F_4, H_4)\}$. This last condition says that the curve lies in the interior of the unit square box, except for the two endpoints of the curve. Consider the sequence of sets $\{A_n = W^{\circ n}(A_0)\}_{n=0}^{\infty}$ where $W : \mathcal{H}(\blacksquare) \to \mathcal{H}(\blacksquare)$ is defined by
$$W(B) = \cup_{n=1}^{4} w_n(B) \text{ for all } B \in \mathcal{H}(\blacksquare).$$

It follows from Theorem 7.1 in Chapter III that the sequence converges to $\blacksquare$ in the Hausdorff metric. The reader should verify that, for each $n = 1, 2, \ldots, A_n$ is a

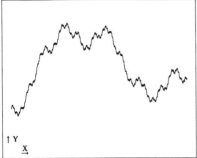
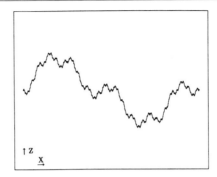
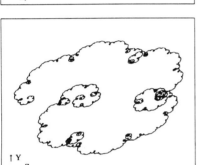

Figure VI.178. This figure shows three projections of the graph of a generalized fractal interpolation function $f : [0, 1] \rightarrow \mathbb{R}^2$. The projections are into the (x, y) plane, the (x, z) plane, and the (y, z) plane. G is the attractor of an IFS of the form $\{\mathbb{R}^3; w_1, w_2\}$, where w_1 and w_2 are affine transformations. Can you find w_1 and w_2?

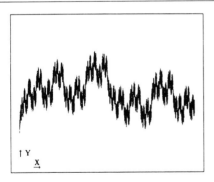
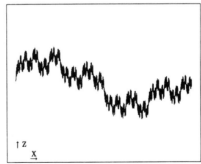
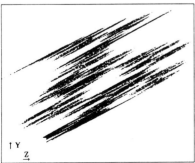

Figure VI.179. Three orthogonal projections of the graph of a generalized fractal interpolation function. The fractal dimension here is higher than for Figure VI.178.

Figure VI.180. Collage of ■ using four similitudes of scaling factor 0.5. The map w_n is chosen so that $w_n(F_0, H_0) = (F_{n-1}, H_{n-1})$ and $w_n(F_4, H_4) = (F_n, H_n)$ for $n = 1, 2, 3, 4$.

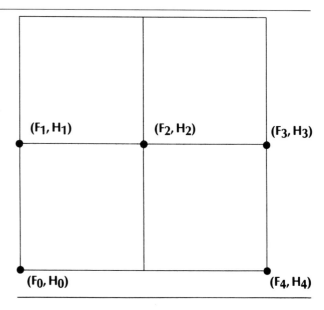

(F_1, H_1) (F_2, H_2) (F_3, H_3)

(F_0, H_0) (F_4, H_4)

Table VI.1.   IFS code for ■, constrained to yield a space-filling curve.

w	a	b	c	d	e	f	p
1	0	0.5	0.5	0	0	0	0.25
2	0.5	0	0	0.5	0	0.5	0.25
3	0.5	0	0	0.5	0.5	0.5	0.25
4	0	-0.5	-0.5	0	1	0.5	0.25

simple curve that connects the points (F_0, H_0) to the point (F_4, H_4). Sequences of such curves are illustrated in Figures VI.181–VI.184.

We use the IFS defined in the previous paragraph to construct a continuous function $f : [0, 1] \to$ ■ such that $f([0, 1]) =$ ■. We achieve this by exploiting a hidden variable fractal function constructed in a special way. We use ideas presented in Chapter VI, section 4. Consider the IFS $\{\mathbb{R}^3; w_n, n = 1, 2, \ldots, N\}$, where the map $w_n : \mathbb{R}^3 \to \mathbb{R}^3$ is the affine transformation

$$w_n \begin{bmatrix} x \\ y \\ z \end{bmatrix} = \begin{bmatrix} 0.25 & 0 & 0 \\ 0 & a_n & b_n \\ 0 & c_n & d_n \end{bmatrix} \begin{bmatrix} x \\ y \\ z \end{bmatrix} + \begin{bmatrix} (n-1)/4 \\ e_n \\ f_n \end{bmatrix} \quad \text{for } n \in \{1, 2, 3, 4\}.$$

The constants $a_n, b_n, c_n, d_n, e_n,$ and f_n are defined in Table VI.1. This IFS satisfies Theorem 4.2, corresponding to the set of data

$$\{(0, F_0, H_0), (0.25, F_1, H_1), (0.5, F_2, H_2), (0.75, F_3, H_3), (1, F_4, H_4)\}.$$

It follows that the attractor of the IFS is the graph, G, of a continuous function $f : [0, 1] \to \mathbb{R}^2$. What is the range of this function? It is

Figure VI.181. A sequence of curves "converging to" a space-filling curve. These are obtained by application of the Deterministic Algorithm to the IFS code in Table VI.1, starting from a curve A_0, which connects $(0, 0)$ to $(1, 0)$ and lies in ■.

Figure VI.182. A higher-resolution view of one of the panels in Figure VI.184. How long is the shortest path from the lower left corner to the lower right corner?

Figure VI.183. A sequence of sets "converging to" a ■. These are obtained by application of the Deterministic Algorithm to the IFS code in Table VI.1, starting from the set A_0 in the lower left panel. How fascinating they are!

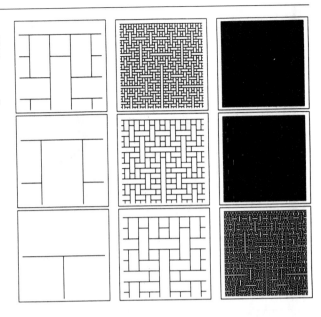

Figure VI.184. A sequence of curves "converging to" a space-filling curve. These are obtained by application of the Deterministic Algorithm to the IFS code in Table VI.1, starting from a curve A_0, which connects $(0, 0)$ to $(1, 0)$ and lies in ■.

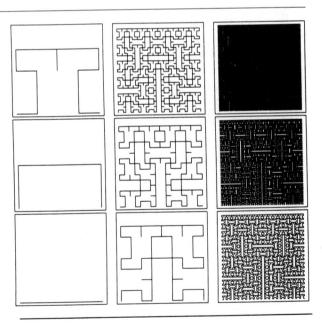

$$G_{yz} = \{(y, z) \in \mathbb{R}^2 : (x, y, z) \in G\},$$

namely the projection of G into the (y, z) plane. It is straightforward to prove that G_{yz} is the attractor $A = \blacksquare$ of the IFS defined by the IFS code in Table VI.1. It follows that $f([0, 1]) = \blacksquare$. So we have our space-filling curve!

We have something else very exciting as well. The attractor of the three-dimensional IFS is the graph of a function from $[0, 1]$ to $\blacksquare$. The projections G_{xy} and G_{xz}, in the obvious notation, are graphs of hidden-variable fractal functions, while $G_{yz} = \blacksquare$. What does G look like from other points of view? Various views of the attractor are illustrated in Figures VI.185 and VI.186. We conclude that G is a curious, complex, three-dimensional object. It would be wonderful to have a three-dimensional model of G made out of very thin strong wire.

The following theorem summarizes what we have just learned.

Theorem 5.1 *Let $A \subset \mathbb{R}^2$ be a nonempty pathwise-connected compact set, such that the following conditions hold. Let N be an integer greater than 1. Let there be a hyperbolic IFS $\{\mathbb{R}^2; M_n, n = 1, 2, \ldots, N\}$ such that A is the attractor of the IFS. Let there be a set of distinct points $\{(F_i, G_i) \in A : i = 0, 1, 2, \ldots, N\}$ such that*

$$M_n(F_0, H_0) = (F_{n-1}, H_{n-1}) \text{ and } w_n(F_N, H_N) = (F_n, H_n) \text{ for } n = 1, 2, \ldots, N.$$

Then there is a continuous function $f : [0, 1] \to \mathbb{R}^2$ such that $f([0, 1]) = A$. One such function is the one whose graph is the attractor of the IFS

$$\{\mathbb{R}^3; w_n(x, y, z) = (\frac{1}{N}x + \frac{n-1}{N}, M_n(y, z)), n = 1, 2, \ldots, N\}.$$

Examples & Exercises

5.1. Let $\triangle$ denote the Sierpinski triangle with vertices at the points $(0, 0)$, $(0, 1)$, and $(1, 0)$. Find an IFS of the form $\{\mathbb{R}^3; w_1, w_2, w_3\}$, where the maps are affine, such that the attractor of the IFS is the graph of a continous function $f : [0, 1] \to \mathbb{R}^2$ such that $f([0, 1]) = \triangle$. Four projections of such an attractor are shown in Figure VI.187.

5.2. Find an IFS $\{\mathbb{R}^3; w_1, w_2, w_3, w_4\}$, where the transformations are affine, whose attractor is the graph of a continuous function $f : [0, 1] \to \mathbb{R}^2$ such that $f([0, 1]) = A$, where A is the set represented in Figure VI.188.

Figure VI.185. Various views of the attractor of a certain IFS. From some points of view we see that it is the graph of a function. From one point of view it is the graph of a space-filling curve!

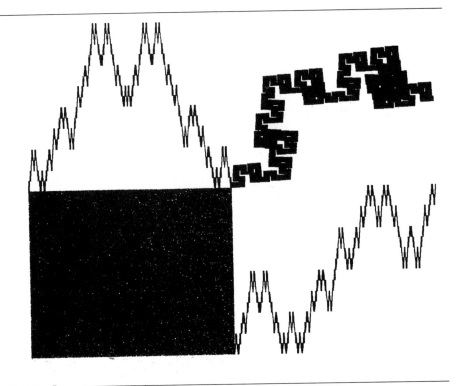

Figure VI.186. Higher-resolution view of the lower right panel of Figure VI.185.

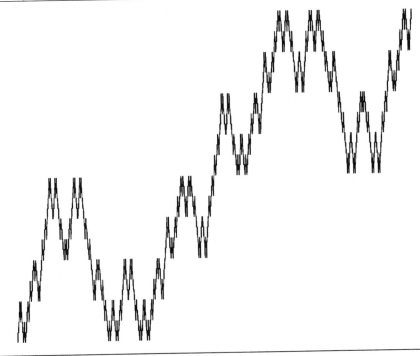

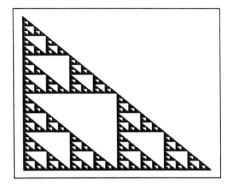

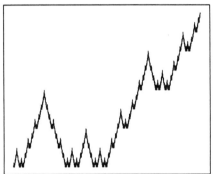
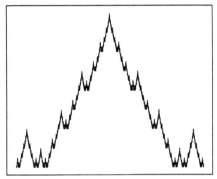

Figure VI.187. Four views of the attractor of an IFS. This attractor is the graph of a continuous function $f : [0, 1] \to \mathbb{R}^2$ such that $f([0, 1])$ is a Sierpinski triangle. This function provides a "space-filling" curve, where space is a fractal!

Figure VI.188. Find an IFS $\{\mathbb{R}^3; w_1, w_2, w_3, w_4\}$, where the transformations are affine, whose attractor is the graph of a continuous function $f : [0, 1] \to \mathbb{R}^2$ such that $f([0, 1]) = A$, where A is the set represented here.

Chapter VII

Julia Sets

1 The Escape Time Algorithm for Computing Pictures of IFS Attractors and Julia Sets

Let us consider the dynamical system $\{\mathbb{R}^2; f\}$, where $f : \mathbb{R}^2 \to \mathbb{R}^2$ is defined by

$$f(x, y) = \begin{cases} (2x, 2y - 1) & \text{if } y > 0.5, \\ (2x - 1, y) & \text{if } x > 0.5 \text{ and } y \leq 0.5, \\ (2x, 2y) & \text{otherwise.} \end{cases}$$

This dynamical system is related to the IFS $\{\mathbb{R}^2; w_1, w_2, w_3\}$, where

$$w_1(x, y) = (0.5x, 0.5y + 0.5),$$
$$w_2(x, y) = (0.5x + 0.5, 0.5y),$$
$$w_3(x, y) = (0.5x, 0.5y)\}.$$

The attractor of the IFS is a Sierpinski triangle $\triangle$ with vertices at $(0, 0)$, $(0, 1)$, and $(1, 0)$. The relationship between the dynamical system $\{\mathbb{R}^2; f\}$ and the IFS $\{\mathbb{R}^2; w_1, w_2, w_3\}$ is that $\{\triangle; f\}$ is a shift dynamical system associated with the IFS. (Shift dynamical systems are discussed in Chapter IV, section 4.) One readily verifies that f restricted to $\triangle$ satisfies

$$f(x, y) = \begin{cases} w_1^{-1}(x, y) & \text{if } (x, y) \in w_1(\triangle) \setminus \{(0, 0.5), (0.5, 0.5)\}, \\ w_2^{-1}(x, y) & \text{if } (x, y) \in w_2(\triangle) \setminus \{(0.5, 0)\}, \\ w_3^{-1}(x, y) & \text{if } (x, y) \in w_3(\triangle). \end{cases}$$

In particular, f maps $\triangle$ onto itself. The dynamical system $\{\mathbb{R}^2; f\}$ is an extension of the shift dynamical system $\{\triangle; f\}$ to $\mathbb{R}^2$. The situation is illustrated in Figure VII.189.

Let d denote the Euclidean metric on $\mathbb{R}^2$. The shift dynamical system $\{\mathbb{R}^2; f\}$ is "expanding": for any pair of points x_1, x_2 lying in any one of the three domains associated with f, we have

$$d(f(x_1), f(x_2)) = 2d(x_1, x_2),$$

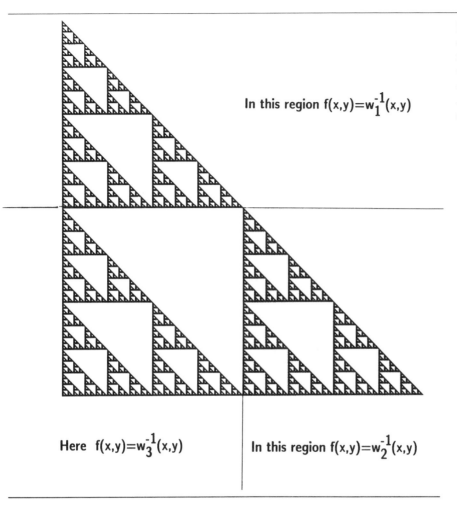

In this region $f(x,y)=w_1^{-1}(x,y)$

Here $f(x,y)=w_3^{-1}(x,y)$

In this region $f(x,y)=w_2^{-1}(x,y)$

Figure VII.189. The dynamical system $\{\mathbb{R}^2, f\}$ is obtained by extending the definition of a shift dynamical system on a Sierpinski triangle to all of $\mathbb{R}^2$.

One can prove that the orbit $\{f^{\circ n}(x)\}_{n=0}^{\infty}$ diverges toward infinity if x does not belong to $\mathbb{A}$. That is

$$d(O, f^{\circ n}(x)) \to \infty \text{ as } n \to \infty \text{ for any point } x \in \mathbb{R}^2 \setminus \mathbb{A}.$$

What happens if we compute numerically the orbit of a point $x \in \mathbb{A}$? Recall that the fractal dimension of $\mathbb{A}$ is $\log(3)/\log(2)$. This tells us that $\mathbb{A}$ is "thin" compared to $\mathbb{R}^2$. Hence, although $f(\mathbb{A}) = \mathbb{A}$, errors in a computed orbit are likely to produce points that do not lie on $\mathbb{A}$. This means that, in practice, most numerically computed orbits will diverge, regardless of whether or not the initial point lies on $\mathbb{A}$. The Sierpinski triangle $\mathbb{A}$ is an "unstable" invariant set for the transformation $f : \mathbb{R}^2 \to \mathbb{R}^2$. It is a "repulsive" fixed point for the transformation $f : \mathcal{H}(\mathbb{R}^2) \to \mathcal{H}(\mathbb{R}^2)$. It is an attractive fixed point for the transformation $W : \mathcal{H}(\mathbb{R}^2) \to \mathcal{H}(\mathbb{R}^2)$, where $W = w_1 \cup w_2 \cup w_3$ is defined in the usual manner.

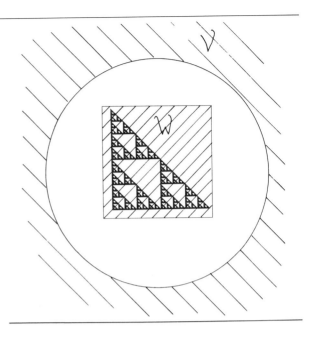

Figure VII.190. How long do orbits of points in $\mathcal{W}$ take to arrive in $\mathcal{V}$? We expect that the number of iterations required should tell us something about the structure of $\mathbb{A}$.

Intuitively, we expect that orbits of the dynamical system $\{\mathbb{R}^2; f\}$ that start close to $\mathbb{A}$ should "take longer to diverge" than those that start far from $\mathbb{A}$. How fast do different orbits diverge? Here we describe a numerical, computergraphical experiment to compare the number of iterations required for the orbits of different points to escape from a ball of large radius, centered at the origin. Let (a, b) and (c, d), respectively, denote the coordinates of the lower left corner and the upper right corner of a closed, filled rectangle $\mathcal{W} \subset \mathbb{R}^2$. Let M denote a positive integer, and define an array of points in $\mathcal{W}$ by

$$x_{p,q} = (a + p\frac{(c-a)}{M}, b + q\frac{(d-b)}{M}) \text{ for } p, q = 0, 1, 2, \ldots, M.$$

In the experiment these points will be represented by pixels on a computer graphics display device. We compare the orbits $\{f^{\circ n}(x_{p,q}) :\}_{n=0}^{\infty}$ for $p, q = 0, 1, 2, \ldots, M$.

Let R be a positive number, sufficiently large that the ball with center at the origin and radius R contains both $\mathbb{A}$ and $\mathcal{W}$. Define

$$\mathcal{V} = \{(x, y) \in \mathbb{R} : x^2 + y^2 > R\}.$$

A possible choice for the rectangle $\mathcal{W}$ and the set $\mathcal{V}$, in relation to $\mathbb{A}$, is illustrated in Figure VII.190. In order that the comparison of orbits provides information about $\mathbb{A}$, one should choose $\mathcal{W}$ so that $\mathcal{W} \cap \mathbb{A} \neq \emptyset$.

Let *numits* denote a positive integer. The following program computes a finite set of points

$$\{f^{\circ 1}(x_{p,q}), f^{\circ 2}(x_{p,q}), f^{\circ 3}(x_{p,q}), \ldots, f^{\circ n}(x_{p,q})\}$$

belonging to the orbit of $x_{p,q} \in \mathcal{W}$, for each $p, q = 1, 2, \ldots, M$. The total number of points computed on an orbit is at most *numits*. If the set of computed points of the orbit of $x_{p,q}$ does not include a point in $\mathcal{V}$ when $n = numits$, then the computation passes to the next value of (p, q). Otherwise the pixel corresponding to $x_{p,q}$ is rendered in a color indexed by the first integer n such that $f^{\circ n}(x_{p,q}) \in \mathcal{V}$, and then the computation passes to the next value of (p, q). This provides a computergraphical method for comparing how long the orbits of different points in $\mathcal{W}$ take to reach $\mathcal{V}$.

The program is written in BASIC. It runs without modification on an IBM PC with Enhanced Graphics Adaptor and Turbobasic. On any line the words preceded by a ' are comments: they are not part of the program.

Program 1. ((Example of the Escape Time Algorithm))

```
numits=20: a=0 : b=0: c=1 : d=1 : M=100   'Define viewing
                                           window, W, and numits.

R=200   'Define the region V.

screen 9: cls   'Initialize graphics.

for p=1 to M

for q=1 to M

x = a + (c-a)*p/M : y = b + (d-b)*q/M   'Specify the initial
                                        point of an orbit, x(p,q).

for n=1 to numits   'Compute at most numits points on the orbit
                    of x(p,q).

  if y > 0.5 then             'Evaluate $f$ applied to the
                              previous point on the orbit.

  x = 2*x : y = 2*y - 1

  elseif x > 0.5 then
                              'THE FORMULA FOR THE
  x = 2*x - 1 : y = 2*y        FUNCTION f(x)

  else

  x = 2*x : y = 2*y

  end if

150 if x*x + y*y > R then    'If the most recently computed
                             point lies in V then...
```

```
160 pset(p,q),n : n = numits   '...render the pixel x(p,q)
                                   in color n, and go to the next
(p,q).

170 end if

if instat then end   'Stop computing if any key is pressed!

next n : next q : next p

end
```

Color Plate 6 shows the result of running a version of Program 1 on a Masscomp 5600 workstation with Aurora graphics.

In Figure VII.191 we show the result of running a version of Program 1, but this time in black and white. A point is plotted in black if the number of iterations required to reach $\mathcal{V}$ is an odd integer, or if the orbit of the point does not reach $\mathcal{V}$ during the first *numits* iterations.

In Figure VII.192 we show the result of running a version of Program 1, with $(a, b) = (0, 0)$, $(c, d) = (5 \times 10^{-18}, 5 \times 10^{-18})$, and *numits* $= 65$. This viewing window is minute. See also Color Plate 7. Now you should be convinced that Δ is not simplified by magnification.

The dynamical system $\{\mathbb{R}^2; f\}$ contains deep information about the "repelling" set Δ. Some of this information is revealed by means of the Escape Time Algorithm. The orbits of points that lie close to Δ do indeed appear to take longer to escape from $\mathbb{R}^2 \setminus \mathcal{V}$ than those of points which lie further away.

Examples & Exercises

1.1. Let $\{\mathbb{R}^2, f\}$ denote the dynamical system defined at the start of this chapter and let Δ denote the associated Sierpinski triangle. Prove that the orbit $\{f^{\circ n}(x)\}_{n=0}^{\infty}$ diverges, for each $x \in \mathbb{R}^2 \setminus \Delta$. That is, prove that $d(O, f^{\circ n}(x)) \to \infty$ as $n \to \infty$ for each $x \in \mathbb{R}^2 \setminus \Delta$.

1.2. Rewrite Program 1 in a form suitable for your own computergraphical environment, then run it and obtain hardcopy of the output.

1.3. If the Escape Time Algorithm is applied to the dynamical system

$$\{\mathbb{R}^2; f(x, y) = (2x, 2y)\},$$

what will be the general appearance of resulting colored regions?

1.4. By changing the window size in Program 1, obtain images of "zooms" on the Sierpinski triangle. For example, use the following windows: $(0, 0) - (0.5, 0.5)$; $(0, 0) - (0.25, 0.25)$; $(0, 0) - (0.125, 0.125)$; .... How must the total number of iterations, *numits*, be adjusted as a function of window size in order

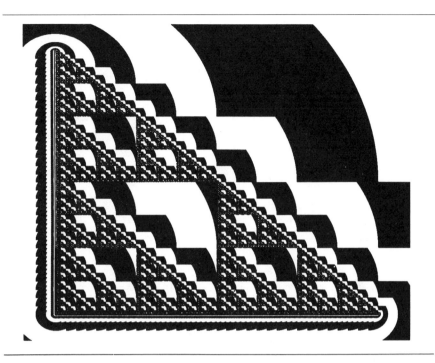

Figure VII.191. Output from a modified version of Program 1. A pixel is rendered in black if either the number of iterations required to reach $\mathcal{V}$ is an odd integer, or the orbit does not reach $\mathcal{V}$ during the first numits iterations.

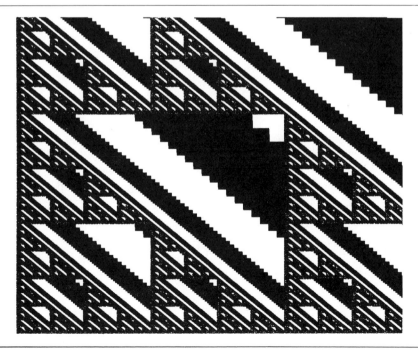

Figure VII.192. Here we show the result of running a version of Program 1, with $(a, b) = (0, 0)$, $(c, d) = (5 \times 10^{-18}, 5 \times 10^{-18})$, and *numits* = 65. This viewing window is minute, yet the computation time was not significantly increased. If we did not know it before, we are now convinced that Δ is not simplified by magnification.

that (approximately) the quality of the images remains uniform? Make a graph of the total number of iterations against the window size. Is there a possible relationship between the behavior of *numits* as a function of window size, and the fractal dimension of the Sierpinski triangle? Make a hypothesis and test it experimentally.

Here we construct another example of a dynamical system whose orbits "try to escape" from the attractor of an IFS. This time we treat an IFS whose attractor has a nonempty interior. Consider the hyperbolic IFS $\{\mathbb{R}^2; w_1, w_2\}$, where

$$w_1 \begin{bmatrix} x \\ y \end{bmatrix} = \begin{bmatrix} 0 & -s^{-1} \\ s^{-1} & 0 \end{bmatrix} \begin{bmatrix} x \\ y \end{bmatrix} + \begin{bmatrix} 1 \\ 0 \end{bmatrix}$$

$$w_2 \begin{bmatrix} x \\ y \end{bmatrix} = \begin{bmatrix} 0 & -s^{-1} \\ s^{-1} & 0 \end{bmatrix} \begin{bmatrix} x \\ y \end{bmatrix} - \begin{bmatrix} 1 \\ 0 \end{bmatrix}$$

and $s = \sqrt{2}$.

The attractor of this IFS is a closed, filled rectangle, which we denote here by ■. This attractor is the union of two copies of itself, each scaled by a factor $1/\sqrt{2}$, rotated about the origin anticlockwise through $90°$, and then translated horizontally, one copy to the left and one to the right. The inverse transformations are

$$w_1^{-1} \begin{bmatrix} x \\ y \end{bmatrix} = \begin{bmatrix} 0 & s \\ -s & 0 \end{bmatrix} \begin{bmatrix} x \\ y \end{bmatrix} + \begin{bmatrix} 0 \\ s \end{bmatrix}, \text{ and}$$

$$w_2^{-1} \begin{bmatrix} x \\ y \end{bmatrix} = \begin{bmatrix} 0 & s \\ -s & 0 \end{bmatrix} \begin{bmatrix} x \\ y \end{bmatrix} - \begin{bmatrix} 0 \\ s \end{bmatrix}.$$

Define $f : \mathbb{R}^2 \to \mathbb{R}^2$ by

$$f(x, y) = \begin{cases} w_1^{-1}(x, y) & \text{when } x > 0 \\ w_2^{-1}(x, y) & \text{when } x \leq 0. \end{cases}$$

Then the dynamical system $\{\mathbb{R}^2; f\}$ is an extension of the shift dynamical system $\{■; f\}$ to $\mathbb{R}^2$.

What happens when we apply the Escape Time Algorithm to this dynamical system? To see, one can replace the function $f(x)$ in Program 1 by

```
if x > 0 then

newx = s*y : newy = -s*x + s

else

newx = s*y : newy = -s*x - s
                                    'THE FORMULA FOR THE
end if                               FUNCTION f(x)

x = newx : y = newy
```

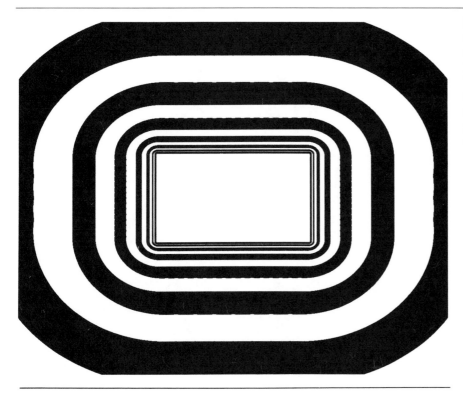

Figure VII.193. An image of an IFS attractor computed using the Escape Time Algorithm. This time the attractor of the IFS is a filled rectangle and the computed orbits of points in ■ seem never to escape.

Results of running Program 1, thus modified, with the window $\mathcal{W}$ and the escape region $\mathcal{V}$ chosen appropriately, are shown in Figure VII.193.

It appears that the orbits of points in the interior ■ do not escape. This is not suprising. The fractal dimension of the attractor of the IFS is the same as the fractal dimension of $\mathbb{R}^2$, so small computational errors are unlikely to knock the orbit off the invariant set. It also appears that the orbits of points that lie in $\mathbb{R}^2 \setminus$ ■ reach $\mathcal{V}$ after fewer and fewer iterations, the farther away from ■ they start.

Again we see that the Escape Time Algorithm provides a means for the computation of the attractor of an IFS. Indeed, we have here the bare bones of a new algorithm for computing images of the attractors of some hyperbolic IFS on $\mathbb{R}^2$. Here are the main steps. (a) Find a dynamical system $\{\mathbb{R}^2; f\}$ which is an extension of a shift dynamical system associated with the IFS, and which tends to transform points off the attractor of the IFS to new points that are farther away from the attractor. (This is always possible if the IFS is totally disconnected. The tricky part is to find a formula for $f(x)$, one which can be input conveniently into a computer. In the case of affine transformations in $\mathbb{R}^2$, one can often define the extensions of the domains of the inverse transformations with the aid of straight lines.) (b) Apply the Escape Time Algorithm, with $\mathcal{V}$ and $\mathcal{W}$ chosen appropriately, but plot only those points whose numerical orbits require sufficiently many iterations before they reach $\mathcal{V}$. For example,

Figure VII.194. Images of an IFS attractor computed using the Escape Time Algorithm. Only points whose orbits have not escaped from $\mathbb{R} \setminus \mathcal{V}$ after numits iterations are plotted. The value for numits must be chosen not too large, as in (a); and not too small, as in (b); but just right, as in (c).

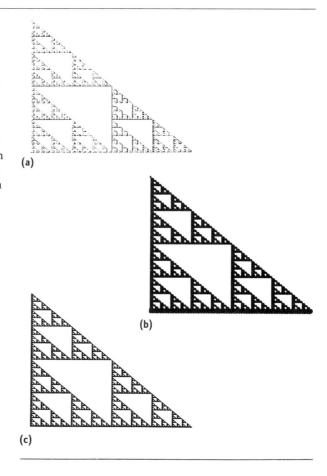

(a)

(b)

(c)

in Program 1 as it stands, one can replace the three lines 150, 160, and 170 by the two lines

```
150 if n = numits then pset(p,q),1
```

```
160 if x*x + y*y > R then n = numits
```

and define *numits* = 10. If the value of *numits* is too high, then very few points will not escape from $\mathcal{W}$ and a poor image of Δ will result. If the value of *numits* is too low, then a coarse image of the Δ will be produced. An image of an IFS attractor computed using the Escape Time Algorithm, modified as described here, is shown in Figure VII.194.

Color Plates 8–12 show the results of applying the Escape Time Algorithm to the dynamical system associated with various hyperbolic IFS in $\mathbb{R}^2$. In each case the maps are affine, and the shift dynamical system associated with the IFS has been extended to $\mathbb{R}^2$.

Examples & Exercises

1.5. Modify your version of Program 1 to compute images of the attractor of the IFS $\{C; w_1(z) = re^{i\theta}z - 1, w_2(z) = re^{i\theta}z + 1\}$, when $r = 1/\sqrt{2}$ and $\theta = \pi/2$.

1.6. Show that it is possible to define a dynamical system $\{C; f\}$ which extends to C the shift dynamical system associated with the IFS

$$\{C; w_1(z) = re^{i\theta}z - 1, w_2(z) = re^{i\theta}z + 1\},$$

for any $\theta \in [0, 2\pi)$, provided that the positive real number r is chosen sufficiently small. Note that this can be done in such a way that f is continuous.

1.7. Let $\{A; f\}$ denote the shift dynamical system associated with a totally disconnected hyperbolic IFS in $\mathbb{R}^2$. A denotes the attractor of the IFS. Show that there are many ways to define a dynamical system $\{\mathbb{R}^2; g\}$ so that $f(x) = g(x)$ for all $x \in A$.

The Escape Time Algorithm can be applied, often with interesting results, to any dynamical system of the form $\{\mathbb{R}^2; f\}$, $\{C; f\}$, or $\{\hat{C}; f\}$. One needs only to specify a viewing window W and a region V, to which orbits of points in W might escape. The result will be a "picture" of W wherein the pixel corresponding to the point z is colored according to the smallest value of the positive integer n such that $f^{\circ n}(z) \in V$. A special color, such as black, may be reserved to represent points whose orbits do not reach V before ($numits +1$) iterations.

What would happen if the Escape Time Algorithm were applied to the dynamical system $f : \hat{C} \to \hat{C}$ defined by $f(z) = z^2$? This transformation can be expressed $f(x, y) = (x^2 - y^2, 2xy)$. From the discussion of the quadratic transformation in Chapter III, section 4, we know that the orbits of points in the complement of the unit disk $F = \{z \in C : |z| \le 1\}$ converge to the point at infinity. Orbits of points in the interior of F converge to the origin. So if W is a rectangle that contains F and if the radius R, which defines V, is sufficiently large, then we expect that the Escape Time Algorithm would yield pictures of F surrounded by concentric rings of different colors. The reader should verify this!

F is called the *filled Julia set* associated with the polynomial transformation $f(z) = z^2$. The boundary of F is called the *Julia set* of f, and we denote it by J. It consists of the circle of radius 1 centered at the origin. One can think of J on the Riemann Sphere as being represented by the equator on a globe. This Julia set separates those points whose orbits converge to the point at Infinity from those whose orbits converge to the origin. Orbits of points on J itself cannot escape, either to infinity or to the origin. In fact $J \in \mathcal{H}(\hat{C})$ and $f(J) = J = f^{-1}(J)$. It is an "unstable" fixed point for the transformation $f : \mathcal{H}(\hat{C}) \to \mathcal{H}(\hat{C})$.

Definition 1.1 *Let $f : \hat{C} \to \hat{C}$ denote a polynomial of degree greater than 1. Let F_f denote the set of points in C whose orbits do not converge to the point at infinity. That is,*

$$F_f = \{z \in C : \{|f^{\circ n}(z)|\}_{n=0}^{\infty} \text{ is bounded }\}.$$

Figure VII.195. Illustration showing what is going on in the proof of Theorem 1.1. This illustrates the increasing sequence of sets $\{\mathcal{V}_n\}$ of the point at infinity. It also shows the decreasing sequence of sets, K_n, the complements of the latter, which converge to the filled Julia set F_f. In general the origin, O, need not belong to F_f.

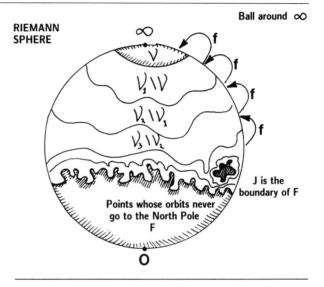

This set is called the filled Julia set *associated with the polynomial* f. *The boundary of* F_f *is called the* Julia set *of the polynomial* f, *and it is denoted by* J_f.

Theorem 1.1 *Let* $f : \hat{C} \to \hat{C}$ *denote a polynomial of degree greater than 1. Let* F_f *denote the filled Julia set of* f *and let* J_f *denote the Julia set of* f. *Then* F_f *and* J_f *are nonempty compact subsets of* C; *that is,* $F_f \in \mathcal{H}(C)$ *and* $J_f \in \mathcal{H}(C)$. *Moreover,* $f(J_f) = J_f = f^{-1}(J_f)$ *and* $f(F_f) = F_f = f^{-1}(F_f)$. *The set* $\mathcal{V}_\infty = \hat{C} \setminus F_f$ *is pathwise-connected.*

Proof We outline the proof for the one-parameter family of transformations $f_\lambda : \hat{C} \to \hat{C}$ defined by

$$f_\lambda(z) = z^2 - \lambda, \qquad \text{where } \lambda \in C \text{ is the parameter.}$$

The general case is treated in [Blanchard 1984], [Brolin], [Fatou 1919–20], and [Julia 1918], for example. This outline proof is constructed to provide information about the relationship between the theorem and the Escape Time Algorithm. Some of the ideas and notation used here are illustrated in Figure VII.195.

Let J_λ denote the Julia set for f_λ and let F_λ denote the filled Julia set for f_λ. Let d denote the Euclidean metric on C and let

$$R > 0.5 + \sqrt{0.25 + |\lambda|}.$$

Then it is readily verified that

$$d(O, f(z)) > d(O, z) \qquad \text{for all } z \text{ such that } d(O, z) \geq R.$$

Define

$$\mathcal{V} = \{z \in C : |z| > R\} \cup \{\infty\}.$$

Then it follows that

$$f(V) \subset V.$$

One can prove that the orbit $\{f^{\circ n}(z)\}$ converges to ∞ for all $z \in V$. No bounded orbit intersects V. It follows that

$$F_\lambda = \{z \in \hat{C} : f^{\circ n}(z) \notin V \text{ for each finite positive integer } n\}.$$

That is, F_λ is the same as the set of points whose orbits do not intersect V.

Now consider the sequence of sets

$$V_n = f^{\circ -n}(V) \qquad \text{for } n = 0, 1, 2, \ldots.$$

For each nonnegative integer n, V_n is an open connected subset of $(\hat{C}, \text{Spherical})$. V_n is open because V is open and f is continuous. V_n is connected because of the geometry of the quadratic transformation, described in Chapter III, section 4: The inverse image of a path that joins the point at infinity to any other point on the sphere is a path that contains the point at infinity.

Since $f(V) \subset V$ it follows that $V \subset f^{-1}(V)$. This implies that

$$V = V_0 \subset V_1 \subset V_2 \subset V_3 \subset \cdots \subset V_n \subset \cdots. \tag{1}$$

For each nonnegative integer n,

$$V_n = \left\{z \in \hat{C} : \{z, f^{\circ 1}(z), f^{\circ 2}(z), f^{\circ 3}(z), \cdots, f^{\circ n}(z)\} \cap V \neq \emptyset\right\}.$$

That is, V_n is the set of points whose orbits require at most n iterations to reach V. Let

$$K_n = \hat{C} \setminus V_n \qquad \text{for } n = 0, 1, 2, 3, \ldots.$$

Then K_n is the set of points whose orbits do not intersect V during the first n iterations. That is,

$$K_n = \left\{z \in \hat{C} : \{z, f^{\circ 1}(z), f^{\circ 2}(z), f^{\circ 3}(z), \ldots, f^{\circ n}(z)\} \cap V = \emptyset\right\}.$$

For each nonnegative integer n, K_n is a nonempty compact subset of the metric space $(\hat{C}, \text{Spherical})$. How do we know that K_n is nonempty? Because we can calculate that f possesses a fixed point $z_f \in C$, by solving the equation

$$f(z_f) = z_f^2 - \lambda = z_f.$$

The orbit of z_f converges to z_f. Hence it cannot belong to V_n for any nonnegative integer n. Hence $z_f \in K_n$ for each nonnegative integer n.

Equation 1 implies that

$$K_0 \supset K_1 \supset K_2 \supset K_3 \supset \cdots \supset K_n \supset \cdots.$$

It follows that $\{K_n\}$ is a Cauchy sequence in $\mathcal{H}(\hat{C})$. It follows that $\{K_n\}$ converges to a point in $\mathcal{H}(\hat{C})$. The limit is the set of points whose orbits do not intersect V. Hence

$$F_f = \lim_{n \to \infty} K_n = \cap_{n=0}^{\infty} K_n,$$

and we deduce that F_f belongs to $\mathcal{H}(\hat{C})$.

The equation

$$K_{n+1} = f^{\circ -1}(K_n) \qquad \text{for } n = 0, 1, 2, \ldots$$

now implies, as in the proof of Theorem 4.1, that

$$F_\lambda = f^{\circ -1}(F_\lambda).$$

Applying f to both sides of this equation, we obtain

$$f(F_\lambda) = F_\lambda.$$

Let us now consider the boundary of F_λ, namely the Julia set J_λ for the dynamical system $\{\hat{C}; f_\lambda\}$. Let $z \in \text{interior}(F_\lambda)$. Then the continuity of f implies $f^{-1}(z) \subset \text{interior}(F_\lambda)$. Hence $F_\lambda \supset f^{-1}(\partial F_\lambda) \supset \partial F_\lambda$. Now suppose that $z \in f^{-1}(\partial F_\lambda)$. Let $\mathcal{O}$ be any open ball that contains z. Since f is analytic, $f(\mathcal{O})$ is an open ball, and it contains $f(z) \in \partial F_\lambda$. Hence $f(\mathcal{O})$ contains a point whose orbit converges to the point at infinity. It follows that $\mathcal{O}$ contains a point whose orbit converges to the point at infinity. Thus $f^{-1}(\partial F_\lambda) \subset \partial F_\lambda$. We conclude that $f^{-1}(\partial F_\lambda) = \partial F_\lambda$ and in particular that $f(\partial F_\lambda) = \partial F_\lambda$. This completes the proof of the theorem.

We summarize some of what we discovered in the course of this proof. The filled Julia set F_λ is the limit of a decreasing sequence of compact sets. Its complement, which we denote by V_∞, is the limit of an increasing sequence $\{V_n\}$ of open pathwise-connected sets in $(\hat{C}, \text{spherical})$. That is,

$$V_\infty = \lim_{n \to \infty} V_n = \cup_{n=0}^{\infty} V_n.$$

The latter is called the *basin of attraction* of the point at infinity under the polynomial tranformation f_λ. It is connected because each of the sets V_n is connected. We have

$$\hat{C} = F_\lambda \cup V_\infty.$$

V_∞ is open, connected, and nonempty. F_λ is compact and nonempty.

The Escape Time Algorithm provides us with a means for "seeing" the filled Julia sets F_λ, as well as the sequences of sets $\{V_n\}$ and $\{K_n\}$ referred to in the theorem. Let us look at what happens in the case $\lambda = 1.1$. Define V by choosing $R = 4$, and let $W = \{(x, y) : -2 \leq x \leq 2, -2 \leq y \leq 2\}$. The function $f_{\lambda=1.1} : C \to C$ is given by the formula

$$f_{\lambda=1.1}(x, y) = (x^2 - y^2 - 1.1, 2xy) \qquad \text{for all } (x, y) \in C.$$

An example of the result of running the Escape Time Algorithm, with V, W and $f : \hat{C} \to \hat{C}$ thus defined, is shown in Figure VII.196. The black object represents the filled Julia set $F_{\lambda=1.1}$. The contours separate the regions $\mathcal{V}_{n+1} \setminus \mathcal{V}_n$, for some sucessive values of n. These contours also represent the boundaries of the regions K_n referred to in the proof of the theorem. We refer to them as escape time contours. Points in $\mathcal{V}_{n+1} \setminus \mathcal{V}_n$ have orbits that reach $\mathcal{V}$ in exactly $(n+1)$ iterations. In Color Plate 13 we show another example of running the Escape Time Algorithm to produce an image of the same set. The regions $\mathcal{V}_{n+1} \setminus \mathcal{V}_n$ are represented by different colors.

Figure VII.197 shows a zoom on an interesting piece of $F_{\lambda=1.1}$, including parts of some escape time contours. This image was computed by choosing $\mathcal{W}$ to be a small rectangular subset of the window used in Figure VII.196.

Figures VII.198(a)–(e) shows pictures of the filled Julia sets F_λ for a set of real values of λ. These pictures also include a number of the escape time contours, to help indicate the location of F_λ. F_0 is a filled disk. As λ increases, the set becomes more and more pinched together until, when $\lambda = 2$, it is the closed interval $[-2, 2]$. For some values of $\lambda \in [0, 2]$, it appears that F_λ has no interior, and is "tree-like"; for other values it seems to possess a roomy interior. It also appears that F_λ is connected for all $\lambda \in [0, 2]$, and totally disconnected when $\lambda > 2$. In the latter case F_λ may be described as a "Cantor-like" set, or as a "dust." The transition between the totally disconnected set and the connected, bubbly set as the parameter λ is varied reminds us of the transition between the the Cantor set and the Sierpinski triangle, discussed in connection with Figure IV.118.

Examples & Exercises

1.8. Modify your version of the Escape Time Algorithm to allow you to compute pictures of filled Julia sets for the family of quadratic polynomials $z^2 - \lambda$ for

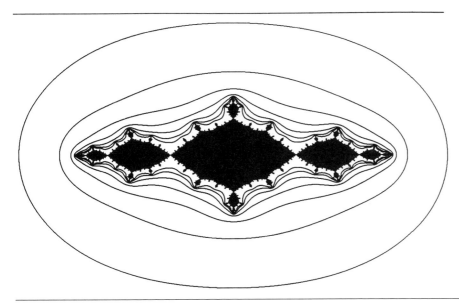

Figure VII.196. The Escape Time Algorithm provides us with a means for "seeing" the filled Julia sets F_λ, as well as the sequences of sets $\{\mathcal{V}_n\}$ and $\{K_n\}$ referred to in Theorem 1. In this illustration, $\lambda = 1.1$. The black object represents the filled Julia set $F_{\lambda=1.1}$. The contours separate the regions $\mathcal{V}_{n-1} \setminus \mathcal{V}_n$, for some successive values of n. These contours also represent the boundaries of the regions K_n referred to in the proof of the theorem.

Figure VII.197. Zoom in on an interesting piece of Figure VII.196.

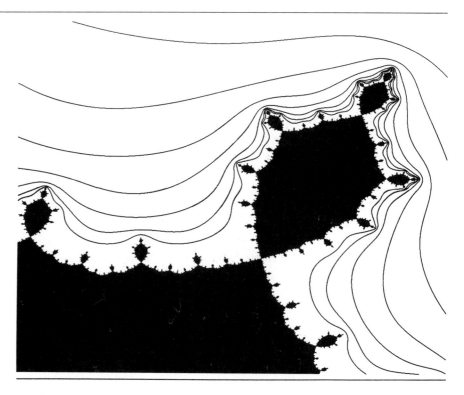

complex values of λ. Compute a picture of the filled Julia set for $\lambda = i$ and obtain hardcopy of the output.

1.9. Give the iteration formulas, and find a suitable value for R in terms of $|\lambda|$, so that the Escape Time Algorithm can be applied to the complex polynomial $z^3 - \lambda$.

1.10. Study web diagrams associated with $x^2 - \lambda$, for increasing values of $\lambda \in [0, 3]$. Speculate on the relation of these diagrams to the corresponding filled Julia sets.

1.11. Let $\lambda \in [0, 0.7] \cup [0.8, 1.2]$. Let $\mathcal{V}$ to be an open ball of radius 0.00001 centered at the origin. Run the Escape Time Algorithm to the dynamical system $\{C, z^2 - \lambda\}$ with this choice of $\mathcal{V}$. Obtain computergraphical data in support of the hypothesis that, in this case, the algorithm yields approximate pictures of pieces of the closure of $C \setminus F_\lambda$. Design an escape region $\mathcal{V}$ so that, for $\lambda \in [0, 0.7] \cup [0.8, 1.2]$, the Escape Time Algorithm yields approximate pictures of J_λ.

1.12. The Escape Time Algorithm introduces numerical errors in the computation of orbits. These errors should lead to inaccuracies in the computed pictures of Julia sets and IFS attractors. Consider the application to the filled Julia set for $z^2 - 1$. By means of computergraphical experiments, determine the importance of these errors in the images you compute. One way to proceed is to choose successively smaller

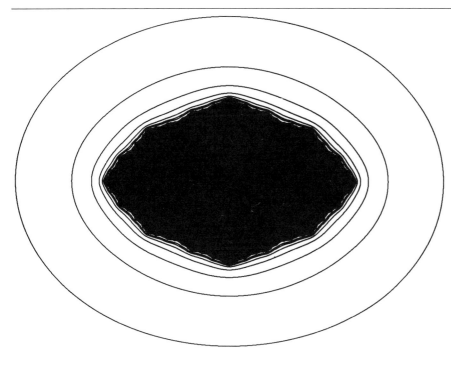

Figure VII.198. (a)–(e) A sequence of Julia set images, as in Figure VII.196, for an increasing sequence of values λ in the range 0 to 3. In (d) and (e) the filled Julia set is the same as the Julia set: the filled Julia set has no interior, so it equals its boundary. In (d) the Julia set is "tree-like." In (e) the Julia set is totally disconnected.

Figure VII.198. (b)

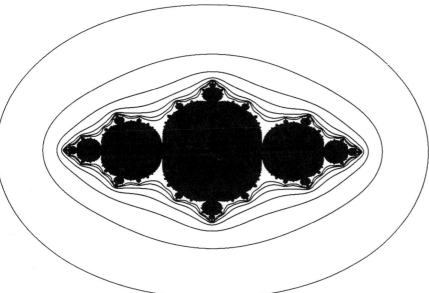

Figure VII.198. (c)

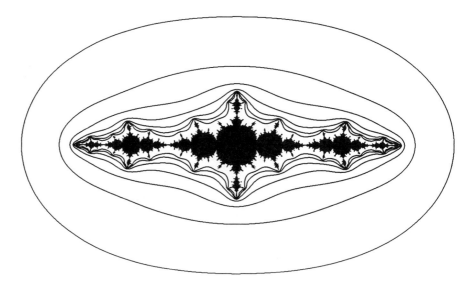

Figure VII.198. (d)

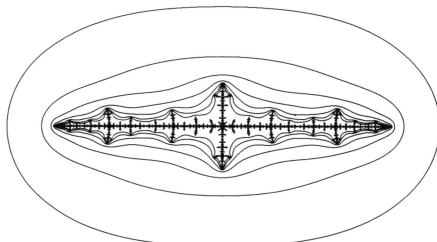

Figure VII.198. (e)

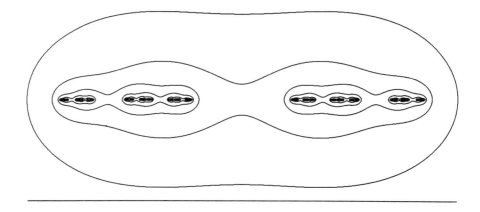

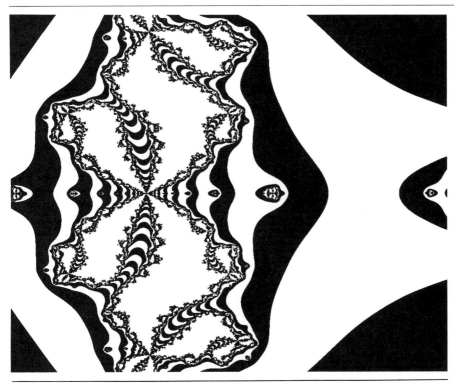

Figure VII.199. This image was computed by applying the Escape Time Algorithm to the dynamical system $\{C; f(z) = z^4 - z - 0.78\}$. The viewing window is $\mathcal{W} = \{(x, y) : -1 \le x \le 1, -1 \le y \le 1\}$. Can you determine the escape region $\mathcal{V}$?

windows $\mathcal{W}$, which intersect the apparent boundary of the filled Julia set, and to seek the window size at which the quality of computed images seems to deteriorate. (You will need to increase the maximum number of iterations, M, as you zoom.) Can you give evidence to show that the apparently deteriorated images are not, in fact, correct?

1.13. Figure VII.199 was computed by applying the Escape Time Algorithm to the dynamical system $\{C; f(z) = z^4 - z - 0.78\}$. The viewing window is $\mathcal{W} = \{(x, y) : -1 \le x \le 1, -1 \le y \le 1\}$. Determine the escape region $\mathcal{V}$. Also, you might like to try magnifying one of the little faces in this image.

1.14. The images in Figure VII.200 (a),(b),(c), and (d) represent the nontrivially distinct attractors of all IFS of the form

$$\{\blacksquare; w_1, w_2, w_3\},$$

where the maps are similitudes of scaling factor one-half, and rotation angles in the set $\{0°, 90°, 180°, 270°\}$. The three translations $(0, 0)$, $(1, 0)$, and $(0, 1)$ are used. These IFS are all just-touching. For $i \ne j$ the set $w_i(A) \cap w_j(A)$ is contained in one of the two straight lines $x = 1$ or $y = 1$. Show that, as a result, it is easy to compute these images using the Escape Time Algorithm.

Here are some observations about this "group" of images. Many of them contain

Figure VII.200. (a)–
(d) The images in (a),
(b), (c), and (d) represent
the nontrivially distinct
attractors of all IFS of
the form $\{\blacksquare;\, w_1,\, w_2,\, w_3\}$,
where the maps are simil-
itudes of scaling factor
one-half, and the rota-
tion angles are in the
set $\{0°, 90°, 180°, 270°\}$.
The three translations
$(0, 0)$, $(1, 0)$, and $(0, 1)$
are used. These IFS are all
just-touching. For $i \neq j$
the set $w_i(A) \cap w_j(A)$ is
contained in one of the
two straight lines $x = 1$ or
$y = 1$. Hence it is easy to
compute images of these
attractors using the Escape
Time Algorithm.

Figure VII.200. (b)

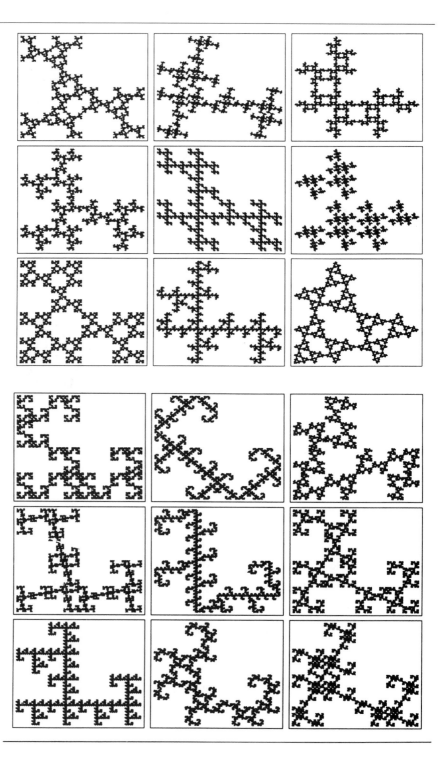

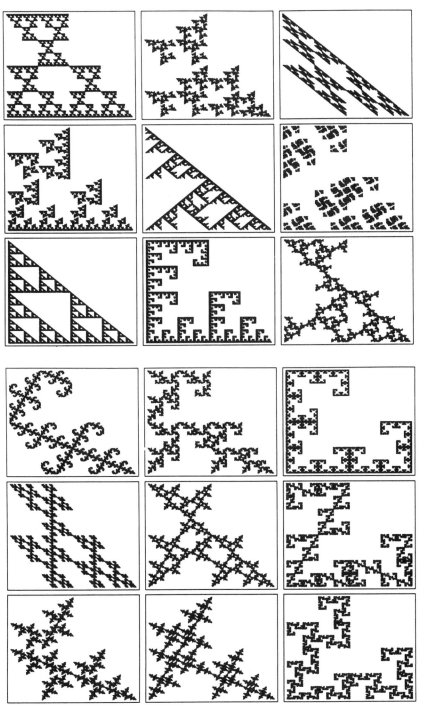

Figure VII.200. (c)

Figure VII.200. (d)

straight lines. They all have the same fractal dimension. They all use approximately the same amount of ink. Many of them are connected. Make some more observations. Can you formalize and prove some of these observations?

1.15.   Verify computationally that a "snowflake" curve is a basin boundary for the dynamical system $\{\mathbb{R}^2; f\}$, where for all $(x, y) \in \mathbb{R}^2$,

$$f(x, y) = (0, -1) \text{ if } y < 0;$$

$$f(x, y) = (3x, 3y) \text{ if } y \geq 0 \text{ and } x < -y/\sqrt{3} + 1;$$

$$f(x, y) = ((9 - 3x - 3\sqrt{3}y)/2, (3\sqrt{3} - 3\sqrt{3}x + 3)/2)$$
$$\text{if } y \geq 0 \quad \text{and} \quad -y/\sqrt{3} + 1 \leq x < 3/2;$$

$$f(x, y) = ((3x - 3\sqrt{3}y)/2, (3\sqrt{3}x + 3y - 6\sqrt{3})/2)/2),$$
$$\text{if } y \geq 0 \quad \text{and} \quad 3/2 \leq x < y/\sqrt{3} + 2;$$

$$f(x, y) = (9 - 3x, 3y), \text{ if } y \geq 0, \text{ and } x \geq y/\sqrt{3} + 2.$$

2   Iterated Function Systems Whose Attractors Are Julia Sets

In section 1 we learned how to define some IFS attractors and filled Julia sets with the aid of the Escape Time Algorithm applied to certain dynamical systems. In this section we explain how the Julia set of a quadratic transformation can be viewed as the attractor of a suitably defined IFS.

The Escape Time Algorithm compares how fast different points in $\mathcal{W}$ escape to $\mathcal{V}$, under the action of a dynamical system. Which set repels the orbits? From where do the escaping orbits originate? In the case of the dynamical systems considered at the start of section 1, orbits were "escaping from" the attractor of the IFS.

Let $\lambda \in \mathbb{C}$ be fixed. Which set repels the orbits, in the case of the dynamical system $\{\hat{\mathbb{C}}; f_\lambda(z) = z^2 - \lambda\}$? To find out let us consider the inverse of $f_\lambda(z)$. This is provided by a pair of functions, $f^{-1}(z) = \{+\sqrt{z + \lambda}, -\sqrt{z + \lambda}\}$, where, for example, the positive square root of a complex number is that complex root that lies on the nonnegative real axis or in the upper half plane. Explicitly, $\sqrt{z} = \sqrt{x_1 + ix_2} = (a(x_1, x_2), b(x_1, x_2))$ with

$$a(x_1, x_2) = \sqrt{\sqrt{\frac{x_1^2 + x_2^2} + x_1}{2}} \qquad \text{when } x_2 \geq 0,$$

$$a(x_1, x_2) = -\sqrt{\sqrt{\frac{x_1^2 + x_2^2} + x_1}{2}} \qquad \text{when } x_2 < 0,$$

$$b(x_1, x_2) = \sqrt{\sqrt{\frac{x_1^2 + x_2^2} - x_1}{2}}.$$

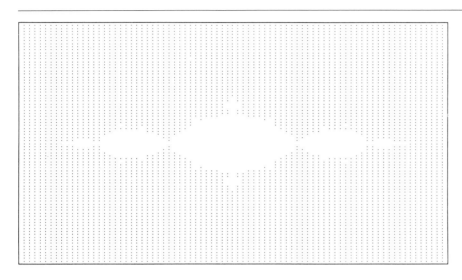

Figure VII.201. (a) and (b) The attractor of the IFS $\{\hat{\mathbb{C}}; w_1(z) = \sqrt{z+1}, w_2(z) = -\sqrt{z+1}\}$ is the Julia set for the transformation $f(z) = z^2 - 1$. (a) illustrates points whose orbits "escape" when the Escape Time Algorithm is applied. (b) shows the results of applying the Random Iteration Algorithm to the IFS, superimposed on (a).

To find the "repelling" set, we must try to run the dynamical system backwards. This leads us to study the IFS

$$\{\hat{\mathbb{C}}; w_1(z) = \sqrt{z+\lambda}, w_2(z) = -\sqrt{z+\lambda}\}.$$

The natural idea is that this IFS has an attractor. This attractor is the set from which points try to flee, under the action of the dynamical system $\{\hat{\mathbb{C}}; z^2 - \lambda\}$.

A few computergraphical experiments quickly suggest a wonderful idea: they suggest that the the IFS indeed possesses an attractor, namely the Julia set $J_\lambda = \partial F_\lambda$ for $f_\lambda(z)$. Consider, for example, the case $\lambda = 1$. Figure VII.201 (a) illustrates points in the window $W = \{z = (x, y) \in \mathbb{C} : -2 \leq x \leq 2, -2 \leq y \leq 2\}$ whose orbits diverge. It was computed using the Escape Time Algorithm. Figure VII.201 (b) shows the results of applying the Random Iteration Algorithm to the above IFS, with $\lambda = 1$ and the same screen coordinates, superimposed on (a). The boundary of the region $F_{\lambda=1}$ is outlined by points on the attractor of the IFS.

Figures VII.202(a)–(d) show the results of applying the Random Iteration Algorithm to the IFS $\{\hat{\mathbb{C}}; w_1(z) = \sqrt{z+\lambda}, w_2(z) = -\sqrt{z+\lambda}\}$ for various $\lambda \in [0, 3]$. In all cases it appears that the IFS possesses an attractor, and this attractor is the Julia set J_λ.

Perhaps

$$\{\hat{\mathbb{C}}; w_1(z) = \sqrt{z+\lambda}, w_2(z) = -\sqrt{z+\lambda}\}$$

is a hyperbolic IFS with J_λ as its attractor? No, it is not, because $\hat{\mathbb{C}} = w_1(\hat{\mathbb{C}}) \cup w_2(\hat{\mathbb{C}})$. The IFS is not associated with a unique fixed point in the space $\mathcal{H}(\hat{\mathbb{C}})$. In order to make the IFS have a unique attractor, we need to remove some pieces from $\hat{\mathbb{C}}$, to produce a smaller space on which the IFS acts.

Figure VII.201. (b)

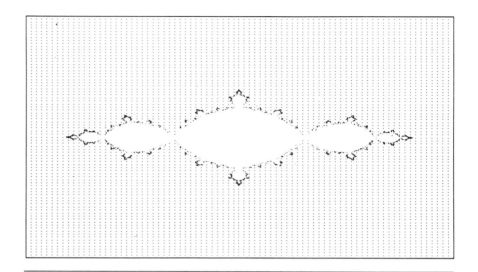

Theorem 2.1 *Let $\lambda \in \mathbb{C}$. Suppose that the dynamical system $\{\hat{\mathbb{C}}; f(z) = z^2 - \lambda\}$ possesses an attractive cycle $\{z_1, z_2, z_3, \ldots, z_p\} \subset \mathbb{C}$. Let ϵ be a very small positive number. Let $\mathbf{X}$ denote the Riemann Sphere $\hat{\mathbb{C}}$ with $(p + 1)$ open balls of radius ϵ removed. (The radius is measured using the spherical metric.) One ball is centered at each point of the cycle, and one ball is centered at the point at infinity, as illustrated in Figure VII.203. Define an IFS by*

$$\{\mathbf{X}; w_1(z) = \sqrt{z + \lambda}, w_2(z) = -\sqrt{z + \lambda}\}.$$

Then the transformation W on $\mathcal{H}(\mathbf{X})$, defined by

$$W(B) = w_1(B) \cup w_2(B) \text{ for all } B \in \mathcal{H}(\mathbf{X}),$$

maps $\mathcal{H}(\mathbf{X})$ into itself, continuously with respect to the Hausdorff metric on $\mathcal{H}(\mathbf{X})$. Moreover $W : \mathcal{H}(\mathbf{X}) \to \mathcal{H}(\mathbf{X})$ possesses a unique fixed point, J_λ, the Julia set for $z^2 - \lambda$. Also

$$\lim_{n \to \infty} W^{\circ n}(B) = J_\lambda \text{ for all } B \in \mathcal{H}(\mathbf{X}).$$

These conclusions also hold if the orbit of the origin, $\{f^{\circ n}(O)\}$, converges to the point at infinity, and $\mathbf{X} = \hat{\mathbb{C}} \setminus B(\infty, \epsilon)$.

Sketch of proof The fact that W takes $\mathcal{H}(\mathbf{X})$ continuously into itself follows from Theorem 4.1. To apply Theorem 4.1, three conditions must be met. These conditions are (i), (ii) and (iii), stated next. f is analytic on $\hat{\mathbb{C}}$ so (i) it is continuous, *and* (ii) it maps open sets to open sets. The way in which $\mathbf{X}$ is constructed ensures that, for small enough ϵ, (iii) $f(\mathbf{X}) \supset \mathbf{X}$. (The latter implies $W(\mathbf{X}) = f^{-1}(\mathbf{X}) \subset \mathbf{X}$.)

To prove that W possesses a unique fixed point we again make use of Theorem 4.1. Consider the limit $A \in \mathcal{H}(\mathbf{X})$ of the decreasing sequence of sets, $\{W^{\circ n}(\mathbf{X})\}$, namely,

$$A = \cap_{n=1}^\infty f^{\circ(-n)}(\mathbf{X}) = \lim_{n \to \infty} W^{\circ n}(\mathbf{X}).$$

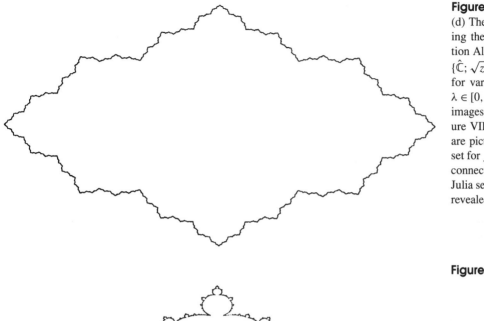

Figure VII.202. (a)–(d) The results of applying the Random Iteration Algorithm to the IFS $\{\hat{\mathbb{C}}; \sqrt{z + \lambda}, -\sqrt{z + \lambda}\}$ for various values of $\lambda \in [0, 3]$. Compare these images with those in Figure VII.198. The results are pictures of the Julia set for $f_\lambda(z) = z^2 - \lambda$. The connection between these Julia sets and IFS theory is revealed!

Figure VII.202. (b)

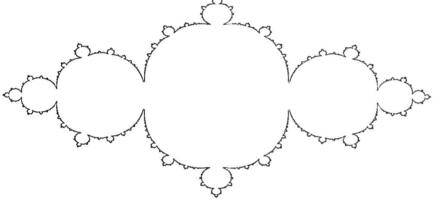

Figure VII.202. (c)

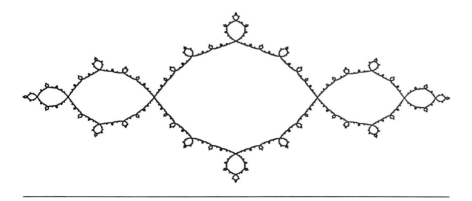

Figure VII.202.   (d)

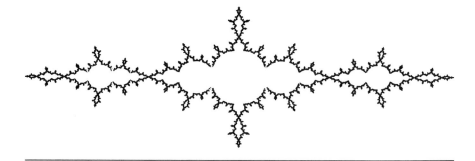

Figure VII.203.   The Riemann Sphere $\hat{\mathbb{C}}$ with a number of very small open balls of radius ϵ removed. One ball is centered at each of the points $\{z_p \in \mathbb{C}\}$ belonging to an attractive cycle of the transformation $f_\lambda(z) = z^2 - \lambda$. One ball is centered at the point at infinity.

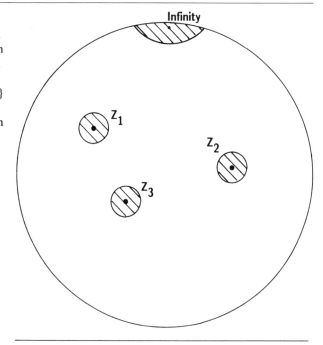

This obeys $W(A) = A$. It follows from [Brolin 65], Lemma 6.3, that $A = J_\lambda$, the Julia set. This completes the sketch of the proof.

Theorem 2.1 can be generalized to apply to polynomial tranformations f : $\hat{\mathbb{C}} \to \hat{\mathbb{C}}$ of degree N greater than 1. Here is a rough description: let $f^{-1}(z) = \{w_1(z), w_2(z), \ldots, w_N(z)\}$ denote a definition of branches of the inverse of f. Then consider the IFS $\{\hat{\mathbb{C}}; w_1(z), w_2(z), \ldots, w_N(z)\}$. This IFS is not hyperbolic: the "typical" situation is that the associated operator $W : \mathcal{H}(\hat{\mathbb{C}}) \to \mathcal{H}(\hat{\mathbb{C}})$ possesses a finite number of fixed points, all except one of which are "unstable." The one "stable" fixed point is J_f, and $W^{\circ n}(A) \to J_f$ for "most" $A \in \mathcal{H}(\hat{\mathbb{C}})$. In principle, J_f can be computed using the Random Iteration Algorithm.

Results like Theorem 2.1 are concerned with what are known as *hyperbolic* Julia sets. The Julia set of a rational transformation $f : \hat{\mathbb{C}} \to \hat{\mathbb{C}}$ is hyperbolic if, whenever $c \in \hat{\mathbb{C}}$ is a critical point of f, the orbit of c converges to an attractive cycle of f. The Julia set for $z^2 - 0.75$ is an example of a nonhyperbolic Julia set. We refer to [Peitgen 1986] as a good source of further information about Julia sets from the dynamical systems' point of view.

Explicit formulas for the inverse maps, $\{w_n(z) : n = 1, 2, \ldots, N\}$, for a polynomial of degree N, are not generally available. So the Random Iteration Algorithm cannot usually be applied. Pictures of Julia sets and filled Julia sets are often computed with the aid of the Escape Time Algorithm. The case of quadratic transformations is somewhat special, because both algorithms can be used. The Random Iteration Algorithm can also be applied to compute Julia sets of cubic and quartic polynomials, and of special polynomials of higher degree such as $z^n + \lambda$ where $n = 5, 6, 7, \ldots$, and $\lambda \in \mathbb{C}$.

Examples & Exercises

2.1. Consider the dynamical system $\{\hat{\mathbb{C}}; f(z) = z^2\}$. The origin, O, is an attractive cycle of period 1: indeed $f(O) = O$ and $|f'(O)| = 0 < 1$. Notice that $\lim_{n \to \infty} f^{\circ n}(z) = O$ for all $z \in B(O, 0.99999999)$. Let $\overset{\circ}{B}(z, r)$ denote the open ball on $\hat{\mathbb{C}}$, with center at z and radius r. Theorem 2.1 tells us that the IFS

$$\{\mathbf{X} = \hat{\mathbb{C}} \setminus \{\overset{\circ}{B}(O, 0.0000001) \cup \overset{\circ}{B}(\infty, 0.0000001)\}; w_1(z) = \sqrt{z}, w_2(z) = -\sqrt{z}\}$$

possesses a unique attractor. The attractor is actually the circle of radius 1 centered at the origin. It can be computed by means of the Random Iteration Algorithm. Notice that if we extend the space $\mathbf{X}$ to include O, then $O \in \mathcal{H}(\mathbf{X})$ and $O = W(O) = w_1(O) \cup w_2(O)$. If we extend $\mathbf{X}$ to include $\overset{\circ}{B}(O, 0.0000001)$, then the filled Julia set F_0 belongs to $\mathcal{H}(\mathbf{X})$ and obeys $F_0 = W(F_0)$. If we take $\mathbf{X}$ to be all of $\hat{\mathbb{C}}$ then $\hat{\mathbb{C}} = W(\hat{\mathbb{C}})$. In other words, if the space on which the IFS acts is too large then uniqueness of the "attractor" of the IFS is lost.

Can you find two more nonempty compact subsets of $\hat{\mathbb{C}}$ that are fixed points of W, in the case $\mathbf{X} = \hat{\mathbb{C}}$?

Establish that, for all $\lambda \in (-0.25, 0.75)$, the point $z_0 = 0.5 - \sqrt{0.25 + \lambda}$ is an attractive cycle of period 1 for $\{\hat{\mathbb{C}}; z^2 - \lambda\}$. Deduce that the corresponding IFS, acting on a suitably chosen space $\mathbf{X}$, possesses a unique attractor.

2.2. Let $\lambda \in (0.75, 1.25)$. Consider the dynamical system $\{\hat{\mathbb{C}}; f(z) = z^2 - \lambda\}$. Let $z_1, z_2 \in \mathbb{R}$ denote the two solutions of the equation $z^2 + z + (1 - \lambda) = 0$. Show $f(z_1) = z_2, f(z_2) = z_1, |(f^{\circ 2})'(z_1)| = |(f^{\circ 2})'(z_2)| < 1$ and hence that $\{z_1, z_2\}$ is an attractive cycle of period 2. Deduce that the IFS

$$\{\hat{\mathbb{C}} \setminus \{\overset{\circ}{B} (z_1, \epsilon) \cup \overset{\circ}{B} (z_2, \epsilon) \cup \overset{\circ}{B} (\infty, \epsilon)\}; +\sqrt{z + \lambda}, -\sqrt{z + \lambda}\}$$

possesses a unique attractor when ϵ is sufficiently small.

2.3. The Julia set J_λ for the polynomial $z^2 - \lambda$ is a union of two "copies" of itself. Identify these two copies for various values of λ. Explain how, when $\lambda = 1$, the two inverse maps $w_1^{-1}(z)$ and $w_2^{-1}(z)$ rip the Julia set apart, and the set map $W = w_1 \cup w_2$ puts it back together again. Where is the rip? Describe the geometry of what is going on here.

2.4. Consider the one-parameter family of polynomials $f(z) = z^3 - \lambda$, where $\lambda \in \mathbb{C}$ is the parameter. Give explicit formulas for the real and imaginary parts of three inverse functions $w_1(z)$, $w_2(z)$, and $w_3(z)$ such that $f^{-1}(z) = \{w_1(z), w_2(z), w_3(z)\}$ for all $\lambda \in \mathbb{C}$. Compute images of the filled Julia set for $f(z)$ for $\lambda = 0.01$ and $\lambda = 1$. Compare these images with those obtained by applying the Random Iteration Algorithm to the IFS $\{C; w_1(z), w_2(z), w_3(z)\}$.

2.5. Consider the dynamical system $\{\hat{\mathbb{C}}; f(z) = z^2 - \lambda\}$ for $\lambda > 2$. Show that $\{f^{\circ n}(O)\}$ converges to the point at infinity. Deduce that the IFS

$$\{\mathbf{X} = \hat{\mathbb{C}} \setminus B(\infty, \epsilon); +\sqrt{z + \lambda}, -\sqrt{z + \lambda}\}$$

possesses a unique attractor $A(\lambda)$. $A(\lambda)$ is a generalized Cantor set. Compute some pictures of $A(3)$. Use the Collage Theorem to help find a pair of affine transformations $w_i : \mathbb{R} \to \mathbb{R}$, $i = 1, 2$, such that the attractor of the IFS $\{\mathbb{R}; w_1, w_2\}$ is an approximation to $A(3)$. Define $\tilde{f} : \mathbb{R} \to \mathbb{R}$ by $\tilde{f}(x) = w_1^{-1}(x)$ when $x < 0$ and $\tilde{f}(x) = w_2^{-1}(x)$ when $x \geq 0$. Compare the graphs of the functions $f(x) = x^2 - 3$ and $\tilde{f}(x)$ for, say, $x \in [-4, 4]$. Compare one-dimensional "images" obtained by applying the Escape Time Algorithm in a similar manner to both $\{\mathbb{R}; f\}$ and $\{\mathbb{R}; \tilde{f}\}$.

One can sometimes obtain a *hyperbolic* IFS associated with a Julia set, if the domains and ranges of the inverse transformations are defined carefully. The following theorem provides such an example.

Theorem 2.2   *Let $\lambda \in [-0.249, 0.749]$, and let ϵ be a very small positive number. Let $a = 0.5 - \sqrt{0.25 + \lambda}$, an attractive fixed point of the dynamical system $\{\hat{\mathbb{C}}; f(z) = z^2 - \lambda\}$. Let $\tilde{\mathbf{X}} = \hat{\mathbb{C}} \setminus \{\overset{\circ}{B} (a, \epsilon) \cup \overset{\circ}{B} (\infty, \epsilon) \cup (0, \infty)\}$. That is, $\tilde{\mathbf{X}}$ consists of the Riemann Sphere with a small open ball centered at a, a small open ball centered at ∞, and the open interval $(0, \infty)$, removed. (This space is not compact because the edges of the lips of the cut, from O to ∞, are missing.) To each lip let there be attached copies of the pieces of the real interval $(0, \infty)$ which were removed, to provide a compact space $\mathbf{X}$, as illustrated in Figure VII.204. The distance $d(z_1, z_2)$ between a pair of points z_1 and $z_2 \in \mathbf{X}$ is the length (measured using the spherical metric) of the shortest path that lies in $\mathbf{X}$ and connects z_1 to z_2. (Paths in $\mathbf{X}$ cannot cross the cut, they have to go around it.) $(\mathbf{X}, d)$ is a compact metric space. Define $w_1 : \mathbf{X} \to \mathbf{X}$ by $w_1(z) = \sqrt{z + \lambda}$, the root that lies in the "upper half*

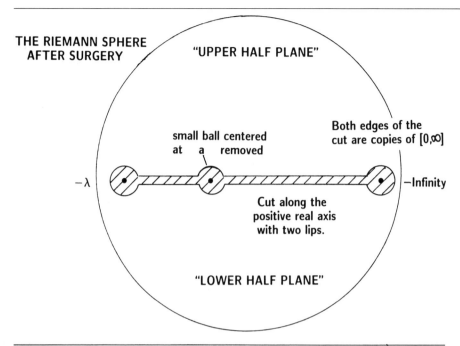

THE RIEMANN SPHERE
AFTER SURGERY

"UPPER HALF PLANE"

small ball centered
at   a   removed

Both edges of the
cut are copies of [0,∞]

$-\lambda$

$-$Infinity

Cut along the
positive real axis
with two lips.

"LOWER HALF PLANE"

Figure VII.204. Construction of a compact metric space $\mathbf{X}$ for IFS $\{\mathbf{X}; \sqrt{z+\lambda}, -\sqrt{z+\lambda}\}$ with $\lambda \in (-0.25, 0.75)$, used in Theorem 2.2. Both sides of the "slit" from $-\lambda$ to infinity belong to the space. The distance between a pair of points on $\mathbf{X}$ is the length of the shortest path that connects the points without crossing the slit. The distance between points may be much greater than it looks.

plane." For z on the upper edge of the cut, $w_1(z)$ is also on the upper edge. For z on the lower edge of the cut, $w_1(z)$ lies on the negative real axis. Define $w_2 : \mathbf{X} \to \mathbf{X}$ by $w_2(z) = -\sqrt{z+\lambda}$, the root that lies in the "lower half plane." For z on the the upper edge of the cut, $w_2(z)$ lies on the negative real axis; and, for z located on the lower edge of the cut, $w_2(z)$ lies on the lower edge of the cut.

Then there is a metric on $\mathbf{X}$, equivalent to the metric d, such that the IFS $\{\mathbf{X}; w_1, w_2\}$ is hyperbolic. The attractor is the Julia set J_λ for $z^2 - \lambda$, where the real point $0.5 + \sqrt{0.25 + \lambda}$ is repeated on both the upper and the lower edges of the cut.

Sketch of proof Let $e = (e_1, e_2, \ldots, e_n, \ldots) \in \Sigma$, the code space on the two symbols $\{1, 2\}$. Define a sequence of nonempty compact subsets of $\mathbf{X}$ by

$$X_n(e) = w_{e_1} \circ w_{e_2} \circ w_{e_3} \circ w_{e_4} \circ w_{e_5} \circ w_{e_6} \circ \cdots \circ w_{e_n}(\mathbf{X}) \text{ for } n = 1, 2, 3, \ldots.$$

It follows, using [Brolin], Theorem 6.2, and Lemma 6.3, that the sequence $\{X_n \in \mathcal{H}(\mathbf{X})\}$ converges to a singleton, say $\{\phi(e)\}$, where $\phi(e) \in J_\lambda$ and

$$\cup_{e \in \Sigma} \phi(e) = J_\lambda.$$

A beautiful theorem of Elton [Elton 1988] applies under just these conditions and provides the conclusion of the theorem. This completes the outline of the proof.

In those situations where the IFS $\{\mathbf{X}; +\sqrt{z+\lambda}, -\sqrt{z+\lambda}\}$ is hyperbolic one can use the associated code space to discuss both the Julia set and the associated shift dynamical system $\{J_\lambda; f(z) = z^2 - \lambda\}$. Here we give some of the flavor of such a discussion. More details can be found in [Barnsley 1984].

For the remainder of this section let $\lambda \in (-0.25, 0.75)$ and consider the IFS $\{\mathbf{X}; w_1(z) = \sqrt{z + \lambda}, w_2(z) = -\sqrt{z + \lambda}\}$, as defined in Theorem 2.2. Let Σ denote the code space on the two symbols $\{1, 2\}$, and let $\phi : \Sigma \to J_\lambda$ denote the associated code space map, introduced in Theorem 2.1 in Chapter IV. If $e = (e_1, e_2, \ldots, e_n, \ldots) \in \Sigma$, then

$$\phi(e) = \lim_{n \to \infty} w_{e_1} \circ w_{e_2} \circ w_{e_3} \circ w_{e_4} \circ w_{e_5} \circ w_{e_6} \circ \cdots \circ w_{e_n}(z).$$

Replace the symbol "1" by the symbol "+" and replace the symbol "2" by the symbol "−". Then the point $\phi(e)$ on the Julia set J_λ can be represented by the formula

$$\phi(e) = e_1 \sqrt{\lambda e_2 \sqrt{\lambda e_3 \sqrt{\lambda e_4 \sqrt{\lambda e_5 \sqrt{\lambda e_6 \sqrt{\lambda e_7 \sqrt{\lambda e_8 \ldots e_n \sqrt{\lambda} \ldots}}}}}}},$$

where $e_i \in \{+, -\}$ for each positive integer i. The set J_λ itself can be represented by the collection of formulas

$$\pm \sqrt{\lambda \pm \sqrt{\lambda \pm \sqrt{\lambda \pm \sqrt{\lambda \pm \sqrt{\lambda \pm \sqrt{\lambda \pm \cdots \pm \sqrt{\lambda} \cdots}}}}}}, \tag{1}$$

where all possible sequences of plus and minus signs are permitted. A particular sequence of signs, corresponding to a point in J_λ, is an address of the point.

In Figure VII.205 we show the Julia set for $z^2 - 0.7$, with the addresses of various points marked on it. Some points on $J_{0.7}$ have multiple addresses while others have single addresses. It appears that the the IFS is just-touching.

The shift dynamical system associated with the IFS is $\{J_\lambda; f(z) = z^2 - \lambda\}$. Notice how the set of points represented by the formulas in equation 1 is mapped into itself by the function that "squares" a formula and subtracts λ from the result. A point on a cycle of period 2 is represented by

$$+ \sqrt{\lambda - \sqrt{\lambda + \sqrt{\lambda - \sqrt{\lambda + \sqrt{\lambda - \sqrt{\lambda + \sqrt{\lambda - \cdots + \sqrt{\lambda} \cdots}}}}}}}.$$

The other point on this cycle is obtained by squaring the formula and subtracting λ.

In Theorem 2.4 in Chapter IV, we learned that the set of periodic points of the shift dynamical system associated with a hyperbolic IFS is dense in the attractor of the IFS. Here this tells us that the set of periodic points of the dynamical system

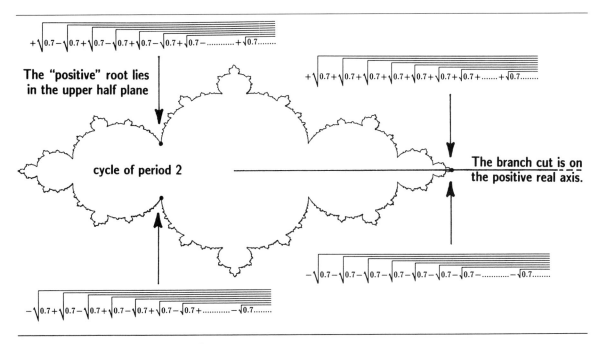

Figure VII.205. The Julia set for $z^2 - 0.7$, labelled with various addresses. Chaotic dynamics takes place on the Julia set and orderly dynamics takes place off it. Boundaries of a fractal character often separate regions where the dynamical system behaves differently. The behavior of the dynamical system on such a boundary may then be indecisive, and in some way chaotic.

$\{J_\lambda;\ f(z) = z^2 - \lambda\}$ is dense in J_λ. In fact, a related idea was the starting point of Julia's original investigations. He considered dynamical systems of the form $\{\hat{\mathbb{C}};\ f(z)\}$, where $f(z)$ is analytic. He defined the (Julia) set to be the closure of the set of repulsive cycles of f.

Following Theorem 8.1 in Chapter IV we explained the sense in which the shift dynamical system associated with a hyperbolic IFS is chaotic. In the present context we learn that the dynamical system $\{J_\lambda;\ z^2 - \lambda\}$ is chaotic.

One can think of the dynamical system $\{\hat{\mathbb{C}};\ z^2 - \lambda\}$ as being the union of two dynamical systems, a chaotic one $\{J_\lambda;\ z^2 - \lambda\}$ and an orderly one $\{\hat{\mathbb{C}} \setminus J_\lambda;\ z^2 - \lambda\}$. The orbit of any point in the latter system converges to a fixed point of the transformation. The orbits of "most" points in the former system are wild. In practice they are usually so wild they cannot be constrained to remain on the repelling set J_λ. They escape and thereafter behave in a rather predictable manner.

An example of chaotic dynamics on a Julia set is provided by the dynamical system $\{[0, 1];\ f(x) = 4x(1 - x)\}$. The interval $[0, 1]$ is exactly the Julia set for the transformation. This system is close to the "chaotic" one illustrated in Figure IV.108.

Examples & Exercises

2.6. The Julia set for $z^2 - 2$ is the interval $[-2, 2]$. Show that the shift dynamical system associated with the IFS $\{[-2, 2]; +\sqrt{z+2}, -\sqrt{z+2}\}$ is precisely the dynamical system $\{[-2, 2]; z^2 - 2\}$. Use a chain of square roots to locate a cycle of minimal period 3.

2.7. Verify numerically that for various choices of $\pm$ on each square root, and for various complex numbers λ such that $|\lambda|$ is very small, the expression below evaluates approximately to a complex number that lies on the unit circle centered at the origin, if enough square roots are taken. ($+\sqrt{z}$ means the solution, w, of the equation $w^2 = z$, which lies either on the nonnegative real axis or in the upper half plane.) Make a hand-waving explanation of why this is, in terms of Julia set theory.

$$\pm\sqrt{\lambda \pm \sqrt{\lambda \pm \sqrt{\lambda \pm \sqrt{\lambda \pm \sqrt{\lambda \pm \sqrt{\lambda \pm \sqrt{\lambda \pm \cdots \cdots \cdots \pm \sqrt{\lambda}}}}}}}}$$

2.8. Design an IFS with condensation such that its attractor looks like an infinite nested chain of square root signs.

$$\sqrt{\sqrt{\sqrt{\sqrt{\cdots}}}}$$

2.9. Figure VII.206 represents a sequence of sets $\{A_n\}$ which converges to the Julia set of $f(z) = z^2 - 1$. A_0 denotes the union of the two largest faces and $A_n = f^{\circ(-n)}(A_0)$. Identify the set A_2.

3 The Application of Julia Set Theory to Newton's Method

We are familiar, since our first course in calculus, with Newton's method for computing solutions of the equation $F(x) = 0$. Or are we?

Consider the polynomial $F(z) = z^4 - 1$ for $z \in \mathbb{C}$. There are four distinct complex numbers, a_i ($i = 1, 2, 3, 4$) such that $F(a_i) = 0$. These are called the roots, or the zeros, of the polynomial $F(z)$. Newton's method provides a means to compute them. Pretend that we do not know that $a_1 = 1$, $a_2 = -1$, $a_3 = i$, and $a_4 = -i$. Then Newton tells us to consider the dynamical system

$$\{\hat{\mathbb{C}}; f(z) = z - \frac{F(z)}{F'(z)}\}.$$

We call $f(z)$ the *Newton transformation* associated with the function $F(z)$. The general expectation is that a typical orbit $\{f^{\circ n}(z_0)\}$, which starts from an initial

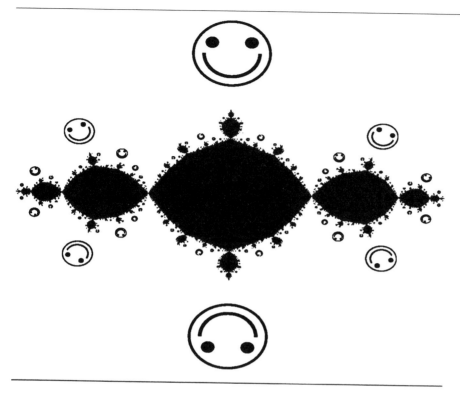

Figure VII.206. Let $f : \hat{\mathbb{C}} \to \hat{\mathbb{C}}$ denote the polynomial $z^2 - 1$. Let A_0 be the union of the two largest smiley faces. Define a transformation $W : \mathcal{H}(\hat{\mathbb{C}}) \to \mathcal{H}(\hat{\mathbb{C}})$ by $W(B) = f^{-1}(B)$ for all $B \in \mathcal{H}(\hat{\mathbb{C}})$ and let $A_n = W^{\circ n}(A_0)$ for $n = 0, 1, 2, \ldots$. Then the sequence of sets $\{A_n\}$ converges to the Julia set $z^2 - 1$. Can you identify A_2?

"guess" $z_0 \in \mathbb{C}$, will converge to one of the roots of $F(z)$. In the present example, the Newton transformation is given by

$$f(z) = \frac{3z^4 + 1}{4z^3}.$$

We expect the orbit of z_0 to converge to one of the numbers a_1, a_2, a_3, or a_4. If we choose z_0 close enough to a_i, then it is readily proved that

$$\lim_{n \to \infty} f^{\circ n}(z_0) = a_i, \qquad \text{for } i = 1, 2, 3, 4.$$

If, on the other hand, z_0 is far away from all of the a_i's, then what happens? Perhaps the orbit of z_0 converges to the root of $F(z)$ closest to z_0? Or perhaps the orbit does not settle down, but wanders, hopelessly, forever?

Let us make a computergraphical experiment to help answer these questions. We use the Escape Time Algorithm to produce a picture of those points $z_0 \in \hat{\mathbb{C}}$ whose orbits converge to a_1. Define $\mathcal{W} = \{(x, y) \in \mathbb{C} : -2 \leq x \leq 2, -2 \leq y \leq 2\}$ and $\mathcal{V} = \{z \in \mathbb{C} : |z - a_1| \leq 0.0001\}$. The real and imaginary parts of $f(x + iy)$ are given by

$$f(x + iy) = \frac{(ce + df)}{(e^2 + f^2)} + i \frac{(de - cf)}{(e^2 + f^2)},$$

where $a = x^2 - y^2$, $b = 2xy$, $c = 3a^2 - 3b^2 + 1$, $d = 6ab$, $e = 4(xa - yb)$, and $f = 4(xb + ya)$. Program 7.1.1 is modified accordingly. Pixels corresponding to points in $\mathcal{W}$ whose orbits reach $\mathcal{V}$ in less than a fixed number of iterations are plotted. A picture resulting from such an experiment is shown in Figure VII.207. See also Figures VII.208 and VII.209.

Color Plate 14 shows the output from another such experiment. This time Mercator's projection is used to represent the Riemann sphere, and points whose orbits converge to the different points a_1, a_2, a_3, and a_4 are plotted in different colors.

The following definition is equivalent to Definition 1.1 in the case of polynomials, [Brolin].

Definition 3.1 *The Julia set of a rational function $f : \hat{\mathbb{C}} \to \hat{\mathbb{C}}$, of degree greater than 1, is the closure of the set of repulsive periodic points of the dynamical system $\{\hat{\mathbb{C}}; f\}$.*

For the rational function $f(z)$ considered above, one can prove that the Julia set J is the same as the set of points whose orbits do not converge to any one of the points a_1, a_2, a_3, a_4. In Figure VII.207, $J \cap \mathcal{W}$ is represented by the boundary between the black and white regions. In Color Plate 14, $J \cap \mathcal{W}$ is the place where the four colors meet. The complement of the Julia set consists of four open sets, the *basins of attraction* of the four attractive fixed points of the Newton iteration scheme. In Color Plate 14 the dark green region represents part of the basin of attraction of a_1. The black regions in the color plate are caused by (a) rounding errors, and (b) the fact that only 100 points on each orbit are tested, for convergence to one of the points a_1, a_2, a_3, a_4.

The Julia set J is the part of $\hat{\mathbb{C}}$ on which chaotic dynamics occurs. It can be characterized as the closure of the set of points whose orbits wander, hopelessly, forever. Orderly, slightly boring motion takes place on $\hat{\mathbb{C}} \setminus J$. J is the boundary of the white region. It is the boundary of the dark green region. It is a bonafide fractal, yet nobody knows its fractal dimension.

There is a beautiful theorem of Sullivan, which can be illustrated using the "petals" in Color Plate 14. The complement of the Julia set is the union of a countable collection of connected open sets, which we call petals. If P is a petal then $f(P)$ is another petal. The Non-Wandering Domain Theorem [Sullivan 1982] says that no connected component of the complement of the Julia set wanders, hopelessly, forever. It always settles into a periodic orbit of petals. If P is a petal in the present example, then one can prove that there is a positive integer S so that $f^{\circ S}(P) = f^{\circ(S+1)}(P) = f^{\circ(S+2)}(P) = f^{\circ(S+3)}(P) = \cdots$. The final petal $f^{\circ S}(P)$ is one of the connected components of the complement of the Julia set that contains one of the points a_1, a_2, a_3, a_4. Each petal is eventually periodic. The orbit of a petal ends up in a cycle of petals of period 1.

How are we to think about this fabulous Julia set? IFS theory provides a simple point of view, as we show next. We begin by defining the inverse map associated

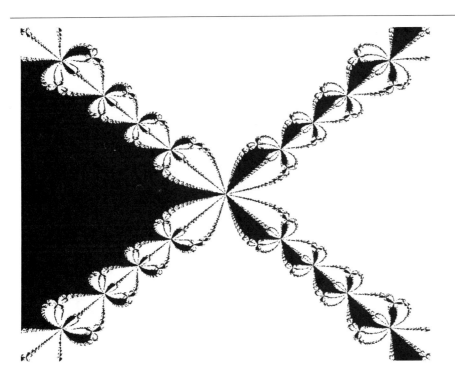

Figure VII.207. The Escape Time Algorithm is applied to analyze Newton's method for finding the complex roots of the polynomial $z^4 - 1$. The boundary of this region represents the Julia set for the rational function $f(z) = (3z^4 + 1)/4z^3$. The points plotted black are those points $z = x + iy$ with $-2 \leq x \leq 2$, and $-2 \leq y \leq 2$, whose orbits intersect $\mathcal{V} = \{z \in \mathbb{C} : |z + 1| \leq 0.0001\}$ in less than 1,000 iterations.

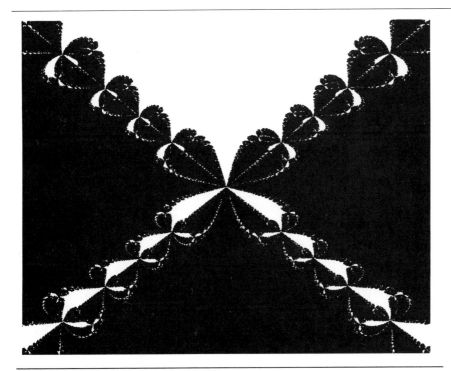

Figure VII.208. The boundary of this region represents the Julia set for the rational function $f(z) = (3z^4 + 1)/4z^3$. The two shades of gray, black and white, correspond to the basins of attraction of the four attractive fixed points of $f(z)$. To which point in $\mathbb{C}$ do the orbits of points in the white region converge?

Figure VII.209. Mercator's projection of the Riemann sphere showing the basins of attraction of the four attractive fixed points of the Newton transformation of $z^4 - 1$. The top of the rectangle corresponds to the point at infinity, and the bottom of the box corresponds to the origin. The points -1, $+i$, $+1$, and -1 lie on the equator. The shading follows the same convention as in (a) and (b). Which point on $\hat{\mathbb{C}}$ is represented by the midpoint of this image?

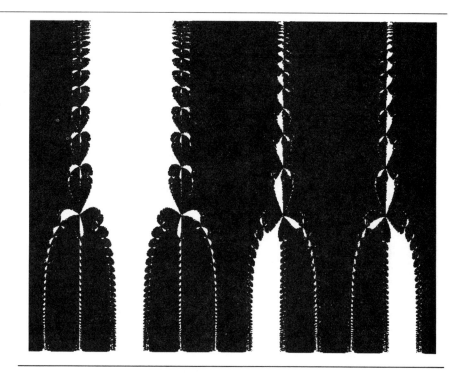

with f. Let $z \in \hat{\mathbb{C}}$ be given and solve

$$z = \frac{3w^4 + 1}{4w^3},$$

to find w in terms of z. This leads to the quartic equation

$$3w^4 - 4zw^3 + 1 = 0.$$

This has four solutions, when we count solutions according to their multiplicities. We can organize these solutions to provide four functions; that is, we write $f^{-1}(z) = \{w_1(z), w_2(z), w_3(z), w_4(z)\}$. Then the Julia set is the "attractor" for the IFS $\{\hat{\mathbb{C}}; w_i, i = 1, 2, 3, 4\}$. However, as in the case of quadratic transformations on $\hat{\mathbb{C}}$, this statement must be treated cautiously: for example, clearly this IFS admits more than one invariant set.

Theorem 3.1 *Let $f : \hat{\mathbb{C}} \to \hat{\mathbb{C}}$ be the Newton transformation associated with the polynomial $z^4 - 1$. Let $\epsilon > 0$ be very small. Let $X = \hat{\mathbb{C}} \setminus \cup_{i=1}^{4} \overset{\circ}{B}(a_i, \epsilon)$ where $a_1 = 1$, $a_2 = -1$, $a_3 = i$, and $a_4 = -i$. As above, define $W : \mathcal{H}(X) \to \mathcal{H}(X)$ by*

$$W(B) = \cup_{i=1}^{4} w_i(B) = f^{-1}(B) \qquad \text{for all } B \in \mathcal{H}(X).$$

Then W is continuous, possesses a unique fixed point, J, the Julia set of f, and

$$\lim_{n \to \infty} W^{\circ n}(B) = J \qquad \text{for all } B \in \mathcal{H}(X).$$

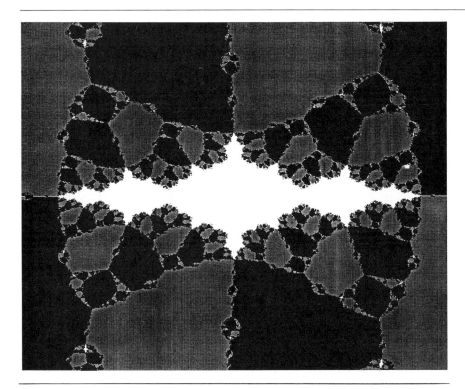

Figure VII.210. The Escape Time Algorithm is applied to a Newton transformation $f(z)$ associated with a cubic polynomial. $f(z)$ possesses an attractive cycle $\{b_1, b_2\}$ of minimal period 2. The basin of attraction of the two-cycle is represented in white. Points whose orbits arrive within a distance of 0.01 of the cycle, prior to 100 iterations, are plotted in white. Does the basin of attraction of the cycle look familiar?

Sketch of proof This is essentially the same as the sketch of the proof of Theorem 2.1.

The Newton transformation associated with a polynomial may possess an attractive cycle of minimal period greater than 1. This cycle may not be directly related to the roots of the polynomial. As as example consider the Newton transformation $f(z)$ associated with the polynomial

$$F(z) = z^3 + (\lambda - 1)z + 1.$$

$\lambda \in \mathbb{C}$ can be chosen so that $f(z)$ possesses an attractive cycle $\{b_1, b_2\}$ of minimal period 2. Figure VII.210 illustrates the basin of attraction of the cycle. The Escape Time Algorithm was used to obtain this image. Points whose orbits arrive within a distance 0.01 of the cycle, prior to 100 iterations, are plotted in black. Accordingly, the escape region is $\mathcal{V} = B(b_1, 0.00001) \cup B(b_2, 0.00001)$. Notice the resemblance of the basin of attraction of $\{b_1, b_2\}$ to the filled Julia set for $z^2 - 1$. This similarity is not accidental. It can be explained using the theory of "polynomial-like" mappings [Douady 1985].

Some interesting computergraphical experiments involving Julia sets for Newton's method are described in [Curry 1983], [Peitgen 1986], and [Vrscay 1986].

Examples & Exercises

3.1. Verify that $z = 1$ is an attractive fixed point for the Newton transformation associated with $F(z) = z^4 - 1$.

3.2. The Newton transformation associated with the polynomial $F(z) = z^2 + 1$ is

$$f(z) = \frac{1}{2}\left(z - \frac{1}{z}\right).$$

Show the corresponding IFS is $\{\hat{\mathbb{C}}; w_1(z) = z + \sqrt{z^2 + 1}, w_2(z) = z - \sqrt{z^2 + 1}\}$, where the square root is defined appropriately. Verify that $A = \mathbb{R} \cup \{\infty\}$ is an attractor of the IFS. Prove, or give evidence to show, that A is the Julia set for $f(z)$. (Hint: see exercise 3.7.) How could the space $\hat{\mathbb{C}}$ be modified so that the IFS has a unique attractor? Notice that numerically computed orbits of points on A, under the dynamical system $\{\hat{\mathbb{C}}; f\}$, can be constrained from escaping from A by keeping imaginary parts equal to zero. Verify numerically that the dynamics of $\{A; f\}$ are wild.

3.3. Find the Newton transformation $f(z)$ associated with the polynomial $F(z) = z^3 - 1$. Use the Escape Time Algorithm to obtain an image, analogous to Figure VII.207, which illustrates this Julia set. Discuss the dynamics of the "petals" in the image.

3.4. In this example we speculate on the application of fractal geometry to biological modelling. Let $F_\lambda(z) = (z - i\lambda)(z - 1)(z + 1)$, where λ is a real parameter, and let $f_\lambda(z)$ denote the associated Newton transformation. Let J_λ denote the Julia set for $f_\lambda(z)$. In Figure VII.211 we show images relating to J_λ, for an increasing sequence of values of λ. These images were computed by applying the Escape Time Algorithm to f_λ.

These images show complex blobs that are reminiscent of something small, biological, and organic. They make one think of the nuclei of cells; of collections of cells during the early stages of development of an embryo; of the process of cell division; and of protozoans. As we track the images we see that the blobs pass through one another. Somehow they do so while preserving their complex geometries. Their geometries seem to interact with one another. Such images suggest that fractal geometry can do more than provide a means for modelling static biological structures, such as ferns: it appears feasible to construct deterministic fractal models, which describe the processes of physiological change that occur during the growth, metamorphosis, and movement of living organisms.

3.5. Find the Newton transformation $f(z)$ associated with the function $F(z) = e^z - 1$. What are the attractive fixed points of the dynamical system $\{\mathbb{C}; f\}$? Figure VII.212 was computed using the Escape Time Algorithm applied to $f(z)$, with $W = \{(x, y) \in \mathbb{R}^2 : -2.5 \leq x \leq 2.5, -2.5 \leq y \leq 2.5\}$. Describe the main features of the image. Explain, roughly, the causes of some of these features.

3.6. What are the "petals" in the case of the Julia set for $z^2 - 1$? Use a picture of

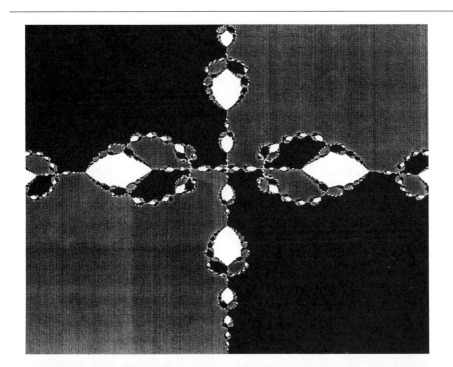

Figure VII.211. (a)–(h) Julia sets associated with a one-parameter family of dynamical systems. Can such systems be used to model biological processes such as meiosis?

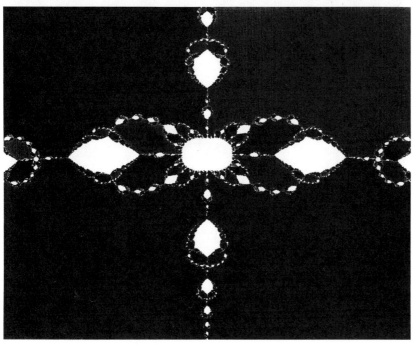

Figure VII.211. (b)

Figure VII.211. (c)

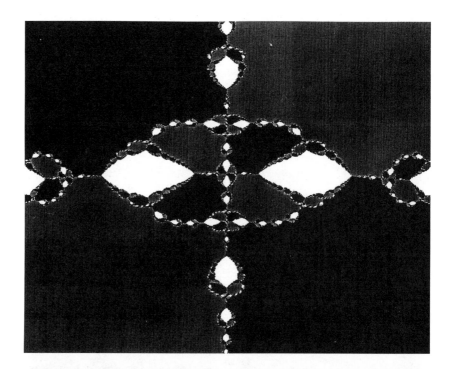

Figure VII.211. (d)

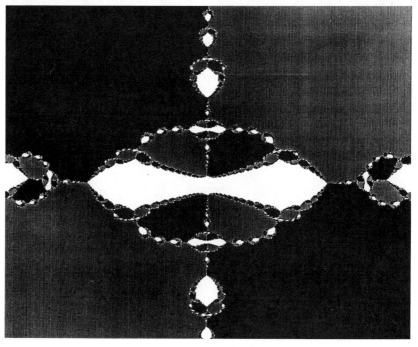

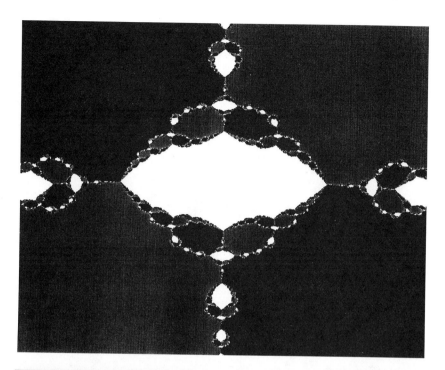

Figure VII.211. (e)

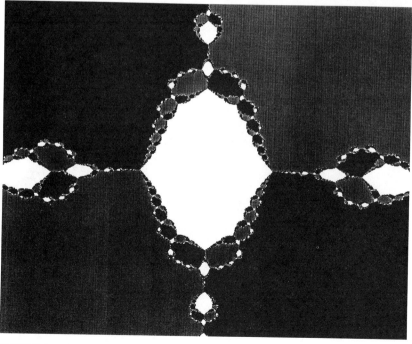

Figure VII.211. (f)

Figure VII.211. (g)

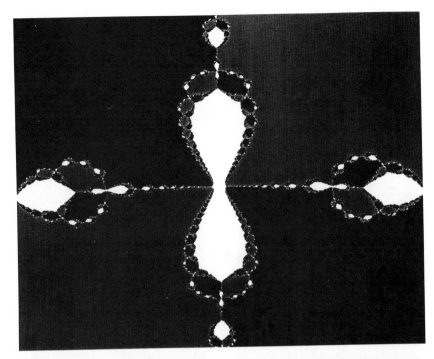

Figure VII.211. (h)

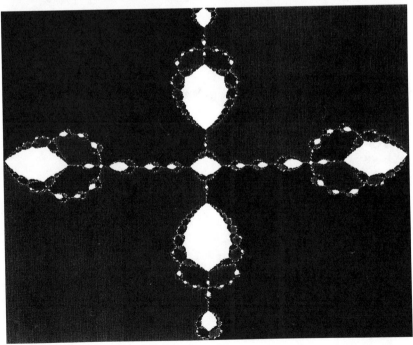

Figure VII.212. This image was computed using the Escape Time Algorithm applied to the Newton transformation associated with $f(z) = e^z - 1$. The viewing window is $\mathcal{W} = \{(x, y) \in \mathbb{R}^2 : -2.5 \le x \le 2.5, -2.5 \le y \le 2.5\}$. Can you work out what "escape region" was used?

the Julia set for $z^2 - 1$ to illustrate the orbit of a tiny petal that is eventually periodic with minimal period 2.

3.7. By making an explicit change of the coordinates, using a Möbius transformation, show that the following two dynamical systems are equivalent:

$$\{\hat{\mathbb{C}}; f(z) = \frac{1}{2}\left(z - \frac{1}{z}\right)\} \text{ and } \{\hat{\mathbb{C}}; f(z) = z^2\}.$$

4   A Rich Source for Fractals: Invariant Sets of Continuous Open Mappings

Let f be a transformation that acts on a space $\mathbf{X}$. Recall that a set A is invariant under f if $f^{-1}(A) = A$, and that this implies $f(A) = A$. We are interested in invariant sets of f that belong to $\mathcal{H}(\mathbf{X})$. The following theorem is a theoretical tool that provides both the existence and a means for the computation of invariant sets. Most of the material in this chapter is based on it.

Theorem 4.1   *Let* $(\mathbf{Y}, d)$ *be a metric space. Let* $\mathbf{X} \subset \mathbf{Y}$ *be compact and nonempty. Let* $f : \mathbf{X} \to \mathbf{Y}$ *be continuous and such that* $f(\mathbf{X}) \supset \mathbf{X}$. *Then, (1) a transformation* $W : \mathcal{H}(\mathbf{X}) \to \mathcal{H}(\mathbf{X})$ *is defined by*

$$W(A) = f^{-1}(A) \qquad \text{for all } A \in \mathcal{H}(\mathbf{X}).$$

(2) W possesses a fixed point $A \in \mathcal{H}(\mathbf{X})$, given by

$$A = \cap_{n=0}^{\infty} f^{\circ(-n)}(\mathbf{X}) = \lim_{n \to \infty} W^{\circ n}(\mathbf{X}).$$

Suppose f obeys the additional condition that $f(\mathcal{O})$ is an open subset of the metric space $(f(\mathbf{X}), d)$ whenever $\mathcal{O} \subset \mathbf{X}$ is an open subset of the metric space $(\mathbf{X}, d)$. Then (3) W is a continuous transformation from the metric space $(\mathcal{H}(\mathbf{X}), h(d))$ into itself.

Proof of (1) and (2). The proof of (3) can be found in (Barnsley 1988c).. (1) We begin by proving that W maps $\mathcal{H}(\mathbf{X})$ into $\mathcal{H}(\mathbf{X})$. Let $B \in \mathcal{H}(\mathbf{X})$. The condition $f(\mathbf{X}) \supset \mathbf{X}$ implies that $f^{-1}(B) \subset \mathbf{X}$, and that $f^{-1}(B)$ is nonempty. B is compact, so it is a closed set in the metric space $(\mathbf{X}, d)$. It follows that $\mathbf{X} \setminus B$ is open. The continuity of f implies that $f^{-1}(\mathbf{X} \setminus B)$ is open. Since $f(\mathbf{X}) \supset \mathbf{X} \supset B$, it follows that $f^{-1}(B) = \mathbf{X} \setminus f^{-1}(\mathbf{X} \setminus B)$. Hence $f^{-1}(B)$ is closed in the metric space $(\mathbf{X}, d)$. Since $\mathbf{X}$ is compact it follows that $f^{-1}(B)$ is compact. This completes the proof of (1).

(2) Since $f(\mathbf{X}) \supset \mathbf{X}$, it follows that $\mathbf{X} \supset f^{\circ(-1)}(\mathbf{X})$. Application of $f^{\circ(-n)}$ to both sides of the latter equation yields

$$\mathbf{X} \supset f^{\circ(-1)}(\mathbf{X}) \supset f^{\circ(-2)}(\mathbf{X}) \supset f^{\circ(-3)}(\mathbf{X}) \cdots \supset f^{\circ(-n)}(\mathbf{X}) \supset \cdots.$$

It follows that $\{f^{\circ(-n)}(\mathbf{X})\}$ is a Cauchy sequence in $\mathcal{H}(\mathbf{X})$, and it possesses a limit $A \in \mathcal{H}(\mathbf{X})$, given by

$$A = \cap_{n=0}^{\infty} f^{\circ(-n)}(\mathbf{X}) = \lim_{n \to \infty} W^{\circ n}(\mathbf{X}).$$

It remains to be proved that A is a fixed point of f. We need to show that $f^{\circ(-1)}(\cap_{n=0}^{\infty} A_n) = \cap_{n=0}^{\infty} A_n$ where $A_n = f^{\circ(-n)}(\mathbf{X})$ for $n = 1, 2, \ldots$. First we prove that $f^{\circ(-1)}(\cap_{n=0}^{\infty} A_n) \supset \cap_{n=0}^{\infty} A_n$. Suppose $x \in f^{\circ(-1)}(\cap_{n=0}^{\infty} A_n)$. Then there is $y \in \cap_{n=0}^{\infty} A_n$ such that $x \inf^{\circ(-1)}(y)$, and $y \in A_n$ for $n = 0, 1, 2, \ldots$. It follows that $x \in f^{\circ(-1)}(A_n) = A_{n+1}$ for $n = 0, 1, 2, \ldots$. It follows that $x \in \cap_{n=0}^{\infty} A_n$. To prove the inclusion the other way around, suppose that $x \in \cap_{n=0}^{\infty} A_n$. Then $x \in A_{n+1} = f^{-1}(A_n)$ for $n = 0, 1, 2, 3, \ldots$. It follows that there is $y_n \in A_n$ such that $f(y_n) = x$ for $n = 0, 1, 2, 3 \ldots$. The sequence $\{y_n\}$ possesses a convergent subsequence. Let y denote the limit of this subsequence. Then $y \in A_n$ for $n = 0, 1, 2, \ldots$, and so $y \in \cap_{n=0}^{\infty} A_n$. Since f is continuous it follows that $f(y) = x$. Hence $x \in f^{\circ(-1)}(\cap_{n=0}^{\infty} A_n)$. We have shown that $f^{\circ(-1)}(\cap_{n=0}^{\infty} A_n) \subset \cap_{n=0}^{\infty} A_n$. This completes the proof of (2).

The invariant set A referred to in Theorem 4.1 can be expressed

$$A = \{x \in \mathbf{X} : f^{\circ n}(x) \in \mathbf{X} \text{ for all } n = 1, 2, 3, \ldots\}.$$

That is, A is the set of points whose orbits do not escape from $\mathbf{X}$. Also, A is the complement of the set of points whose orbits do escape. If $\mathbf{X} \subset \mathbb{R}^2$, then pictures of A can be computed by using the Escape Time Algorithm.

The last statement in the theorem expresses a desirable property for a transformation W on $\mathcal{H}(\mathbf{X})$. If W is not continuous, yet $A_0 \in \mathcal{H}(\mathbf{X})$ and $\{W^{\circ n}(A_0)\}$ converges to $A \in \mathcal{H}(\mathbf{X})$, one cannot conclude that $W(A) = A$. Without continuity of W, one should not trust the results of applying the Escape Time Algorithm. For example, slight numerical errors may mean that a computed sequence of sets $\{\tilde{A}_n \approx W^{\circ n}(\mathbf{X})\}$ is not decreasing. One may still wish to define $\tilde{A} = \lim \tilde{A}_n$. Without continuity one cannot suppose that $f^{-1}(\tilde{A}) = \lim W(\tilde{A}_n) = \tilde{A}$, even approximately.

Analytic transformations map open sets to open sets. Hence their inverses act continuously on the space $\mathcal{H}(\mathbf{X})$, where $\mathbf{X} \subset \hat{\mathbb{C}}$ is chosen appropriately. To help visualize this, look back at Figure III.52. If the Sierpinski triangle ABO is deformed or moved, its inverse image $POQ \cup \tilde{P}O\tilde{Q}$ will move continuously with it.

The Hausdorff metric is the metric of perception: what we call a small change in the appearance of a picture is probably a small change in a Hausdorff distance. When one talks about continuous motion in the context of graphics, continuous growth in the context of botany, or continuous change in the context of a chemical system, the word "continuous" can often be replaced, pedantically, by "continuous in the Hausdorff metric." Theorem 4.1 suggests that one could use continuous *open* maps to model such systems.

Examples & Exercises

*4.1.   Let $\lambda \in [-1, 1]$. Define a transformation $f : \mathbb{R}^2 \to \mathbb{R}^2$ by

$$f(x, y) = \begin{cases} (x^2 - y^2 - 1, 2xy) & \text{when } x > 0, \\ (x^2 - y^2 - 1 + \lambda x, 2xy) & \text{when } x \leq 0. \end{cases}$$

Show that f is continuous. Show that if $\mathbf{X}$ denotes a ball, centered at the origin, of sufficiently large radius, then $f(\mathbf{X}) \supset \mathbf{X}$. Also verify that if $\lambda \in [-1, 0]$ then f maps open sets into open sets. Show that this is not the case for $\lambda = 1$. (Hint: look at what the map does to a very small disk centered at the origin.)

Figure VII.213 shows the result of applying the Escape Time Algorithm to f when $\lambda = 1$. The inner region, bounded by an ellipse, actually represents a disk $\mathbf{X}$ such that $f(\mathbf{X}) \supset \mathbf{X}$. Different scales have been used in the x and y directions. $f(\mathbf{X})$ is the region bounded by the outer curve. The image of a point that goes once around the inner ellipse is a point that goes twice around the origin, following the outer curve, which looks like a folded figure eight. Different "escape times" of orbits of points in $\mathbf{X}$ are represented by different gray tones. A magnified version of $\mathbf{X}$, painted by escape times, is shown in Figure VII.214. Roughly speaking, regions closest to the outside escape fastest. Points in the white region also escape. So where is the invariant set A? It is right in the middle. It appears to be a branching, connected, tree-like set, with no interior.

Figure VII.215 shows the result of applying the Escape Time Algorithm to f when $\lambda = 0$. This time we see that the invariant set A, in the center, in white, is just the filled Julia set for $z^2 - 1$.

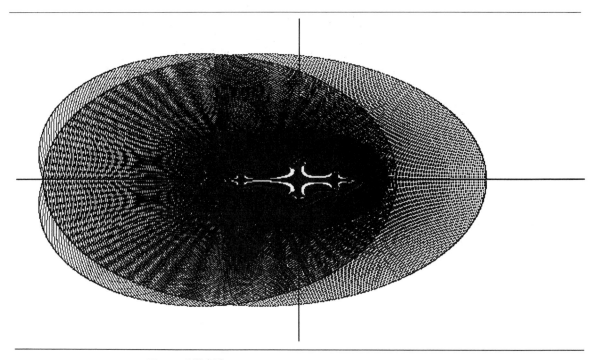

Figure VII.213.   The result of applying the Escape Time Algorithm to the function f in example 4.1 with $\lambda = 1$. This function is continuous and is such that $f(\mathbf{X}) \supset \mathbf{X}$, where $\mathbf{X}$ denotes the Fabergé egg in the middle. $f(\mathbf{X})$ is the region bounded by the outer curve. Different "escape times" of orbits of points in $\mathbf{X}$ are represented by different gray tones.

What happens if we choose $\lambda = -1$? This time we obtain the image shown in Figure VII.216. However, this time things may not be as simple as they appear to be. The inner "layers" that surround the apparent invariant set A are highly irregular and unstable. That is, points that are very close together seem to have orbits that have very different escape times. Could it be that, although the mapping $W : \mathcal{H}(\mathbf{X}) \to \mathcal{H}(\mathbf{X})$ is continuous, it has a very poor modulus of continuity?

4.2.   Construct a function $f : \mathbf{X} \to \mathbb{R}^2$, where $\mathbf{X} \subset \mathbb{R}^2$, which obeys the conditions of Theorem 4.1. Use the Escape Time Algorithm to analyze the associated invariant set A described in the statement of the theorem. Your example should be interesting and of a different character from those specifically described in this chapter.

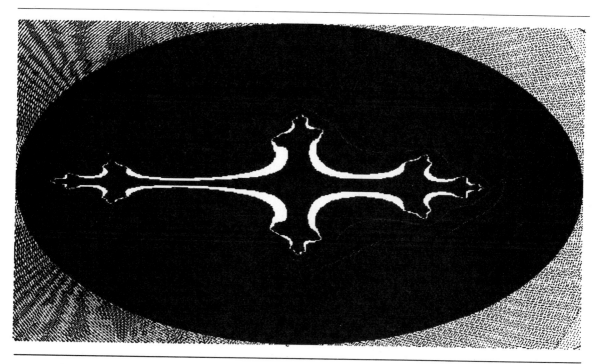

Figure VII.214.   A magnified version of the region **X** in Figure VII.213. Approximately, orbits of points closest to the outside escape the fastest. Points in the white region also escape. The invariant set *A* is in the middle. It appears to be a branching, connected, tree-like set. It is surrounded by layers, just as the center of a real tree is surrounded by layers of growth.

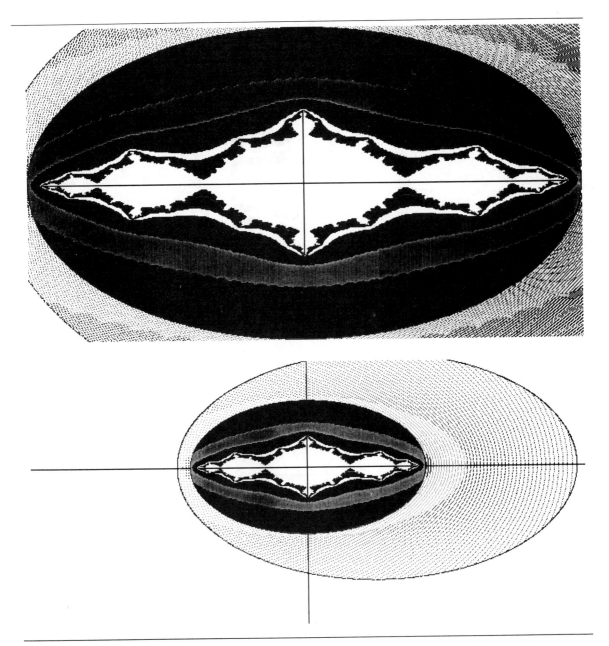

Figure VII.215. The result of applying the Escape Time Algorithm to f in example VII.213 with $\lambda = 0$. This time we see that the invariant set A, in the center, in white, is just the filled Julia set for $z^2 - 1$.

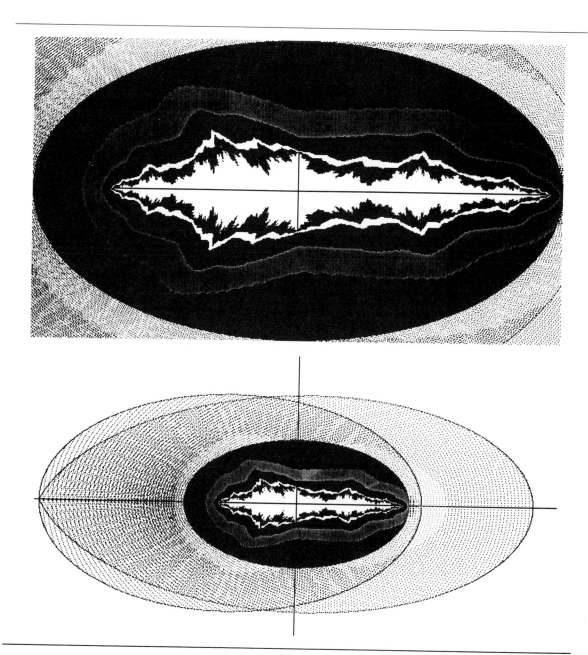

Figure VII.216. This image was computed using the Escape Time Algorithm applied to the Newton transformation associated with $f(z) = e^z - 1$. The viewing window is $\mathcal{W} = \{(x, y) \in \mathbb{R}^2 : -2.5 \leq x \leq 2.5, -2.5 \leq y \leq 2.5\}$. Can you work out what "escape region" was used?

Chapter VIII

Parameter Spaces and Mandelbrot Sets

1 The Idea of a Parameter Space: A Map of Fractals

A map with nothing marked on it is practically useless. A map of a 1000×1000 square mile region containing the British Isles is shown in Figure VIII.217; it does not convey much information. However, as a concept it is quite exciting. Each location on the map corresponds to somewhere on Earth. For example, the dot with coordinates (750, 227.3) represents the town of Maidstone. A point on the map may represent a certain grain of soil in a ploughed field, or a molecule of flotsam on the top of some foam on the surface of the sea. Nearby points in the map correspond to nearby points on the Earth. Connected sets with interiors correspond to physical regions.

How could a perfect map be made? Ideally, it should specify locations on the Earth's surface at a certain instant. The coordinates would be relative to some absolute coordinate system, perhaps determined by reference to the fixed stars. Moreover, the surface of the Earth would have to be defined precisely, up to the last molecule of water, soil, and plant matter; for this purpose one can imagine using a straight line from the center of the Earth as suggested in Figure VIII.218. Of course, maps are not made like this, but the goal is the same: to have an accurate correspondence between points on the physical surface of the Earth and the physical surface of the paper.

We must be careful how we interpret a map. Geographical maps are complicated by the real number system and the unphysical notion of infinite divisibility. Mathematically, the map is an abstract place. A point on the map cannot represent a certain physical atom in the real world, not just because of inaccuracies in the map, but because of the dual nature of matter: according to current theories one cannot know the exact location of an atom at a given instant.

Fractal geometers avoid this problem by pretending that the surface of the Earth is an abstract place too; we imagine, once again, that matter is infinitely divisible, and

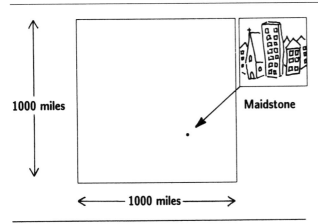

Figure VIII.217. A map with no information marked on it is quite exciting. A very unhelpful map of the British Isles. The dot represents Maidstone. The map is accurate up to dot size.

that we can address every point. In the same spirit, we presume that we can model trees, clouds, horizons, churning seas, and infinitely finely defined coastlines. Then, for example, we can define the Hausdorff-Besicovitch dimension of the coastline of the British Isles.

For a map to be useful it must have information marked on it, such as heights above sea level, population densities, roads, vegetation, rainfall, types of underlying rock, ownership, names, incidence of volcanoes, malarial infestation, and so on. A good way of providing such information is with colors. For example, if we use blue for water and green for land then we can "see" the land on the map and we can understand some geometrical relationships. We can estimate overland distances between points, land areas of islands, the shortest sea passage from LLanellian Bay

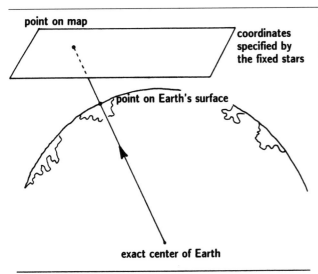

Figure VIII.218. How Figure VIII.217 might have been made.

Figure VIII.219. In this map points corresponding to land have been shaded. A fascinating entity, the coastline, is revealed.

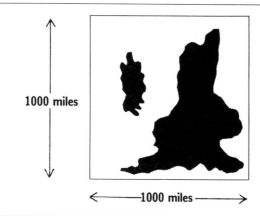

1000 miles

← ————1000 miles ————→

to Amylwch Harbour, the length of the coastline, etc. All this is achieved through the device of marking some colors on a blank map!

Let us consider the boundary of the shaded region in Figure VIII.219. It is here that the map conveys extra information. In the interior of the shaded region we learn no more about the surface of the Earth than that "there is land there." However, on the boundary we learn not only that "there are land and sea there," but also, if the map is accurate enough, a feature that we will actually "see" on the surface, namely the local shape of the coastline.

The latter idea can be extended. If we include more colors on a geographical map, to provide more information about properties of the Earth's surface, we produce more boundaries on the map. These boundaries can give information about local geometry. For example, a map finely colored according to elevation reveals the shapes of the bases of the mountains, the paths of rivers, and – if we look closely enough—the outlines of buildings. Such a map, made abstract and perfect, placed in a metric space, would contain much detailed information about what, at each point, the local observer would see.

Examples & Exercises

1.1. Study an atlas that contains maps colored according to diverse criteria, such as rainfall, population density, vegetation, and elevation. Discuss to what extent these maps provide information about the local geometry of the surface of the Earth.

We turn attention to making colored maps of parameterized families of fractals. We consider families of iterated function systems and families of dynamical systems that depend on two real parameters. The collection of possible parameter values defines a *parameter space* associated with the family. We use the notation P to denote a parameter space. Typically P is a subspace of $(\mathbb{R}^2, \text{Euclidean})$ such as ∎, a closed ball, or $\mathbb{R}^2$.

An example of a parameter space is $P = \{(\lambda_1, \lambda_2) \in \mathbb{R}^2 : |\lambda_1|, |\lambda_2| < 2^{-0.5}\}$. This

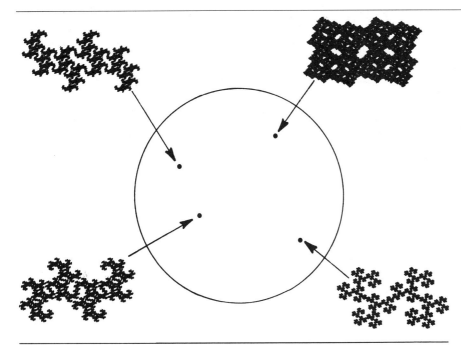

Figure VIII.220. An example of a parameter space. Each point λ in the space corresponds to a fractal, $A(\lambda)$. This is a poor map, because very little has been marked on it. It is like the map of the British Isles shown in Figure VIII.217. It needs coloring.

is a parameter space for the family of hyperbolic IFS $\{\mathbb{C}; (\lambda_1 + i\lambda_2)z + 1, (\lambda_1 + i\lambda_2)z - 1\}$. Each point $\lambda = (\lambda_1, \lambda_2) \in P$ corresponds to an IFS. Each IFS possesses a unique attractor, say $A(\lambda)$. Hence each point of P corresponds to a single fractal. We can think of P as representing part of $\mathcal{H}(\mathbb{C})$, a space of fractals. A map of P, with a few points marked on it, is shown in Figure VIII.220. Each point in P corresponds to a single fractal. Nearby points in P correspond to nearby fractals, that is, points in $\mathcal{H}(\mathbb{C})$ whose Hausdorff distance apart is small.

Another example is $P = \mathbb{C}$, which provides a parameter space for the family of dynamical systems $\{\hat{\mathbb{C}}; f_\lambda(z) = z^2 - \lambda\}$. Each point in the parameter space corresponds to a different dynamical system. Each dynamical system is associated with a unique Julia set, $J(\lambda)$. The collection of fractals $\{J(\lambda) : \lambda \in P\}$, associated with the parameter space, is vast and diverse.

Let $\mathbf{X}$ be a two-dimensional metric space such as $\mathbb{R}^2$ or $\hat{\mathbb{C}}$. Let P denote a parameter space corresponding to a family of fractals $\{A(\lambda) \in \mathcal{H}(\mathbf{X}) : \lambda \in P\}$. Can we provide the explorer of fractals, who wishes to investigate this collection of fractals, with a colored map? This map should give her information about the sets $A(\lambda)$ to be found at different points on P.

Suppose $P = \blacksquare$. To make a map, let us represent P by the set of pixels on the screen of a computer graphics monitor. The idea is to color the pixel λ according to some property of $A(\lambda)$. (We write λ = pixel = point in P, without repeatedly explaining that λ is a point in the small rectangle in P that corresponds to the pixel.)

Suitable properties could relate to the connectivity of $A(\lambda)$, the fractal dimension of $A(\lambda)$, the escape time near a special point on $A(\lambda)$ under an associated dynamical system, the number of holes in $A(\lambda)$, or the presence of straight lines in $A(\lambda)$, for example.

If we make a good selection of the properties to associate with colors, the result will be a useful map containing various differently colored regions. This map will be a ready reference for the explorer of fractals. It will tell him something about what to expect at he travels about P. He might be suprised nonetheless.

The boundaries of the colored regions can provide additional geometrical information to the explorer, over and above the information that the map was originally designed to convey. It sometimes occurs that the local shapes of the boundaries in the map reflect the shapes of the corresponding fractals. There is a deep principle here, which we shall not pin down as a theorem, but which we will illustrate in a number of cases.

Examples & Exercises

1.2. It is often useful, when establishing the Lipshitz continuity condition in Theorem 11.1 in Chapter III, to remove dependencies on $x \in \mathbf{X}$ by restricting the domain of the IFS to a compact set B that contains the attractors $A(\lambda)$ for all values of $\lambda \in P$. Since this does not change the attractors themselves, continuous dependency on λ in B then guarantees it for the original IFS. This can be done in practice by means of the following lemma:

Lemma 1.1 *Let $B \in \mathcal{H}(\mathbf{X})$ such that for all $p \in P$, $W(p, B) \subset B$. Then for all $p \in P$, $A(p) \subset B$.*

Prove the lemma.

1.3. Let $P = \{(\lambda_1, \lambda_2) \in \mathbb{R}^2 : |\lambda_1|, |\lambda_2| \leq 0.9\}$. The family of IFS $\{\mathbb{R}; \lambda_1 x, \lambda_2 x + 1 - \lambda_2, 0.5x + 0.5\}$ is hyperbolic, with contractivity factor $s = 0.9$, for all $\lambda \in P$. Use Theorem 3.11.1 and the lemma of the previous problem to prove that the attractor $A(\lambda)$ depends continuously on λ.

1.4. The family of IFS $\{[0, 1]; \lambda_1 x^2, \lambda_2 x + (1 - \lambda_2)\}$ is hyperbolic, with contractivity factor $s = 0.9$, for all λ in the parameter space

$$P = \{(\lambda_1, \lambda_2) \in \mathbb{R}^2 : 0 \leq \lambda_1 \leq 0.45, 0 \leq \lambda_2 \leq 0.9\}.$$

Since we have

$$d(w_{1_p}(x), w_{1_q}(x)) = |x^2| |\lambda_{1_p} - \lambda_{1_q}| \leq |\lambda_{1_p} - \lambda_{1_q}|$$
$$d(w_{2_p}(x), w_{2_q}(x)) = |x - 1| |\lambda_{2_p} - \lambda_{2_q}| \leq |\lambda_{2_p} - \lambda_{2_q}|,$$

we can satisfy the Lipshitz condition with $k = 1$ in both cases. The attractor depends continuously on λ.

1.5. Sometimes the Lipshitz condition is not that convenient to prove directly. In this case, we can go directly to the continuity statement that it is used to guarantee.

Suppose we can show directly that $w_i(p, x)$ is a continuous function from $P \times \mathbf{X} \to$ $\mathbf{X}$. That is, *without holding either p or x fixed*, given $\epsilon > 0$, there is a $\delta_i > 0$ for each $w_i(p, x)$, such that

$$d(w_i(p, x), w_i(q, y)) < \epsilon \qquad \text{whenever} \quad d((p, x), (q, x)) < \delta_i.$$

Then the conclusions of Theorem 11.1 in Chapter III hold. This does not mean that the Lipshitz condition (or something like it) is unnecessary; such conditions guarantee the statement we have just made. If we can guarantee it some other way, it is equivalent.

1.6. An example of a parameter space is $(P,\text{Euclidean})$, where $P = \{(\lambda_1, \lambda_2) \in \mathbb{R}^2 : |\lambda_1|, |\lambda_2| \le 0.999\}$. This space can be used to represent the family of hyperbolic IFS $\{\mathbb{R}^2; w_1, w_2\}$, where

$$w_1 \begin{pmatrix} x \\ y \end{pmatrix} = \begin{pmatrix} \lambda_1 & 0 \\ 0 & \lambda_2 \end{pmatrix} \begin{pmatrix} x \\ y \end{pmatrix} + \begin{pmatrix} 0 \\ 1 \end{pmatrix},$$

$$w_2 \begin{pmatrix} x \\ y \end{pmatrix} = \begin{pmatrix} 0.3 & -0.2 \\ 0.1 & 0.4 \end{pmatrix} \begin{pmatrix} x \\ y \end{pmatrix}.$$

Since (using $(z, w) = (\lambda_1, \lambda_2)$ to emphasize familiarity)

$$w_1(x, y, z, w) = (zx, wy + 1)$$

is continuous from basic calculus, Theorem 11.1 in Chapter III holds, and the attractor varies continuously with λ.

2 Mandelbrot Sets for Pairs of Transformations

Let $P \subset \mathbb{R}^2$ be a parameter space corresponding to a family of fractals. That is, we have a function $A : P \to \mathcal{H}(\mathbf{X})$, so that each point $\lambda \in P$ corresponds to a set $A(\lambda) \in \mathcal{H}(\mathbf{X})$. One way to make a map is to color the parameter space according to whether or not $A(\lambda)$ is connected.

Theorem 2.1 *Let $\{\mathbf{X}; w_1, w_2\}$ be a hyperbolic IFS with attractor A. Let w_1 and w_2 be one-to-one on A. If*

$$w_1(A) \cap w_2(A) = \emptyset,$$

then A is totally disconnected. If

$$w_1(A) \cap w_2(A) \ne \emptyset,$$

then A is connected.

Proof Suppose that $w_1(A) \cap w_2(A) = \emptyset$. Let Σ denote the code space map associated with the IFS. By Theorem 2.2 in Chapter IV the code space map $\phi : \Sigma \to A$ is invertible. ϕ is also a continuous transformation between two compact metric spaces.

Hence, by Theorem 8.5 in Chapter II, ϕ is a homeomorphism. Hence A is homeomorphic to code space, which is totally disconnected. (Recall that code space on two or more symbols is metrically equivalent to a classical Cantor set.) It follows that A is totally disconnected.

Suppose that $w_1(A) \cap w_2(A) \neq \emptyset$. Then there is at least one point $x \in w_1(A) \cap w_2(A)$. This point x has two addresses, say

$$x = \phi(\zeta) = \phi(\sigma), \qquad \text{where } \zeta_1 = 1 \text{ and } \sigma_1 = 2.$$

Let us see what happens if we additionally suppose that "A is not connected." Then, since A is compact, we can find two nonempty compact sets E and F so that

$$A = E \cup F, \qquad E \cap F = \emptyset.$$

Using compactness, there is a positive real number δ so that

$$d(e, f) \geq \delta \qquad \text{for all } e \in E, f \in F.$$

Let π and ψ be a pair of codes that agree through the first K symbols, for some positive integer K. That is $\pi_i = \psi_i$ for $i = 1, 2, \ldots, K$. Then

$$d(\phi(\pi), \phi(\psi)) \leq s^K \text{diam}(A),$$

where $\text{diam}(A) = \text{Max}\{d(x, y) : x, y \in A\}$, and $s \in [0, 1)$ is a contractivity factor for the IFS. Suppose also that $\phi(\pi) \in E$ and $\phi(\psi) \in F$. Then

$$\delta \leq d(\phi(\pi), \phi(\psi)).$$

Combining the latter two inequalities we discover $\delta \leq s^K \text{diam}(A)$, which implies

$$K \leq \frac{\log(\delta/\text{diam}(A))}{\log(s)}.$$

We conclude that if $e \in E$ and $f \in F$ then the number of successive agreements between an address of e and an address of f cannot exceed the number on the right-handside here. It follows that there is a maximum number, M, of initial agreements between the address of a point $e \in E$ and a point $f \in F$; and this maximum is achieved on some pair of points, say e and f. Then we can find $\rho_i \in \{1, 2\}$ for $i = 1, 2, \ldots, M$ such that

$$\phi(\rho_1, \rho_2, \rho_3, \ldots, \rho_M, 1, \ldots) = e \in E$$

and

$$\phi(\rho_1, \rho_2, \rho_3, \ldots, \rho_M, 2, \ldots) = f \in F.$$

Now consider the point $z \in A$, which has the two addresses

$$z = \phi(\rho_1, \rho_2, \rho_3, \ldots, \rho_M, 1, \zeta_2, \zeta_3, \zeta_4, \ldots)$$
$$= \phi(\rho_1, \rho_2, \rho_3, \ldots, \rho_M, 2, \sigma_2, \sigma_3, \sigma_4, \ldots).$$

Suppose $z \in E$. Then its address agrees with that of $f \in F$ through $(M + 1)$ initial symbols. Hence $z \in F$. But then its address agrees with that of $e \in E$ through $(M +$

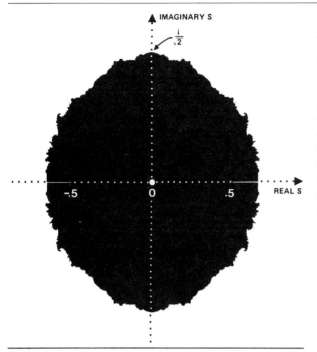

IMAGINARY S

$\frac{i}{\sqrt{2}}$

-.5   0   .5   REAL S

Figure VIII.221. A map of the family of IFS $\{\mathbb{C}; \lambda z - 1, \lambda z + 1\}$, where the parameter space is $P = \{\lambda = (\lambda_1, \lambda_2) \in \mathbb{C} : |\lambda| < 1\}$. This picture of a parameter space is obtained by "painting" black where the attractor of the IFS is disconnected and light where it is connected. The Mandelbrot set is the light region, the sea. It contains a dragon at $\lambda = (0.5, 0.5)$.

1) initial symbols, which is not possible. We have a contradiction. Hence "A is *not* disconnected." It follows that A is connected. This completes the proof of the theorem.

Definition 2.1 *Let* $\{\mathbf{X}; w_1, w_2\}$ *be a family of hyperbolic IFS that depends on a parameter* $\lambda \in P \subset \mathbb{R}^2$. *Let* $A(\lambda)$ *denote the attractor of the IFS. The set of points* $\mathcal{M} \subset P$ *defined by*

$$\mathcal{M} = \{\lambda \in P : A(\lambda) \text{ is connected}\}$$

is called the Mandelbrot set *for the family of IFS.*

For the rest of this section we consider the family of IFS

$$\{\mathbb{C}; \lambda z - 1, \lambda z + 1\},$$

where the parameter space is

$$P = \{\lambda = (\lambda_1, \lambda_2) \in \mathbb{C} : \lambda_1^2 + \lambda_2^2 < 1\}.$$

Figure VIII.221 shows a picture of the associated Mandelbrot set, $\mathcal{M}$. This is a map for the collection of fractals associated with the IFS. It has been colored dark where the attractor is totally disconnected and light where it is connected.

Here is an outline of an algorithm to compute images of the Mandelbot set $\mathcal{M}$ associated with the family $\{\mathbb{C}; w_1(z) = \lambda z - 1, w_2(z) = \lambda z + 1\}$. It is based on Theorem 2.1.

Algorithm 2.1 Example of Method for Making Pictures of the Mandelbrot Set of a Family of IFS.

(1) Choose a positive integer, L, corresponding to the amount of computation one is able to do. The greater the value of L, the more accurate the resulting map image will be.

(2) Represent the parameter space $P = \{\lambda \in \mathbb{C} : |\lambda| < 1\}$ by an array of pixels. Carry out the following steps for each λ in the array.

(3) Calculate a number R, so that the attractor is contained in a ball of radius R, centered at the origin; that is, choose $R > 0$ so that $A(\lambda) \subset B(0, R)$.

(4) Compute the number

$$H = \min\{d(x, y) : x \in w_1(W^{\circ L}(\{0\})), y \in w_2(W^{\circ L}(\{0\}))\},$$

where $W = w_1 \cup w_2$. If $H \leq 2|\lambda|^{L+1}R$, then the pixel λ is assumed to belong to $\mathcal{M}$ and is colored accordingly.

Step (4) is based on the following observation. The attractor of the IFS is contained in the set $W^{\circ(L+1)}(B(0, R))$, which consists of 2^{L+1} balls of radius $|\lambda|^{L+1}R$. The centers of these balls lie in the union of the two sets $w_1(W^{\circ L}(0))$ and $w_2(W^{\circ L}(0))$. If H is greater than $2|\lambda|^{L+1}R$, then $A(\lambda)$ must be disconnected.

Figure VIII.222 shows the "coastal region" of a quarter of the complement of the Mandelbrot set in Figure VIII.221. It has been laid over a grid in order to help you locate points where interesting fractals lie.

The boundary of $\mathcal{M}$ is complicated and intricate. Close-ups of the "coastline" near the places marked (a), (b), and (c) are shown in Figure VIII.223. Figure VIII.224 shows a zoom on the spiral peninsula in Figure VIII.222.

Now look at Figures VIII.225 and VIII.226, which show pictures of the attractors $A(\lambda)$ for some points λ located near the boundary of the Mandelbrot set. There is a "family resemblance" between the places on the boundary from which the fractals come, and the fractals themselves. To help see this you should look back at the close-ups on the coastline in Figure VIII.223. Figure VIII.227 shows the IFS attractor corresponding to the tip of the peninsula in Figure 224. Notice how it contains spirals, very much like the ones in the peninsula in parameter space. At the end of this chapter we make some comments on why such "family resemblances" occur.

The following theorem provides rigorous bounds on the locations of $\mathcal{M}$ and $\partial\mathcal{M}$. The proof is delightful, because it relies on a fractal dimension estimate.

Theorem 2.2 *[Barnsley 1985c] The attractor $A(\lambda)$ of the IFS $\{\mathbb{C}; \lambda z - 1, \lambda z + 1\}$ is totally disconnected if $|\lambda| < 0.5$ and connected if $1 > |\lambda| > 1/\sqrt{2}$. The boundary of the associated Mandelbrot set is contained in the annulus $1/2 \leq |\lambda| \leq 1/\sqrt{2}$.*

Proof Let A denote the attractor of the IFS and let $D(A)$ denote its fractal dimension. The two maps in the IFS are similitudes of scaling factor $|\lambda|$. This means that Theorem 2.3 in Chapter V can be applied.

Figure VIII.222. This shows the coastal region of a quarter of the complement of the Mandelbrot set in Figure VIII.221. It has been laid over a grid in order to help you locate points where interesting fractals lie. Close-ups of the coast at (a), (b), and (c) are shown in Figure VIII.223. The coordinates of the grid are $(0, 0) - -(0.71, 0.71)$.

Suppose that A is totally disconnected. Then the IFS is totally disconnected and, by Theorem 2.3 in Chapter V,

$$D(A) = \log(1/2)/\log(|\lambda|).$$

By Theorem 2.1 in Chapter V, $D(A) \leq 2$. Hence

$$\log(1/2)/\log(|\lambda|) \leq 2.$$

This implies that $|\lambda| \leq 1/\sqrt{2}$.

Suppose that A is connected. Then it contains a path that connects two distinct points. The fractal dimension of any path is greater than or equal to 1. Hence $D(A) \geq 1$. However, by Theorem 2.3 in Chapter V,

$$D(A) \leq \log(1/2)/\log(|\lambda|).$$

It follows that

$$1 \leq \log(1/2)/\log(|\lambda|).$$

This implies that $|\lambda| \geq 1/2$. This completes the proof of the theorem.

A different point of view on the Mandelbrot set considered above is given by

Figure VIII.223. Close-ups of the boundary of the Mandelbrot set at (a), (b), and (c). The diverse structures in this boundary echo the shapes of the attractors of the corresponding IFS.

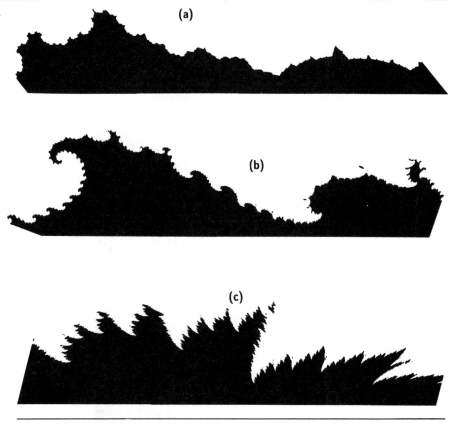

(a)

(b)

(c)

Figure VIII.228. $\mathcal{M}$ has been turned inside-out by making the change of variables $\lambda' = \lambda^{-1}$. The inner white disk is no-man's land; it does not belong to the parameter space. Also included are the two bounds provided by Theorem 2.2, namely the circle $|\lambda'| = 2$ and the circle $|\lambda'| = \sqrt{2}$. The fractal dimension decreases with increasing distance from the origin.

Examples & Exercises

2.1. Sketch the Mandelbrot set for the family of IFS $\{\mathbb{R}; \lambda_1 x + \lambda_2, \lambda_2 x + \lambda_1\}$ where the parameter space is $P = \{(\lambda_1, \lambda_2) : |\lambda_1|, |\lambda_2| < 1\}$.

2.2. Let $\{\mathbf{X}; w_1, w_2\}$ be a family of hyperbolic IFS which depends on a parameter $\lambda \in P \subset \mathbb{R}^2$. Let w_1 and w_2 depend Lipshitz continuously on λ for some $k > 0$ and for fixed $x \in \mathbf{X}$. Assume that the IFS has contractivity factor $s \in [0, 1)$ which is independent of $\lambda \in P$. Then by Theorem 11.1 in Chapter III, the function $A : P \rightarrow \mathcal{H}(\mathbf{X})$ is continuous. Use this continuity to prove that the Mandelbrot set associated with the family of IFS is closed. It is suggested that you begin by showing that the set $S = \{B \in \mathcal{H}(\mathbf{X}) : B \text{ is not connected}\}$ is an open subset of $\mathcal{H}(\mathbf{X})$.

Figure VIII.224. Close-up on the spiral peninsula on the edge of the Mandelbrot set in Figures VIII.221 and VIII.222. What information about the corresponding fractals does this boundary convey?

2.3. Use Figure VIII.222 to determine some values of λ that belong, approximately, to the boundary $\partial\mathcal{M}$ of the Mandelbrot set. Compute images of the corresponding attractors. Compare images corresponding to two points λ_1 and $\lambda_2 \in \partial\mathcal{M}$, with $|\lambda_1| < |\lambda_2|$. Explain why the picture of $A(\lambda_1)$ is more delicate than the picture of $A(\lambda_2)$. Also comment on similarities and differences between your images and the local geography of the parts of $\partial\mathcal{M}$ from which they come.

2.4. The pictures of the Mandelbrot set associated with the family of IFS $\{\mathbb{C}; \lambda z -$

Figure VIII.225. Some of the fractals to be found at various points near the boundary of the Mandelbrot set associated with the parameterized family of IFS $\{\mathbb{C}; \lambda z - 1, \lambda z + 1\}$.

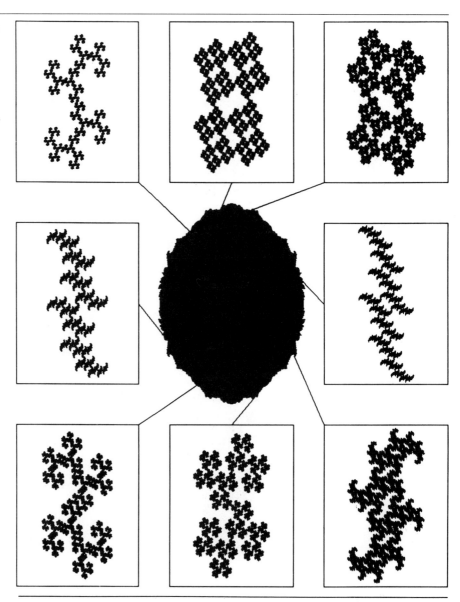

$1, \lambda z + 1\}$ suggest the conjecture that $\mathcal{M}$ is symmetric about the x-axis and about the origin. Prove the conjecture.

2.5. An interesting point in the parameter space for the family $\{\mathbb{C}; \lambda z - 1, \lambda z + 1\}$ is $\lambda = (1/2, 1/2)$. This lies on the circle $1/|\lambda| = |\lambda'| = \sqrt{2}$ in Figure VIII.228. It appears to be located in the interior of the Mandelbrot set, although the IFS is just-touching. It corresponds to the Twin-Dragon Fractal. A picture of it is shown in Figure VIII.229. It is possible to tile the plane with Twin-Dragons. Various other

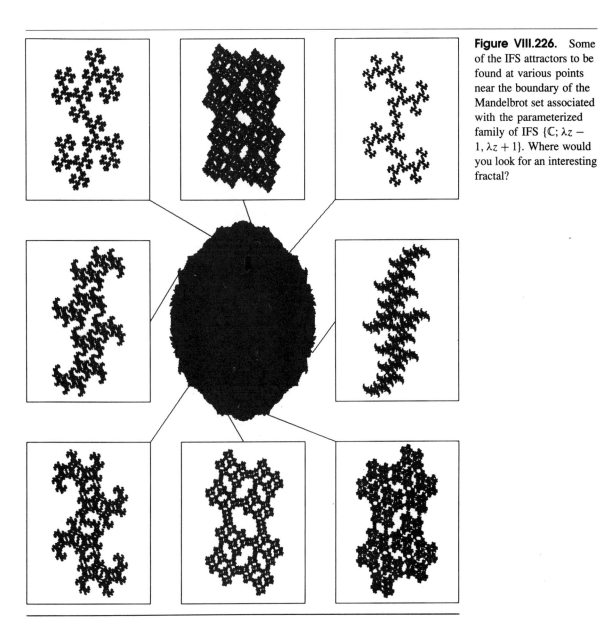

Figure VIII.226. Some of the IFS attractors to be found at various points near the boundary of the Mandelbrot set associated with the parameterized family of IFS $\{\mathbb{C}; \lambda z - 1, \lambda z + 1\}$. Where would you look for an interesting fractal?

values of λ also correspond to tilings of the plane. See [Gilbert 1982]. Show that the attractor at the point $\lambda = (0, 1/\sqrt{2})$ can be used to tile the plane.

2.6. Notice the line segments on the real axis in Figure VIII.228. In [Barnsley 1985c] it is proved that

$$\{\lambda \in \mathbb{C} : 0.5 \le \lambda_1 \le 0.53; \lambda_2 = 0\} \subset \mathcal{M}.$$

Figure VIII.227. Attractor of the IFS $\{\mathbb{C}; \lambda z - 1, \lambda z + 1\}$ corresponding to the value of λ at the tip of the spiral peninsula, shown in Figure VIII.224.

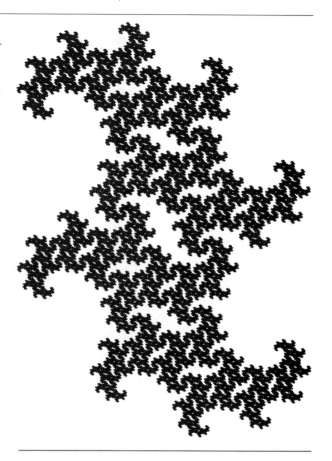

For λ in such a line segment what does the attractor look like? Are you suprised, in view of what you know about maps of coastlines?

2.7. Some of the most delicate attractors of the family $\{\mathbb{C}; \lambda z - 1, \lambda z + 1\}$ are associated with points on $\partial \mathcal{M}$ where it touches the circle $1/|\lambda| = |\lambda'| = 2$. These have the lowest possible fractal dimension while still being connected. Let us call these attractors *tree-like* if $w_1(A) \cap w_2(A)$ is a single point. Argue (or, better yet, prove) that a tree-like attractor A contains no trapped holes; that is, A contains no nontrivial, non–self-intersecting paths that start and finish at the same point. A picture of a tree-like attractor is shown in Figure VIII.230.

2.8. Let $e = e_1 e_2 e_3 \ldots e_n \ldots$ be a point in the code space Σ of two symbols, with $e_n \in \{+1, -1\}$ for all n. Let $\lambda \in \mathbb{C}$. Prove that the series

$$f(\lambda) = e_1 + e_2\lambda + e_3\lambda^2 + e_4\lambda^3 + e_5\lambda^4 + e_6\lambda^5 \ldots + e_n\lambda^n + \ldots$$

has radius of convergence 1. What is the relationship between $f(\lambda)$ and the code space map $\phi : \Sigma \to A(\lambda)$ associated with the family of IFS $\{\mathbb{C}; \lambda z - 1, \lambda z + 1\}$? Let

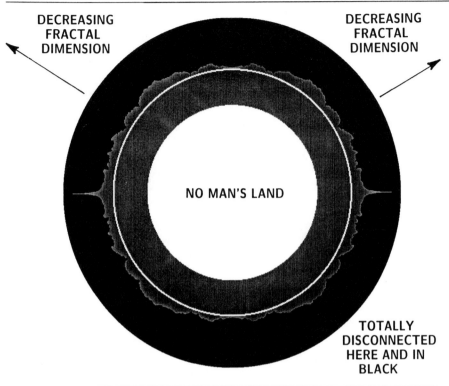

DECREASING
FRACTAL
DIMENSION

DECREASING
FRACTAL
DIMENSION

NO MAN'S LAND

TOTALLY
DISCONNECTED
HERE AND IN
BLACK

Figure VIII.228. Inside-out picture of the Mandelbrot set for $\{\mathbb{C}; \lambda z - 1, \lambda z + 1\}$. It has been turned inside-out by making the change of variables $\lambda' = \lambda^{-1}$. The inner white disk is no-man's land; it does not belong to the parameter space. The figure also includes the two bounds provided by Theorem 2.2, namely the circle $|\lambda'| = 2$ and the circle $|\lambda'| = \sqrt{2}$.

$|\lambda| < 1$. Show that the attractor of the IFS is the union of all points which can be written in the form

$$\pm 1 \pm \lambda \pm \lambda^2 \pm \lambda^3 \pm \lambda^4 \pm \lambda^5 \pm \lambda^6 \pm \lambda^7 \pm \lambda^8 \pm \lambda^9 \pm \lambda^{10} \pm \lambda^{11} \pm \dots.$$

3 The Mandelbrot Set for Julia Sets

In this section we introduce a good method for making maps, such as might be found in an atlas, of families of dynamical systems. The method is based on the use of escape times and is discussed more generally in section 4. Here we restict attention to the family

$$\{\hat{\mathbb{C}}; f_\lambda(z) = z^2 - \lambda\},$$

where the parameter space is $P = \mathbb{C}$. This family is of special importance because it provides a model for the onset of chaotic behavior in physical and biological systems; see [May 1976] and [Feigenbaum 1979]. Moreover, it was the first family of dynamical systems for which a useful computergraphical map was constructed, by Mandelbrot. We concentrate on map making.

Figure VIII.229. The Twin-Dragon Fractal. You can tile the plane with these sets. Although it is just-touching, it appears to lie in the interior of the Mandelbrot set.

The Julia set $J(\lambda)$ associated with $f_\lambda(z)$ is symmetric about the origin, O. We know this because the filled Julia set, of which $J(\lambda)$ is the boundary, is the set of points whose orbits remain bounded. The orbit of $z \in \mathbb{C}$ remains bounded if and only if the orbit of $-z$ remains bounded.

For some values of $\lambda \in P$, O belongs to the filled Julia set, $F(\lambda)$, while for others it is quite far from $F(\lambda)$. This suggests that we try to color the parameter space according to the distance from O to $F(\lambda)$. How can we estimate this distance? An approximate method is to look at the "escape time" of the orbit of O. That is, we can color the parameter space according to the number of steps along the orbit of O that are required before it lands in a ball around the point at infinity, from where we know that all orbits diverge. The intuitive idea is that the longer an orbit of O takes to reach the ball, the closer O must be to $F(\lambda)$. Of course, if an orbit does not diverge then we know that $O \in F(\lambda)$.

Suppose that we want to make a map corresponding to a region $\mathcal{W} \subset P$. Here we choose

$$W = \{\lambda = (\lambda_1, \lambda_2) \in \mathbb{C} : |\lambda_1|, |\lambda_2| \leq 2\}.$$

Let $R > 0$ and define

$$\mathcal{V}(R) = \{z \in \mathbb{C} : |z| > R\} \cup \{\infty\}.$$

Suppose

$$R > 0.5 + 0.25 + |\lambda|.$$

Then it is readily proved that the orbit $\{f_\lambda^{\circ n}(z)\}$ diverges if and only if it intersects $\mathcal{V}(R)$. So if we choose $R = 10$ we are sure that, for all $\lambda \in \mathcal{W}$, the orbit $\{f_\lambda^{\circ n}(O)\}$

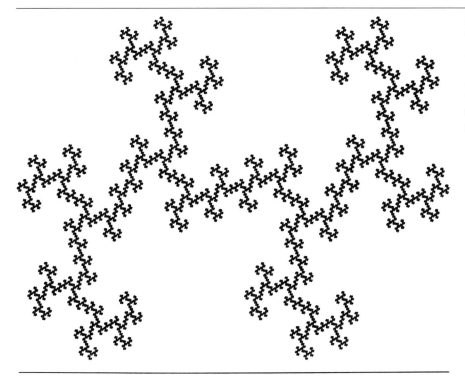

Figure VIII.230. A "tree-like" attractor A from the family $\{\mathbb{C}; w_1(z) = \lambda z - 1, w_2 = \lambda z + 1\}$. The two sets $w_1(A)$ and $w_2(A)$ meet approximately at a single point.

diverges if and only if it intersects $\mathcal{V}(R)$. Let us see what happens if we color the pixels of $\mathcal{W}$ according to the number of iterations required to enter $\mathcal{V}(10)$.

The following program is written in BASIC. It runs without modification on an IBM PC with Enhanced Graphics Adaptor and Turbobasic. On any line the words preceded by ' are comments: they are not part of the program.

Program 1. (Algorithm for Coloring Parameter Space According to an Escape Time)

```
'Define viewing window, W, and numits.
numits = 20 :   a = -2 : b = -2 :   c = 2 : d = 2 : M = 100

R =10   'Define the region V.

screen 9 : cls 'Initialize graphics.

for p =1 to M

for q =1 to M

'Specify the value of lambda (k,l) in  P
k =   a+(c-a)*p/M : l = b+(d-b)*q/M
```

```
'Specify the initial point, O, on the orbit
x =0 : y =0

'Compute at most numits points on the orbit of O
for n = 1 to numits

newx   =   x*x-y*y-k

newy = 2*x*y-l

x = newx : y = newy

'If the most recently computed point lies in V then...
if x*x+y*y > R then

'... render the pixel (p,q) in color n, and
'go to the next (p,q).
pset(p,q),n : n = numits

end if

if instat then end 'Stop computing if any key is pressed!

next n : next q : next p

end
```

Color Plate 16 shows the result of running a version of Program 1 on a Masscomp 5600 workstation with Aurora graphics.

In Figure VIII.231 we show the result of running a version of Program 1, but this time in halftones. The central white object corresponds to values of λ for which the computed orbit of O does not reach V during the first *numits* iterations. It represents the Mandelbrot set (defined below) for the dynamical system $\{\hat{\mathbb{C}}; z^2 - \lambda\}$. The bands of colors (or white and shades of gray) surrounding the Mandelbrot set correspond to different numbers of iterations required before the orbit of O reaches $\mathcal{V}(10)$. The bands farthest away from the center represent orbits that reach O most rapidly. Approximately, the distance from O to $F(\lambda)$ increases with the distance from λ to the Mandelbrot set.

Definition 3.1 *The* Mandelbrot set *for the family of dynamical systems $\{\hat{\mathbb{C}}; z^2 - \lambda\}$ is*

$$\mathcal{M} = \{\lambda \in P : J(\lambda) \text{ is connected}\}.$$

The relationship between escape times of orbits of O and the connectivity of $J(\lambda)$ is provided by the following theorem.

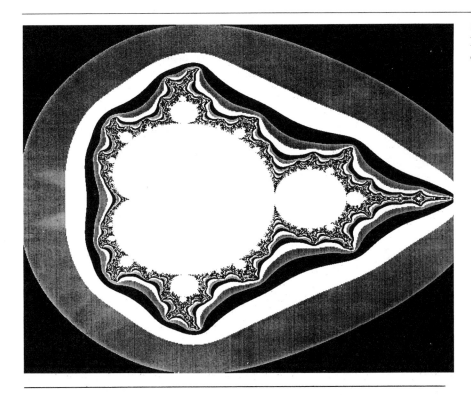

Figure VIII.231. The Mandelbrot set for $z^2 - \lambda$, computed by escape times.

Theorem 3.1 *The Julia set for the family of dynamical systems* $\{\hat{\mathbb{C}}; f_\lambda(z) = z^2 - \lambda\}, \lambda \in P = \mathbb{C}$, *is connected if and only if the orbit of the origin does not escape to infinity; that is*

$$\mathcal{M} = \{\lambda \in \mathbb{C} : |f_\lambda^{\circ n}(0)| \nrightarrow \infty \text{ as } n \to \infty\}$$

Proof This theorem follows from [Brolin], Theorem 11.2, which says that the Julia set of a polynomial, of degree greater than 1, is connected if and only if none of the finite critical points lie in the basin of attraction of the point at infinity. $f_\lambda(z)$ possesses two critical points, 0 and ∞. Hence $J(\lambda)$ is connected if and only if $|f_\lambda^{\circ n}(0)| \nrightarrow \infty$ as $n \to \infty$.

In this paragraph we discuss the relationship between the Mandelbrot set for the family of dynamical systems $\{\hat{\mathbb{C}}; z^2 - \lambda\}$ and the corresponding family of IFS $\{\hat{\mathbb{C}}; \sqrt{z + \lambda}, -\sqrt{z + \lambda}\}$. We know that for various values of λ in $\mathbb{C}$ the IFS can be modified so that it is hyperbolic, with attractor $J(\lambda)$. For the purposes of this paragraph *let us pretend that the IFS is hyperbolic, with attractor* $J(\lambda)$, *for all* $\lambda \in \mathbb{C}$. Then Definition 2.1 would be equivalent to Definition 3.1. By Theorem 2.1, the attractor of the IFS would be connected if and only if $w_1(J(\lambda)) \cap w_2(J(\lambda)) \neq \emptyset$. But $w_1(\mathbb{C}) \cap w_2(\mathbb{C}) = 0$. Then it would follow that the attractor of the IFS is connected if and only if $0 \in J(\lambda)$. In other words: we discover the same criteria for connectivity of $J(\lambda)$ if we argue informally using the IFS point of view, as can be proved using Julia set theory. This completes the discussion.

We return to the theme of coastlines and the possible resemblance between fractal sets corresponding to points on boundaries in parameter space and the local geometry of the boundaries. Figures VIII.232 and VIII.233 show the Mandelbrot set for $z^2 - \lambda$, together with pictures of filled Julia sets corresponding to various points around the boundary. If one makes a very high-resolution image of the boundary of the Mandelbrot set, at a value of λ corresponding to one of these Julia sets, one "usually" finds structures that resemble the Julia set. It is as though the boundary of the Mandelbrot set is made by stitching together microscopic copies of the Julia sets that it represents. An example of such a magnification of a piece of the boundary of $\mathcal{M}$, and a picture of a corresponding Julia set, are shown in Figures VIII.234 and VIII.235.

If you look closely at the pictures of the Mandelbrot set $\mathcal{M}$ considered in this section, you will see that there appear to be some parts of the set that are not connected to the main body. Pictures can be misleading.

Theorem 3.2 *[Mandelbrot-Douady-Hubbard]*
The Mandelbrot set for the family of dynamical systems $\{\hat{\mathbb{C}}; z^2 - \lambda\}$ *is connected.*

Proof This can be found in [Douady 1982].

The Mandelbrot set for $z^2 - \lambda$ is related to the exciting subject of cascades of bifurcations, quantitative universality, chaos, and the work of Feigenbaum. To learn more you could consult [Feigenbaum 1979], [Douady 1982], [Barnsley 1984], [Devaney 1986], [Peitgen 1986], and [Scia 1987].

Examples & Exercises

3.1. Rewrite Program 1 in a form suitable for your own computergraphical environment. Run your program and obtain hardcopy of the output. Adjust the window parameters a, b, c, and d, to allow you to make zooms on the boundary of the Mandelbrot set.

3.2. Figure VIII.236 shows a picture of the Mandelbrot set for the family of dynamical systems $\{\hat{\mathbb{C}}; z^2 - \lambda\}$ corresponding to the coordinates $-0.5 \le \lambda_1 \le 1.5$, $-1.0 \le \lambda_2 \le 1.0$. It has been overlaid on a coordinate grid. The middle of the first bubble has not been plotted, to clarify the coordinate grid. Let $B_0, B_1, B_2, B_3, \ldots$ denote the sequence of bubbles on the real axis, reading from left to right. Verify computationally that when λ lies in the interior of B_n the dynamical system possesses an attractive cycle, located in $\mathbb{C}$, of minimal period 2^n, for $n = 0, 1, 2$, and 3.

3.3. The sequence of bubbles $\{B_n\}_{n=0}^{\infty}$ in exercise 3.2 converges to the *Myreberg point*, $\lambda = 1.40115\ldots$. The ratios of the widths of successive bubbles converge to the *Feigenbaum ratio* $4.66920\ldots$. Make a conjecture about what sort of "attractive cycle" the dynamical system $\{\hat{\mathbb{C}}; z^2 - \lambda\}$ might possess at the Myreberg point. Test your conjecture numerically. You will find it easiest to restrict attention to real orbits.

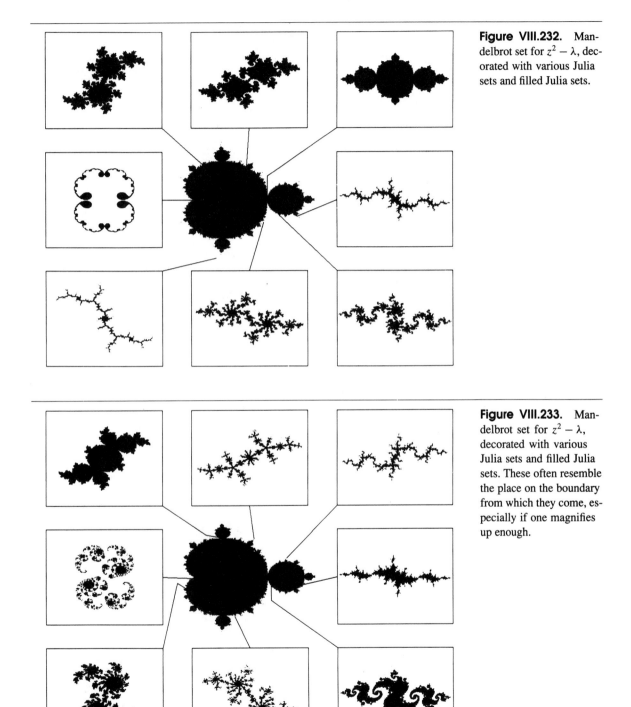

Figure VIII.232. Mandelbrot set for $z^2 - \lambda$, decorated with various Julia sets and filled Julia sets.

Figure VIII.233. Mandelbrot set for $z^2 - \lambda$, decorated with various Julia sets and filled Julia sets. These often resemble the place on the boundary from which they come, especially if one magnifies up enough.

Figure VIII.234. A zoom on a piece of the boundary of the Mandelbrot set for $z^2 - \lambda$.

Figure VIII.235. A filled Julia set corresponding to the piece of the coastline of the Mandelbrot set in Figure VIII.234. Notice the family resemblances.

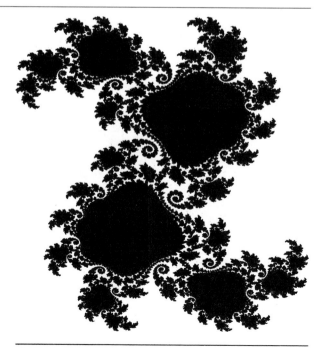

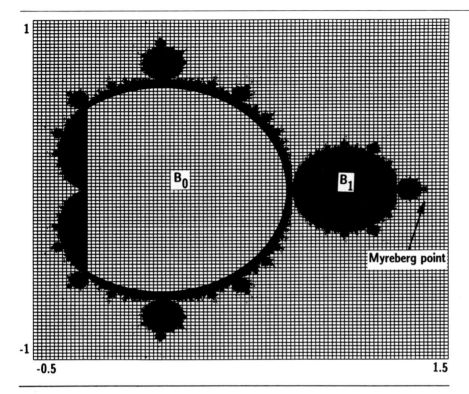

Figure VIII.236. A picture of the Mandelbrot set for the family of dynamical systems $\{\mathbb{C}; z^2 - \lambda\}$. It has been overlaid on a coordinate grid. The middle of the first bubble has not been plotted, to clarify the coordinate grid.

3.4. Make a parameter space map for the family of dynamical systems $\{\hat{\mathbb{C}}; f_\lambda(z)\}$, where f_λ is the Newton transformation associated with the family of polynomials

$$F(z) = z^3 + (\lambda - 1)z + 1, \qquad \lambda \in P = \mathbb{C}.$$

Notice that the polynomial has a root located at $z = 1$, independent of λ. Color your map according to the "escape time" of the orbit of O to a ball of small radius centered at $z = 1$. Use black to represent values of λ for which O does not converge to $z = 1$. Examine some Julia sets of f_λ corresponding to points on the boundary of the black region. Are there resemblances between structures that occur in your map of parameter space, and some of the corresponding collection of Julia sets? (The correct answer to this question can be found in [Curry 1983].)

4   How to Make Maps of Families of Fractals Using Escape Times

We begin by looking at the Mandelbrot set for a certain family of IFS. It is disappointing, and we do not learn much. We then introduce a related family of dynamical systems and color the parameter space using escape times. The result is a map

Figure VIII.237. The complement of the Mandelbrot set $\mathcal{M}_1$ associated with the family of IFS $\{\mathbb{C}; w_1 = \lambda z + 1, w_2 = \lambda^* z - 1\}$. Points in the complement of the Mandelbrot set are colored black. The boundary of $\mathcal{M}_1$ is smooth and does not reveal much information about the family of fractals it represents. The figure also shows attractors of the IFS corresponding to various points on the boundary of $\mathcal{M}_1$. What a disappointing map this is!

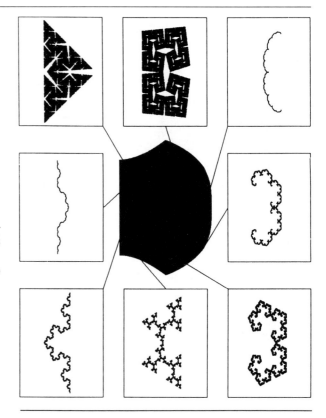

packed with information. We generalize the procedure to provide a method for making maps of other families of dynamical systems. We discover how certain boundaries in the resulting maps can yield information about the appearance of the fractals in the family. That is, we begin to learn to read the maps.

Figures VIII.237 and VIII.238 show the Mandelbrot set $\mathcal{M}_1$ for the family of hyperbolic IFS

$$\{\mathbb{C}; w_1(z) = \lambda z + 1, w_2(z) = \lambda^* z - 1\}, P = \{\lambda \in \mathbb{C} : |\lambda| < 1\}.$$

We use the notation $\lambda^* = (\lambda_1 + i\lambda_2)^* = (\lambda_1 - i\lambda_2)$ for the complex conjugate of λ. The two transformations are similitudes of scaling factor $|\lambda|$. At fixed λ, they rotate in opposite directions through the same angle. The figures also show attractors of the IFS corresponding to various points around the boundary of the Mandelbrot set. What a disappointing map this is! There are no secret bays, jutting peninsulas, nor ragged rocks in the coastline.

Theorem 4.1 (Hardin 1985). *The Mandelbrot set $\mathcal{M}_1$ is connected. Its boundary is the union of a countable set of smooth curves and is piecewise differentiable.*

Proof This can be found in [Barnsley 1988d].

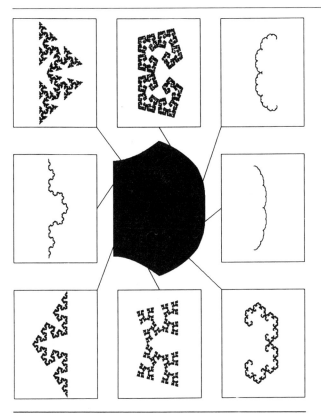

Figure VIII.238. The complement of the Mandelbrot set $\mathcal{M}_1$ associated with the family of IFS $\{\mathbb{C}; w_1 = \lambda z + 1, w_2 = \lambda^* z - 1\}$, together with some of the corresponding fractals. Notice how these have subsets of points that lie on straight lines, like the local structure of $\partial \mathcal{M}_1'$.

Let us try to obtain a better map of this family of attractors. In order to do so we begin by defining an extension of the associated shift dynamical system, for each $\lambda \in P \setminus \mathcal{M}_1$. Let $A(\lambda)$ denote the attractor of the IFS. One can prove that $A(\lambda)$ is symmetric about the y-axis. Hence $A(\lambda) \in \mathcal{M}_1$ if and only if $A(\lambda)$ intersects the y-axis. Define $f_\lambda : \mathbb{C} \to \mathbb{C}$

$$f_\lambda(z) = \begin{cases} w_1^{-1}(z) & \text{if Re } z \geq 0; \\ w_2^{-1}(z) & \text{if Re } z < 0. \end{cases}$$

Then, when λ is such that $A(\lambda)$ is disconnected, $\{A(\lambda); f_\lambda\}$ is the shift dynamical system associated with the IFS; $\{\mathbb{C}; f_\lambda\}$ is an extension of the shift dynamical system to all of $\mathbb{C}$; and $A(\lambda)$ is the "repelling set" of $\{\mathbb{C}; f_\lambda\}$. This system can be used to compute images of $A(\lambda)$ in the just-touching and totally disconnected cases, using the Escape Time Algorithm, as discussed in Chapter VII, section 1.

We make a map of the family of dynamical systems $\{\mathbb{R}^2; f_\lambda\}, \lambda \in P$. To do this we use the following algorithm, which was illustrated in Program 8.3.1. The algorithm applies to any family of dynamical systems $\{\mathbb{R}^2; f_\lambda\}$ that possesses a "repelling set" $A(\lambda)$, and such that P is a two-dimensional parameter space with a nice classical shape, such as a square or a disk.

Algorithm 4.1 Method for Coloring Parameter Space According to an Escape Time.

(i) Choose a positive integer, numits, *corresponding to the amount of computation one is able to do. Fix a point $Q \in \mathbb{R}^2$ such that $Q \in A(\lambda)$ for some, but not all, $\lambda \in P$.*

(ii) Fix a ball $B \subset \mathbb{R}^2$ such that $A(\lambda) \subset B$ for all $\lambda \in P$. Define an escape region to be $\mathcal{V} = \mathbb{R}^2 \setminus B$.

(iii) Represent the parameter space P by an array of pixels. Carry out the following step for each λ in the array.

(iv) Compute $\{f_\lambda^{\circ n}(Q) : n = 0, 1, 2, 3, \dots, \text{numits}\}$. Color the pixel λ according to the least value of n such that $f_\lambda^{\circ n}(Q) \in \mathcal{V}$. If the computed piece of the orbit does not intersect $\mathcal{V}$, color the pixel black.

The result of applying this algorithm to the dynamical system defined above, with $Q = O$, is illustrated in Figures VIII.239 and VIII.240 (a)–(g) and Color Plates 17 and 18.

Figure VIII.239 contains four different regions. The first is a neighborhood of O, surrounded by almost concentric bands of black, gray, and white. The location of this region is roughly the same as that of $P \setminus \mathcal{M}_1$, which corresponds to totally disconnected and just-touching attractors. The second region is the grainy area, which

Figure VIII.239. A map of the family of dynamical systems $\{\mathbb{C}; f_\lambda\}$, where

$$f_\lambda(z) = \begin{cases} (z-1)/\lambda & \text{if } \operatorname{Re} z \geq 0; \\ (z+1)/\lambda & \text{if } \operatorname{Re} z < 0. \end{cases}$$

The parameter space is $P = \{\lambda \in \mathbb{C}; 0 < \lambda_1 < 1, 0 < \lambda_2 \leq 0.75\}$. The map is obtained by applying Algorithm 8.4.1. Pixels are shaded according to the "escape time" of a point $O \in \mathbb{R}^2$. The exciting places where the interesting fractals are to be found are not within the solid bands of black, gray, or white, but within the foggy coastline. This coastline is itself a fractal object, revealing infinite complexity under magnification. In it one finds approximate pictures of some of the connected and "almost connected" repelling sets of the dynamical system. Why are they there?

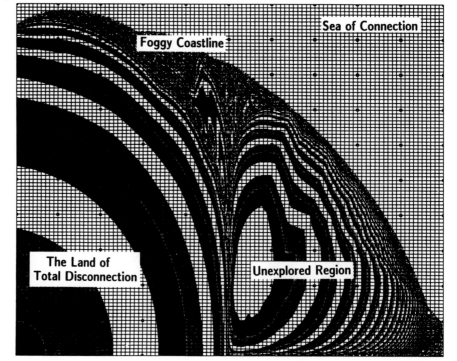

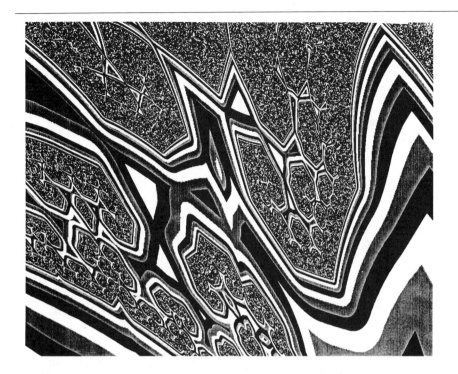

Figure VIII.240. A sequence of zooms on a piece of the foggy coastline in Figure VIII.239. The window coordinates of the highest power zoom are $0.4123 \leq \lambda_1 \leq 0.4139, 0.6208 \leq \lambda_2 \leq 0.6223$. Can you find where each picture lies within the one that precedes it?

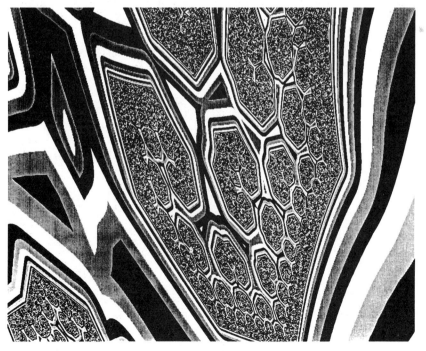

Figure VIII.240. (b)

Figure VIII.240. (c)

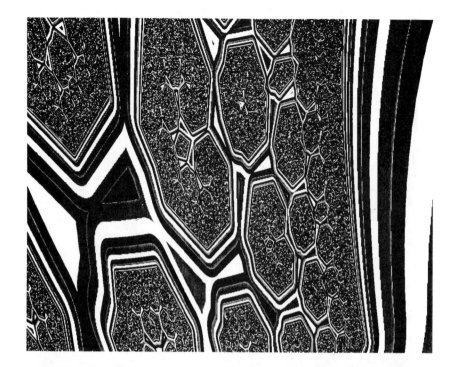

Figure VIII.240. (d)

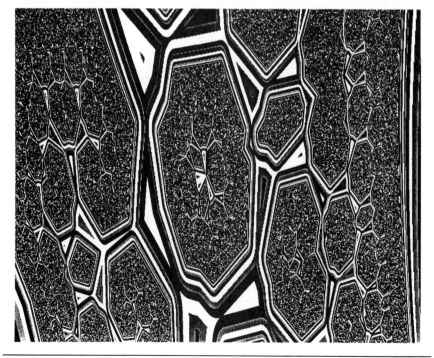

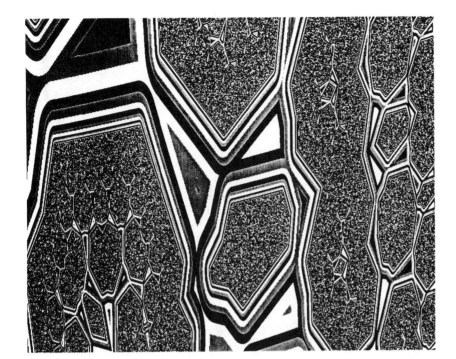

Figure VIII.240. (e)

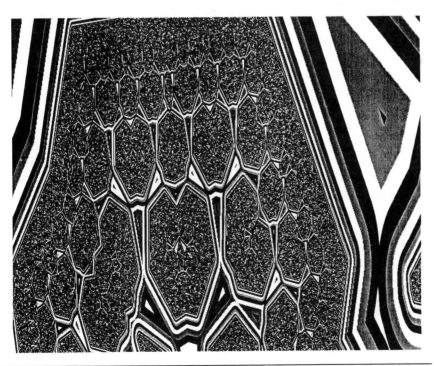

Figure VIII.240. (f)

Figure VIII.240. (g)

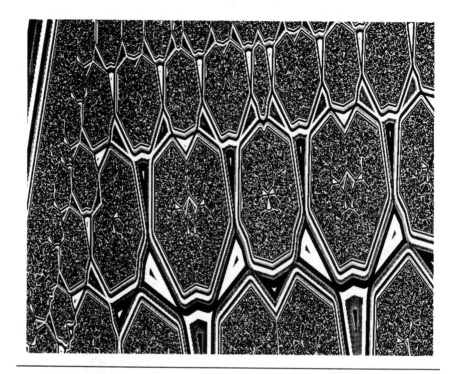

we refer to as the foggy coastline. Here, upon magnification, one finds complex geometrical structures. An example is illustrated in the sequence of zooms in Figure VIII.240 (a)–(g). The structures appear to be subtly different from one another. Early experiments show that if λ is chosen in the vicinity of one of these structures, then images of the "repelling set" of the dynamical system $\{\mathbb{R}^2; f_\lambda\}$, computed using the Escape Time Algorithm, contains similar structures. An example of such an image is shown in Figure VIII.241. The third region, at the lower right in Figure VIII.239, is made up of closed contours of black, grey, and white. Here the map conveys little information about the family of dynamical systems. To obtain information in this region one should examine the orbits of a point Q, different from O. The fourth region, the outer white area in Figure VIII.239, corresponds to dynamical systems for which the orbit of O does not escape. It is likely that for λ in this region, the "repelling set" of the dynamical system possesses an interior.

Our new maps, such as Figure VIII.239, can provide information about the family of IFS

$$\{\mathbb{C}; w_1(z) = \lambda z + 1, w_2(z) = \lambda^* z - 1\}, P = \{\lambda \in \mathbb{C} : |\lambda| < 1\}$$

in the vicinity of the boundary of the Mandelbrot set. For $\lambda \in \partial \mathcal{M}_1$ the attractor of the IFS is the same as the repelling set of the dynamical system. For λ close to $\partial \mathcal{M}_1$ the attractor of the IFS "looks like" the repelling set of the dynamical system.

Figure VIII.242 shows a transverse section through the anther of a lily. We include it because some of the structures in Figure VIII.240 (a)–(g) are reminiscent of cells.

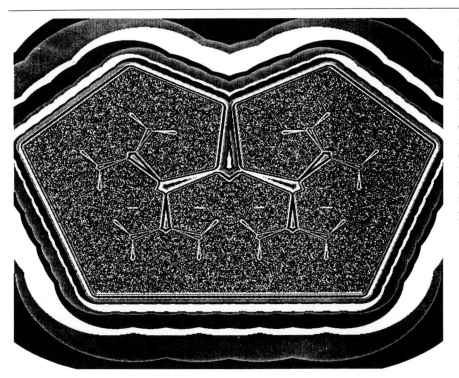

Figure VIII.241. Image of the repelling set for one of the family of the dynamical systems whose parameter space was mapped in Figure VIII.239. This image corresponds to a value of λ that lies within the highest power zoom in Figure VIII.240. Notice how the objects here resemble those in the corresponding position in the parameter space.

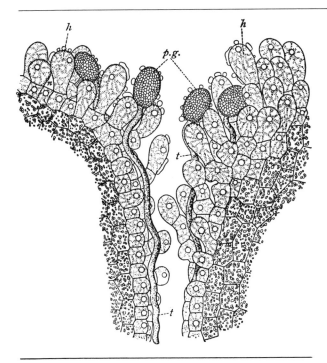

Figure VIII.242. Longitudinal section through part of the stigma of a lily, showing germinating pollen-grains. *h* pappillae of stigma; *p.g.*, pollen grains; *t*, pollen tubes. Highly magnified. (After DodelPort, [Scott 1917].)

Algorithm 1 in section 4 can be applied to families of dynamical systems of the type described in Theorem 4.1 in Chapter VII. For example, let $\{\mathbb{R}^2; f_\lambda\}$, where $\lambda \in P = \blacksquare \subset \mathbb{R}^2$, denote a family of dynamical systems. Let $\mathbf{X} \subset \mathbb{R}^2$ be compact. Let $f_\lambda : \mathbf{X} \to \mathbb{R}^2$ be continuous and such that $f(\mathbf{X}) \supset \mathbf{X}$. Then f_λ possesses an invariant set $A(\lambda) \in \mathcal{H}(\mathbf{X})$, given by

$$A(\lambda) = \cap_{n=0}^{\infty} f_\lambda^{\circ(-n)}(\mathbf{X}).$$

$A(\lambda)$ is the set of points whose orbits do not escape from $\mathbf{X}$. The set of points in P corresponding to which orbit of Q does not escape from $\mathbf{X}$ is

$$\mathcal{M}(Q) = \{\lambda \in P : Q \in A(\lambda)\}.$$

We conclude this chapter by giving an "explanation" of how family resemblances can happen between structures that occur on the boundary of $\mathcal{M}(Q)$ and the sets $A(\lambda)$. (1) Suppose that $A(\lambda)$ is a set in $\mathbb{R}^2$ that looks like a map of Great Britain, translated by λ. Then what does $\mathcal{M}(Q)$ look like? It looks like a map of Great Britain. (2) Suppose that $A(\lambda)$ is a set that looks like a map of Great Britain at time λ_1, translated by λ. We picture the set $A(\lambda)$ varying slowly, perhaps its boundary changing continuously in the Hausdorff metric as λ varies. Now $A(\lambda)$ looks like a deformed map of Great Britain. The local coves and inlets will be accurate representations of those coves at about the time λ_1 to which they correspond in the parameter space map. That is, the boundary of $\mathcal{M}(Q)$ will consist of neighboring bays and inlets at different times stitched together. It will be a map that is microscopically accurate (at some time) and globally inaccurate. (3) Now pretend in addition that the coastline of Great Britain is self-similar at each time λ_1. That is, imagine that little bays look like whole chunks of the coastline, at a given instant. Now what will $\mathcal{M}(Q)$ look like? At a given microscopic location on the boundary, magnified enormously, we will see a picture of a whole chunk of the coastline of Great Britain, at that instant. (4) Now imagine that for some values of λ, Great Britain, in the distant future, is totally disconnected, reduced to grains of isolated sand. It is unlikely that those values of λ belong to $\mathcal{M}(Q)$. As λ varies in a region of parameter space for which $A(\lambda)$ is totally disconnected, it is not probable that $Q \in A(\lambda)$. In these regions we would expect $\mathcal{M}(Q)$ to be totally disconnected.

The families of sets $\{A(\lambda) \in \mathbf{X} : \lambda \in P\}$ considered in this chapter broadly fit into the description in the preceding paragraph. Both P and $\mathbf{X}$ are two-dimensional. The sets $A(\lambda)$ are derived from transformations that behave locally like similitudes. For each $\lambda \in P$, $A(\lambda)$ is either connected or totally disconnected. Finally, the sets $A(\lambda)$ and their boundaries appear to depend continuously on λ.

Examples & Exercises

4.1. In the above section we applied Algorithm 1 in section 4, with $Q = (0, 0)$, to compute a map of the family of dynamical systems

re VIII.243. A
of the family of dy-
cal systems described
ample 4.2, computed
Algorithm 1 in sec-
4. The parameter
is $P = \{\lambda \in \mathbb{C}; 0 <$
1, $0 < \lambda_2 < 1\}$. The
grainy area is the in-
ing region. This is
oastline"; it is itself
tal object, revealing
e complexity under
ification.

e VIII.244. Zoom
small piece of
ggy area in Fig-
III.242. In it one
grainy areas that re-
the repelling sets
corresponding dy-
al systems. At what
f λ does one find
At the value of λ
map where the pic-
u are interested in

$$f_\lambda(z) = \begin{cases} (z-1)/\lambda & \text{if } \operatorname{Re} z \geq 0, \\ (z+1)/\lambda^* & \text{if } \operatorname{Re} z < 0. \end{cases}$$

The resulting map was shown in Figure VIII.239. This map contains an unexplored region. Repeat the computation, but with (a) $Q = 0.5$, and (b) $Q = -0.5$, to obtain information about the unexplored region.

4.2. In this example we consider the family of dynamical systems $\{\mathbb{C}; f_\lambda\}$, where

$$f_\lambda(z) = \begin{cases} (z-1)/\lambda & \text{if } \lambda_2 x - \lambda_1 y \geq 0, \\ (z+1)/\lambda & \text{if } \lambda_2 x - \lambda_2 y < 0. \end{cases}$$

The parameter space is $\lambda \in P = \{\lambda \in \mathbb{C} : 0 < |\lambda| < 1\}$. This family is related to the family of IFS

$$\{\mathbb{C}; w_1(z) = \lambda z + 1, w_2(z) = \lambda z - 1\}.$$

Let $A(\lambda)$ denote the attractor of the IFS and let $\tilde{A}(\lambda)$ denote the "repelling set" associated with the dynamical system. Let

$$S = \{\lambda \in P : \text{the line } \lambda_2 x - \lambda_1 y = 0 \text{ separates } w_1(A(\lambda)) \text{ and } w_2(A(\lambda))\}.$$

If $\lambda \in S$ then $\{A(\lambda); f_\lambda\}$ is the shift dynamical system associated with the IFS, and $A(\lambda) = \tilde{A}(\lambda)$. Even when $\lambda \notin S$ we expect there to be similarities between $A(\lambda)$ and $\tilde{A}(\lambda)$.

In Figures VIII.243, VIII.244, and VIII.245 and Color Plates 19 and 20 we show some results of applying Algorithm 1 in section 4 to the dynamical system $\{\mathbb{C}; f_\lambda\}$.

In Figure VIII.243, the outer white region represents systems for which the orbit of the point O does not diverge, and probably corresponds to "repelling sets" with nonempty interiors. The inner region, defined by the patchwork of gray, black, and white sections, bounded by line segments, represents systems for which the orbit of O diverges and corresponds to totally disconnected "repelling sets." The grainy gray area is the interesting region. This is the "coastline"; it is itself a fractal object, revealing infinite complexity under magnification. Figures VIII.244 and VIII.245 show magnifications at two places on the coastline. The grainy areas revealed by magnification resemble pictures of the repelling set of the dynamical system at the corresponding values of λ.

4.3. This exercise refers to the family of dynamical systems $\{\mathbb{C}; z^2 - \lambda\}$. Use Algorithm 1 in section 4 with $-0.25 \leq \lambda_1 \leq 2$, $-1 \leq \lambda_2 \leq 1$, and $Q = (0.5, 0.5)$ to make a picture of the "Mandelbrot Set" $\mathcal{M}(0.5, 0.5)$. An example of such a set, for a different choice of Q, is shown in Figure VIII.246.

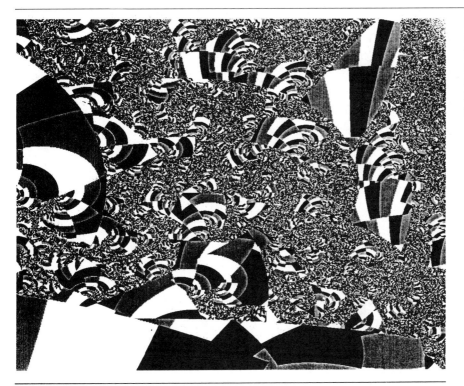

Figure VIII.245. Zoom on a small piece of the foggy area in Figure VIII.242. The grainy areas in this picture here have different shapes from those in Figure VIII.244.

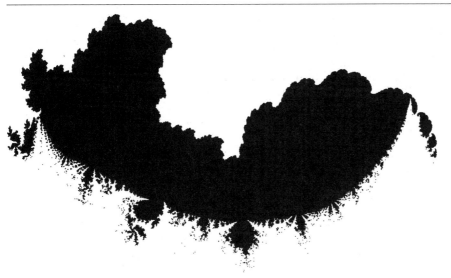

Figure VIII.246. A "Mandelbrot set" $\mathcal{M}(z_0)$ associated with the family of dynamical systems $\{\mathbb{C}; z^2 - \lambda\}$. This was computed using escape times of orbits of the point $z = z_0$ different from the critical point, $z = 0$.

Chapter IX

Measures on Fractals

1 Introduction to Invariant Measures on Fractals

In this section we give an intuitive introduction to measures. We focus on measures that arise from iterated function systems in $\mathbb{R}^2$.

In Chapter III, section 8 we introduced the Random Iteration Algorithm. This algorithm is a means for computing the attractor of a hyperbolic IFS in $\mathbb{R}^2$. In order to run the algorithm one needs a set of probabilities, in addition to the IFS.

Definition 1.1 *An iterated function system* with probabilities *consists of an IFS*

$$\{\mathbf{X}; w_1, w_2, \ldots, w_N\}$$

together with an ordered set of numbers $\{p_1, p_2, \ldots, p_N\}$, *such that*

$$p_1 + p_2 + p_3 + \cdots + p_N = 1 \text{ and } p_i > 0 \qquad \text{for } i = 1, 2, \ldots, N.$$

The probability p_i is associated with the transformation w_i. The nomenclature "IFS with probabilities" is used for "iterated function system with probabilites." The full notation for such an IFS is

$$\{\mathbf{X}; w_1, w_2, \ldots, w_N; p_1, p_2, \ldots, p_N\}.$$

Explicit reference to the probabilities may be suppressed.

An example of an IFS with probabilities is

$$\{\mathbb{C}; w_1(z), w_2(z), w_3(z), w_4(z); 0.1, 0.2, 0.3, 0.4\},$$

where

$$w_1(z) = 0.5z, \, w_2(z) = 0.5z + 0.5,$$
$$w_3(z) = 0.5z + (0.5)i, \, w_4(z) = 0.5z + 0.5 + (0.5)i.$$

It can be represented by the IFS code in Table IX.1. The attractor is the filled square ■, with corners at $(0, 0)$, $(1, 0)$, $(1, 1)$, and $(0, 1)$.

Here is how the Random Iteration Algorithm proceeds in the present case. An initial point, $z_0 \in \mathbb{C}$, is chosen. One of the transformations is selected "at random"

Table IX.1.   IFS code for a measure on ∎.

w	a	b	c	d	e	f	p
1	0.5	0	0	0.5	1	1	0.1
2	0.5	0	0	0.5	50	1	0.2
3	0.5	0	0	0.5	1	50	0.3
4	0.5	0	0	0.5	50	50	0.4

from the set $\{w_1, w_2, w_3, w_4\}$. The probability that w_i is selected is p_i, for $i = 1, 2, 3, 4$. The selected transformation is applied to z_0 to produce a new point $z_1 \in \mathbb{C}$. Again a transformation is selected, in the same manner, independently of the previous choice, and applied to z_1 to produce a new point z_2. The process is repeated a number of times, resulting in a finite sequence of points $\{z_n : n = 1, 2, \ldots, numits\}$, where *numits* is a positive integer. For simplicity, we assume that $z_0 \in$ ∎. Then, since $w_i(∎) \subset ∎$, for $i = 1, 2, 3, 4$, the "orbit" $\{z_n : n = 1, 2, \ldots, numits\}$ lies in ∎.

Consider what happens when we apply the algorithm to the IFS code in Table IX.1. If the number of iterations is sufficiently large, a picture of ∎ will be the result. That is, every pixel corresponding to ∎ is visited by the "orbit" $\{z_n : n = 1, 2, \ldots, numits\}$. The rate at which a picture of ∎ is produced depends on the probabilities. If *numits* $= 10,000$, then we expect that because the images of ∎ are just-touching,

$$\text{the number of computed points in } w_1(∎) \approx 1000,$$

$$\text{the number of computed points in } w_2(∎) \approx 1000,$$

$$\text{the number of computed points in } w_3(∎) \approx 1000,$$

$$\text{the number of computed points in } w_4(∎) \approx 1000.$$

These estimates are supported by Figure IX.247, which shows the result of running a modified version of Program 2 in Chapter III, with the IFS code in Table IX.1, and *numits* $= 100,000$.

In Figure IX.248 we show the result of running a modified version of Program 2 in Chapter III, for the IFS code in Table IX.1, with various choices for the probabilities. In each case we have halted the program after a relatively small number of iterations, to stop the image from becoming "saturated." The results are diverse textures. In each case the attractor of the IFS is the same set, ∎. However, the points produced by the Random Iteration Algorithm "rain down" on ∎ with different frequencies at different places. Places where the "rainfall" is highest appear "darker" or "more dense" than those places where the "rainfall" is lower. In the end all places on the attractor get wet.

The pictures in Figure IX.248 (a)–(c) suggest a wonderful idea. They suggest that associated with an IFS with probabilities there is a unique "density" on the

Figure IX.247. The Random Iteration Algorithm, Program 1 in Chapter III, is applied to the IFS code in Table IX.1, with numits = 100,000. Verify that the number of points that lie in $w_i(\blacksquare)$ is approximately $(numits)p_i$, for $i = 1, 2, 3, 4$.

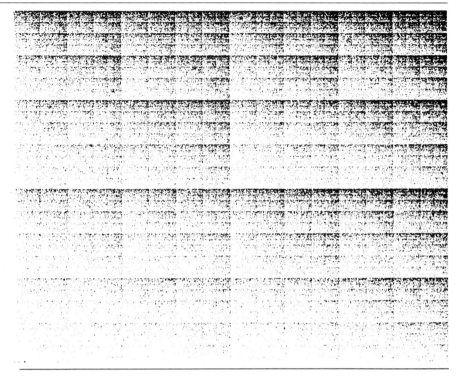

attractor of the IFS. The Random Iteration Algorithm gives one a glimpse of this "density," but one loses sight of it as the number of iterations is increased. This is true, and much more as well! As we will see, the "density" is so beautiful that we need a new mathematical concept to describe it. The concept is that of a *measure*. Measures can be used to describe intricate distributions of "mass" on metric spaces. They are introduced formally further on in this chapter. The present section provides an intuitive understanding of what measures are and of how an interesting class of measures arises from IFS's with probabilities.

As a second example, consider the IFS with probabilities

$$\{\mathbb{C}; w_1(z) = 0.5z + 24 + 24i, w_2(z) = 0.5z + 24i, w_3(z) = 0.5z; 0.25, 0.25, 0.5\}.$$

The attractor is a Sierpinski triangle $\triangle$. The probability associated with w_3 is twice that associated with either w_1 or w_2. In Figure IX.249 we show the result of applying the Random Iteration Algorithm, with these probabilities, to compute 10,000 points belonging to $\triangle$. There appear to be different "densities" at different places on $\triangle$. For example, $w_3(\triangle)$ appears to have more "mass" than either $w_1(\triangle)$ or $w_2(\triangle)$.

In Figure IX.250 we show the result of applying the Random Iteration Algorithm to another IFS with probabilities, for three different sets of probabilities. The IFS is $\{\mathbb{R}^2; w_1, w_2, w_3, w_4\}$, where w_i is an affine transformation for $i = 1, 2, 3, 4$. The attractor of the IFS is a leaf-like subset of $\mathbb{R}^2$. In each case we see a different pattern

Figure IX.248. The Random Iteration Algorithm is applied to the IFS code in Table IX.1, but with various different sets of probabilities. The result is that points rain down on the attractor of the IFS at different rates at different places. What we are seeing are the faint traces of wonderful mathematical entities called *measures*. These are the true fractals. Their supports, the attractors of IFS, are merely sets upon which the measures live.

Figure IX.248.   (b)

Figure IX.248. (c)

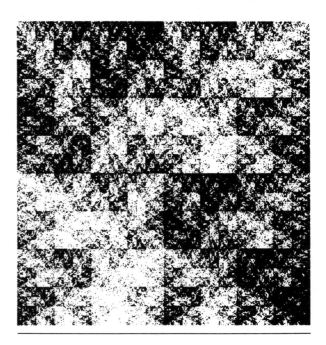

of "mass" on the attractor of the IFS. It appears that each "density" is itself a fractal object.

Examples & Exercises

1.1. Carry out the following numerical experiment. Apply the Random Iteration Algorithm to the IFS code in Table IX.1, for *numits* = 1000, 2000, 3000, ... In each case record the number, $\mathcal{N}$, of computed points that land in $B = \{(x, y) \in \mathbb{R}^2 : (x-1)^2 + (y-1)^2 \leq 1\}$, and make a table of your results. Verify that the ratio $\mathcal{N}/numits$ appears to approach a constant.

1.2. Repeat the computergraphical experiment that produced Figure IX.247. Verify that you obtain "similar-looking" output to that shown in Figure IX.247, even though you (probably) use a different random-number sequence.

1.3. The Random Iteration Algorithm is used to compute 100,000 points belonging to ■, using the IFS code in Table IX.1. How many of these points, do you expect, would belong to $w_1 \circ w_3(■)$? Why?

Let $(\mathbf{X}, d)$ be a complete metric space. Let $\{\mathbf{X}; w_1, \ldots, w_N; p_1, \ldots, p_N\}$ be an IFS with probabilities. Let A denote the attractor of the IFS. Then there exists a thing called the *invariant measure* of the IFS, which we denote here by μ. μ assigns "mass" to many subsets of $\mathbf{X}$. For example, $\mu(A) = 1$ and $\mu(\emptyset) = 0$. That is, the "mass" of the attractor is one unit, and the "mass" of the empty set is zero. Also $\mu(\mathbf{X}) = 1$, which says that the whole space has the same "mass" as the attractor of the IFS; the "mass" is located on the attractor.

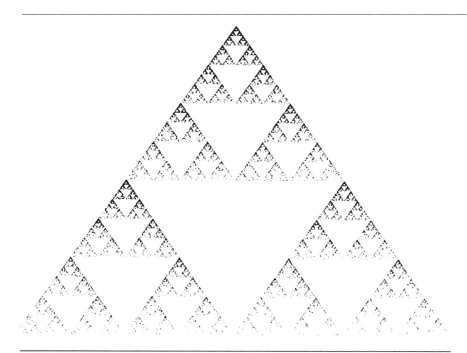

Figure IX.249. The Random Iteration Algorithm is used to compute an image of the Sierpinski triangle $\triangle$. The probability associated with w_3 is twice that associated with w_1 or w_2. One thousand points have been computed. The result is that $w_3(\triangle)$ appears denser than $w_1(\triangle)$ or $w_2(\triangle)$. This appearance is lost when the number of iterations is increased. We are led to the idea of a "mass" or measure that is supported on the fractal.

Not all subsets of $\mathbf{X}$ have a "mass" assigned to them. The subsets of $\mathbf{X}$ that do have a "mass" are called the *Borel subsets* of $\mathbf{X}$, denoted by $\mathcal{B}(\mathbf{X})$. The Borel subsets of $\mathbf{X}$ include the compact nonempty subsets of $\mathbf{X}$, so that $\mathcal{H}(\mathbf{X}) \subset \mathcal{B}(\mathbf{X})$. Also, if $\mathcal{O}$ is an open subset of $\mathbf{X}$, then $\mathcal{O} \in \mathcal{B}(\mathbf{X})$. So there are plenty of sets that have "mass." Let B denote a closed ball in $\mathbf{X}$. Here is how to calculate the "mass" of the ball, $\mu(B)$. Apply the Random Iteration Algorithm to the IFS with probabilities, to produce a sequence of points $\{z_n\}_{n=0}^{\infty}$. Let

$$\mathcal{N}(B, n) = \text{ number of points in } \{z_0, z_1, z_2, z_3, \ldots, z_n\} \cap B, \text{ for } n = 0, 1, 2, \ldots.$$

Then, almost always,

$$\mu(B) = \lim_{n \to \infty} \left\{ \frac{\mathcal{N}(B, n)}{(n + 1)} \right\}.$$

That is, the "mass" of the ball B is the proportion of points, produced by the Random Iteration Algorithm, which land in B. (To be precise we also have to require that the "mass" of the boundary of B is zero; see Corollary 7.1.)

By now you should be bursting with questions. How do we know that this formula "almost always" gives the same answer? What are Borel sets? Why don't all sets have "mass"? Welcome to measure theory!

As an example, we evaluate the measure of some subsets of $\mathbb{C}$, for the IFS with probabilities

Figure IX.250.   The Random Iteration Algorithm is used to compute an image of a leaf. Different sets of probabilities lead to different distributions of "mass" on the leaf.

$$\{\mathbb{C};\, w_1(z) = 0.5z,\, w_2(z) = 0.5z + (0.5)i,\, w_3(z) = 0.5z + 0.5;\, 0.33, 0.33, 0.34\}.$$

The attractor is a Sierpinski triangle $\triangle$ with vertices at 0, i, and 1. We compute the measures of the following sets:

$$B_1 = \{z \in \mathbb{C} : |z| \leq 0.5\}$$
$$B_2 = \{z \in \mathbb{C} : |z - (0.5 + 0.5i)| \leq 0.2\}$$
$$B_3 = \{z \in \mathbb{C} : |z - (0.5 + 0.5i)| \leq 0.5\}$$
$$B_4 = \{z \in \mathbb{C} : |z - (2 + i)| \leq \sqrt{2}\}.$$

The results are presented in Table IX.2.

Figure IX.251 illustrates the ideas introduced here.

Examples & Exercises

1.4. Explain why $\mu(B_4) \approx 0$ in Table IX.2.

1.5. What value, approximately, would have been obtained for $\mu(B_1)$ in Table IX.2, if the probabilities on the three maps had been $p_1 = 0.275$, $p_2 = 0.125$, and $p_3 = 0.5$?

Table IX.2. The measures of some subsets of $\mathbb{A}$ are computed by random iteration.

n	$\mathcal{N}(B_1, n)/n$	$\mathcal{N}(B_2, n)/n$	$\mathcal{N}(B_3, n)/n$	$\mathcal{N}(B_4, n)/n$
5,000	0.3313	0.1036	0.6385	0.0004
10,000	0.3314	0.1050	0.6500	0.0002
15,000	0.3323	0.1041	0.6512	0.0001
20,000	0.3330	0.1030	0.6525	0.0000
50,000	0.3326	0.1041	0.6527	0.0000
100,000	0.3325	0.1054	0.6497	0.0000

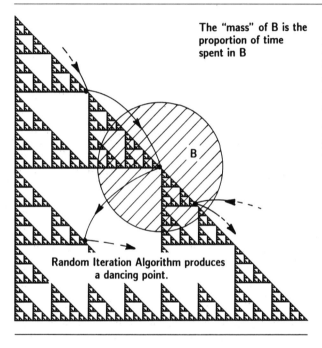

The "mass" of B is the proportion of time spent in B

Random Iteration Algorithm produces a dancing point.

Figure IX.251. Diagram of the Random Iteration Algorithm running, and a dancing point coming and going from the ball B. The "mass" or measure of the ball is $\mu(B)$. It is equal to the proportion of points that land in B.

1.6. Why, do you think, is the phrase "almost always" written in connection with the formula for $\mu(B)$, given above?

2 Fields and Sigma-Fields

Definition 2.1 *Let* $\mathbf{X}$ *be a space. Let* $\mathcal{F}$ *denote a nonempty class of subsets of a space* $\mathbf{X}$*, such that*

(1) $A, B \in \mathcal{F} \Rightarrow A \cup B \in \mathcal{F}$*;*
(2) $A \in \mathcal{F} \Rightarrow \mathbf{X} \setminus A \in \mathcal{F}$*.*

Then $\mathcal{F}$ is called a field.

(In exercise 2.12 you will be asked to prove that $\mathbf{X} \in \mathcal{F}$.)

Theorem 2.1 *Let $\mathbf{X}$ be a space. Let $\mathcal{G}$ be a nonempty set of subsets of $\mathbf{X}$. Let $\mathcal{F}$ be the set of subsets of $\mathbf{X}$ which can be built up from finitely many sets in $\mathcal{G}$ using the operations of union, intersection, and complementation with respect to $\mathbf{X}$. Then $\mathcal{F}$ is a field.*

Proof Elements of $\mathcal{F}$ consist of sets such as

$$\mathbf{X} \setminus (((\mathbf{X} \setminus (G_1 \cup G_2)) \cap G_3) \cup (G_5 \cap G_6)),$$

where G_1, G_2, G_3, G_3, ... denote elements of $\mathcal{G}$. That is, $\mathcal{F}$ is made of all those sets that can be expressed using a finite chain of parentheses, $\setminus$, $\cup$, $\cap$, elements of $\mathcal{G}$, and $\mathbf{X}$. (In fact, using de Morgan's laws one can prove that it is not necessary to use the intersection operation.) If we form the union of any two such expressions we obtain another one. Similarly, if we form the complement of such an expression with respect to $\mathbf{X}$, we obtain another such expression. So conditions (i) and (ii) in Definition 2.1 are satisfied. This completes the proof.

Definition 2.2 *The field referred to in Theorem 2.1 is called the field* generated *by $\mathcal{G}$.*

Examples & Exercises

2.1. Let $\mathbf{X}$ be a space and let $A \subset \mathbf{X}$. Then $\mathcal{F} = \{\mathbf{X}, A, \mathbf{X} \setminus A, \emptyset\}$ is a field.

2.2. Let $\mathbf{X}$ be the set of all leaves on a certain tree and let $\mathcal{F}$ be the set of all subsets of $\mathbf{X}$. Then $\mathcal{F}$ is a field. Let A denote the set of all the leaves on the lowest branch of the tree. Then $A \in \mathcal{F}$. Prove that $\mathcal{F}$ is generated by the leaves.

2.3. Let $\mathbf{X} = [0, 1] \subset \mathbb{R}$. Let $\mathcal{G}$ denote the set of all subintervals (open, closed, half-open) of $[0, 1]$. Let $\mathcal{F}$ denote the field generated by $\mathcal{G}$. Examples of members of $\mathcal{F}$ are $[0.5, 0.6) \cup (0.7, 0.81)$; $[0, 1]$; $[1, 1]$; and $(\frac{1}{2}, 1) \cup (\frac{1}{4}, \frac{1}{3}) \cup \cdots \cup (\frac{1}{100}, \frac{1}{99})$. Show that

$$\cup_{n=1}^{\infty} \left(\frac{1}{(n+1)}, \frac{1}{n} \right) = \left(\frac{1}{2}, 1 \right) \cup \left(\frac{1}{3}, \frac{1}{2} \right) \cup \left(\frac{1}{4}, \frac{1}{3} \right) \cup \cdots$$

is a subset of $\mathbf{X}$ but it is not a member of $\mathcal{F}$.

2.4. Let $\mathbf{X} = \blacksquare \subset \mathbb{R}^2$. Let $\mathcal{G}$ denote the set of closed rectangles contained in $\mathbf{X}$, whose sides are parallel to the coordinate axes and whose corners have rational coordinates. Let $\mathcal{F}$ denote the field generated by $\mathcal{G}$. An example of an element of $\mathcal{F}$ is

$$((\blacksquare \setminus ((\blacksquare \setminus R_1) \cup R_2) \cap R_3) \cup (R_4 \cap (\blacksquare \setminus R_5)),$$

where R_1, R_2, R_3, R_4, and R_5 are rectangles in $\mathcal{G}$. Let $S \in \mathcal{F}$. Prove that the area of S is a rational number. Deduce that $\mathcal{F}$ does not contain the ball $B(O, 1) = \{(x, y) \in \blacksquare : x^2 + y^2 \leq 1\}$.

Plate 23
Compare the simple, green IFS attractor and the corresponding Borel measure rendered using autumnal colors.

Plate 24
Compare the collage for a maple leaf and the corresponding Borel measure.

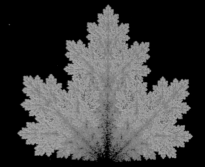

Plate 25

A sequence of frames taken from an IFS encoded movie entitled *A Cloud Study* [Barnsley, 1987a]. This illustates the continous dependence of the invariant measure on parameters in the IFS code. Theorem 11.1 in Chapter III makes movies!

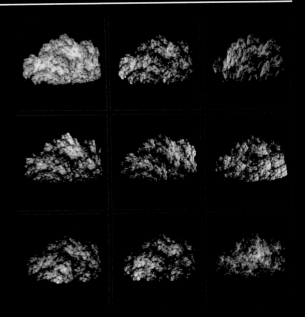

Plate 27

Andes Girl.

Plate 26

Monterey Coast.

Plate 28
Arctic Wolf.

Plate 30
Sunflower Field.

Plate 29
Leaf and Sunflower, condensation sets used in
Color Plate 30.

Plate 31
Black Forest.

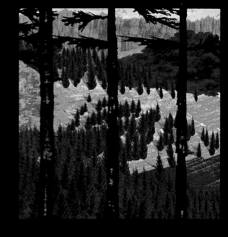

Plate 32
Zoom on Black Forest.

Plate 33
Black Forest in Winter.

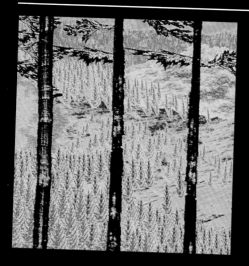

Plate 34
Zoom on Black Forest in Winter.

Plate 35
A rendered Borel measure mimics leaping flames.

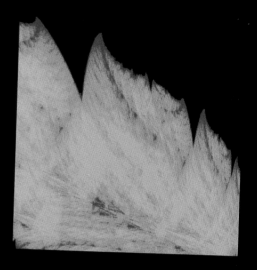

Plate 36
A fern and three progressive zooms. It was constructed using a recurrent IFS. Notice how at one scale the fronds are alternating and at another scale they are opposed.

Plate 37

An image of a root in soil. Notice how the gravel appears at many scales around the root. This plate illustrates a measure $\mu \in P$, which is a fixed point of the Markov operator M associated with a recurrent IFS of affine maps, where the maps have been carefully chosen to create this image. Computed using VRIFS™.

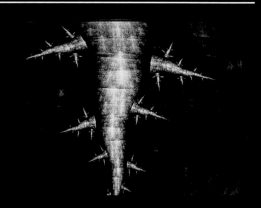

Plate 38

This picture of birds and trees is made up of a number of segments, each of which is made from a vector recurrent IFS with probabilities using VRIFS™. The PRESENTS™ system is then used to map the grayscale values for each segment into color values using a look-up table which is then adjusted interactively to make the image look as realistic as possible.

Plate 39
A fractal tree. Computed using VRIFS™ and colored by PRESENTS™.

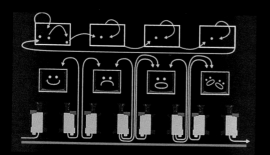

Plate 40
A collection of TV cameras pointing at TV screens illustrates the idea of a Vector Recurrent Iterated Function System. There is a Markov process on each "screen." Some points are spawned and other trajectories die.

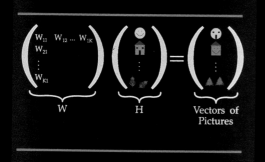

Plate 41
The transformation W is a contraction mapping in the space of vectors of pictures.

Plate 42
An original photograph of a cafe in Valencia, Spain was scanned to create a 24-bit per pixel file that was 1,345,960 bytes in size.

Plate 43
The decompressed image after compression down to 147,557 bytes; a compression ratio of 9:1. Compression and decompression for Plates 43, 44, and 45 were done using the fractal transform as implemented in Images Incorporated™ 2.0.

Plate 44
The decompressed image after compression down to 30,173 bytes; a compression ratio of 44:1.

Plate 45
The decompressed image after compression down to 23,399 bytes; a compression ratio of 58:1.

Figure IX.252. A field whose elements are sets of pixels. Can you find two elements of the field which generate the field?

2.5. Let **X** denote the set of pixels corresponding to a certain computer graphics display device. The set of all monochrome images that can be produced on this device forms a field. Figure IX.252 shows an example of a small field of subsets of **X**. It is generated by the pair of images, G_1 and G_2, in the middle of the second row, together with the set **X**. **X** is represented by the black rectangle. The empty set is represented by a blank screen. Find formulas for all of the images in Figure IX.252, in terms of G_1, G_2, and **X**.

2.6. Let Σ denote the code space on two symbols 1 and 2. Let $n \in \{1, 2, 3, \ldots\}$ and $e_i \in \{1, 2\}$ for $i = 1, 2, \ldots, n$. Let

$$C(e_1, e_2, \ldots, e_n) = \{x \in \Sigma : x_i = e_i \text{ for } i = 1, 2, \ldots, n\}.$$

Any subset of Σ that can be written in this form is called a *cylinder subset* of Σ. Let $\mathcal{F}$ denote the field generated by the cylinder subsets of Σ. Find a subset of Σ that is not in $\mathcal{F}$.

2.7. Let **X** be a space. Let $\mathcal{F}$ denote the set of all subsets of **X**. The customary notation for this field is $\mathcal{F} = 2^{\mathbf{X}}$. Show that $\mathcal{F}$ is a field.

Definition 2.3 *Let $\mathcal{F}$ be a field such that*

$$A_i \in \mathcal{F} \text{ for } i \in 1, 2, 3, \ldots \Rightarrow \cup_{i=1}^{\infty} A_i \in \mathcal{F}.$$

Then $\mathcal{F}$ is called a σ-field (sigma-field).

Given any field, there always is a minimal, or smallest, σ-field which contains it.

Theorem 2.2 *Let $\mathbf{X}$ be a space and let $\mathcal{G}$ be a set of subsets of $\mathbf{X}$. Let $\{\mathcal{F}_\alpha : \alpha \in I\}$ denote the set of all σ-fields on $\mathbf{X}$ which contain $\mathcal{G}$. Then $\mathcal{F} = \cap_\alpha \mathcal{F}_\alpha$ is a σ-field.*

Proof Note that there is at least one σ-field that contains $\mathcal{G}$, namely $2^{\mathbf{X}}$, the field consisting of all subsets of $\mathbf{X}$. We have to show that $\cap_\alpha \mathcal{F}_\alpha$ is a σ-field if each $\mathcal{F}_\alpha$ is a σ-field that contains $\mathcal{G}$. Suppose that $A_i \in \cap_\alpha \mathcal{F}_\alpha$; then, for each α, A_i is an element of the σ-field $\mathcal{F}_\alpha$ and so $\cup_{i=1}^\infty A_i \in \mathcal{F}_\alpha$. Suppose $A \in \cap_\alpha \mathcal{F}_\alpha$; then, for each α, $A \in \mathcal{F}_\alpha$ and so $\mathbf{X} \setminus A \in \mathcal{F}_\alpha$. Hence $\mathbf{X} \setminus A \in \cap_\alpha \mathcal{F}_\alpha$. This completes the proof.

Definition 2.4 *Let $\mathcal{G}$ be a set of subsets of a space $\mathbf{X}$. The minimal σ-field which contains $\mathcal{G}$, defined in Theorem 2.2, is called the σ-field generated by $\mathcal{G}$.*

Definition 2.5 *Let $(\mathbf{X}, d)$ be a metric space. Let $\mathcal{B}$ denote the σ-field generated by the open subsets of $\mathbf{X}$. $\mathcal{B}$ is called the* Borel field *associated with the metric space. An element of $\mathcal{B}$ is called a* Borel subset *of $\mathbf{X}$.*

The following theorem gives the flavor of ways in which the Borel field can be generated.

Theorem 2.3 *Let $(\mathbf{X}, d)$ be a compact metric space. Then the associated Borel field $\mathcal{B}$ is generated by a countable set of balls.*

Proof We prove a more general result first. Let $\mathcal{G} = \{b_n \subset \mathbf{X} : n = 1, 2, 3, \ldots; b_n \text{ open }\}$ be a *countable base* for the open subsets of $\mathbf{X}$. That is, every open set in $\mathbf{X}$ can be written as a union of sets in $\mathcal{G}$. Then $\mathcal{B}$ is generated by $\mathcal{G}$. To see this, let $\tilde{\mathcal{B}}$ denote the σ-field generated by $\mathcal{G}$. Then $\tilde{\mathcal{B}} \subset \mathcal{B}$ because $\mathcal{G}$ is contained in the set of open subsets of $\mathbf{X}$. On the other hand, $\mathcal{B} \subset \tilde{\mathcal{B}}$ because $\tilde{\mathcal{B}}$ contains all the generators of $\mathcal{B}$. Hence $\mathcal{B} = \tilde{\mathcal{B}}$.

It remains to construct a countable base for the open subsets of $\mathbf{X}$ using balls. For $R > 0$ let

$$B(x, R) = \{y \in \mathbf{X} : d(x, y) < R\}.$$

Let n be a positive integer. Then $\mathbf{X} = \cup_{x \in \mathbf{X}} B(x, \frac{1}{n})$. Hence $\{B(x, \frac{1}{n}) : x \in \mathbf{X}\}$ is an open covering of $\mathbf{X}$. Since $\mathbf{X}$ is compact it contains a finite subcovering $\{B(x_m^{(n)}, \frac{1}{n}) : m = 1, 2, \ldots, M(n)\}$ for some integer $M(n)$. We claim that

$$\mathcal{D} = \{B(x_m^{(n)}, \frac{1}{n}) : m = 1, 2, \ldots, M(n); n = 1, 2, 3, \ldots\}$$

is a countable base for the open subsets of $\mathbf{X}$. Let $\mathcal{O}$ be an open subset of $\mathbf{X}$, and let $x \in \mathcal{O}$. Then there is an open ball, of radius $R > 0$, such that $B(x, R) \subset \mathcal{O}$. Let n be large enough that $\frac{1}{n} < \frac{R}{2}$. Then there is $m \in \{1, 2, \ldots, M(n)\}$ so that x is in the ball $B(x_m^{(n)}, \frac{1}{n})$, and this ball is contained in $\mathcal{O}$. Each x in $\mathcal{O}$ is contained in such a ball, belonging to $\mathcal{D}$. Hence $\mathcal{D}$ is indeed a countable base for the open subsets of $\mathbf{X}$. This completes the proof.

Examples & Exercises

2.8. Let $\mathcal{B}$ denote the σ-field generated by the field in exercise 2.4. Then $\mathcal{B}$ contains the ball $B(O, 1)$. Similarly it contains all balls in $\blacksquare \subset \mathbb{R}^2$. Show that $\mathcal{B}$ is the Borel field associated with ($\blacksquare$, Manhattan).

2.9. Let Σ denote the code space on the two symbols $\{0, 1\}$. Show that the Borel field associated with (Σ, code space metric) is generated by the cylinder subsets of Σ, defined in exercise 2.5.

2.10. Let $\triangle \subset \mathbb{R}^2$ denote a Sierpinski triangle. Let $\mathcal{G}$ denote the set of connected components of $\mathbb{R}^2 \setminus \triangle$. Let $\mathcal{F}$ denote the σ-field generated by $\mathcal{G}$. Show that $\mathcal{F}$ is contained in, but not equal to, the Borel field associated with ($\mathbb{R}^2$, Euclidean).

2.11. Let $\mathbf{X}$ be a space and let $\mathcal{G}$ be a set of subsets of $\mathbf{X}$. Let $\mathcal{F}_1$ be the field generated by $\mathcal{G}$, let $\mathcal{F}_2$ be the σ-field generated by $\mathcal{G}$, and let $\mathcal{F}_3$ be the σ-field generated by $\mathcal{F}_1$. Prove that $\mathcal{F}_3 = \mathcal{F}_2$.

2.12. Let $\mathcal{F}$ be a field of subsets of a space $\mathbf{X}$. Prove that $\mathbf{X} \in \mathcal{F}$.

---

3 Measures

A measure is defined on a field. Each member of the field is assigned a nonnegative real number, which tells us its "mass."

Definition 3.1 *A measure μ, on a field $\mathcal{F}$, is a real nonnegative function μ : $\mathcal{F} \to [0, \infty) \subset \mathbb{R}$, such that whenever $A_i \in \mathcal{F}$ for $i = 1, 2, 3, \ldots$, with $A_i \cap A_j = \emptyset$ for $i \neq j$ and $\cup_{i=1}^{\infty} A_i \in \mathcal{F}$, we have*

$$\mu(\cup_{i=1}^{\infty} A_i) = \sum_{i=1}^{\infty} \mu(A_i).$$

(In other texts a measure as defined here is usually referred to as a finite measure.)

Definition 3.2 *Let $(\mathbf{X}, d)$ be a metric space. Let $\mathcal{B}$ denote the Borel subsets of $\mathbf{X}$. Let μ be a measure on $\mathcal{B}$. Then μ is called a* Borel measure.

Some basic properties of measures are summarized below.

Theorem 3.1 *Let $\mathcal{F}$ be a field and let $\mu : \mathcal{F} \to \mathbb{R}$ be a measure. Then*

(1) If $B \supset A$, then $\mu(B) = \mu(B \setminus A) + \mu(B)$, for A, $B \in \mathcal{F}$;
(2) If $B \supset A$, then $\mu(B) \geq \mu(A)$;
(3) $\mu(\emptyset) = 0$;
(4) If $A_i \in \mathcal{F}$ for $i = 1, 2, 3, \ldots$, and $\cup_{i=1}^{\infty} A_i \in \mathcal{F}$, then $\mu\left(\cup_{i=1}^{\infty} A_i\right) \leq \sum_{i=1}^{\infty} \mu(A_i)$;
(5) If $\{A_i \in \mathcal{F}\}$ obeys $A_1 \subset A_2 \subset A_3 \subset \ldots$, and if $\cup_{i=1}^{\infty} A_i \in \mathcal{F}$, then $\mu(A_i) \to \mu(\cup_{i=1}^{\infty} A_i)$.

(6) If $\{A_i \in \mathcal{F}\}$ obeys $A_1 \supset A_2 \supset A_3 \supset \ldots$, and if $\cap_{i=1}^{\infty} A_i \in \mathcal{F}$, then $\mu(A_i) \to \mu(\cap_{i=1}^{\infty} A_i)$.

Proof [Rudin 1966] Theorem 1.19, p. 17. These are fun to prove for yourself!

We are concerned with measures on compact subsets of metric spaces such as $(\mathbb{R}^2$, Euclidean). The natural underlying σ-field is the Borel field, generated by the open subsets of the metric space. The following theorem allows us to work with the restriction of the measure to any field that generates the σ-field.

Theorem 3.2 *[Caratheodory] Let μ denote a measure on a field $\mathcal{F}$. Let $\hat{\mathcal{F}}$ denote the σ-field generated by $\mathcal{F}$. Then there exists a unique measure $\hat{\mu}$ on $\hat{\mathcal{F}}$ such that $\mu(A) = \hat{\mu}(A)$ for all $A \in \mathcal{F}$.*

Sketch of proof The proof can be found in most books on measure theory; see [Eisen 1969] Theorem 5, p. 180, Chapter 6, for example. First μ is used to define an "outer measure" μ^0 on the set of subsets of **X**. μ^0 is defined by

$$\mu^0(A) = \inf \left\{ \sum_{n=1}^{\infty} \mu(A_n) : A \subset \cup_{n=1}^{\infty} A_n, A_n \in \mathcal{F} \forall n \in \mathcal{Z}^+ \right\}.$$

μ^0 is not usually a measure. However, one can show that the class $\mathcal{F}^0$ of subsets A of **X** such that—this was Caratheodory's smart idea—

$$\mu^0(E) = \mu^0(A \cap E) + \mu^0((\mathbf{X} \setminus A) \cap E) \qquad \text{for all } E \in 2^{\mathbf{X}}$$

is a σ-field that contains $\mathcal{F}$. One can also show that μ^0 is a measure on $\mathcal{F}^0$. Note that $\mathcal{F}^0 \supset \hat{\mathcal{F}}$. $\hat{\mu}$ is defined by restricting μ^0 to $\hat{\mathcal{F}}$. Finally one shows that this extension of μ to $\hat{\mathcal{F}}$ is unique. This completes the sketch.

In the above sketch we have discovered how to evaluate the extended measure $\hat{\mu}$ in terms of its values on the original field.

Theorem 3.3 *Let a measure μ on a field $\mathcal{F}$ be extended to a measure $\hat{\mu}$ on the minimal σ-field $\hat{\mathcal{F}}$ that contains $\mathcal{F}$. Then, for all $A \in \hat{\mathcal{F}}$,*

$$\hat{\mu}(A) = \inf \left\{ \sum_{n=1}^{\infty} \mu(A_n) : A \subset \cup_{n=1}^{\infty} A_n, A_n \in \mathcal{F} \forall n = 1, 2, \ldots \right\}.$$

Examples & Exercises

3.1. Consider the field $\mathcal{F} = \{\mathbf{X}, A, \mathbf{X} \setminus A, \emptyset\}$, where $A \neq \mathbf{X}$ and $A \neq \emptyset$. A measure $\mu : \mathcal{F} \to \mathbb{R}$ is defined by $\mu(\mathbf{X}) = 7.2$, $\mu(A) = 3.5$, $\mu(\mathbf{X} \setminus A) = 3.7$, and $\mu(\emptyset) = 0$. $\mathcal{F}$ is also a σ-field. The extension of the measure promised by Caratheodory's theorem is just the measure itself.

3.2. Let $\mathcal{F}$ be the field made of sets of leaves on a certain tree, at a certain instant in time, and let $\mu(A)$ be the number of aphids on all the leaves in $A \in \mathcal{F}$. Then μ is a measure on a finite σ-field.

3.3. Let $X = [0, 1] \subset \mathbb{R}$. Let $\mathcal{F}$ be the field generated by the set of subintervals of $[0, 1]$. Let $a, b \in [0, 1]$ and define $\mu((a, b)) = \mu([a, b]) = b - a$, for $a \leq b$; and more generally let

$\mu(\text{element of } \mathcal{F}) = $ sum of lengths of subintervals which comprise the element.

Show that μ is a measure on $\mathcal{F}$. The σ-field $\hat{\mathcal{F}}$ generated by $\mathcal{F}$ is the Borel field for $([0, 1], \text{Euclidean})$. Show that $S = \{x \in [0, 1] : x \text{ is a rational number }\}$ belongs to $\hat{\mathcal{F}}$ but not to $\mathcal{F}$. Evaluate $\hat{\mu}(S)$, where $\hat{\mu}$ is the extension of μ to $\hat{\mathcal{F}}$.

3.4. Let $X = \Sigma$, the code space on the two symbols 1 and 2. Let $\mathcal{F}$ denote the field generated by the cylinder subsets of Σ, as defined in exercise 2.5. Let $0 \leq p_1 \leq 1$ and $p_2 = 1 - p_1$. Define

$$\mu(C(e_1, e_2, \ldots, e_n)) = p_{e_1} p_{e_2} \cdots p_{e_n},$$

for each cylinder subset $C(e_1, e_2, \ldots, e_n)$ of Σ. Show how μ can be defined on the other elements of $\mathcal{F}$ in such a way as to provide a measure on $\mathcal{F}$. Evaluate

$$\mu(\{x \in \Sigma : x_7 = 1\}) \text{ and } \mu(\Sigma).$$

Extend $\mathcal{F}$ to the field $\hat{\mathcal{F}}$ generated by $\mathcal{F}$, and correspondingly extend μ to $\hat{\mu}$. Show that

$$S = \{x \in \Sigma : x_{\text{odd}} = 1\} \in \hat{\mathcal{F}}$$

and evaluate $\hat{\mu}(S)$.

3.5. This example takes place in the metric space $([0, 1], \text{Euclidean})$. Consider the IFS with probabilities

$$\{[0, 1]; w_1(x) = \frac{1}{3}x, w_2(x) = \frac{1}{3}x + \frac{2}{3}; p_1, p_2\}.$$

Let $\mathcal{F}$ denote the field generated by the set of intervals that can be expressed in the form

$$w_{e_1} \circ w_{e_2} \circ \ldots \circ w_{e_n}([0, 1]),$$

where $n \in \{1, 2, \ldots\}$ and $e_i \in \{1, 2\}$ for each $i = 1, 2, \ldots, n$. Let $0 \leq p_1 \leq 1$ and $p_2 = 1 - p_1$. Show that one can define a measure on $\mathcal{F}$ so that, for every such interval,

$$\mu(w_{e_1} \circ w_{e_2} \circ \cdots \circ w_{e_n}([0, 1])) = p_{e_1} p_{e_2} \cdots p_{e_n}.$$

Let A denote the attractor of the IFS. Evaluate $\mu(A)$, $\mu(X \setminus A)$, and $\mu([\frac{1}{3}, \frac{2}{3}])$.

3.6. What happens in exercise 3.5 if the IFS is replaced by

$$\{[0, 1]; w_1(x) = \frac{1}{2}x, w_2(x) = \frac{1}{2}x + \frac{1}{2}; p_1, p_2\}?$$

For what value of p_1 is the extension of the measure to the σ-field generated by $\mathcal{F}$ the same as the Borel measure defined in exercise 3.3?

Definition 3.3 *Let* $(\mathbf{X}, d)$ *be a metric space, and let* μ *be a Borel measure. Then the* support *of* μ *is the set of points* $x \in \mathbf{X}$ *such that* $\mu(B(x, \epsilon)) > 0$ *for all* $\epsilon > 0$, *where* $B(x, \epsilon) = \{y \in \mathbf{X} : d(y, x) < \epsilon\}$.

The support of a measure is the set on which the measure lives. The following is an easy exercise.

Theorem 3.4 *Let* $(\mathbf{X}, d)$ *be a metric space, and let* μ *be a Borel measure. Then the support of* μ *is closed.*

Examples & Exercises

3.7. Let $(\mathbf{X}, d)$ be a compact metric space and let μ be a Borel measure on $\mathbf{X}$ such that $\mu(\mathbf{X}) \neq 0$. Show that the support of μ belongs to $\mathcal{H}(\mathbf{X})$, the space of nonempty compact subsets of $\mathbf{X}$.

3.8. Prove the following. "Let μ be a measure on a σ-field $\mathcal{F}$, and let $\overline{\mathcal{F}}$ be the class of all sets of the form $A \cup B$ where $A \in \mathcal{F}$ and B is a subset of a set of measure zero. Then $\overline{\mathcal{F}}$ is a σ-field and the function $\overline{\mu} : \overline{\mathcal{F}} \to \mathbb{R}$ defined by $\overline{\mu}(A \cup B) = \overline{\mu}(A)$ is a measure." The measure $\overline{\mu}$ referred to here is called the *completion* of μ. The completion of the measure in exercise 3.3 is called the *Lebesgue* measure on $[0, 1]$.

4   Integration

In the next section we will introduce a remarkable compact metric space. It is a space whose points are measures! In order to define the metric on this space we need to be able to integrate continuous real-valued functions with respect to measures. Can one integrate a continuous function defined on a fractal? How does one evaluate the "average" temperature of the coastline of Sweden? Here we learn how to integrate functions with respect to measures. Let $(\mathbf{X}, d)$ be a compact metric space. Let μ be a Borel measure on $\mathbf{X}$. Let $f : \mathbf{X} \to \mathbb{R}$ be a continuous function. We will explain the meaning of integrals such as

$$\int_{\mathbf{X}} f(x) \, d\mu(x).$$

Definition 4.1 *We reserve the notation* χ_A *for the* characteristic function *of a set* $A \subset \mathbf{X}$. *It is defined by*

$$\chi_A(x) = \begin{cases} 1 & \text{for } x \in A, \\ 0 & \text{for } x \in \mathbf{X} \setminus A. \end{cases}$$

A function $f : \mathbf{X} \to \mathbb{R}$ *is called* simple *if it can be written in the form*

$$f(x) = \sum_{i=1}^{N} y_i \chi_{I_i}(x),$$

where N is a positive integer, $I_i \in \mathcal{B}$ and $y_i \in \mathbb{R}$ for $i = 1, 2, \ldots, N$, $\cup_{i=1}^{N} I_i = \mathbf{X}$, and $I_i \cap I_j = \emptyset$ for $i \neq j$.

The graphs of several simple functions, associated with different spaces, are shown in Figures IX.253 and IX.254.

Definition 4.2 *The integral (with respect to μ) of the simple function f in Definition 4.1, is*

$$\int_{\mathbf{X}} f(x) \, d\mu(x) = \int_{\mathbf{X}} f \, d\mu = \sum_{i=1}^{N} y_i \mu(I_i).$$

Examples & Exercises

4.1. Let $f : [0, 1] \to \mathbb{R}$ be a piecewise constant function, with finitely many discontinuities. Show that f is a simple function. Let μ denote the Borel measure on $[0, 1]$ such that $\mu(I) = $ length of I, when I is a subinterval of $[0, 1]$. Show that

$$\int_0^1 f(x) \, dx = \int_{[0,1]} f(x) \, d\mu(x),$$

where the right-hand side denotes the area under the graph of f.

4.2. This example takes place in the metric space ($\blacksquare$, Euclidean). Let $\mathcal{G}$ denote the set of rectangular subsets of $\blacksquare$. Let $\mathcal{F}$ denote the field generated by $\mathcal{G}$. Show that there is a unique measure μ on $\mathcal{F}$ such that $\mu(A) = $ area of A, for all $A \in \mathcal{G}$. Notice that the σ-field generated by $\mathcal{F}$ is precisely the Borel field $\mathcal{B}$ associated with ($\blacksquare$, Euclidean). Let $\hat{\mu}$ denote the extension of μ to $\mathcal{B}$. Let $\mathbb{A}$ denote a Sierpinski triangle contained in $\blacksquare$. Show that $\mathbb{A} \in \mathcal{B}$, and

$$\int_{\blacksquare} \chi_{\mathbb{A}} \, d\hat{\mu} = \hat{\mu}(\mathbb{A}) = 0.$$

4.3. This example concerns the IFS with probabilities

$$\{\mathbb{C}; w_1(z), w_2(z), w_3(z); p_1 = 0.2, p_2 = 0.3, p_3 = 0.5\},$$

where

$$w_1(z) = 0.5z, \qquad w_2(z) = 0.5z + (0.5)i, \qquad w_3(z) = 0.5z + 0.5.$$

Let $\mathbb{A}$ denote the attractor of the IFS, and $\mathcal{B}$ the Borel subsets of ($\mathbb{A}$, Euclidean). Let μ denote the unique measure on $\mathcal{B}$ such that

$$\mu(\mathbb{A}) = 1$$
$$\mu(w_i(\mathbb{A})) = p_i \qquad \text{for } i \in \{1, 2, 3\};$$
$$\mu(w_i \circ w_j(\mathbb{A})) = p_i p_j \qquad \text{for } i, j \in \{1, 2, 3\};$$
$$\vdots$$
$$\mu(w_i \circ w_j \cdots \circ w_k(\mathbb{A})) = p_i p_j \ldots p_k \qquad \text{for } i, j, \ldots, k \in \{1, 2, 3\};$$

Figure IX.253. The graph of a simple function on a Sierpinski triangle. The domain is a Sierpinski triangle in the (x, y) plane. The function values are represented by the z-coordinates.

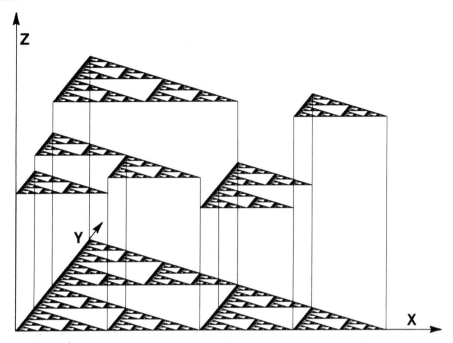

Figure IX.254. The graph of a function whose domain is a fractal fern. If, instead, the function values were represented by colors, a painted fern would replace the graph.

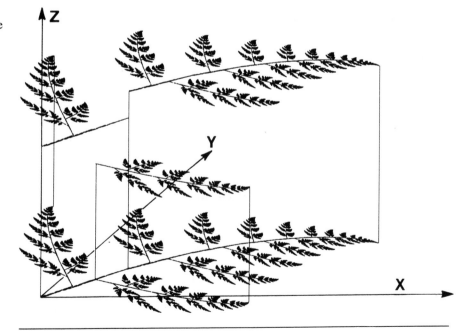

Define a simple function on $\triangle$ by

$$f(x+iy) = \begin{cases} 1 & \text{for } x+iy \in \triangle \text{ and } 1/3 \le x \le 1, \\ -1 & \text{for } x+iy \in \triangle \text{ and } 0 \le x \le 2/3. \end{cases}$$

Calculate $\int_\triangle f(z)d\mu(z)$, accurate to two decimal places.

Based on the ideas in section 1 of this chapter, can you guess a method for calculating the integral that makes use of the Random Iteration Algorithm? Try it!

4.4. Show that if $\alpha, \beta \in \mathbb{R}$ and f, g are simple functions then $\alpha f + \beta g$ is a simple function, and

$$\alpha \int_X f \, d\mu + \beta \int_X g \, d\mu = \int_X (\alpha f + \beta g) \, d\mu.$$

4.5. Black ink is printed to make this page. Let $\blacksquare \subset \mathbb{R}^2$ be a model for the page, and represent the ink by means of a Borel measure μ, so that $\mu(A)$ is the mass of ink associated with the set $A \subset \blacksquare$. Let $\mathcal{A} \in \mathcal{F}$ denote the smallest Borel set that contains all of the letters "a"on the page. Assume that the total mass of ink on the page is one unit. Estimate $\int_\blacksquare \chi_\mathcal{A} d\mu$.

4.6. Let Σ denote code space on two symbols $\{1, 2\}$. $\mathcal{B}$ denotes the Borel field associated with $(\Sigma, \text{code space metric})$. Consider the IFS $\{\Sigma; w_1(x) = 1x, w_2(x) = 2x; p_1 = 0.4, p_2 = 0.6\}$, where "$1x$" means the string "$1x_1x_2x_3\ldots$" and "$2x$" means the string "$2x_1x_2x_3\ldots$." The attractor of the IFS is Σ. Let μ denote the unique measure on $\mathcal{B}$ such that

$$\mu(w_i \circ w_j \cdots \circ w_k(\Sigma)) = p_i p_j \ldots p_k \qquad \text{for } i, j, \ldots, k \in \{1, 2\}.$$

Define sets A and B in $\mathcal{B}$ by

$$A = \{x \in \mathcal{B} : x_1 = 0\} \text{ and } B = \{x \in \mathcal{B} : x_2 = 1\}.$$

Define $f : \Sigma \to \mathbb{R}$ by

$$f(x) = \chi_A(x) + (2.3)\chi_B(x) \qquad \text{for all } x \in \Sigma.$$

Evaluate the integral

$$\int_\Sigma f(x) \, d\mu(x).$$

Definition 4.3 *Let $(\mathbf{X}, d)$ be a compact metric space, and let $\mathcal{B}$ denote the associated Borel field. Let μ be a Borel measure. A partition of $\mathbf{X}$ is a finite set of nonempty Borel sets, $\{A_i \in \mathcal{B} : i = 1, 2, \ldots, M\}$, such that $\mathbf{X} = \cup_{i=1}^M A_i$, and $A_i \cap A_j = \emptyset$ for $i \ne j$. The diameter of the partition is $\max\{\sup\{d(x, y) : x, y \in A_i\} : i = 1, 2, \ldots, M\}$.*

Theorem 4.1 *Let $(\mathbf{X}, d)$ be a compact metric space. Let $\mathcal{B}$ denote the associated Borel field. Let μ be a Borel measure on $\mathbf{X}$. Let $f : \mathbf{X} \to \mathbb{R}$ be continuous. (i) Let n be a positive integer. Then there exists a partition $\mathcal{B}_n = \{A_{n,m} \in \mathcal{B} : m =$*

1, 2, ..., $M(n)$} *of diameter* $1/n$. *(ii) Let* $x_{n,m} \in A_{n,m}$ *for* $m = 1, 2, 3, \ldots$, *and define a sequence of simple functions by*

$$f_n(x) = \sum_{m=1}^{M(n)} f(x_{n,m}) \chi_{A_{n,m}}(x) \qquad \text{for } n = 1, 2, 3, \ldots.$$

Then $\{f_n\}$ *converges uniformly to* $f(x)$. *(iii) The sequence* $\{\int_X f_n d\mu\}$ *converges. (iv) The value of the limit is independent of the particular sequence of partitions, and of the choices of* $x_{n,m} \in A_{n,m}$.

Sketch of proof

(i) Since X is compact it is possible to cover X by a finite set of closed balls of diameter $1/n$, say $b_{n,1}, b_{n,2}, \ldots, b_{n,M(n)}$. We can assume that each ball contains a point not in any of the other balls. Then define $A_{n,1} = b_{n,1}$, and $A_{n,j} = b_{n,j} \setminus \cup_{k=1}^{j-1} A_{n,k}$, for $j = 2, 3, \ldots, M(n)$. Then $B_n = \{A_{n,m} \in B : m = 1, 2, \ldots, M(n)\}$ is a partition of X of diameter $1/n$.

(ii) Let $\epsilon > 0$. f is continuous on a compact space, so it is uniformly continuous. It follows that there exists an integer $N(\epsilon)$ so that if $x, y \in X$ and $d(x, y) \leq 1/N(\epsilon)$ then $|f(x) - f(y)| \leq \epsilon$. It follows that $|f(x) - f_n(x)| \leq \epsilon$ when $n \geq N(\epsilon)$.

(iii) It is readily proved that $\{\int_X f_n d\mu\}$ is a Cauchy sequence. Indeed, for all $n, m \geq N(\epsilon)$ we have

$$\left| \int_X f_n d\mu - \int_X f_m d\mu \right| \leq \int_X |f_n - f_m| d\mu \leq 2\epsilon \mu(X).$$

It follows that the sequence converges.

(iv) Let $\{\tilde{f}_n\}$ be a sequence of simple functions, constructed as above. Then there is an integer $\tilde{N}(\epsilon)$ such that $|f(x) - \tilde{f}_n(x)| \leq \epsilon$ when $n \geq \tilde{N}(\epsilon)$. It follows that for all $n \geq \max\{N(\epsilon), \tilde{N}(\epsilon)\}$,

$$\left| \int_X f_n d\mu - \int_X \tilde{f}_n d\mu \right| \leq \int_X |f_n - \tilde{f}_n| d\mu \leq 2\epsilon \mu(X).$$

This completes the sketch of the proof.

Definition 4.4 *The limit in Theorem 4.1 is called the* integral of f *(with respect to* μ). *It is denoted by*

$$\lim_{n \to \infty} \int_X f_n d\mu = \int_X f \, d\mu.$$

Examples & Exercises

4.7. Let (X, d) be a metric space. Let $a \in X$. Define a Borel measure δ_a by $\delta_a(B) = 1$ if $a \in B$ and $\delta_a(B) = 0$ if $a \notin B$, for all Borel sets $B \subset X$. This measure is referred to as a "a delta function" and "a point mass at a." Let $f : \mathbf{X} \to \mathbb{R}$ be continuous. Show that

$$\int_{\mathbf{X}} f(x) \, d\delta_a(x) = f(a).$$

4.8. This example takes place in the metric space ($\blacksquare$, Euclidean). Let μ be the measure defined in exercise 4.2, and define $f : \blacksquare \to \mathbb{R}$ by $f(x, y) = x^2 + 2xy + 3$. Evaluate

$$\int_{\blacksquare} f \, d\mu.$$

4.9. Make an approximate evaluation of the integral $\int_{\mathbb{A}} x^2 d\mu(x)$ where μ and $\mathbb{A}$ are as defined in exercise 4.3.

4.10. Let $\mathbf{X}$ denote the set of pixels corresponding to a certain computer graphics display device. Define a metric d on $\mathbf{X}$ so that $(\mathbf{X}, d)$ is a compact metric space. Give an example of a Borel subset of $\mathbf{X}$ and of a nontrivial Borel measure on $\mathbf{X}$. Show that any function $f : \mathbf{X} \to \mathbb{R}$ is continuous. Give a specific example of such a function, and evaluate $\int_{\mathbf{X}} f \, d\mu$.

5 The Compact Metric Space $(\mathcal{P}(X), d)$

We introduce the most exciting metric space in the book. It is the space where fractals *really* live.

Definition 5.1 *Let $(\mathbf{X}, d)$ be a compact metric space. Let μ be a Borel measure on $\mathbf{X}$. If $\mu(\mathbf{X}) = 1$, then μ is said to be* normalized.

Definition 5.2 *Let $(\mathbf{X}, d)$ be a compact metric space. Let $\mathcal{P}(\mathbf{X})$ denote the set of normalized Borel measures on $\mathbf{X}$. The* Hutchinson metric d_H *on $\mathcal{P}(\mathbf{X})$ is defined by*

$$d_H(\mu, v) = \sup \left\{ \int_{\mathbf{X}} f \, d\mu - \int_{\mathbf{X}} f \, dv : f : \mathbf{X} \to \mathbb{R} f \text{ continuous,} \right.$$

$$\left. |f(x) - f(y)| \leq d(x, y) \forall x, y \in \mathbf{X} \right\},$$

for all $\mu, v \in \mathcal{P}(\mathbf{X})$.

Theorem 5.1 *Let $(\mathbf{X}, d)$ be a compact metric space. Let $\mathcal{P}(\mathbf{X})$ denote the set of normalized Borel measures on $\mathbf{X}$ and let d_H denote the Hutchinson metric. Then $(\mathcal{P}(\mathbf{X}), d_H)$ is a compact metric space.*

Sketch of proof A direct proof, using the tools in this book, is cumbersome. It is straightforward to verify that d_H is a metric. It is most efficient to use the concept of the "weak topology" on $\mathcal{P}(\mathbf{X})$ to prove compactness. One shows that this topology is the same as the one induced by the Hutchinson metric, and then applies Alaoglu's theorem. See [Hutchinson 1981] and [Dunford 1966].

Examples & Exercises

5.1. Let K be a positive integer. Let $\mathbf{X} = \{(i, j) : i, j = 1, 2, \ldots, K\}$. Define a metric on $\mathbf{X}$ by $d((i_1, j_1), (i_2, j_2)) = |i_1 - i_2| + |j_1 - j_2|$. Then $(\mathbf{X}, d)$ is a compact metric space. Let $\mu \in \mathcal{P}(\mathbf{X})$ be such that $\mu((i, j)) = (i + j)/(K^3 + K^2)$ and let $\nu \in \mathcal{P}(\mathbf{X})$ be such that $\nu(i, j) = 1/K^2$, for all $i, j \in \{1, 2, \ldots, N\}$. Calculate $d_H(\mu, \nu)$.

5.2. Let $(\mathbf{X}, d)$ be a compact metric space. Let $\mu \in \mathcal{P}(\mathbf{X})$. Prove that the support of μ belongs to $\mathcal{H}(\mathbf{X})$.

---

6   A Contraction Mapping on $\mathcal{P}(\mathbf{X})$

Let $(\mathbf{X}, d)$ denote a compact metric space. Let $\mathcal{B}$ denote the Borel subsets of $\mathbf{X}$. Let $w : \mathbf{X} \to \mathbf{X}$ be continuous. Then one can prove that $w^{-1} : \mathcal{B} \to \mathcal{B}$. It follows that if ν is a normalized Borel measure on $\mathbf{X}$ then so is $\nu \circ w^{-1}$. In turn, this implies that the function defined next indeed takes $\mathcal{P}(\mathbf{X})$ into itself.

Definition 6.1 *Let $(\mathbf{X}, d)$ be a compact metric space and let $\mathcal{P}(\mathbf{X})$ denote the space of normalized Borel measures on $\mathbf{X}$. Let*

$$\{\mathbf{X}; w_1, w_2, \ldots, w_N; p_1, p_2, \ldots, p_N\}$$

be a hyperbolic IFS with probabilities. The Markov *operator associated with the IFS is the function $M : \mathcal{P}(\mathbf{X}) \to \mathcal{P}(\mathbf{X})$ defined by*

$$M(\nu) = p_1 \nu \circ w_1^{-1} + p_2 \nu \circ w_2^{-1} + \cdots + p_N \nu \circ w_N^{-1}$$

for all $\nu \in \mathcal{P}(\mathbf{X})$.

Lemma 6.1 *Let M denote the Markov operator associated with a hyperbolic IFS, as in Definition 6.1. Let $f : \mathbf{X} \to \mathbb{R}$ be either a simple function or a continuous function. Let $\nu \in \mathcal{P}(\mathbf{X})$. Then*

$$\int_{\mathbf{X}} f d(M(\nu)) = \sum_{i=1}^{N} p_i \int_{\mathbf{X}} f \circ w_i \, d\nu.$$

Proof Suppose that $f : \mathbf{X} \to \mathbb{R}$ is continuous. By Theorem 5.1 we can find a sequence of simple functions $\{f_n\}$ which converges uniformly to f. Let n be a positive integer. It is readily verified that

$$\int_{\mathbf{X}} f_n d(M(\nu)) = \sum_{i=1}^{N} p_i \int_{\mathbf{X}} f_n \, d\nu \circ w_i^{-1}$$

$$= \sum_{i=1}^{N} p_i \int_{w_i(\mathbf{X})} f_n \, d\nu \circ w_i^{-1}$$

$$= \sum_{i=1}^{N} p_i \int_{\mathbf{X}} f_n \circ w_i \, d\nu.$$

The sequence $\{\int f_n d(M(\nu))\}$ converges to $\int f d(M(\nu))$.

For each $i \in \{1, 2, \ldots, N\}$ and each positive integer n, $f_n \circ w_i$ is a simple function. The sequence $\{f_n \circ w_i\}_{n=1}^{\infty}$ converges uniformly to $f \circ w_i$. It follows that $\{\int f_n \circ w_i d\nu\}_{n=1}^{\infty}$ converges to $\int f \circ w_i$. It follows that $\{\sum_{i=1}^{N} p_i \int f_n \circ w_i d\nu\}_{n=1}^{\infty}$ converges to $\sum_{i=1}^{N} p_i \int f \circ w_i d\nu$. This completes the proof.

Theorem 6.1 *Let* $(\mathbf{X}, d)$ *be a compact metric space. Let*

$$\{\mathbf{X}; w_1, w_2, \ldots, w_N; p_1, p_2, \ldots, p_N\}$$

be a hyperbolic IFS with probabilities. Let $s \in (0, 1)$ *be a contractivity factor for the IFS. Let* $M : \mathcal{P}(\mathbf{X}) \to \mathcal{P}(\mathbf{X})$ *be the associated Markov operator. Then* M *is a contraction mapping, with contractivity factor* s, *with respect to the Hutchinson metric on* $\mathcal{P}(\mathbf{X})$. *That is,*

$$d_H(M(\nu), M(\mu)) \leq s d_H(\nu, \mu) \qquad \text{for all } \nu, \mu \in \mathcal{P}(\mathbf{X}).$$

In particular, there is a unique measure $\mu \in \mathcal{P}(\mathbf{X})$ *such that*

$$M\mu = \mu.$$

Proof Let L denote the set of continuous functions $f : \mathbf{X} \to \mathbb{R}$ such that $|f(x) - f(y)| \leq d(x, y) \forall x, y \in \mathbf{X}$. Then

$$d_H(M(\nu), M(\mu)) = \sup\{\int f d(M(\mu)) - \int f d(M(\nu)) : f \in L\}$$

$$= \sup\{\int \sum_{i=1}^{N} p_i f \circ w_i \, d\mu - \int \sum_{i=1}^{N} p_i f \circ w_i \, d\nu : f \in L\}.$$

Let $\tilde{f} = s^{-1} \sum_{i=1}^{N} p_i f \circ w_i$. Then $\tilde{f} \in L$. Let $\tilde{L} = \{\tilde{f} \in L : \tilde{f} = s^{-1} \sum_{i=1}^{N} p_i f \circ w_i$, some $f \in L\}$. Then we can write

$$d_H(M(\nu), M(\mu)) = \sup\{s \int \tilde{f} d\mu - s \int \tilde{f} d\nu : \tilde{f} \in \tilde{L}\}.$$

Since $\tilde{L} \subset L$, it follows that

$$d_H(M(\nu), M(\mu)) \leq s d_H(\nu, \mu).$$

This completes the proof.

Definition 6.2 *Let μ denote the fixed point of the Markov operator, promised by Theorem 6.1. μ is called the* invariant measure *of the IFS with probabilities.*

We have arrived at our goal! This invariant measure is the object we discussed informally in section 1 of this chapter. *Now* we know what fractals are.

Examples & Exercises

6.1. Verify that the Markov operator associated with a hyperbolic IFS on a compact metric space indeed maps the space into itself.

6.2. This example uses the notation in the proof of Theorem 6.1. Let $f \in L$ and let $\tilde{f} = s^{-1} \sum_{i=1}^{N} p_i f \circ w_i$. Prove that $\tilde{f} \in L$.

6.3. Consider the hyperbolic IFS

$$\{\blacksquare \subset \mathbb{R}^2; w_1, w_2, w_3, w_4; p_1, p_2, p_3, p_4\}$$

corresponding to the collage in Figure IX.255(a). Let M be the associated Markov operator. Let $\mu_0 \in \mathcal{P}(\mathbf{X})$, so that $\mu_0(\blacksquare) = 1$. For example, μ_0 could be the uniform measure, for which $\mu_0(S)$ is the area of $S \in \mathcal{P}(\blacksquare)$. We look at the sequence of measures $\{\mu_n = M^{\circ n}(\mu_0)\}$. The measure $\mu_1 = M(\mu_0)$ is such that $\mu(w_i(\blacksquare)) = p_i$ for $i = 1, 2, 3, 4$, as illustrated in Figure IX.255(b). It follows that $\mu_2 = M^{\circ 2}(\mu_0)$ obeys $\mu(w_i \circ w_j(\blacksquare)) = p_i p_j$ for $i, j = 1, 2, 3, 4$, as illustrated in Figure IX.255(c). We quickly get the idea. When the Markov operator is applied, the "mass" in a cell $\blacksquare_{ij...k} = w_i \circ w_j \circ \cdots \circ w_k(\blacksquare)$ is redistributed among the four smaller cells $w_1(\blacksquare_{ij...k})$, $w_2(\blacksquare_{ij...k})$, $w_3(\blacksquare_{ij...k})$, and $w_4(\blacksquare_{ij...k})$. Also, mass from other cells is mapped into subcells of $\blacksquare_{ij...k}$ in such a way that the total mass of $\blacksquare_{ij...k}$ remains the same as before the Markov operator was applied. In this manner the distribution of "mass" is defined on finer and finer scales as the Markov operator is repeatedly applied. What a wonderful idea. We have also illustrated this idea in Figures IX.256 and IX.257.

6.4. Apply the Random Iteration Algorithm to an IFS of the form considered in example 6.3. Choose the probabilities p_1, p_2, p_3, and p_4 so as to obtain a "picture" of the invariant measure that would occur at the end of the sequence that commences in Figure IX.257(a), (b), (c), and (d).

6.5. Consider the IFS

$$\{[0, 1] \subset \mathbb{R}; w_1(x) = (0.5)x, w_2(x) = (0.7)x + 0.3; p_1 = 0.45, p_2 = 0.55\}.$$

The attractor of the IFS is $[0, 1]$. Let M denote the associated Markov operator. Let $\mu_0 \in \mathcal{P}([0, 1])$ be the uniform measure on $[0, 1]$. In Figure IX.258(a), μ_0 is represented by a rectangle, whose base is $[0, 1]$ and whose area is 1. The successive iterates $M(\mu_0)$, $M^{\circ 2}(\mu_0)$, $M^{\circ 3}(\mu_0)$ are represented in Figure IX.258(b), (c) and (d). Each measure is represented by a collection of rectangles whose bases are contained in the interval $[0, 1]$. The area of a rectangle equals the measure of the base of the rectangle. Although the sequence of measures converges $\{M^{\circ n}(\mu_0)\}$ in the metric

(a)

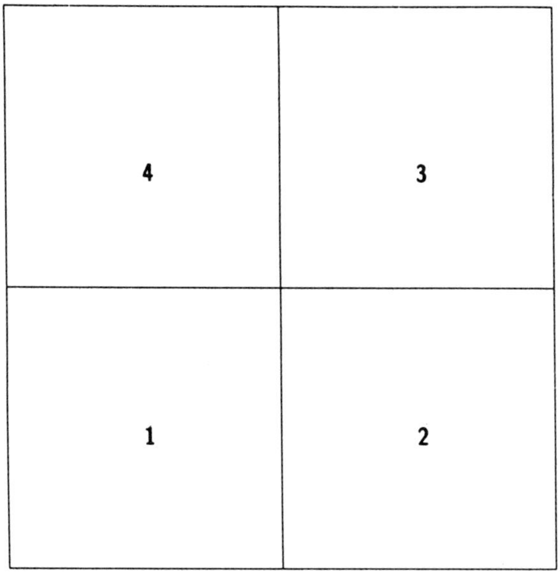

Figure IX.255. A collage for an IFS of four maps. The attractor of the IFS is ■, and the probability of the map w_i is p_i for $i = 1, 2, 3, 4$. Let M denote the associated Markov operator. Let $\mu_0 = 1$. Then $\mu_1 = M(\mu_0)$ is a measure such that $\mu(w_i(\blacksquare)) = p_i$ for $i = 1, 2, 3, 4$, as illustrated in (b). The measure $\mu_2 = M^{\circ 2}(\mu_0)$ is such that $\mu(w_i \circ w_j(\blacksquare)) = p_i p_j$ for $i, j = 1, 2, 3, 4$, as illustrated in (c). See also Figures IX.256 and IX.257.

(b)

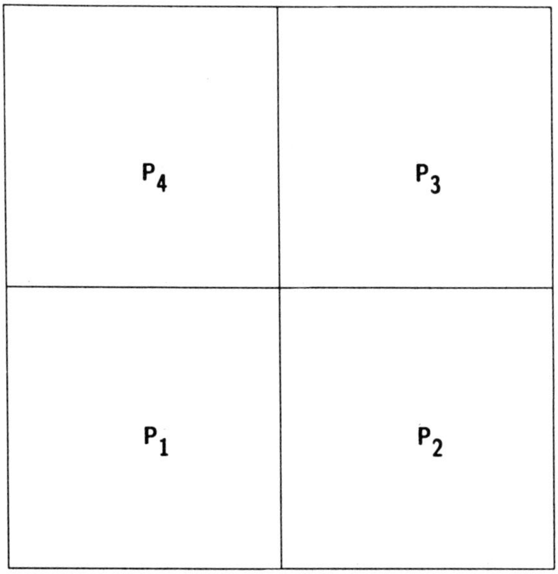

Figure IX.255.   (b)

Figure IX.255. (c)

(c)			
P_4P_4	P_4P_3	P_3P_4	P_3P_3
P_4P_1	P_4P_2	P_3P_1	P_3P_2
P_1P_4	P_1P_3	P_2P_4	P_2P_3
P_1P_1	P_1P_2	P_2P_1	P_2P_2

space$\{\mathcal{P}([0, 1], d_H\}$, some of the rectangles would become infinitely tall as n tends to infinity.

6.6. Make a sequence of figures, analagous to Figure IX.258(a)–(d), to represent the Markov operator applied to the uniform measure μ_0, for each of the following IFS's with probabilities:

$(i)\{[0, 1] \subset \mathbb{R}; w_1(x) = (0.5)x, w_2(x) = (0.5)x + 0.5; p_1 = 0.5, p_2 = 0.5\};$

$(ii)\{[0, 1] \subset \mathbb{R}; w_1(x) = (0.5)x, w_2(x) = (0.5)x + 0.5;$
$\quad p_1 = 0.99, p_2 = 0.01\};$

$(iii)\{[0, 1] \subset \mathbb{R}; w_1(x) = (0.9)x, w_2(x) = (0.9)x + 0.1; p_1 = 0.45, p_2 = 0.55\}.$

In each case describe the associated invariant measure.

6.7. Let $\mathbf{X} = \{A, B, C\}$ denote a space that consists of three points. Let $\mathcal{B}$ denote the σ-field that consists of all subsets of $\mathbf{X}$. Consider the IFS with probabilities

$$\{\mathbf{X}; w_1, w_2; p_1 = 0.6, p_2 = 0.4\},$$

where $w_1 : \mathbf{X} \to \mathbf{X}$ is defined by $w_1(A) = B$, $w_1(B) = B$, $w_1(C) = B$, and $w_2 : \mathbf{X} \to \mathbf{X}$ is defined by $w_2(A) = C$, $w_2(B) = A$, and $w_2(C) = C$. Let $\mathcal{P}(\mathbf{X})$ denote the set of normalized measures on $\mathcal{B}$. Let $\mu_0 \in \mathcal{P}(\mathbf{X})$ be defined by $\mu_0(A) = \mu_0(B) = \mu_0(C) = \frac{1}{3}$. Let M denote the Markov operator associated with the IFS, and let $\mu_n = M^{\circ n}(\mu_0)$ for $n = 1, 2, 3, \ldots$. Determine real numbers $a, b, c, d, e, f, g, h, i$ such that for each n,

$$\begin{bmatrix} \mu_n(A) \\ \mu_n(B) \\ \mu_n(C) \end{bmatrix} = \begin{bmatrix} a & b & c \\ d & e & f \\ g & h & i \end{bmatrix} \begin{bmatrix} \mu_{n-1}(A) \\ \mu_{n-1}(B) \\ \mu_{n-1}(C) \end{bmatrix}.$$

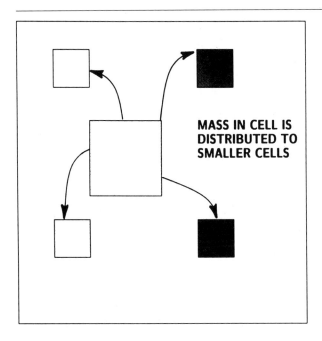

MASS IN CELL IS
DISTRIBUTED TO
SMALLER CELLS

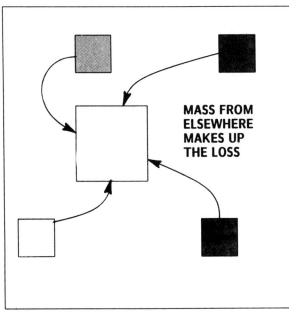

MASS FROM
ELSEWHERE
MAKES UP
THE LOSS

Figure IX.256.   This illustrates the action of the Markov operator on one of the sequence of measures $\{M^{\circ n}(\mu_0)\}$, where $\mu_0(\blacksquare) = 1$. When the Markov operator is applied, the "mass" in a cell $\blacksquare_{ij\ldots k} = w_i \circ w_j \circ \cdots \circ w_k(\blacksquare)$ is redistributed among the four cells $w_1(\blacksquare_{ij\ldots k})$, $w_2(\blacksquare_{ij\ldots k})$, $w_3(\blacksquare_{ij\ldots k})$, and $w_4(\blacksquare_{ij\ldots k})$. Also, mass from other cells is mapped into subcells of $\blacksquare_{ij\ldots k}$ in such a way that the total mass of $\blacksquare_{ij\ldots k}$ remains the same as before the Markov operator was applied. In this manner the distribution of "mass" is defined on finer and finer scales as the Markov operator is repeatedly applied.

Figure IX.257. This
sequence of figures represents successive measures
produced by iterative ap-
plications of a Markov
operator of the type con-
sidered in Figures IX.255
and IX.256. The result
of one application of the
operator to the uniform
measure on ■ is repre-
sented in (a). Figures (b),
(c), and (d) show the re-
sults of further successive
applications of the Markov
operator. The measures are
represented in such a way
as to keep the total number
of dots constant. The measure of a set corresponds
approximately to the num-
ber of dots it contains.
This represents the first
few of a sequence of mea-
sures that converges in the
metric space $(\mathcal{P}(\blacksquare), d_H)$
to the invariant measure of
the IFS.

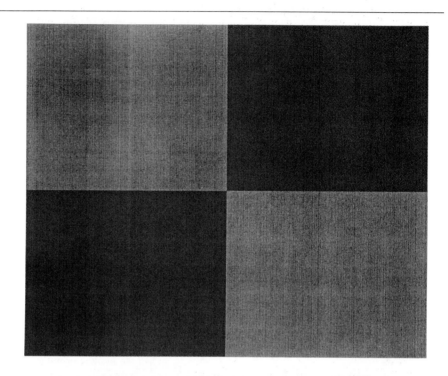

Figure IX.257. (b)

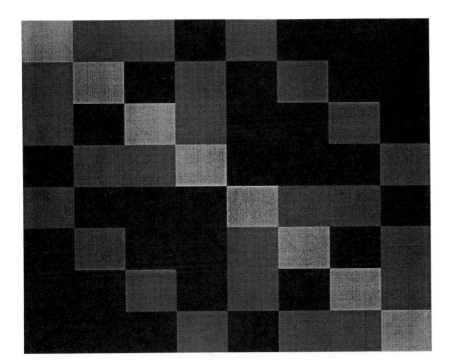

Figure IX.257.   (c)

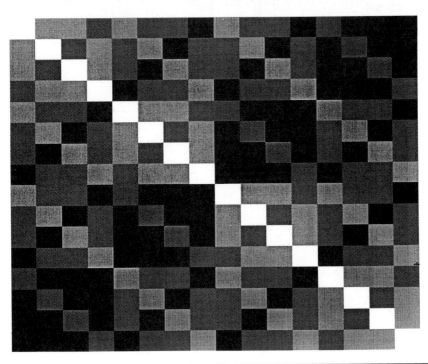

Figure IX.257.   (d)

Figure IX.258. This sequence of images relates to the IFS $\{[0, 1] \subset \mathbb{R}; w_1(x) = (0.5)x, w_2(x) = (0.7)x + 0.3, p_1 = 0.45, p_2 = 0.55\}$. The attractor of the IFS is $[0, 1]$. Let M denote the associated Markov operator. Let $\mu_0 \in \mathcal{P}([0, 1])$ be the uniform measure on $[0, 1]$. The successive iterates $M(\mu_0)$, $M^{\circ 2}(\mu_0)$, $M^{\circ 3}(\mu_0)$, and $M^{\circ 4}(\mu_0)$ are represented in parts (a),(b),(c), and (d). Each measure is represented by a collection of rectangles whose bases are contained in the interval $[0, 1]$. The area of a rectangle equals the measure of the base of the rectangle.

(a)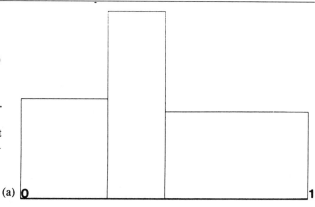

(b)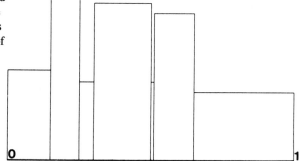

(c)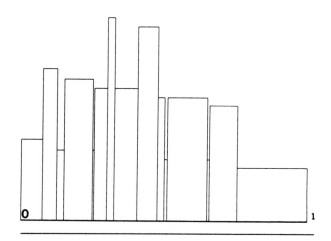

Figure IX.258. (d)

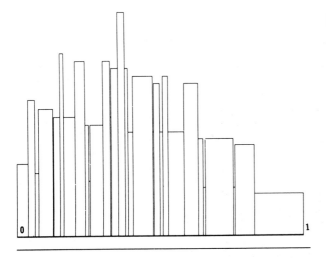

Let $\tilde{M}$ denote the 3×3 matrix here. Explain how $\tilde{M}$ is related to M, and show that the invariant measure of the IFS can be described in terms of an eigenvector of $\tilde{M}$.

6.8. Let

$$\{\mathbf{X}; w_1, w_2, \ldots, w_N; p_1, p_2, \ldots, p_N\}$$

be a hyperbolic IFS with probabilities. Let μ denote the associated invariant measure. Let A denote the attractor of the IFS. Let $\mu_0 \in \mathcal{P}(\mathbf{X})$ be such that $\mu_0(A) = 1$. By considering the sequence of measures $\{\mu_n = M^{\circ n}(\mu_0)\}$, prove that

$$\mu(w_i \circ w_j \circ \cdots \circ w_k(A)) \geq p_i p_j \ldots p_k, \qquad \text{for all } i, j, \ldots, k \in \{1, 2, \ldots, N\}.$$

Show that if the IFS is totally disconnected then the equality sign holds.

Theorem 6.2 *Let $(\mathbf{X}, d)$ be a compact metric space. Let*

$$\{\mathbf{X}; w_1, w_2, \ldots, w_N; p_1, p_2, \ldots, p_N\}$$

be a hyperbolic IFS with probabilities. Let μ be the associated invariant measure. Then the support of μ is the attractor of the IFS $\{\mathbf{X}; w_1, w_2, \ldots, w_N\}$.

Proof Let B denote the support of μ. Then B is a nonempty compact subset of $\mathbf{X}$. Let A denote the attractor of the IFS. Then

$$\{A; w_1, w_2, \ldots, w_N; p_1, p_2, \ldots, p_N\}$$

is a hyperbolic IFS. Let ν denote the invariant measure of the latter. Then ν is also an invariant measure for the original IFS. So, since μ is unique, $\nu = \mu$. It follows that $B \subset A$.

Let $a \in A$. Let $\mathcal{O}$ be an open set that contains a. We will use the notation of Theorem 2.1 in Chapter IV. Let Σ denote the code space associated with the IFS and let $\sigma \in \Sigma$ denote the address of a. It follows from Theorem 2.1 in Chapter IV that $\lim_{n \to \infty} \phi(\sigma, n, A) = a$, where the convergence is in the Hausdorff metric. It follows that there is a positive integer n so that $\phi(\sigma, n, A) \subset \mathcal{O}$. But

$$\mu(\phi(\sigma, n, A)) \geq p_{\sigma_1} p_{\sigma_2} \cdots p_{\sigma_n} > 0.$$

It follows that $\mu(\mathcal{O}) > 0$. It follows that a is in the support of μ. It follows that $a \in B$. It follows that $A \subset B$. This completes the proof.

Theorem 6.3 The Collage Theorem for Measures.. *Let*

$$\{\mathbf{X}; w_1, w_2, \ldots, w_N; p_1, p_2, \ldots, p_N\}$$

be a hyperbolic IFS with probabilities. Let μ be the associated invariant measure. Let $s \in (0, 1)$ be a contractivity factor for the IFS. Let $M : \mathcal{P}(\mathbf{X}) \to \mathcal{P}(\mathbf{X})$ be the associated Markov operator. Let $\nu \in \mathcal{P}(\mathbf{X})$. Then

$$d_H(\nu, \mu) \leq \frac{d_H(\nu, M(\nu))}{(1 - s)}.$$

Proof This is a corollary of Theorem 6.1.

We conclude this section with a description of the application of Theorem 6.3 to an inverse problem. The problem is to find an IFS with probabilities whose invariant measure, when represented by a set of dots, looks like a given texture.

A measure supported on a subset of $\mathbb{R}^2$ such as ■ can be represented by a lot of black dots on a piece of white paper. Figures IX.248 and IX.250 provide examples. The dots may be granules of carbon attached to the paper by means of a laser printer. The number of dots inside any circle of radius $\frac{1}{2}$ inch, say, should be approximately proportional to the measure of the corresponding ball in $\mathbb{R}^2$. A gray-tone image in a newspaper is made of small dots and can be thought of as representing a measure.

Let two such images, each consisting of the same number of points, be given. Then we expect that the degree to which they look alike corresponds to the Hutchinson distance between the corresponding measures. Let such an image, L, be given. We imagine that it corresponds to a measure ν. Theorem 6.3 can be used to help to find a hyperbolic IFS with probabilities whose invariant measure, represented with dots, approximates the given image. Let N be a positive integer. Let $w_i : \mathbb{R}^2 \to \mathbb{R}^2$ be an affine transformation, for $i = 1, 2, \ldots, N$. Let

$$\{\mathbb{R}^2; w_1, w_2, \ldots, w_N; p_1, p_2, \ldots, p_N\}$$

denote the sought-after IFS. Let M denote the associated Markov operator.

Let $p_i \& L$ mean the set of dots L after the "density of dots" has been decreased by a factor p_i. For example $0.5 \& L$ means L after "every second dot" in L has been removed. The action of the Markov operator on ν is represented by $\cup_{i=1}^{N} w_i(p_i \& L)$. This set consists of approximately the same number of dots as L. Then we seek contractive affine transformations and probabilities such that

$$\cup_{i=1}^{N} w_i(p_i \& L) \approx L. \tag{1}$$

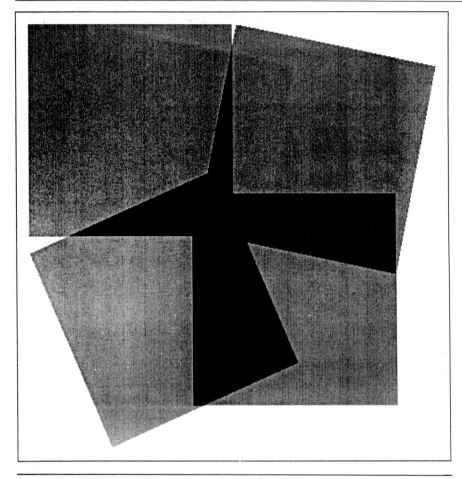

Figure IX.259. This illustration relates to the Collage Theorem for Measures. The shades of gray "add up" in the overlapping regions.

That is, the coefficients that define the affine transformations and the probabilities must be adjusted so that the left-hand side "looks like" the original image.

Suppose we have found an IFS with probabilities so that equation 1 is true. Then generate an image $\tilde{L}$ of the invariant measure of the IFS, containing the same number of points as L. We expect that

$$\tilde{L} \approx L. \tag{2}$$

If the maps are sufficiently contractive, then the meaning of "$\approx$" should be the same in both equations 1 and 2. These ideas are illustrated in Figure IX.259.

Examples & Exercises

6.9. Use the Collage Theorem for Measures to help find an IFS with probabilities for each of the images in Figures IX.260, IX.261, and IX.262.

Figure IX.260. Can you find the IFS and probabilities corresponding to this texture?

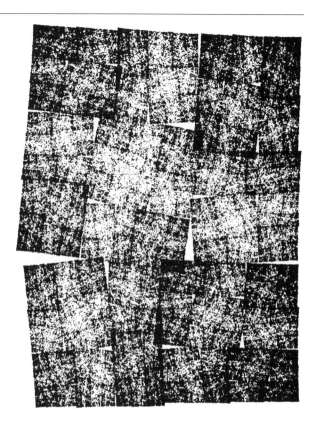

Figure IX.261. Determine the IFS and probabilities for this cloud texture.

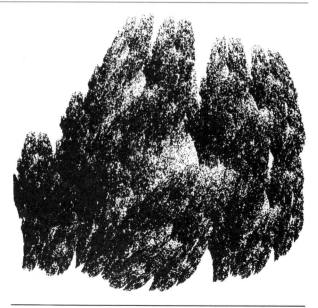

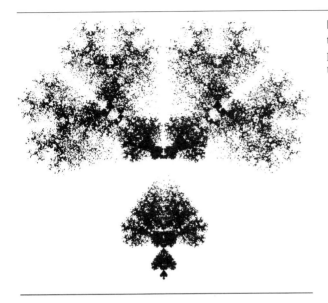

Figure IX.262. Find the four affine maps and probabilities for this texture.

6.10. Estimate the probabilities and transformations used to make each part of Figure IX.248.

6.11. Let

$$\{\mathbf{X}; w_1, w_2, \ldots, w_N; p_1, p_2, \ldots, p_N\}$$

be a hyperbolic IFS. Let μ denote the invariant measure. Let A denote the attractor. Let Σ denote the associated code space on the N symbols $\{1, 2, \ldots, N\}$. Let $T_i : \Sigma \to \Sigma$ be defined by $T_i(\sigma) = i\sigma$, for all $\sigma \in \Sigma$, for $i = 1, 2, 3, 4$. Let ρ denote the invariant measure for the hyperbolic IFS

$$\{\Sigma; T_1, T_2, T_3, T_4; p_1, p_2, p_3, p_4\}.$$

Let $\phi : \Sigma \to A$ denote the continuous map between code space and the attractor of the IFS intoduced in Theorem 4.2.1. Prove that $\rho(\phi^{-1}(B)) = \mu(B)$ for all Borel subsets B of $\mathbf{X}$.

6.12. Figure IX.263 depicts the invariant measure for the IFS $\{[0, 1] \subset \mathbb{R}; w_1(x) = a_1 x, w_2(x) = a_2 x + e_2; p_1,$
$p_2\}$, where a_1, a_2, and e_2 are real constants such that the attractor is contained in $[0, 1]$. The measure of a Borel subset of $[0, 1]$ is approximately the amount of black that lies "vertically" above it. Find a_1, a_2, and e_2.

Figure IX.263. This figure depicts the invariant measure for the IFS $\{[0, 1] \subset \mathbb{R}; w_1(x) = a_1x, w_2 = a_2x + e_2; p_1, p_2\}$, where a_1, a_2, and e_2 are real constants such that the attractor is contained in $[0, 1]$. The measure of a Borel subset of $[0, 1]$ is approximately the amount of black that lies "vertically" above it. Can you find a_1, a_2, and e_2?

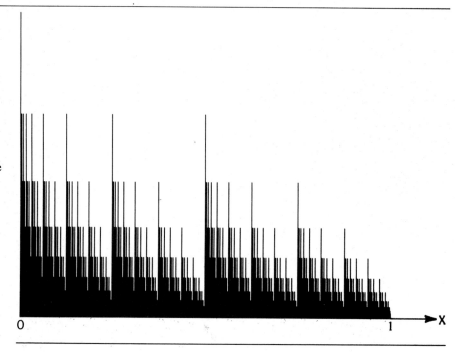

7   Elton's Theorem

Both the following theorem and its corollary claim that certain events occur "with probability one." Although this has a very precise technical meaning, it is fine to interpret it in the same way as you would interpret the statement "There is a 100% chance of rain tomorrow." After the statements we mention the mathematical framework used for dealing with probabilistic statements. To go further we recommend reading parts of [Eisen 1969].

The theorem below is actually true when the p_i's are functions of x, the w_i's are only contraction mappings "on the average," and the space is "locally" compact.

Theorem 7.1 *Let* $(\mathbf{X}, d)$ *be a compact metric space. Let*

$$\{\mathbf{X}; w_1, w_2, \ldots, w_N; p_1, p_2, \ldots, p_N\}$$

be a hyperbolic IFS with probabilities. Let $(\mathbf{X}, d)$ *be a compact metric space. Let* $\{x_n\}_{n=0}^{\infty}$ *denote an orbit of the IFS produced by the Random Iteration Algorithm, starting at x_0. That is,*

$$x_n = w_{\sigma_n} \circ w_{\sigma_{n-1}} \circ \cdots \circ w_{\sigma_1}(x_0),$$

where the maps are chosen independently according to the probabilities

$$p_1, p_2, \ldots, p_N, \qquad \text{for } n = 1, 2, 3, \ldots.$$

Let μ be the unique invariant measure for the IFS. Then with probability one (that is, for all code sequences $\sigma_1, \sigma_2, \ldots$ except for a set of sequences having probability zero),

$$\lim_{n \to \infty} \frac{1}{n+1} \sum_{k=0}^{n} f(x_k) = \int_{\mathbf{X}} f(x) \, d\mu(x)$$

for all continuous functions $f : \mathbf{X} \to \mathbb{R}$ and all x_0.

Proof See [Elton 1987].

Corollary 7.1 *Let B be a Borel subset of $\mathbf{X}$ and let $\mu(\text{boundary of } B) = 0$. Let*

$\mathcal{N}(B, n) = $ number of points in $\{x_0, x_1, x_2, x_3, \ldots, x_n\} \cap B$, for $n = 0, 1, 2, \ldots$.

Then, with probability one,

$$\mu(B) = \lim_{n \to \infty} \left\{ \frac{\mathcal{N}(B, n)}{(n+1)} \right\}$$

for all starting points x_0. That is, the "mass" of B is the proportion of iteration steps, when running the Random Iteration Algorithm, which produce points in B.

Let's explain more deeply the context of the statement "with probability one." Let Σ denote the code space on the N symbols $\{1, 2, \ldots, N\}$. Let ρ denote the unique Borel measure on Σ such that

$$\rho(C(\sigma_1, \sigma_2, \ldots, \sigma_m)) = p_{\sigma_1} p_{\sigma_2} \cdots p_{\sigma_m}$$

for each positive integer m and all $\sigma_1, \sigma_2, \ldots, \sigma_m \in \{1, 2, \ldots, N\}$, where

$$C(\sigma_1, \sigma_2, \ldots, \sigma_m) = \{\omega \in \Sigma : \omega_1 = \sigma_1, \omega_2 = \sigma_2, \ldots, \omega_m = \sigma_m\}.$$

Then $\rho \in \mathcal{P}(\Sigma)$. This measure provides a means for assigning probabilities to sets of possible outcomes of applying the Random Iteration Algorithm. Let us see how this works.

When the Random Iteration Algorithm is applied, an infinite sequence of symbols $\omega_1, \omega_2, \omega_3, \ldots$, namely a code $\omega = \omega_1 \omega_2 \omega_3 \ldots \in \Sigma$, is generated. Provided that we keep $x_0 \in \mathbf{X}$ fixed, we can describe the probabilities of orbits $\{x_n\}$ in terms of the probabilities of codes ω. So we examine how probabilities are associated to sets of codes.

The Random Iteration Algorithm is applied and produces a code $\omega \in \Sigma$. What is the probability that $\omega_1 = 1$? Clearly it is $p_1 = \rho(C(1))$. What is the probability that $\omega_1 = \sigma_1, \omega_2 = \sigma_2, \ldots$, and $\omega_n = \sigma_n$? Because the symbols are chosen independently, it is

$$\rho(C(\sigma_1, \sigma_2, \ldots, \sigma_m)) = p_{\sigma_1} p_{\sigma_2} \cdots p_{\sigma_m}.$$

Let B denote a Borel subset of Σ. What is the probability that the Random Iteration Algorithm produces a code $\sigma \in B$? It is at least intuitively reasonable that it is

$\rho(B)$. This can be formalized, see, for example [Eisen 1969]. The measure ρ provides a means of describing the probabilities of outcomes of the Random Iteration Algorithm.

Here is a heavy way of stating the central part of Theorem 7.1. "... Let $x_0 \in \mathbf{X}$. Let $B \subset \Sigma$ denote the set of codes $\sigma \in \Sigma$ such that

$$\lim_{n \to \infty} \frac{1}{n+1} \sum_{k=0}^{n} f(x_k) = \int_{\mathbf{X}} f(x) \, d\mu(x),$$

for all $x_0 \in \mathbf{X}$ and all continuous functions $f : \mathbf{X} \to \mathbb{R}$, where

$$x_n = w_{\sigma_n} \circ w_{\sigma_{n-1}} \circ \ldots \circ w_{\sigma_1}(x_0).$$

Then B is a Borel subset of Σ and $\rho(B) = 1$." A similar heavy restatement of the corollary can be made.

Examples & Exercises

7.1. This example concerns the IFS

$$\{[0, 1]; \frac{1}{2}x, \frac{1}{2}x + \frac{1}{2}; 0.5, 0.5\}.$$

Show that the invariant measure μ is such that $\mu([x, x + \delta]) = \Delta$ when $[x, x + \delta]$ is a subinterval of $[0, 1]$. Deduce that if $f : [0, 1] \to \mathbb{R}$ is a continuous function then

$$\int_0^1 f(x) \, dx = \int_{[0,1]} f \, d\mu.$$

Let $f(x) = 1 + x^2$. Compute approximations to the latter integral with the aid of Elton's theorem and the Random Iteration Algorithm. Compare your results with the exact value $\frac{4}{3}$.

7.2. This example concerns the IFS

$$\{\blacksquare \subset \mathbb{R}^2; w_1, w_2, w_3, w_4; 0.25, 0.25, 0.25, 0.25\}$$

corresponding to the collage in Figure IX.255(a). Let μ denote the invariant measure. Argue that μ is the uniform measure that assigns "measure" $dxdy$ to an infinitesimal rectangular cell of side lengths dx and dy. Use Elton's theorem and the Random Iteration Algorithm to evaluate approximations to

$$\int_{\blacksquare} (x^2 + 2xy + 3y^2) \, dxdy.$$

Compare your approximations with the exact value.

7.3. This example concerns the IFS

$$\{\blacktriangle \subset \mathbb{R}^2; w_1, w_2, w_3; \frac{1}{3}, \frac{1}{3}, \frac{1}{3}\},$$

where

$$w_1(x, y) = (\frac{1}{2}x, \frac{1}{2}y), \quad w_2(x, y) = (\frac{1}{2}x, \frac{1}{2}y + \frac{1}{2}), \quad w_3(x, y) = (\frac{1}{2}x + \frac{1}{2}, \frac{1}{2}y),$$

and $\mathbb{A}$ is the attractor of the IFS, our old friend. Let μ denote the invariant measure of the IFS. Argue that μ provides a good concept of a "uniform" measure on $\mathbb{A}$. Use Elton's theorem and the Random Iteration Algorithm to compute approximations to

$$\int_{\mathbb{A}} (x^2 + 2xy + 3y^2)\, dx dy.$$

In Chapters II, III, and IV, we introduced the space Σ_N of shifts on N symbols. It was mentioned in passing in Chapter IV that any invertible mixing function could be represented by a baker's transformation with "uneven cutting and stretching." We are now in a position to show how this comes about using an example involving two simple IFS. The same model with some necessary refinements yields the code space mixing model used to justify the representations. It is easier to visualize without the refinements, as we present it here. It is one of the most important properties involved with the modelling of physical chaos.

We begin with perhaps the most simple of all IFS with probabilities. On the interval $[0, 1]$, we define the just-touching IFS with N maps and with probabilities

$$\{[0, 1];\ w_1, w_2, \ldots, w_N;\ p_1, p_2, \ldots, p_N\},$$

where

$$w_1(x) = \frac{1}{N}x$$

$$w_2(x) = \frac{1}{N}x + \frac{1}{N}$$

$$w_3(x) = \frac{1}{N}x + \frac{3}{N}$$

$$\vdots$$

$$w_N(x) = \frac{1}{N}x + \frac{N-1}{N},$$

and the probabilities are arbitrary, subject to the usual condition

$$\sum_{i=1}^{N} p_i = 1.$$

Associated with this IFS there is an invariant measure on $[0, 1]$, which we denote by ν.

Now we define another IFS on $[0, 1]$, this time without probabilities, using the p_i from above. On $[0, 1]$ define the IFS

$$\{[0, 1];\ v_1, v_2, \ldots, v_N\},$$

where

$$v_1(x) = p_1 x$$
$$v_2(x) = p_2 x + p_1$$
$$v_3(x) = p_3 x + (p_1 + p_2)$$

$$\vdots$$

$$v_k(x) = p_k x + \sum_{i=1}^{k-1} p_i$$

$$\vdots$$

$$v_N(x) = p_N x + \sum_{i=1}^{N-1} p_i.$$

This IFS is also just-touching by construction and, because the probabilities from the first IFS sum to one, has as its attractor the interval $[0, 1]$ as well. We are going to use it to define an equivalent metric on $[0, 1]$ as follows:

Each point has a unique address under this IFS in code space, except the points $v_i(A)$, whose multiple addresses correspond to

$$\sigma = i\overline{N-1} = (i+1)\overline{0}.$$

These are precisely the points in a base N expansion of a real number which are equated to form the real line. We denote the *value* of a point x with address $x_1 x_2 x_3 \ldots$ in this new metric space to be the real number with N-ary expansion $.x_1 x_2 x_3 \ldots$. In effect we have given each point the numeric value that would correspond to having measured its distance from say 0 with a ruler on which the spacing of the tick marks had been made uneven in a very specific way by the IFS.

With these values, the space is still $[0, 1]$, but we put a metric on it by assigning the distance between two real numbers to be the distance measured with a "normal" ruler. Another way to put it is that we take the normal interval $[0, 1]$ and assign the distance between two points to be the distance between the addresses corresponding to their N-ary expansions in the above IFS. Thus if $N = 10$ for instance, the distance between $.251$ and $.137$ is not $.251 - .137$, but rather the distance between the points with addresses $251\overline{0}$ and $137\overline{0}$ in the IFS $\{[0, 1]; v_1, v_2, \ldots, v_{10}\}$. We will call this space $[0, 1]_p$, and the distance function d_p to avoid confusion.

We have a metric space, so we will now assign a Borel measure to it by defining $\mu([a, b]) = \mu((a, b)) = d_p(a, b)$, which is uniform for this metric space. And to proceed with the example, we need a function, $f : [0, 1] \to [0, 1]_p$ which we define by $f(x) =$(point with value x in $[0, 1]_p$). Because the definition was very careful to preserve the ordering of the real line and its conventions about multiple addressing, f is both a homeomorphism and a metric equivalence. Because it is continuous, it is also what is called a *measurable function* in that if $A \in \mathcal{B}([0, 1]_p)$ then $f^{-1}(A) \in \mathcal{B}([0, 1])$.

7.4. Show that f is *measure-preserving* with respect to the invariant measure on v

associated with

$$\{[0, 1]; w_1, w_2, \ldots, w_N; p_1, p_2, \ldots, p_N\};$$

that is, that for any Borel subset $A \in [0, 1]$, we have $\nu(A) = \mu(f(A))$.

We now have the machinery to cast the Random Iteration Algorithm entirely in terms of IFS with no recourse to randomness. It is really a deterministic model, with the random part coming in to help when a very simple statement made all the time in mathematics turns out to be something a computer cannot do.

The exact transfer of the Random Iteration Algorithm to the model using the space $([0, 1]_p, d_p)$ looks like this: define the function $g : [0, 1] \to [0, 1]_p$ by

$$d_p(g(x), 0) = x.$$

Define the map $h : [0, 1]_p \to \{0, 1, 2, \ldots, N - 1\}$ defined by $h(p) = [Np]$ where $[\cdot]$ is the greatest integer function. In other words, take the first N-ary digit of the value of the point $p \in [0, 1]_p$. Define the map $y : [0, 1] \to [0, 1]$ given by $y(x) = Nx \bmod N$. Then the Random Iteration Algorithm is precisely the iteration of the map $R : [0, 1] \times \mathbf{X}$ given by

$$R(p, x) = (y(p), w_{h \circ g(p)}(x)).$$

Where does the random part of the algorithm come in? We need it to "pick a real number." One can think of the random number at each iteration as a function to get the next digit of the real number we "picked." In the above expression, we get a random number and find out which function to use via $h(g(p))$. Then we iterate the IFS using $w_{h \circ g(p)}(x)$, and in order to have a new "random number" the next time, we advance p to the next digit using $y(p)$.

Now, consider the space $[0, 1]_p \times [0, 1]$. Think of it as a square with coordinates spaced unevenly in the x direction and evenly in the y direction. Your "usual" point in the square (where here usual means with probability one) has an N-ary expansion for y in which every digit occurs with equal probability, while the x value has an N-ary expansion in which 0 occurs with probability p_1, 1 occurs with probability p_2, etc.

7.5. Draw a diagonal from $(0, 0)$ to $(1, 1)$ on this square. Show that this statement is still true if we pick a "usual point" from this diagonal.

7.6. Draw a smooth curve from $(0, 0)$ to $(1, 1)$ on the square. Then the statement is still true if we pick a "usual point" from this curve.

By using the diagonal in exercise 7.5, we can take a point x in $[0, 1]$ and map it to a new point $\tilde{x}$, by putting x along the vertical coordinate and reading the horizontal coordinate like a web diagram. In terms of all the functions we have defined, this operation is $\tilde{x} = f^{-1}(g(x))$. Under the original IFS with probabilities, this new point will, with probability 1, have an orbit under the shift dynamical system $\{A; S\}$ with a distribution of dots identical to the one we would get by using the Random Iteration Algorithm with probabilities $\{p_1, p_2, \ldots, p_N\}$.

There seems to be a lot of mileage in this square with the strange coordinates. There should be; the uneven coordinates correspond to future cuts and stretches for the baker's transformation with uneven stretches and cuts. (A real baker's transformation would not use the just-touching IFS used here, but it's easier to visualize, and for the most general case N is allowed to be infinite.) It is a mixing function, so it automatically satisfies the equation that results from Elton's theorem (a property called *ergodicity*). The theorem takes care of how little "hyperbolicity" an IFS can have and still retain this property. Alternatively, Elton's theorem can be viewed as a set of minimal requirements on the w_i such that the baker's transformation as set up here accurately reflects the behavior of the IFS on addresses.

8 Application to Computer Graphics

We begin by illustrating how a color image of the invariant measure of an IFS with probabilities can be produced. The idea is very simple. We start from an IFS such as

$$\{\mathbb{C}; 0.5z + 24 + 24i, 0.5z + 24i, 0.5z; 0.25, 0.25, 0.5\}.$$

A viewing window and a corresponding array of pixels P_{ij} is specified. The Random Iteration Algorithm is applied to the IFS, to produce an orbit $\{z_n : n = 0, 1, \ldots,$ *numits*$\}$, where *numits* is the number of iterations. For each (i, j) the number of points, $\mathcal{N}(P_{ij})$, which lie in the pixel P_{ij} are counted. The pixel P_{ij} is assigned the value $\mathcal{N}(P_{ij})/$*numits*. By Elton's theorem, if *numits* is large, this value should be a good approximation to the measure of the pixel. The pixels are plotted on the screen in colors determined from their measures.

The following program implements this procedure. It is is written in BASIC. It runs without modification on an IBM PC with Enhanced Graphics Adaptor and Turbobasic.

Program 1. (Uses the Random Iteration Algorithm to Make a "Picture" of the Invariant Measure Associated with an IFS with Probabilities)

```
screen 9 : cls   'Initialize graphics.

dim s(51,51)   'Allocate array of pixels.

'IFS code for a Sierpinski triangle.
a(1)=0.5 : b(1)=0 : c(1)=0 : d(1)=0.5 : e(1)=24 : f(1)=24

a(2)=0.5 : b(2)=0 : c(2)=0 : d(2)=0.5 : e(2)=0 : f(2)=24
```

```
a(3)=0.5 : b(3)=0 : c(3)=0 : d(3)=0.5 : e(3)=0 : f(3)=0

'Probabilities for the IFS; they must add to one!
p(1)=0.25 : p(2)=.25 : p(3)=.5

mag=1 'Magnification factor.

'Increase the number of iterations as you magnify.
numits=5000

factor =100 'Scales pixel values to color values.

'This is the number of colors you are able to use.
numcols=8

for n=1 to numits 'Random iteration begins!

r=rnd : k=1 'Pick a number in [0,1] at random.

if r > p(1) then k=2

if r > p(1)+p(2) then k=3

'Map k is picked with probability p(k).
newx=a[k]*x + b[k]*y + e[k]
newy=c[k]*x + d[k]*y + f[k]

x=newx : y=newy

i=int(mag*x) : j=int(mag*y) 'Scale by magnification factor.

if (((i < 50) and (i>=0))and((0=<j) and (j<50))) then
'If the scaled value is

s(i,j)=s(i,j)+1
'...in the array add one to pixel (i,j).

end if

pset(i,j) 'Plot the point.

if instat then end 'Stop if a key is pressed.

next
```

```
for i=0 to 49 'Normalize values in pixel array, and plot...

for j=0 to 49 '...in colors corresponding to the normalized...

'...values of the numbers s(i,j).
col=s(i,j)*numcols*factor*mag*mag/numits

'Plot the pixel (i,j) in the color determined by...
pset(i,j),col

next j '...its measure.

next i

end
```

The program allows the user to zoom in on a piece of the rendered measure by altering the value of the magnification parameter *mag*. The result of running an adaptation of this program on a Masscomp workstation and then printing the contents of the graphics screen is shown in Figure IX.264.

Rendered invariant measures for IFS's acting in $\mathbb{R}^2$ are also shown in Figure IX.265.

By carrying out some simple computergraphical experiments, using a program such as the one above, we discover that "pictures" of invariant measures of IFS's possess a number of properties. (i) Once the viewing window and color assignments have been fixed, the image produced is stable with respect to the number of iterations, provided that the number of iterations is sufficiently large. (ii) Images vary consistently with respect to translation and rotation of the viewing window, and with respect to changes in resolution. In particular they vary consistently when they are magnified. (iii) The images depend continuously on the IFS code, including the probabilities. Property (i) ensures that the images are well defined. The properties in (ii) are also true for views of the real world seen through the viewfinder of a camera. Property (iii) means that images can be controlled interactively. These properties suggest that IFS theory is applicable to computer graphics.

We should, if we have done our measure theory homework, understand the reasons for (i) and (ii). They are consequences of corresponding properties of Borel of measures on $\mathbb{R}^2$. Property (iii) follows from a theorem by Withers [Withers 1987].

Examples & Exercises

8.1. Rewrite Program 1, section 8, in a form suitable for your own computer environment. Adjust *numits* and *factor* to ensure that a stable image results. Then make experiments to verify that the conditions (i)–(iii) above are verified. For example, to test the consistency of images with respect to changes in resolution you should try *mag* = 0.5, 1, and 1.5. Unless you have a very powerful system, do not make ex-

Figure IX.264. The result of running a modified version of Program 9.8.1 and then printing the contents of the graphics screen in gray tones. A rendered picture of a measure is the result.

treme adjustments. For example, do not choose *mag* too small, otherwise you will need a very large value for *numits*.

Applications of fractal geometry to computer graphics have been investigated by a number of authors including Mandelbrot [Mandelbrot 1982], Kawaguchi [Kawaguchi 1982], Oppenheimer [Oppenheimer 1986], Fournier *et al.* [Fournier 1982], Smith [Smith 1984], Miller [Miller 1986], and Amburn *et al.* [Amburn 1986]. In all cases the focus has been on the modelling of natural objects and scenes. Both deterministic and random geometries have been used. The application of IFS theory to computer graphics was first reviewed in [Demk 85]. It provides a single framework that can reach an unlimited range of images. It is distinguished from other fractal approaches because it is the only one that uses measure theory.

The modelling of natural scenes is an important area of computer graphics. Photographs of natural scenes contain redundant information in the form of subtle patterns and variations. There are two characteristic features: (i) the presence of complex geometrical structure and distributions of color and brightness at many scales; and (ii) the hierarchical layout of objects. (i) Natural boundaries and textures are not

Figure IX.265. Rendered invariant measures for IFS's of two maps.

(a)

smoothed out under magnification; they preserve some degree of geometrical complexity. (ii) Natural scenes are organized in hierarchical structures. For example, a forest is made of trees; a tree is a collection of boughs and limbs along a trunk; on each branch there are clusters of leaves; and a single leaf is filled with veins and covered with fine hairs. It appears often in a natural scene that a recognizable entity is built up from numerous near repetitions of some smaller structure. These two observations can be integrated into systems for modelling images using IFS theory.

Examples & Exercises

8.2. Examine a good-quality color photograph of a natural scene, such as can be found in a *Sierra Club* calender, or an issue of *National Geographic*. Discuss the extent to which (i) and (ii) are true for that photograph. Be specific.

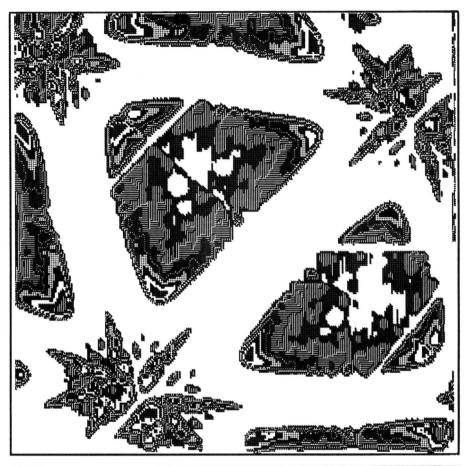

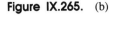

Figure IX.265.   (b)

In [Barnsley 1988a] it is reported that IFS theory can be used efficiently to model photographs of clouds, mountains, ferns, a field of sunflowers, a forest, seascapes and landscapes, a hat, the face of a girl, and a glaring arctic wolf.

There are two parts to making any computer graphics image: geometrical modelling and rendering. Consider an architect making a computergraphical house: first she defines the dimensions of the floor, the roof, the windows, the shapes of the gables, and so on, to produce the geometrical model. Traditionally this is specified in terms of polygons, circles, and other classical geometrical objects that can be conveniently input to the computer. This model is not a picture. To make a picture, the model must be projected into two dimensions from a certain point of view and distance, discretized so that it can be represented with pixels, and finally rendered in colors on a display device.

Here we describe briefly the software system designed by the author, Alan Sloan, and Laurie Reuter, which was used to produce the color images that accompany this section. More details can be found in [Reuter 1987] and [Barnsley 1988a]. The system consists of two subsystems known as *Collage* and *Seurat*. *Collage* is used for geometrical modelling, while *Seurat* is used for rendering.

Collage and *Seurat* process IFS structures of the form

$$\{\mathbb{R}^2; w_1, w_2, \ldots, w_N; p_1, p_2, \ldots, p_N : n = 1, 2, \ldots, N\},$$

where the maps are affine transformations in $\mathbb{R}^2$. An IFS is represented by a file that consists of an IFS code, where each coefficient is written with a fixed number of bits Let μ denote the invariant measure of such an IFS and let A denote the attractor. The pair (A, μ) is referred to as an *underlying model*. The attractor A carries the geometry while μ carries the rendering information. One can think of the IFS code, or equivalently (A, μ), as being analagous to the plans of an architect. It corresponds to many different pictures.

Collage is a geometrical modelling system used to determine the coefficients of the affine transformations $w_1, w_2, \ldots, w_N$. It is based on the Collage Theorem. *Seurat* is a software system for rendering images starting from an IFS code. An image is produced once a viewing window, color table, and resolution have been specified. This is achieved using the Random Iteration Algorithm. Its mathematical basis is Elton's Theorem. *Seurat* is also used in an interactive mode to determine the probabilities and color values.

The input to *Collage* is a target image, which we denote here by T. For example, T may be a polygonal approximation to a leaf. We suppose that

$$T \subset \blacksquare = \{(x, y) \in \mathbb{R}^2 : 0 \le x \le 1, 0 \le y \le 1\},$$

and that the screen of the computer display device corresponds to $\blacksquare$. T is rendered on the graphics workstation monitor. An affine transformation

$$w_1(x, y) = \begin{pmatrix} a_1 & b_1 \\ c_1 & d_1 \end{pmatrix} \begin{pmatrix} x \\ y \end{pmatrix} + \begin{pmatrix} e_1 \\ f_1 \end{pmatrix} = A_1 x + t_1$$

is introduced, with coefficients initialized at $a_1 = d_1 = 0.25$, $b_1 = c_1 = e_1 = f_1$. The image $w_1(T)$ is displayed on the monitor in a different color from T. $w_1(T)$ is a quarter-sized copy of T, centered closer to the point $(0, 0)$. The user now interactively adjusts the coefficients with a mouse or some other interaction technique, so that the image $w_1(T)$ is variously translated, rotated, and sheared on the screen. The goal of the user is to transform $w_1(T)$ so that it lies over part of T. It is important that the dimensions of $w_1(T)$ are smaller than those of T, to ensure that w_1 is a contraction. Once $w_1(T)$ is suitably positioned, it is fixed, and a new subcopy of the target, $w_2(T)$, is introduced. w_2 is adjusted until $w_2(T)$ covers a subset of those pixels in T that are not in $w_1(T)$. Overlap between $w_1(T)$ and $w_2(T)$ is allowed, but in general it should be made as small as possible, for efficiency. New maps are added and adjusted until $\cup_{j=1}^N w_j(T)$ is a good approximation to T. The output from *Collage* is the resulting IFS code. The probability p_j is chosen proportional to $|a_j d_j - b_j c_j|$ if this number is nonzero, and equal to a small positive number if the determinant of A_j equals zero.

The input to *Seurat* is one or more IFS codes generated by *Collage*. The viewing window and the number of iterations are specified by the user. The measures of the pixels are computed. The resulting numbers are multiplied by the inverse of the maximum value so that all of them lie in [0, 1]. Colors are assigned to numbers in [0, 1] using a color assignment function. The default is a gray scale where the intensity is proportional to the number, such as 0 corresponds to black and 1 corresponds to brightest white. The coloring and texture of the image can be controlled through the probabilities and the color assignment function. Although one does not explicitly use it, Theorem 9.6.3 lies in the background and can help in the adjustment of the probabilities.

Color Plate 21 shows some smoking chimneys in a landscape. We obtained the IFS codes for the elements of this image we obtained using *Collage*. Different color assignment functions are associated to different elements in the image. The image was rendered using *Seurat*.

The consistency of images with respect to changes in resolution is illustrated in Color Plate 22, which shows a zoom on one of the smokestacks in Color Plate 21. The number of iterations must be increased with magnification to keep the number of points landing within the viewing window constant. This requirement ensures the consistency of the textures in an image throughout the magnification process.

Color Plates 23 and 24 show various renderings of leaves produced by *Seurat*.

Color Plate 25 shows a sequence of frames taken from an IFS encoded movie entitled *A Cloud Study* [Barnsley 1987]. The smooth transition from frame to frame is a consequence of the continuous dependence on parameters of the invariant measure of the IFS for the cloud.

Color Plates 26, 27, and 28 were encoded from color photographs. Segmentation according to color was performed on the originals to define textured pieces. IFS codes for these components were obtained using *Collage*. The IFS data base contained less than 180 maps for the Monterey seascape, and less than 160 maps for the Andes Indian girl.

The two primitives, a leaf and a flower, in Color Plate 29 were used as condensation sets in the picture *Sunflower Field*, Color Plate 30. Here we see the hierarchical structure: the leaf is itself the attractor of an IFS; and the flower is an overlay of four IFS attractors. The leaf is a condensation set for the IFS that generates all of the leaves. The flower is a condensation set to an IFS that generates many flowers, converging to the horizon. In the pictures *Sunflower Field* and *Black Forest*, shown in Color Plates 31–34, the primitives were displayed from back to front. The data bases for the *Sunflower Field* and *Black Forest* contain less than 100 and 120 maps, respectively. Notice the shadows behind the little trees in the background in Color Plate 32. The winter forest pictures were obtained by adjusting the color assignment function. The important point is that once the adjustment has been made, the image and the zoom are consistent.

Examples & Exercises

8.3. Use the Collage Theorem to help you find an IFS code for a leaf. Adjust your version of Program 1 in section 8 to allow you to render images of associated invariant measures. Assign colors in the range from red through orange to green. Adjust the probabilities. Obtain a spectacular color picture of the leaf showing the veins. Make a color slide of the output. To photograph a picture on the screen of a computergraphics monitor, use a telephoto lens. Mount the camera on a tripod, and take the photograph in a darkened room, on Ectachrome 64 ASA color slide film, 0.1 sec exposure, f-stop 5.6. For possible publication, submit the color slide, together with a letter of copyright assignment, to Michael Barnsley, Iterated Systems, Inc., 5550-A Peachtree Parkway, Suite 650, Norcross GA 30092 USA. Include a self-addressed envelope.

8.4. Obtain a very powerful computer with good graphics. Find the heirarchical IFS codes for the *Sunflower Field*. Replace the sunflowers by roses. Fly into your picture, to explore forever that scent-filled horizon. You are on your own.

Chapter X

Recurrent Iterated Function Systems

1 Fractal Systems

The goal of this chapter is to describe some general systems that can be used to construct deterministic fractals. We are concerned with the invention of mathematical machines for designing and controlling fractals.

We use the name "recurrent iterated function systems" loosely to encompass the systems introduced in this chapter. Examples include "IFS's" in which the application of transformations to points depends on the "addresses" of the points. Other examples include "IFS's with probabilities" which have invariant measures that can be computed by means of algorithms of the "Chaos Game" type, wherein, instead of each transformation being applied with a fixed probability, there are different probabilities that depend on which transformation was applied previously. A unified framework for presenting IFS, recurrent IFS, and future developments in the theory of deterministic fractals is provided by the idea of *fractal systems*.

In this section we introduce fractal systems, their associated fractals, and the objects or models which they can be used to approximate. We do this in such a way as to help the reader understand how to invent a fractal system appropriate to the type of model she or he is describing, without having first to develop a new theory from the bottom up. The ideas are loosely stated in order to convey most simply the underlying principles, which seem to be of general applicability and to be very powerful. From this author's experience, it seems that any framework for a fractal system, once decided upon, can be readily fleshed out with the appropriate definitions and logical infrastructure to provide a full and clean theory.

Roughly speaking, a fractal system consists of a means for generating deterministic fractals, associated with an underlying space $\mathbb{X}$, and a set or space of objects or *models*, $\mathbb{Y}$, which these fractals can be used to approximate. Throughout this chapter we use $\square$ to denote a closed bounded subset of $\mathbb{R}^2$ such as $\{(x, y) \in \mathbb{R}^2 : 0 \leq x, y \leq 1\}$.

Already in this book we have looked at several very different fractal systems, as described in the following examples.

Example (i). One example of a fractal system is defined by IFS's of contractive affine transformations acting on $\mathbb{X} = \square \subset \mathbb{R}^2$: the underlying space is $\square$; the fractals of the system are attractors of all of the IFS's; and the space of models, $\mathbb{Y}$, which these fractals can be used to approximate, consists of all (compact) subsets of $\square$, namely $\mathbb{Y} = \mathcal{H}(\mathbb{X})$. In this example the models may represent monochrome pictures, with the sets representing the black parts of the pictures, and their complements representing the (background) white parts. The fractal system itself is a device for approximating the models by means of the fractals that it generates.

Example (ii). Another example of a fractal system is defined by the IFS's of contractive affine transformations on $\square \subset \mathbb{R}^2$, with probabilities. The fractals of this system are the attractors of the associated Markov operators, as described in Chapter IX. They lie in the space $\mathbb{Y} = \mathcal{P}(\square)$, the set of all normalized Borel measures on $\square \subset \mathbb{R}^2$. Here the underlying space $\mathbb{X}$ is $\square \subset \mathbb{R}^2$; its points do not belong to the fractals of the system because they are measures, not sets of points. In this example the models may represent grayscale pictures, with the amount of light reflected by a subset of the picture represented by the measure of that subset. Again, the fractal system provides a means for approximating the models, using the fractals that it generates.

Example (iii). A third example of a fractal system involves the space $\mathbb{X} = [0, 1] \subset \mathbb{R}^2$, with $\mathbb{Y} = C[0, 1]$, the continuous real valued functions on $[0, 1]$. In this system the fractals are the fractal interpolation functions on $[0, 1]$ that can be generated by affine transformations, as described in Chapter VI; and the the models may represent time series data or stock market prices.

Further examples of fractal systems are introduced in this chapter, and involve recurrent iterated function systems. Yet others, involving local iterated function systems and appropriate to the modelling of real world images, are described in [BH].[1] All of these fractal systems use the same basic ingredients, which we describe next.

(1) *An underlying metric space* $(\mathbb{X}, d)$ *needed to define the fractals and the models in the system.* For example, $\mathbb{X}$ may be $\mathbb{R}^2$, $\mathbb{R}^3$, or a subset of one of these spaces, such as $\square \subset \mathbb{R}^2$. Typically $(\mathbb{X}, d)$ is complete, and bounded subsets of it are compact.

(2) *A space of models* $\mathbb{Y} = \mathbb{Y}(\mathbb{X})$. Each point of $\mathbb{Y}$ is a model, and models are defined with the aid of the space $\mathbb{X}$. The fractals generated by the fractal system

---

[1] M. F. Barnsley and L. P. Hurd, Fractal Image Compression, (1993), A. K. Peters, Wellesley, Massachusetts.

also belong to $\mathbb{Y}$. Examples of $\mathbb{Y}$ include spaces of sets, function spaces, and measure spaces. We also need *a metric h on the space* $\mathbb{Y}$, *such that* $(\mathbb{Y}, h)$ *is a complete metric space.* Examples of $(\mathbb{Y}, h)$ include $(\mathcal{H}(\mathbb{X})$, the Hausdorff distance generated by $d)$, $(C[0, 1], h(f, g) = \max\{|f(x) - g(x)| : x \in [0, 1]\})$, and $(\mathcal{P}(\mathbb{X}), d_h)$ where d_h is the Hutchinson distance between measures.

(3) *A contractive operator O which acts on the space* $(\mathbb{Y}, h)$. That is, the operator O is such that there exists a real number s with $0 \leq s < 1$, and

$$h(O(\phi), O(\psi)) \leq s \cdot h(\phi, \psi) \text{ for all } \phi, \psi \in \mathbb{Y}.$$

Typically, the operator O is constructed with the aid of elementary contractive functions that act upon the underlying space $\mathbb{X}$.

For the examples of fractal systems, (i), (ii), and (iii) begun above, the operators O are set up as follows:

Example (i) (continued). $\mathbb{X} = \square \subset \mathbb{R}^2$, $w_i : \square \to \square$ is a contractive transformation for each $i = 1, 2, \ldots, N$, $\mathbb{Y} = \mathcal{H}(\mathbb{X})$, and $O : \mathbb{Y} \to \mathbb{Y}$ is defined by

$$O(\mathbb{Y}) = W(\mathbb{Y}) = \bigcup\{w_i(x) : x \in \mathbb{Y}\}.$$

Remark that O is built up using contractive transformations acting on the underlying space.

Example (ii) (continued). $\mathbb{X} = \square \subset \mathbb{R}^2$, $\{\mathbb{X}; w_1, \ldots, w_N; p_1, \ldots, p_N\}$ are IFS's with probabilities, as in section 6 in Chapter IX, with the w_i's as in (i) above, and $\mathbb{Y} = \mathcal{P}(\mathbb{X})$; we can then define a corresponding fractal system by taking O to be the associated Markov operator as defined in section 6 in Chapter IX; that is, $O : \mathbb{Y} \to \mathbb{Y}$ is defined by

$$O(\nu) = M(\nu) = p_1 \nu \circ w_1^{-1} + p_2 \nu \circ w_2^{-1} + \cdots + p_N \nu \circ w_N^{-1}.$$

We know from the theory of Chapter IX that O is contractive in the Hutchinson metric. Again, remark that O is built up using contractive transformations, the w_i's, acting on the underlying space.

Example (iii) (continued). In this example fractal interpolation functions, as described in Chapter VI, are used. $\mathbb{X} = [0, 1] \subset \mathbb{R}$, and $\mathbb{Y} = \mathcal{F} = C[0, 1]$. Then $O : \mathbb{Y} \to \mathbb{Y}$ is the operator $T : \mathcal{F} \to \mathcal{F}$ defined in the proof of Theorem 2.2 in Chapter VI. In the notation of section 2 in Chapter VI, this is, for $f \in \mathcal{F}$,

$$O(f) = (Tf)(x) = c_n l_n^{-1}(x) + d_n f(l_n^{-1}(x)) + f_n$$
$$\text{for } x \in [x_{n-1}, x_n], \text{ for } n = 1, \ldots, N.$$

Then O is a contraction mapping with respect to the supremum distance on $C[0, 1]$. Again, notice how the operator O is built up from contractive transformations acting on the underlying space.

In practice, the contractive operator O is constructed from finite collections of contractive *affine* transformations; for example, affine transformations acting in two dimensions. Furthermore, the coefficients of these constituent functions can be restricted to lie in finite sets, defined for example by rounding. Then O itself can be described by finite sets of discrete coefficients, the full set of which represents the "code" for the operator O.

The consequences of having the ingredients 1, 2, and 3 can be summarized in the following Theorems and Expections:

Theorem 1.1 Existence of Attractors. *Since* O *is contractive, and the metric space* $\mathbb{Y}$ *is complete, there exists an unique attractor* $\phi \in \mathbb{Y}$ *such that*

$$O(\phi) = \phi$$

Proof See Theorem 6.1 in Chapter III.

Expectation 1.1 Fractal Character of Attractors. *We anticipate that* ϕ *is a fractal: that is, we expect that* ϕ *has a resolution-independent, infinitely-magnifiable character. This is because of the contractivity of the functions from which* O *is constructed: the whole invariant set is the same as a sum or union of contractions applied to it, and thus it is made of shrunken copies of (parts of) itself. Depending on the way in which the contractions act, the focus may be on contractivity of various spatial dimensions, and/or measure theoretic contractivity, and we expect that the attractor* ϕ *will inherit corresponding fractal characteristics.*

Theorem 1.2 Computation of Attractors. *To compute* ϕ*, we can use the fact that if* $\psi \in \mathbb{Y}$ *then the result of repeatedly applying* O *to* ψ *converges to the attractor* ϕ*; that is,*

$$\lim_{n \to \infty} O^{\circ n}(\psi) = \phi.$$

Moreover, if there exists a real constant C *such that* $h(\phi_1, \phi_2) < C$ *for all* $\phi_1, \phi_2 \in \mathbb{Y}$*, then we have the error estimate*

$$h(O^{\circ n}(\psi), \phi) \leq s^n C.$$

Proof See Theorem 6.1 in Chapter III.

The latter equation tells us that the fixed point, or attractor, can be computed by algorithms of the "Photocopy Machine" type. The error estimate allows one to predict the number of iterations required to achieve a given accuracy.

Theorem 1.3 General Collage Theorem Estimate. *The distance between* $\psi \in \mathbb{Y}$ *and the attractor* ϕ *of* O *is bounded by the estimate*

$$h(\phi, \psi) \leq \frac{h(\psi, O(\psi))}{(1 - s)}.$$

Proof See Lemma 11.1 in Chapter III.

The set $O(\psi)$ is called a *collage of* ψ, while the distance $h(\psi, O(\psi))$ is called the corresponding *collage error*. The theorem says that if we wish to find an operator

O whose attractor ϕ is approximately ψ, then we have only to solve the problem of choosing O such that application of O to ψ does not change ψ very much, as discussed in Chapters III and IX.

This latter simple prescription is of central importance, yet easy to miss. We repeat it in slightly different terms. Suppose that we have a fractal system, defined by $\mathbb{Y}(\mathbb{X})$ and $\mathcal{O}$, where $\mathcal{O}$ is a set of operators O as above. We seek an operator $O \in \mathcal{O}$ whose attractor ϕ approximates $\psi \in \mathbb{Y}$. To do so, we adjust O so that the collage error $h(\psi, O(\psi))$ is as small as possible. Then O defines our fractal approximation ϕ. To store ϕ it suffices to store O. In the end, one thinks of the controllable family of operators $\mathcal{O}$ as the fractal system.

Whenever we have the ingredients 1, 2, and 3, we have a fractal system. A *fractal system* is a mechanism for finding fractal approximations ϕ for models $\psi \in \mathbb{Y}$. Examples (i), (ii), and (iii) above constitute interactively controllable systems for fractal modelling applied to diverse problems: in (i), sets of points are approximated by fractals made of sets of points; in (ii) measures are approximated by fractal measures; and in (iii), continuous functions are approximated by fractals that are graphs of continuous functions. In each case the parameters that control the problem are embedded in the operator O, which is selected and tailored to produce an apparatus for fractal approximation, a fractal system, appropriate to the modelling problem under consideration.

We note that fractal systems provide deterministic fractals rather than stochastic fractals. Once the operator O has been selected, its attractor, the fractal that it defines, has been fixed once and for all. There is only one fractal associated with each operator. These fractals are typically low information content members of a larger space, the space of models in which they lie; in much the same way as the rational numbers are finite information content members of the information rich space of all real numbers.

In the following sections we develop more elaborate fractal systems, based on recurrent IFS. Their introduction was begun in section 1 of this chapter. These generalize earlier systems in an elegant manner. They can be used to design machines to do elaborate fractal modelling, and illustrate well the design principles enunciated above.

2   Recurrent Iterated Function Systems

Recurrent iterated function systems were introduced in [BEH],[2] where more information can be found.

---

[2] M. Barnsley, J. Elton, D. Hardin, Recurrent Iterated Function Systems, Constructive Approximation (1989) 5, 3–31.

Table X.1.   Example of a recurrent IFS code. Not all transitions are possible!

w_i	a_i	b_i	c_i	d_i	e_i	f_i	P_{i1}	P_{i2}	P_{i3}
1	0.5	0	0	0.5	0	0	0.3	0.7	0
2	0.5	0	0	0.5	0	128	0	0.6	0.4
3	0.5	0	0	0.5	128	128	0.5	0	0.5

Definition 2.1 *A recurrent iterated function system* consists of an IFS $\{\mathbb{X}; w_1,$
$w_2, \ldots, w_N\}$ *together with a matrix* $\{p_{ij} \in [0, 1] : i, j = 1, 2, \ldots, N\}$, *such that (i)*

$$p_{i1} + p_{i2} + p_{i3} + \cdots + p_{iN} = 1 \quad \text{for } i = 1, 2, \ldots, N$$

and (ii) such that for any i and j, there exists a finite sequence of integers $k, l, \ldots,$
$m \in \{1, 2, \ldots, N\}$ *so that*

$$p_{ik} p_{kl} \cdots p_{mj} > 0$$

The transition probability for a certain discrete time Markov process is p_{ij} , which gives the chance, or probability, of moving from state i to state j, given that the process is in state i. Condition (i) says that the process is row stochastic: whichever state the system is in, a set of probabilities is available that sum to 1, and that describe the possible subsequent states to which the system can transition at the next step. Condition (ii) says that if the system is in state i then there is a finite probability of reaching state j in a finite number of steps, for any pair of integers $i, j \in \{1, 2, \ldots, N\}$.

An IFS with probabilities provides a simple example of a recurrent IFS. The IFS with probabilities

$$\{\mathbb{X}; w_1, w_2, \ldots, w_N; p_1, p_2, \ldots, p_N\}$$

is in many ways the same as the recurrent IFS

$$\{\mathbb{X}; w_1, w_2, \ldots, w_N; (p_{ij})\}$$

when $p_{ij} = p_j$ for all $i, j \in \{1, 2, \ldots, N\}$.

We will say that a recurrent IFS is hyperbolic if the associated IFS is hyperbolic, that is, contractive with contractivity factor $0 \leq s < 1$. We restrict attention to hyperbolic recurrent IFS.

Examples of *codes* for recurrent IFS's are presented in Tables X.1 and X.2. In each case the space is $\mathbb{X} = \mathbb{R}^2$, the transformations are affine, and we use the usual notation

$$w_i \begin{pmatrix} x \\ y \end{pmatrix} = \begin{pmatrix} a_i & b_i \\ c_i & d_i \end{pmatrix} \begin{pmatrix} x \\ y \end{pmatrix} + \begin{pmatrix} e_i \\ f_i \end{pmatrix}.$$

Figures X.266 and X.267 provide diagrams of the associated Markov processes. We say that the system is in *state i* if the last transformation to be applied was w_i. For the the recurrent IFS in Table X.1 it is never possible to apply transformation

Table X.2. Example of a recurrent IFS code. All transitions can occur!

w_i	a_i	b_i	c_i	d_i	e_i	f_i	P_{i1}	P_{i2}	P_{i3}
1	0.5	0	0	0.5	0	0	0.3	0.6	0.1
2	0.5	0	0	0.5	0	128	0.1	0.5	0.4
3	0.5	0	0	0.5	128	128	0.4	0.4	0.2

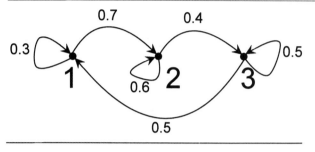

Figure X.266. Markov process for the recurrent IFS given in Table X.1. There is no path from state 1 to state 3, state 2 to state 1, or state 3 to state 2.

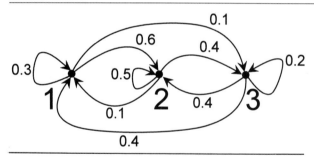

Figure X.267. Markov process for the recurrent IFS given in Table X.2. Paths exist between all possible states.

number 3 if the system is in state 1. It is also not possible to make a transition from state 2 to state 1. For the recurrent IFS in Table X.2, any transformation can follow any other one; that is, the system can transition from any state to any other state, in one step.

A recurrent IFS possesses a unique measure theoretic attractor, [BEH]. This is an invariant measure that can be computed by means of a generalized "Chaos Game" algorithm. It can also be computed using a wierd "Photocopy Machine."

Example of the Chaos Game Algorithm for a Recurrent IFS. We present this algorithm for the case of the the recurrent IFS codes in Tables X.1 and X.2. The algorithm yields an orbit $\{(x_n, y_n) : n = 0, 1, 2, \ldots\}$ that, with probability one, describes the unique invariant measure associated with the recurrent IFS. The relationship between the orbit and the measure is much the same as that described in Chapter IX for the case of standard IFS with probabilities.

(1) Choose a starting point $(x_0, y_0) \in \mathbb{R}^2$. It is desirable that the starting point lies

as close as possible to the attractor. Clearly this lies in $[0, 256] \times [0, 256]$. Thus, we might choose $(x_0, y_0) = (0, 0)$.

(2) Choose a initial state $s_0 = \{1, 2, 3\}$. Any other state in the set $\{1, 2, 3\}$ could be used. The Markov process associated with the transition matrix (p_{ij}) possesses a unique stationary vector (m_1, m_2, m_3) with $m_i > 0$, such that

$$\sum_{i=1}^{3} m_i p_{ij} = m_j, \quad j = 1, 2, 3.$$

If one knows this vector, then it is a good idea to choose the initial state to be the one corresponding to the largest m_j.

(3) Select $s_1 \in \{1, 2, 3\}$ with probability $p_{s_0 j}$ associated to the choice $s_1 = j$.
(4) Compute $(x_1, y_1) = w_{s_1}(x_0, y_0)$.
(5) Select $s_2 \in \{1, 2, 3\}$ with probability $p_{s_1 j}$ associated to the choice $s_2 = j$.
(6) Compute $(x_2, y_2) = w_{s_2}(x_1, y_1)$.

· · ·

$(2n + 1)$ Select $s_n \in \{1, 2, 3\}$ with probability $p_{s_{n-1} j}$ associated to the choice $s_n = j$.

$(2n + 2)$ Compute $(x_n, y_n) = w_{s_n}(x_{n-1}, y_{n-1})$.

· · ·

The result of the above calculation is supposed to be a long orbit

$$Q(\text{numits}) = \{(x_n, y_n) : n = 1, 2, .., \text{numits}\}.$$

(With probability one) the value of the invariant measure $\mu \in \mathcal{P}([0, 256] \times [0, 256])$ of the recurrent IFS is given by the formula

$$\lim_{\text{numits} \to \infty} \mu(S) = \frac{\text{number of points in } \{S \cap Q(\text{numits})\}}{\text{numits}}$$

for any measureable subset $S \subset \mathbb{R}^2$.

In Figure X.268 we illustrate this algorithm for the case of the recurrent IFS in Table X.1. In Figure X.269 we give the results of a computation, showing a long orbit. The invariant measure can be pictured by observing the density of the points.

In Figure X.270 we illustrate this algorithm for the case of the recurrent IFS in Table X.2. In Figure X.271 we give the results of a computation, showing a long orbit. Again, one gets some idea of the invariant measure by looking at the density of the points.

In Figure X.272 we show the structure and the attractor for the recurrent IFS in Table X.3. This involves four transformations which take □ into itself. The range of w_i is the quadrant of □ labeled i, so that the transformations provide a just-touching collage of the square. The transition probability p_{24} is zero, which means that no points on the attractor lie in the subquadrant labeled 24.

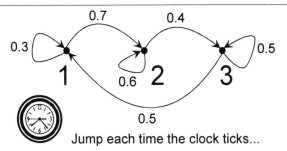

Figure X.268. Chaos Game for a recurrent IFS in which not all transitions are possible. The recurrent IFS code is given in Table X.1.

Jump each time the clock ticks...

IN STATE 1	IN STATE 2	IN STATE 3
Jump to State 1 with Prob 0.3	Jump to State 2 with Prob 0.6	Jump to State 3 with Prob 0.5
Jump to State 2 with Prob 0.7	Jump to State 3 with Prob 0.4	Jump to State 1 with Prob 0.5

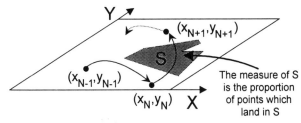

The measure of S is the proportion of points which land in S

In the (x,y) plane, when the system jumps to State j, apply the map w_j to the current point (x_N, y_N) to obtain $(x_{N+1}, y_{N+1}) = w_j(x_N, y_N)$.

Thus produce a dancing point.

Figure X.273 shows the attractor for the maps in Table X.3, for six different sets of transition probabilities. These sets of transition probabilites used were:

(1) $p_{11} = 0.3$, $p_{12} = 0.5$, $p_{13} = 0.1$, $p_{14} = 0.1$,
$p_{21} = 0.1$, $p_{22} = 0.5$, $p_{23} = 0.4$, $p_{24} = 0.0$,
$p_{31} = 0.3$, $p_{32} = 0.3$, $p_{33} = 0.2$, $p_{34} = 0.2$,
$p_{41} = 0.25$, $p_{42} = 0.25$, $p_{43} = 0.25$, $p_{44} = 0.25$;

(2) $p_{11} = 0.3$, $p_{12} = 0.5$, $p_{13} = 0.1$, $p_{14} = 0.1$,
$p_{21} = 0.1$, $p_{22} = 0.5$, $p_{23} = 0.4$, $p_{24} = 0.0$,
$p_{31} = 0.3$, $p_{32} = 0.3$, $p_{33} = 0.2$, $p_{34} = 0.2$,
$p_{41} = 0.25$, $p_{42} = 0.25$, $p_{43} = 0.5$, $p_{44} = 0.0$;

(3) $p_{11} = 0.3$, $p_{12} = 0.5$, $p_{13} = 0.1$, $p_{14} = 0.1$,
$p_{21} = 0.1$, $p_{22} = 0.5$, $p_{23} = 0.4$, $p_{24} = 0.0$,
$p_{31} = 0.0$, $p_{32} = 0.3$, $p_{33} = 0.5$, $p_{34} = 0.2$,
$p_{41} = 0.0$, $p_{42} = 0.5$, $p_{43} = 0.5$, $p_{44} = 0.0$.

(4) $p_{11} = 0.0$, $p_{12} = 0.5$, $p_{13} = 0.2$, $p_{14} = 0.3$,
$p_{21} = 0.1$, $p_{22} = 0.4$, $p_{23} = 0.4$, $p_{24} = 0.1$,

Figure X.269. Some points on the orbit produced by the machinery in Figure X.268.

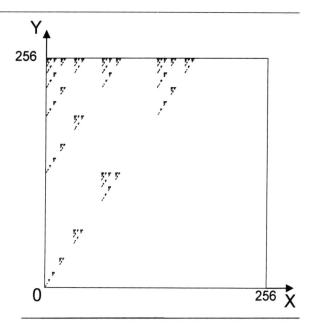

$p_{31} = 0.25$, $p_{32} = 0.25$, $p_{33} = 0.25$, $p_{34} = 0.25$,
$p_{41} = 0.4$, $p_{42} = 0.4$, $p_{43} = 0.1$, $p_{44} = 0.1$.

(5) $p_{11} = 0.25$, $p_{12} = 0.25$, $p_{13} = 0.25$, $p_{14} = 0.25$,
$p_{21} = 0.25$, $p_{22} = 0.25$, $p_{23} = 0.25$, $p_{24} = 0.25$,
$p_{31} = 0.25$, $p_{32} = 0.25$, $p_{33} = 0.25$, $p_{34} = 0.25$,
$p_{41} = 0.25$, $p_{42} = 0.25$, $p_{43} = 0.25$, $p_{44} = 0.25$.

(6) $p_{11} = 0.0$, $p_{12} = 0.34$, $p_{13} = 0.33$, $p_{14} = 0.33$,
$p_{21} = 0.25$, $p_{22} = 0.25$, $p_{23} = 0.25$, $p_{24} = 0.25$,
$p_{31} = 0.25$, $p_{32} = 0.25$, $p_{33} = 0.25$, $p_{34} = 0.25$,
$p_{41} = 0.25$, $p_{42} = 0.25$, $p_{43} = 0.25$, $p_{44} = 0.25$.

Can you work out which set of the above transition probabilities corresponds to each of the images in Figure X.273?

Example of the Photocopy Machine Algorithm for realizing the invariant measure of a recurrent IFS. We illustrate this algorithm for the case of the IFS code in Table X.1. It is illustrated in Figure X.274.

The mathematical machine in Figure X.274 corresponds to the recurrent IFS in Table X.1, wherein there are three affine transformations w_1, w_2, and w_3, and wherein the transition probabilities p_{13}, p_{21}, and p_{32} are all zero. The states of the system are represented by the three copy units, one of which is labeled IN-PUT1/COPY1; the others are labeled INPUT2/COPY2 and INPUT3/COPY3. Each unit has two lens/filter components. The first copy unit is made of two lens/filter components; one of these components applies affine transformation w_1 to INPUT1,

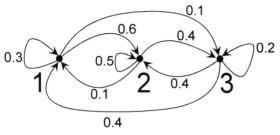

Figure X.270. Random walk generator associated with the recurrent IFS in Table X.2.

Tic, Tic, Tic, Tic ......

IN STATE 1	IN STATE 2	IN STATE 3
Jump to State 1 with Prob 0.3	Jump to State 1 with Prob 0.1	Jump to State 1 with Prob 0.4
Jump to State 2 with Prob 0.6	Jump to State 2 with Prob 0.5	Jump to State 2 with Prob 0.4
Jump to State 3 with Prob 0.1	Jump to State 3 with Prob 0.4	Jump to State 3 with Prob 0.2

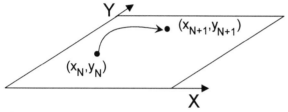

Apply map w_j when system jumps to State j.

attenuates the brightness of the image by factor p_{11}, and directs its output to COPY1; the other of these components applies affine transformation w_2 to the input image, attenuates the brightness of the image by a factor p_{12}, and directs its output to COPY2. The second copy unit has two lens/filter components; the first applies affine transformation w_2 to the input image INPUT2, attenuates the brightness of the image by factor p_{22}, and directs its output to COPY2; the other of these components applies affine transformation w_3 to the input image INPUT2, attenuates the brightness of the image by a factor p_{23}, and directs its output to COPY3. The third copy unit has two lens/filter components; the first applies affine transformation w_3 to the input image INPUT3, attenuates the brightness of the image by factor p_{33}, and directs its output to COPY3; the other of these components applies affine transformation w_1 to the input image INPUT3, attenuates the brightness of the image by a factor p_{31}, and directs its output to COPY1.

It is important to note that the three inputs and the three outputs are set up so that they all represent the same subset $\square$ of $\mathbb{R}^2$. The range of transformation w_1 lies in COPY1, the range of w_2 lies in COPY2, and the range of w_3 lies in COPY3. Thus,

Figure X.271. "Picture" of the point-set attractor corresponding to the fractal system represented in Figure X.270.

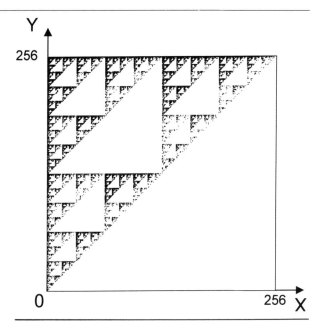

Table X.3. Recurrent IFS code. See Figure X.272.

w_i	a_i	b_i	c_i	d_i	e_i	f_i	P_{i1}	P_{i2}	P_{i3}	P_{i4}
1	0.5	0	0	0.5	0	0	0.3	0.5	0.1	0.1
2	0.5	0	0	0.5	0	128	0.1	0.5	0.4	0.0
3	0.5	0	0	0.5	128	128	0.3	0.3	0.2	0.2
4	0.5	0	0	0.5	128	0	0.25	0.25	0.25	0.25

in the first copy unit transformation $w_2 : \square \rightarrow \square$ in fact takes INPUT1 to COPY2, whilst in the second copy unit it takes INPUT2 to COPY2.

The images that are projected onto the copy screens are superposed; places where images come in from several inputs add up. For example, the total brightness of the image in COPY1 is p_{11} times the total brightness of the image in INPUT1, plus p_{31} times the total brightness of the image in INPUT3.

Copy images are recorded on "photographic paper" and put back on their respective inputs, as indicated in the figure. The system is cycled numerous times until the output images cease to change; then symbolically

$$\text{OUTPUT1} = p_{11} \cdot \text{OUTPUT1} + p_{31} \cdot \text{OUTPUT3}$$
$$\text{OUTPUT2} = p_{12} \cdot \text{OUTPUT1} + p_{22} \cdot \text{OUTPUT2}$$
$$\text{OUTPUT3} = p_{23} \cdot \text{OUTPUT2} + p_{33} \cdot \text{OUTPUT3}.$$

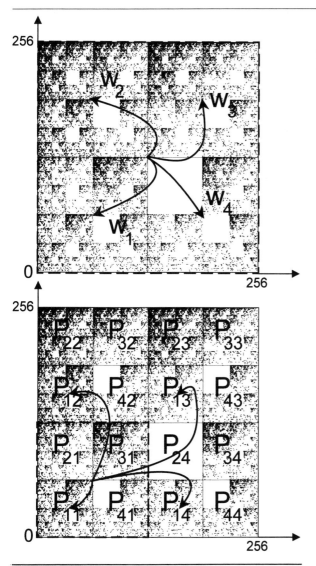

Figure X.272. Four affine transformations provide a just-touching collage of □. A recurrent IFS (Table X.3) made of these transformations is such that $p_{24} = 0$. As a consequence, the attractor has no points in the square whose address is 24.

Finally, the images OUTPUT1, OUTPUT2 and OUTPUT3 are superposed. Symbolically,

$$\text{ATTRACTOR} = \text{OUTPUT1} + \text{OUTPUT2} + \text{OUTPUT3}.$$

Color Plate 37 illustrates a measure theoretic attractor of a related Photocopy Machine. Color Plate 40 illustrates another way of approaching the computation of the invariant measure of a recurrent IFS, using cameras and screens.

Figure X.273.   All of these attractors correspond to different choices for the transition probabilities in Table X.3.

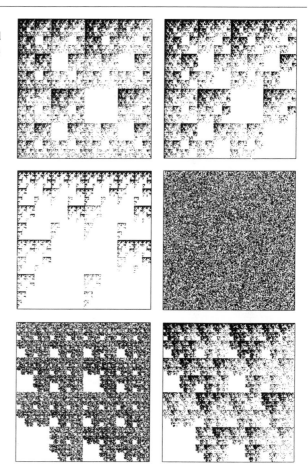

3   Collage Theorem for Recurrent IFS

In this section we are concerned with the construction of fractal systems, based on recurrent IFS, with an emphasis on point-set attractors, that is, attractors that are sets of points rather than measures. Our goal is to describe how a collage theorem and associated machinery can be set up to enable fractal modelling using the point-set attractors of certain recurrent IFS.

A recurrent IFS can described in terms of a Markov chain which acts on a code space built from the symbols $\{1, 2, \ldots, N\}$, as illustrated in Figures X.268 and X.270. The numbers $p_{ij} \geq 0$, $\sum_{j=1}^{N} p_{ij} = 1$, give the probabilities of transfer among the symbols. One can imagine a particle moving from symbol to symbol following the discrete time Markov process. The process is defined, strictly speaking, to be *recurrent* if there is a finite probability of being able to move, on the directed graph,

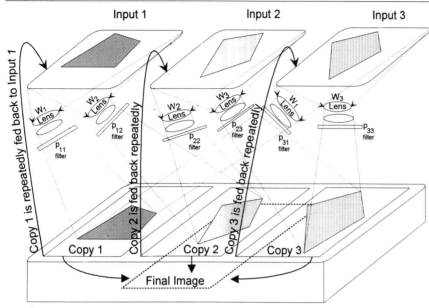

Figure X.274. Photocopy Machine for realizing the invariant measure of the recurrent IFS in Table X.1.

Copies 1, 2 and 3 are combined (superposed) to create Final Image

from any given symbol to any given symbol. A good source of information on Markov chains is [Feller, Ch. 15].[3]

We are concerned with the *hyperbolic* case, namely,

$$d(w_i(x), w_i(y)) \leq s d(x, y) \quad \forall i, \forall x, y \in \mathbb{X}$$

where d is the distance function on $\mathbb{X}$ and $0 \leq s < 1$. In this case, we know that there exists a unique attractive invariant probability measure μ, which describes the random walk on $\mathbb{X}$ produced for example by the "Chaos Game" algorithm described in the previous section. Our focus in this section is on the structure of the support of μ, which we call $A \in \mathbb{X}$, the attractor of the recurrent IFS. This depends only on which p_{ij} are nonzero, the *connection structure* of the Markov chain, and otherwise not on the values of the p_{ij}'s.

We begin by recalling how the analysis proceeds in the case of IFS theory; this provides the simplest framework for explaining the recurrent case. Let $(\mathbb{X}, d)$ denote a compact metric space $\mathbb{X}$, with distance function d. Let $\mathcal{H}$ denote the set of all nonempty compact subsets of $\mathbb{X}$.

Definition 3.1 $d(x, B) = \min_{y \in B} d(x, y) \ \forall x \in \mathbb{X}, \forall B \in \mathcal{H}$. *Note that*

$$B \subset C \Rightarrow d(x, C) \leq d(x, B). \tag{1}$$

---

[3] W. Feller, An Introduction to Probability Theory and its Applications (Wiley, London), 1957.

Definition 3.2 $d(A, B) = \max_{x \in A} d(x, B) \ \forall A, B \in \mathcal{H}.$

Note that this set distance is not symmetric. It has the following properties

(i) $B \subset C \Rightarrow d(A, C) \leq d(A, B)$. This follows at once from (1).
(ii) $d(A \cup B, C) = d(A, C) \vee d(B, C)$, where $x \vee y = \max\{x, y\}$. This follows
from the observation that

$$d(A \cup B, C) = \max_{x \in A \cup B} d(x, C) = \max_{x \in A} d(x, C) \vee \max_{x \in B} d(x, C).$$

Definition 3.3 *For all $A, B \in \mathcal{H}$, the Hausdorff distance is defined by*

$$h(A, B) = d(A, B) \vee d(B, A).$$

We recall that $(\mathcal{H}, h)$ is a compact metric space.

Lemma 3.1 *For all $A, B, C, D \in \mathcal{H}$,*

$$h(A \cup B, C \cup D) \leq h(A, C) \vee h(B, D).$$

Proof

$$d(A \cup B, C \cup D) = d(A, C \cup D) \vee d(B, C \cup D) \quad \text{by (ii)}$$
$$\leq d(A, C) \vee d(B, D) \quad \text{by (i)}$$
$$\leq h(A, C) \vee h(B, D).$$

The same argument yields $d(C \cup D, A \cup B) \leq d(C, A) \vee d(B, D) \leq h(A, C) \vee h(B.D)$.

Let $\{\mathbb{X}; w_j, j = 1, 2, \ldots, N\}$ be a hyperbolic IFS, with $d(w_j x, w_j y) \leq s d(x, y)$
$\forall x, y \in \mathbb{X}$, and $0 \leq s < 1$. Define

$$W : \mathcal{H} \to \mathcal{H}$$

by

$$W(A) = \bigcup_{j=1}^{N} w_j(A).$$

Theorem 3.1 $W : \mathcal{H} \to \mathcal{H}$ *is a contraction, with contractivity factor s, with respect to the Hausdorff metric; that is,*

$$h(W(A), W(B)) \leq s h(A, B) \quad \forall A, B \in \mathcal{H}.$$

Proof For any $A, B \in \mathcal{H}$,

$$h(W(A), W(B)) = h \left(\bigcup_{j=1}^{N} w_j(A), \bigcup_{j=1}^{N} w_j(B) \right)$$

$$\leq \bigvee_{j=1}^{N} h\left(w_j(A), w_j(B)\right) \quad \text{(by Lemma 3.1)}$$

$$= \bigvee_{j=1}^{N} \{ d\left(w_j(A), w_j(B)\right) \vee d\left(w_j(B), w_j(A)\right) \}$$

$$\leq \bigvee_{j=1}^{N} \{ sd(A, B) \vee sd(B, A) \}$$

$$= sh(A, B)$$

We leave it as an exercise for the reader to show that the following extension of the above result is true. We will need this fact later on. If $(\mathbb{X}_1, d_1)$ and $(\mathbb{X}_2, d_2)$ are metric spaces, then $(\mathcal{H}_1, h_1)$ and $(\mathcal{H}_2, h_2)$ are the corresponding spaces of compact nonempty subsets, if

$$w_j : \mathbb{X}_1 \to \mathbb{X}_2 \quad \text{for } j = 1, 2, \ldots, N,$$

obeys

$$d_2(w_j x, w_j y) \leq s d_1(x, y) \quad \forall x, y \in \mathbb{X}_1,$$

then

$$W : \mathcal{H}_1 \to \mathcal{H}_2$$

defined by

$$W(A) = \bigcup_{j=1}^{N} w_j(A)$$

obeys

$$h_2(W(A), W(B)) \leq s h_1(A, B).$$

Corollary 3.1 *There is a unique set $A \in \mathcal{H}$ such that $W(A) = A$.*

A is the attractor of the IFS.

Corollary 3.2 Collage Theorem for IFS. *If $B \in \mathcal{H}$ obeys*

$$h(B, W(B)) \leq \epsilon > 0$$

then

$$h(B, A) \leq \frac{\epsilon}{(1 - s)},$$

where A denotes the attractor of the IFS.

Proof By the contraction mapping theorem

$$h(A, B) = h\left(B, \lim_{n\to\infty} W^{\circ n}(B)\right) = \lim_{n\to\infty} h\left(B, W^{\circ n}(B)\right)$$

where $W^{\circ 0}(B) = W(B)$ and we inductively define

$$W^{\circ(n+1)}(B) = W(W^{\circ n}(B)), \quad n = 0, 1, 2, \ldots$$

But, by the triangle inequality,

$$h\left(B, W^{\circ n}(B)\right) \leq \sum_{m=1}^{n} h\left(W^{\circ(m-1)}(B), W^{\circ m}(B)\right)$$

$$= \sum_{m=1}^{n} h\left(W^{\circ(m-1)}(B), W^{\circ(m-1)}(W(B))\right)$$

$$\leq \sum_{m=1}^{n} s^{m-1} h\left(B, W(B)\right)$$

$$\leq (1-s)^{-1} h\left(B, W(B)\right)$$

We are now ready to extend the structure to the case of recurrent IFS. We actually make a generalization of the recurrent IFS structure to multiple spaces and set maps, suitable for the hyperbolic case where one is concerned with point-set topology issues. We are only concerned here with the connection structure of the chain. Let $(\mathbb{X}_j, d_j)$ be compact metric spaces, $j \in \{1, 2, \ldots, N\}$. Let $(\mathcal{H}_j, h_j)$ denote the associated metric spaces of non-empty compact subsets which use the Hausdorff metrics. Let there be defined maps $W_{ij} : \mathcal{H}_j \to \mathcal{H}_i \forall (i, j) \in I$ where I is some set of pairs of indices with the property that for each $i \in \{1, 2, \ldots, N\}$ there is a $j \in \{1, 2, \ldots, N\}$ with $(i, j) \in I$. That is, $I(i) = \{j | (i, j) \in I\} \neq \emptyset$ for each $i \in \{1, 2, \ldots, N\}$. Furthermore let

$$h_i\left(W_{ij}(A), W_{ij}(B)\right) \leq s_{ij} h_j(A, B)$$

for some number s_{ij}, $\forall (i, j) \in I$, $\forall A, B \in \mathcal{H}_j$. By the remark following Theorem 3.1 such maps can be built up from point maps taking $\mathbb{X}_j$ to $\mathbb{X}_i$.

Suppose $w_i : \mathbb{X} \to \mathbb{X}$ are contractive maps, where $\mathbb{X}$ is compact metric, and (p_{ij}) is row stochastic. Define $(\mathbb{X}_j, d_j) = (\mathbb{X}, d)$ for each j, and define $W_{ij}(S) = \{w_i(x) : x \in S\}$, $i, j = 1, \ldots, N$. Let $I(i) = \{j : p_{ji} > 0\}$. This embeds us in the more general setup that we now study.

Let

$$\tilde{\mathcal{H}} = \mathcal{H}_1 \times \mathcal{H}_2 \times \mathcal{H}_3 \times \cdots \times \mathcal{H}_N,$$

and endow $\tilde{\mathcal{H}}$ with the metric $\tilde{h}$ defined by

$$\tilde{h}\left((A_1, A_2, \ldots, A_N), (B_1, B_2, \ldots, B_N)\right) = \max\left\{h_j(A_j, B_j) | j = 1, 2, \ldots, N\right\}.$$

Then it is readily demonstrated by the enthusiatic reader that $(\tilde{\mathcal{H}}, \tilde{h})$ is a compact metric space.

We think of $\tilde{\mathcal{H}}$ as consisting of a stack of clipped planes $\mathbb{X}_1, \mathbb{X}_2, \ldots, \mathbb{X}_N$ with a point in $\tilde{\mathcal{H}}$ being the N-tuple of one image in each plane. This idea is illustrated in Color Plate 41. Define

$$\tilde{W} : \tilde{\mathcal{H}} \to \tilde{\mathcal{H}}$$

by

$$\tilde{W}(A_1, A_2, \ldots, A_N) = \left(\bigcup_{j \in I(1)} w_{1j}(A_j), \bigcup_{j \in I(2)} w_{2j}(A_j), \ldots, \bigcup_{j \in I(N)} w_{Nj}(A_j) \right)$$

For example, in the case $N = 2$ such a mapping might be symbolized

$$\tilde{W} \begin{pmatrix} A_1 \\ A_2 \end{pmatrix} = \begin{pmatrix} \varnothing & W_{12} \\ W_{21} & W_{22} \end{pmatrix} \begin{pmatrix} A_1 \\ A_2 \end{pmatrix} = \begin{pmatrix} W_{12}(A_2) \\ W_{21}(A_1) \cup W_{22}(A_2) \end{pmatrix}$$

We have now arrived at our goal; we are able to characterize $\tilde{W}$ as a contraction mapping on $\tilde{\mathcal{H}}$; we thus have all of the ingredients of a fractal system.

Theorem 3.2 $\tilde{W} : \tilde{\mathcal{H}} \to \tilde{\mathcal{H}}$ *obeys*

$$\tilde{h}\left(\tilde{W}(A), \tilde{W}(B) \right) \le s \tilde{h}(A, B) \quad \forall A, B \in \mathcal{H},$$

where $s = \max\{s_{ij}, (i, j) \in I\}$.

Proof To keep the notation succinct we suppose

$$\tilde{W} = \begin{pmatrix} W_{11} & W_{12} \\ W_{21} & W_{22} \end{pmatrix}$$

Then if $A = (A_1, A_2)$ and $B = (B_1, B_2)$ we have

$$\tilde{h}\left(\tilde{W}(A), \tilde{W}(B) \right)$$

$$= \tilde{h}\left((W_{11}(A_1) \cup W_{12}(A_2), W_{21}(A_1) \cup W_{22}(A_2)) , \right.$$
$$\left. (W_{11}(B_1) \cup W_{12}(B_2), W_{21}(B_1) \cup W_{22}(B_2)) \right)$$
$$= \max\{ h_1(W_{11}(A_1) \cup W_{12}(A_2), W_{11}(B_1) \cup W_{12}(B_2)),$$
$$h_2(W_{21}(A_1) \cup W_{22}(A_2), W_{21}(B_1) \cup W_{22}(B_2)) \}$$
$$\le \max\{ h_1(W_{11}(A_1), W_{11}(B_1)) \vee h_1(W_{12}(A_2), W_{12}(B_2)),$$
$$h_2(W_{21}(A_1), W_{21}(B_1)) \vee h_2(W_{22}(A_2), W_{22}(B_2)) \} \quad \text{(by Lemma 3.1)}$$
$$\le \max\{ s_{11}h_1(A_1, B_1) \vee s_{12}h_2(A_2, B_2), s_{21}h_1(A_1, B_1) \vee s_{22}h_2(A_2, B_2) \}$$
$$\le s(h_1(A_1, B_1) \vee h_2(A_2, B_2)) = s\tilde{h}((A_1, A_2), (B_1, B_2))$$
$$= s\tilde{h}(A, B).$$

The fact that W is a contraction mapping on $\tilde{\mathcal{H}}$ means that we have a fractal system as described in the first section of this chapter. Conditions 1, 2, and 3, of section 1 apply. In this case the underlying space is $\mathbb{X}$. Multiple copies of $\mathbb{X}$ are used to make up a space of models $\mathbb{Y} = \tilde{\mathcal{H}}$. (It may be that we only care about one "plane" in $\tilde{\mathcal{H}}$.) A contraction mapping $\mathrm{O} = \tilde{W}$ on $\tilde{\mathcal{H}}$ is built up out of elementary transformations acting on $\mathbb{X}$. Fixed points of such operators O provide the fractal approximations to the models. In the present case the specific structure is as follows.

Corollary 3.3 *When $s < 1$ there is a unique element*

$$A = (A_1, A_2, \ldots, A_N) \in \tilde{\mathcal{H}}$$

such that

$$A_i = \cup_{j \in I(i)} W_{ij}(A_j) \quad \text{for } i = 1, 2, \ldots, N.$$

That is to say, $\tilde{W}(A) = A$. A is the attractor of the recurrent IFS.

Corollary 3.4 A Collage Theorem for recurrent IFS. *If $B \in \tilde{\mathcal{H}}$ obeys*

$$\tilde{h}\left(B, \tilde{W}(B)\right) \leq \epsilon > 0$$

then

$$h(B, A) \leq \frac{\epsilon}{(1-s)},$$

where A denotes the attractor of the recurrent IFS.

To connect this with the original single-space point-map recurrent IFS, we have

Corollary 3.5 *Let $\{\mathbb{X}; w_i, (p_{ij}), i, j = 1, \ldots, N\}$ be a recurrent IFS with $\mathbb{X}$ compact and the w_i's uniform contractions. Let A be the support of the unique stationary measure μ mentioned in section 2. Then there exist unique compact sets $A_i \subset A, \ i = 1, \ldots, N$ with $A = \cup_{i=1}^{N} A_i$ such that*

$$A_i = \cup_{j:p_{ji}>0} w_i(A_j), \quad i = 1, \ldots, N.$$

In terms of the random walk produced by the Chaos Game Algorithm, as illustrated in section 2, the A_i's may be characterized as follows: for all x, $x \in A_i$ iff for every neighborhood G of x, for almost all trajectories $x_0, w_{i_1}(x_0), w_{i_2}(w_{i_1}(x_0)), \ldots,$ we have $i_n = i$ and $w_{i_n}(\ldots (w_{i_1} x_0) \ldots) \in G$ for infinitely many n. In other words, to "see" A_i, just look at the points along a trajectory that end in map w_i.

Let us use the above ideas to construct a simple machine (fractal system) for modelling some binary subsets of $\square$. The machine is illustrated in Figure X.275. It uses affine transformations $w_i : \square \to \square, i = 1, 2, 3$. Place a copy of the subset $T \subset \square$ on (1), and place *adjustable* subsets $U, V \subset \square$ on (2) and (3) respectively. Adjust the coefficients $a_i, b_i, c_i, d_i, e_i, f_i$ and the two subsets U and V, so that

$$T \simeq w_1(T) \cup w_2(U) \cup w_3(V)$$
$$U \simeq w_1(T) \cup w_2(U)$$
$$V \simeq w_1(T) \cup w_3(V)$$

Assuming that the system can be adjusted so that these equations hold approximately, while the transformations remain suitably contractive, we know that the attractor for the system, our fractal model, which will be represented by a vector of sets $(\hat{T}, \hat{U}, \hat{V})$, will be such that $\hat{T} \simeq T$. This simple machine generates an 18 parameter family of fractal models for subsets of $\square$.

A simple machine for modelling some binary subsets of $\square$.

$$w_i \begin{pmatrix} x \\ y \end{pmatrix} = \begin{pmatrix} a_i & b_i \\ c_i & d_i \end{pmatrix} \begin{pmatrix} x \\ y \end{pmatrix} + \begin{pmatrix} e_i \\ f_i \end{pmatrix}$$

for i = 1, 2, 3

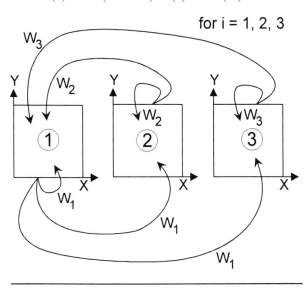

Figure X.275. Imagine that you have control knobs so that you can adjust the affine coefficients.

Another simple point-set fractal system is suggested by the diagram in Figure X.276. This type of machine generates recurrent fractal interpolation functions. The coefficients of the transformations are constrained to make sure that the resulting attractors are graphs of continuous functions, as follows:

$$w_1(x_0, y_0) = (x_0, y_0), \qquad w_3(x_2, y_2) = (x_3, y_3), \qquad w_5(x_2, y_2) = (x_4, y_4)$$
$$w_1(x_2, y_2) = (x_1, y_1), \qquad w_3(x_0, y_0) = (x_2, y_2), \qquad w_5(x_4, y_4) = (x_5, y_5)$$
$$w_2(x_2, y_2) = (x_1, y_1), \qquad w_4(x_4, y_4) = (x_3, y_3), \qquad w_6(x_4, y_4) = (x_5, y_5)$$
$$w_2(x_4, y_4) = (x_2, y_2), \qquad w_4(x_6, y_6) = (x_4, y_4), \qquad w_6(x_6, y_6) = (x_6, y_6).$$

One example of a commercial fractal system, useful for educational purposes and for desktop publishing applications, is the *Desktop Fractal Design System,* [DFDS].[4] This is based on simple recurrent IFS theory using affine transformations in two dimensions. This can be used to illustrate some aspects of modelling using point-set attractors of recurrent IFS built up of affine transformations acting in two dimensions.

---

[4] The Desktop Fractal Design System, Version 2.0, for MacIntosh and IBM PC. Published by Academic Press, 1992.

Figure X.276. A fractal system constrained so that its attractors are graphs of continuous functions.

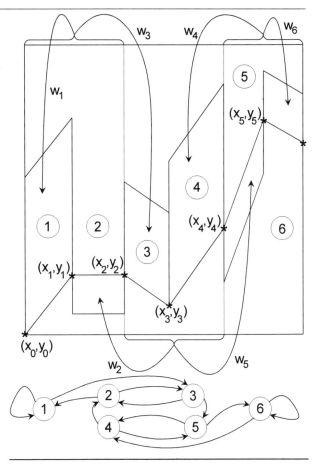

Consider the IFS for a fern, as introduced in Chapter III. We refer to this as Fern 1. Its IFS code is given in Table X.4 and it is shown in Figure X.277.

In Figure X.278 we illustrate how one can create the same picture of Fern 1 by using a duplicate copy of the space □ with the same IFS acting on that space, and allowing a transformation to map from one copy of the space to the other.

Look closely at the structure of Fern 1. Notice how the main fronds alternate as we go up the main stem. As we saw in Chapter III, this same structure is repeated at all scales. If we zoom in on the tiniest frond, we will find that all the frondlets alternate up their respective stems.

Now suppose we modify our more complex recurrent system by adjusting the IFS for the first copy of the fern, that is we adjust the affine coefficients so that Fern 1 on the first copy of □ becomes Fern 2, as illustrated in Figure X.279 and Table X.5.

Notice that the fronds in Fern 2 are opposed instead of alternating. Finally, in Figure X.280, we show the effect of the modifications on the attractor as a whole. The original copy of the fern now has opposed fronds, some of which are made of

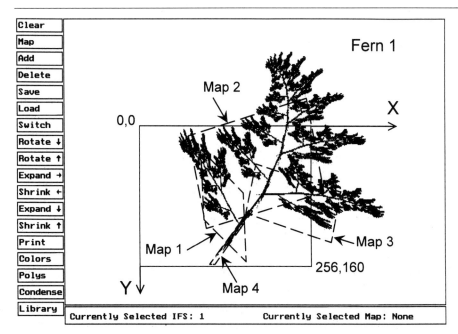

Figure X.277. A binary fern image is the point-set attractor for a single fractal system consisting of an IFS of affine maps acting in $\mathbb{R}^2$. It is closely related to more complex fractals based on recurrent IFS theory.

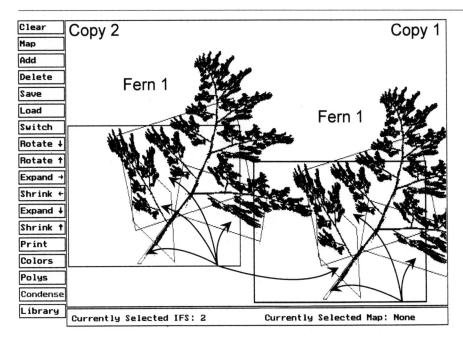

Figure X.278. This shows how the attractor in Figure X.277 is also the attractor of a recurrent fractal system, involving two copies of □ instead of one. The attractor of the IFS appears as *part of* the attractor of a recurrent IFS.

Figure X.279. The IFS acting on the first copy of □ is modified to create a new fern, Fern 2.

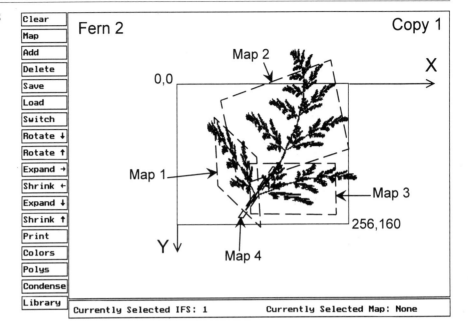

Figure X.280. The result of the modification of the first copy of the fern is to fundamentally alter the structure of the part of the attractor appearing on the first copy of □.

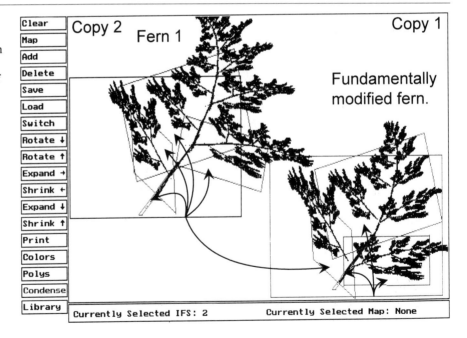

Table X.4.   IFS code for Fern 1.

w	a	b	c	d	e	f
1	-0.02	0.39	-0.31	0.30	96	108
2	0.68	0.17	-0.17	0.68	73	11
3	0.05	-0.80	-0.22	-0.19	286	131
4	0.02	-0.36	0.00	0.38	160	97

Table X.5.   IFS code for Fern 2.

w	a	b	c	d	e	f
1	-0.02	0.39	-0.31	0.30	64	116
2	0.64	0.16	-0.16	0.64	67	15
3	0.00	-0.80	-0.22	0.01	237	147
4	0.20	-0.33	0.00	0.35	144	96

smaller ferns with alternating fronds. We are learning to control a recurrent fractal model!

Finally what happens if we feed the first copy back into some of the fronds on the second copy? The results of such an adjustment are shown in Figure X.281. In fact, in this example, the bottom-most fronds on each fern are affine transformations of the other fern.

Color Plate 36 shows a fern image, and zooms on it, produced by a fractal system. In this case the fractal system involves four copies of □ and a recurrent structure of affine transformations of the type discussed in this section. In this section each of three successive zooms on one of the ferns reveals new structure.

Another example of an attractor, consisting of four point-sets at once, in shown in Figure X.282. It is good to think of the vector of sets as a single entity.

4   Fractal Systems with Vectors of Measures as Their Attractors

In this section we are concerned with the construction of fractal systems, using recurrent IFS, whose attractors are vectors of measures. Our goal is to show how the collage theorem and associated machinery can be set up to enable fractal modelling using the measure theoretic attractors of certain recurrent IFS. That is, recurrent IFS can be used to make fractal systems suitable for modelling measures.

Recall what happens in the case of an IFS with probabilities: for example, suppose $\mathbb{X} = \square \subset \mathbb{R}^2$, $\{\mathbb{X}; w_1, ..., w_N; p_1, \ldots, p_N\}$ is an IFS with probabilities, as in section 6 in Chapter IX, with the w_i's contractive affine, and $\mathbb{Y} = \mathcal{P}(\mathbb{X})$; then we can define a corresponding fractal system by taking O to be the associated Markov

Figure X.281. A fractal system, based on a recurrent IFS structure, produces two ferns whose structures are woven together. Neither could exist without the other!

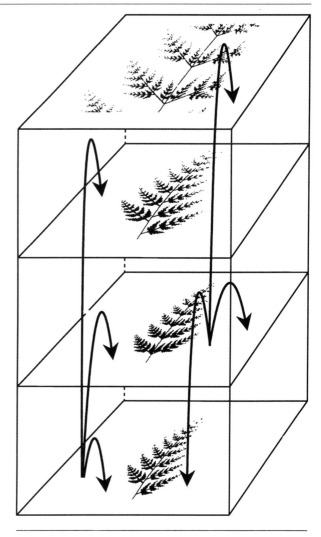

operator as defined in section 6 in Chapter IX; that is $O : \mathbb{Y} \to \mathbb{Y}$ is defined by

$$O(v) = M(v) = p_1 v \circ w_1^{-1} + p_2 v \circ w_2^{-1} + \cdots + p_N v \circ w_N^{-1}.$$

Then we know from the theory of Chapter IX that O is contractive in the Hutchinson metric.

What is the analagous structure in the case of recurrent IFS? We use the notation of the previous section, but restrict attention to the case of measures supported on three copies of $\mathbb{X}$. Instead of using the measure space $\mathcal{P}(\mathbb{X})$, we work with the space of normalized Borel measures $\tilde{\mathcal{P}}$ that are defined on Borel subsets $\tilde{S}$ of $\mathbb{X} \times \mathbb{X} \times \mathbb{X}$ of the special form

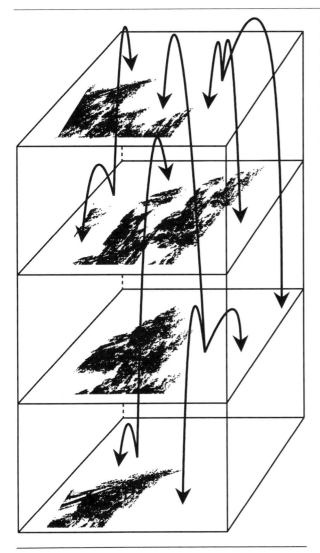

Figure X.282. Affine maps, linking four copies of □, lead to a recurrent structure consisting of four linked sets. Adjustments of the coefficients lead to a wide variety of fractal modes in this fractal system.

$$\tilde{S} = (S_1, S_2, S_3) \in \mathcal{B}(\mathbb{X}) \times \mathcal{B}(\mathbb{X}) \times \mathcal{B}(\mathbb{X}).$$

where $\mathcal{B}(\mathbb{X})$ denotes the Borel subsets of $\mathbb{X}$. A point in $\tilde{\mathcal{P}}$ is described by a vector of measures

$$\tilde{\mu} = (\mu_1, \mu_2, \mu_3),$$

where each μ_i is a Borel measure on $\mathbb{X}$, and

$$\mu_1(\mathbb{X}) + \mu_2(\mathbb{X}) + \mu_3(\mathbb{X}) = 1.$$

Then the measure of vector of subsets $\tilde{S} = (S_1, S_2, S_3)$ is

$$\tilde{\mu}(\tilde{S}) = \mu_1(S_1) + \mu_2(S_2) + \mu_3(S_3).$$

We define a generalized Hutchinson metric $\tilde{d}$ on $\tilde{\mathcal{P}}$ by

$$\tilde{d}(\tilde{\mu}, \tilde{v}) = \sup\left\{\sum_i \left(\int_{\mathbb{X}} f_i \, d\mu_i - \int_{\mathbb{X}} f_i \, dv_i\right) : \tilde{f} = (f_1, f_2, f_3) \in L\right\},$$

for all $\tilde{\mu}, \tilde{v} \in \tilde{\mathcal{P}}$, where

$$L = \left\{\tilde{f} = (f_1, f_2, f_3) : f_i : \mathbb{X} \to \mathbb{R} \text{ continuous,}\right.$$

$$\left.\text{and } |f_i(x) - f_i(y)| \le d(x, y) \quad \forall x, y \in \mathbb{X}, i = 1, 2, 3\right\}.$$

Then, from the fact that $(\mathcal{P}(\mathbb{X}), d_H)$ is a compact metric space whenever $(\mathbb{X}, d)$ is a compact metric space, it follows that $(\tilde{\mathcal{P}}, \tilde{d})$ is a compact metric space too.

We are now ready to describe generalized Markov operators $\tilde{M}$,

$$\tilde{M} : \tilde{\mathcal{P}} \to \tilde{\mathcal{P}},$$

that are associated with recurrent IFS. We illustrate the idea for recurrent IFS such as the one in Table X.1, which is the subject of Figures X.268 and X.274. We simply define

$$\tilde{M}(\tilde{\mu}(\tilde{S})) = \left(\sum_i p_{i1}\mu_i\left(w_1^{-1}(S_1)\right), \sum_i p_{i2}\mu_i\left(w_2^{-1}(S_2)\right), \sum_i p_{i3}\mu_i\left(w_3^{-1}(S_3)\right)\right)$$

Our general expectation is that such operators are contractions on $(\tilde{\mathcal{P}}, \tilde{d})$ when the transformations w_i are contractions.

We demonstrate that $\tilde{M}$ has contractivity factor s when $p_{ij} = p_j$, and the w_i's are contractive with contractivity factor $0 \le s < 1$. We will need the identity

$$\int_{\mathbb{X}} f_j d\left(\tilde{M}(\tilde{v})_j\right) = \sum_i p_{ij} \int_{\mathbb{X}} f_j\left(w_j(x)\right) dv_i(x),$$

which follows from a change of variables. Then

$$\tilde{d}\left(\tilde{M}(\tilde{v}), \tilde{M}(\tilde{\mu})\right)$$

$$= \sup\left\{\sum_j \left(\int_{\mathbb{X}}(f_j d(\tilde{M}(\tilde{\mu})_j)) - \int_{\mathbb{X}} f_j d(\tilde{M}(\tilde{v})_j)\right) : \tilde{f} \in L,\right.$$

$$= \sup\left\{\int_{\mathbb{X}} \sum_{ij} p_{ij} f_j(w_j(x)) \, d\mu_i - \int_{\mathbb{X}} \sum_{ij} p_{ij} f_j(w_j(x)) \, dv_i : \tilde{f} \in L,\right.$$

$$= \sup\left\{\int_{\mathbb{X}} \sum_{ij} p_j f_j(w_j(x)) \, d\mu_i - \int_{\mathbb{X}} \sum_{ij} p_j f_j(w_j(x)) \, dv_i : \tilde{f} \in L\right\}.$$

Let, for $x \in \mathbb{X}$,

$$\hat{f}(x) = \left(s^{-1} \sum_j p_j f_j(w_j(x)), \, s^{-1} \sum_j p_j f_j(w_j(x)), \, s^{-1} \sum_j p_j f_j(w_j(x)) \right).$$

Then $\hat{f} \in L$. Let

$$\hat{L} = \{\hat{f} \in L : \hat{f} \text{ is of the special form of } \hat{f} \text{ above, for some } f \in L\}.$$

Then we can write

$$\tilde{d}(\tilde{M}(\tilde{\nu}), \tilde{M}(\tilde{\mu})) = \sup \left\{ s \int \tilde{f} \, d\tilde{\mu} - s \int \tilde{f} \, d\tilde{\nu} : \tilde{f} \in \hat{L} \right\}.$$

Since $\hat{L} \subset L$, it follows that

$$\tilde{d}(\tilde{M}(\tilde{\nu}), \tilde{M}(\tilde{\mu})) \leq s\tilde{d}(\tilde{\nu}, \tilde{\mu}).$$

This completes the demonstration of the contractivity of the recurrent "multilevel" Markov operator, in the special case $p_{ij} = p_j$.

However, in the general case where the transition matrix $P = (p_{ij})$ is not constrained as above, the contractivity of the Markov operator is controlled not only by the contractivities of the w_i's but also by the behavior of the transition matrix. To understand this, consider what happens when the space $\mathbb{X}$ consists of a single point. Then $\tilde{\mathcal{P}}$ becomes the set

$$\{\tilde{\mu} = (\mu_1, \mu_2, \mu_3) \in \mathbb{R}^3 : \mu_1 + \mu_2 + \mu_3 = 1, \, \mu_1 \geq 0, \, \mu_2 \geq 0, \, \mu_3 \geq 0\}.$$

That is, a typical measure $\tilde{\mu} \in \tilde{\mathcal{P}}$ is described by a vector of length three and total mass one. The Hutchinson distance between two such vectors is simply the sum of the absolute values of the differences between the components as follows:

$$\tilde{d}\,(\tilde{\mu} = (\mu_1, \mu_2, \mu_3), \tilde{\nu} = (\nu_1, \nu_2, \nu_3)) = |\mu_1 - \nu_1| + |\mu_2 - \nu_2| + |\mu_3 - \nu_3|.$$

Then the Markov operator $\tilde{M} : \tilde{\mathcal{P}} \to \tilde{\mathcal{P}}$ is given by

$$\tilde{M}(\tilde{\mu}) = (\mu_1, \mu_2, \mu_3) P = (\mu_1 p_{11} + \mu_2 p_{21} + \mu_3 p_{31}, \, \mu_1 p_{12} + \mu_2 p_{22} + \mu_3 p_{32},$$
$$\mu_1 p_{13} + \mu_2 p_{23} + \mu_3 p_{33}),$$

where

$$P = \begin{bmatrix} p_{11} & p_{12} & p_{13} \\ p_{21} & p_{22} & p_{23} \\ p_{31} & p_{32} & p_{33} \end{bmatrix}.$$

It follows that

$$\tilde{d}\left(\tilde{M}(\tilde{\nu}), \tilde{M}(\tilde{\mu}) \right) \leq \tilde{d}(\tilde{\nu}, \tilde{\mu}).$$

However we cannot, in the general case, assert a contractivity factor smaller than one, as can be seen by considering the considering the case

$$P = \begin{bmatrix} 1 - 2\epsilon & \epsilon & \epsilon \\ \epsilon & 1 - 2\epsilon & \epsilon \\ \epsilon & \epsilon & 1 - 2\epsilon \end{bmatrix},$$

where ϵ is very small and positive. For this transition matrix, we find that

$$\tilde{d}\left(\tilde{M}((1,0,0)), \tilde{M}((0,0,1))\right) = (1 - 3\epsilon)\tilde{d}\left((1,0,0), (0,0,1)\right).$$

However, so long as the Markov operator has a unique fixed point, there always exist metrics such that the operator is contractive, in fact, with an arbitrarily small contractivity factor. The problem is the identification of such a metric, one that is simultaneously intuitively or visually accessible.

The operator $M : \mathcal{P}(\mathbb{X}) \rightarrow \mathcal{P}(\mathbb{X})$, associated with an IFS with probabilities,

$$M(v) = p_1 v \circ w_1^{-1} + p_2 v \circ w_2^{-1} + \cdots + p_N v \circ w_N^{-1},$$

is an example of a measure preserving transformation on a metric space, constructed from contractive transformations.

Given a set of spaces, $\{\mathbb{X}_k : k = 1, 2, \ldots, N\}$, we can construct measure preserving transformations $M_{ij} : \mathcal{P}(\mathbb{X}_i) \rightarrow \mathcal{P}(\mathbb{X}_j)$, according to

$$M_{ij}(v) = p_{ij1} v \circ w_{ij1}^{-1} + p_{ij2} v \circ w_{ij2}^{-1} + \cdots + p_{ijN_{ij}} v \circ w_{N_{ij}}^{-1},$$

where $w_{ijn} : \mathbb{X}_i \rightarrow \mathbb{X}_j$ for $n \in \{1, 2, \ldots, N_{ij}\}$ is contractive between the metric spaces $(\mathbb{X}_i, d_i)$ and $(\mathbb{X}_j, d_j)$ according to

$$d_j\left(w_{ijn}(x), w_{ijn}(y)\right) \leq s \cdot d_i(x, y) \quad \text{for all } x, y \in \mathbb{X}_i,$$

and the p_{ijn}'s being positive numbers such that

$$\sum_n p_{ijn} = 1.$$

(One can think of $\{\mathbb{X}_i, \mathbb{X}_j; w_{ijn}, p_{ijn}, n = 1, 2, \ldots, N_{ij}\}$ as being an "IFS" that acts between the spaces $\mathbb{X}_i$ and $\mathbb{X}_j$.) We now introduce an $N \times N$ transition matrix $P = (p_{ij})$, that is row stochastic, and gives the probability of transiti n from space $\mathbb{X}_i$ to $\mathbb{X}_j$. Then we can define a Markov operator $\tilde{M} : \tilde{\mathcal{P}} \rightarrow \tilde{\mathcal{P}}$, where $\tilde{\mathcal{P}} = \mathcal{P}(\mathbb{X}_1) \times \mathcal{P}(\mathbb{X}_2) \times \cdots \times \mathcal{P}(\mathbb{X}_N)$, by

$$\tilde{M}(\mu_1, \mu_2, \ldots, \mu_N) = \left(\sum_i p_{i1} M_{i1}(\mu_i), \sum_i p_{i2} M_{i2}(\mu_i), \ldots, \sum_i p_{iN} M_{iN}(\mu_i)\right).$$

In general we expect such operators to be contractive, and to provide the basis for useful fractal systems for modelling vectors of measures, and in particular, grayscale photographs. An example of such a system is VRIFS™,[5] which is an interactive image modelling system that uses a fractal system built from operators of the type of $\tilde{M}$

---

[5] VRIFS™, which stands for Vector Recurrent Iterated Function System, is an interactive measure theoretic fractal system for modelling images in two dimensions. It runs on Sun workstations and uses four-dimensional vectors of measures under affine control. It is available from Iterated Systems Inc., Commercial Division, 5550-A Peachtree Parkway, Norcross, Georgia, 30092, USA, telephone number (404) 840 0633.

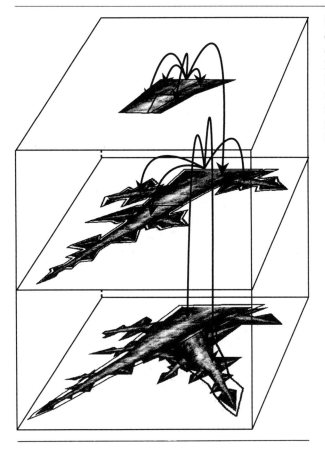

Figure X.283. A special Markov operator acts on a vector of measures. Its attractor is a stack of images, including a split root. Fractal systems of this type are good for fractal modelling of grayscale images.

above, with all transformations being two dimensional affine transformations. An example of an image generated by VRIFS™is shown in Figure X.283. See also Color Plates 37, 38, and 39.

5 References

This section comprises a list of references to fractal systems, new to the second edition of this book.

L. Anson and M. Barnsley, Graphics Compression Technology, Sunworld (October 1991), 42–52.

M. Barnsley, (1) Fractals and (2) Chaos, BCS Conference Documentation Displays Group, State of the Art Seminar, Fractals and Chaos, 6th–7th December 1989, London (The British Computer Society, 13 Mansfield Street, London W1M 0BP).

M. Barnsley, J. Elton, and D. Hardin, Recurrent Iterated Function Systems, *Constructive Approximation (1989) 5*: 3–31.

M. Barnsley, Desktop Fractal Design System, Academic Press, Boston, Version 1.0 (1989), Version 2.0 (1992).

M. Barnsley and A. Sloan, Method and Apparatus for Processing Digital Data, United States Patent #5,065,447.

M. Barnsley, L. P. Hurd, Fractal Image Compression, A. K. Peters, Wellesley, Massachusetts, (1993).

J. M. Beaumont, Image Data Compression Using Fractal Techniques, *B.T. Journal*, Vol. 9, Number 4, (1991), 93–108.

M. M. Cabre, Fractal Series and Infinite Products, Preprint, Documento de Trabajo 9021, Facultad de Ciencias Economicas Y Empresariales, - Universidad Complutense Campus de Somosaguas. 28023 - Madrid.

A. J. Crilly, R. A. Earnshaw, H. Jones, Fractals and Chaos, Springer Verlag, London (1991).

K. Culik II and S. Dube, Rational and Affine Expressions for Image Description, Discrete Applied Mathematics to appear. *Tech. Report TR 90001*, Univ. of South Carolina (1990).

K. Culik II and S. Dube, Affine Automata and Related Techniques for Generation of Complex Images, *Theoretical Computer Science*, to appear.

K. Culik and S. Dube, Balancing Order and Chaos in Image Generation, Proceeding on the 18th International Colloquium on Automata, Languages and Programming, Madrid Spain, July 1991, in: *Lecture Notes in Computer Science 510*, 600–614, Springer-Verlag (1991).

K. Culik II and S. Dube, New Techniques for Image Generation and Compression, Conference on New Trends and Results in Computer Science, Graz, Austria, June 1992, in: *Lecture Notes in Computer Science 555*, 69–90, Springer-Verlag (1991).

F. Dudbridge, Image Approximation by Self Affine Fractals, Ph.D. Thesis, Department of Computing, Imperial College of Science, Technology, and Medicine, London SW7 2BZ, England.

A. Jacquin, Image Coding Based on a Fractal Theory of Iterated Contractive Image Transformations, *IEEE Transactions on Image Processing, 1* (1992) 18–30.

Y. Fisher, R. D. Boss, E. W. Jacobs, Fractal Image Compression, to appear in Data Compression, R. Storer (ed.) Kluwer Academic Publishers, Norwell, MA.

G. A. Edgar, Measure, Topology, and Fractal Geometry, Springer-Verlag (1990).

J. S. Geronimo, D. Hardin, Fractal Interpolation Surfaces and a Related 2-D Multiresolution Analysis, 1990 Preprint, School of Mathematics, Georgia Institute of Technology, Atlanta, Georgia, USA.

J. C. Hart, Thomas A. DeFanti, Efficient Antialiased Rendering of 3-D Linear Fractals, *Computer Graphics, 25* (July 1991), 91–100.

A. Horn, IFSs and the Interactive Design of Tiling Structures, in Fractals and Chaos (Edited by A.J. Crilly, R.A. Earnshaw, H. Jones) Springer-Verlag, London 1991.

Images Incorporated, by Iterated Systems, Inc., Norcross, Georgia (1992).

T. Kaijser, On a new contraction condition for random systems with complete connections,

Rev Roum Math Pure Appl, Vol 24, sid 383–412 (1981). A limit theorem for Markov chains in compact metric spaces with applications to products of random matrices, *Duke Math Journal*, vol 45, sid 311–349 (1978).

R. D. Maudlin, S. C.Williams, Hausdorff dimension in graph directed constructions, *Trans. Amer. Math. Soc. 309* (1988), 811–829.

Microsoft Encarta, by Microsoft Corporation, Bellview, Washington (1993).

D. M. Monroe, F. Dudbridge, Fractal Approximation of Image Blocks, ICASSP-92, Volume 3, Multidimensional Signal Processing, 1992 IEEE International Conference on Acoustics, Speech and Signal Processing.

D. M. Monroe and F. Dudbridge, Fractal Block Coding of Images, *Electronics Letters*, 21st May 1992, Volume 28, No. 11, 1053.

D. Oliver, Fractal Vision, Sams Publishing (A division of Prentice Hall Computer Publishing, Carmel, Indiana), 1992.

H. O. Peitgen, H. Jürgens, and D. Saupe, Chaos and Fractals (New Frontiers in Science), Springer Verlag, London, 1992.

E. R. Vrscay and C. J. Roehrig, Iterated Function Systems and the Inverse Problem of Fractal Construction Using Moments, in Computers and Mathematics (New York, 1989), E. Kaltofen and S. M. Watt, Eds., Springer-Verlag, 250–259.

J. Waite, A Review of Iterated Function System Theory for Image Compression, Preprint, British Telecom Research Laboratories, Martlesham Heath, U.K. (1992).

References

[Amburn 1986] Amburn, P., Grant, E., and Whitted, T., "Managing Geometric Complexity with Enhanced Procedural Methods," *Computer Graphics 20*(4) (August 1986).

[Aono 1984] Aono, M., and Kunti, T. L., "Botanical Tree Image Generation," *IEEE Computer Graphics and Applications 4*(5): 10–33 (May 1984).

[Barnsley 1984] Barnsley, M. F., Geronimo, J. S., and Harrington, A. N., "Geometry and Combinatorics of Julia Sets for Real Quadratic Maps," *Journal of Statistical Physics 37*: 51–92 (1984).

[Barnsley 1985a] Barnsley, M. F. and Demko, S., "Iterated Function Systems and the Global Construction of Fractals," *The Proceedings of the Royal Society of London A399*: 243–275 (1985).

[Barnsley 1985b] Barnsley, M. F., Ervin, V. Hardin, D., and Lancaster, J., "Solution of an Inverse Problem for Fractals and Other Sets," *Proceedings of the National Academy of Science 83*: 1975–1977 (April 1985).

[Barnsley 1985c] Barnsley, M. F., and Harrington, A. N., "A Mandelbrot Set for Pairs of Linear Maps," *Physica D 15*: 421–432 (1985).

[Barnsley 1986] Barnsley, M. F., "Fractal Functions and Interpolation," *Constructive Approximation 2*: 303–329 (1986).

[Barnsley 1987] Barnsley, M. F., Jaquin, A., Reuter, L., and Sloan, A. D., "A Cloud Study." A videotape produced by the Computergraphical Mathematics Laboratory at Georgia Institute of Technology.

[Barnsley 1988a] Barnsley, M. F., Jacquin, A., Reuter, L., and Sloan, A. D., "Harnessing Chaos for Image Synthesis," *Computer Graphics,* SIGGRAPH 1988 Conference Proceedings.

[Barnsley 1988b] Barnsley, M. F., and Sloan, A. D., "A Better Way to Compress Images," *Byte Magazine*: 215–223, January 1988.

[Barnsley 1988c] Barnsley, M. F., Cain, G., and Kasriel, R. K., "The Escape Time Algorithm," Preprint, School of Mathematics, Georgia Institute of Technology, 1988.

[Barnsley 1988d] Barnsley, M. F., Hardin, D. P., "A Mandlebrot Set Whose Boundary Is Piecewise Smooth," to appear in the *Transactions of the American Mathematical Society,* 1988.

[Barnsley 1988e] Barnsley, M. F., and Elton, J., "A New Class of Markov Processes for Image Encoding," *Journal of Applied Probability 20*: 14-32 (1988).

[Barnsley 1988f] Barnsley, M. F., Elton, J., Hardin, D., and Massopust, P., "Hidden Variable Fractal Interpolation Functions," Georgia Institute of Technology Preprint, July 1986, to appear in *SIAM Journal of Analysis* (1988).

[Bedford 1986] Bedford, T. J., "Dimension and Dynamics for Fractal Recurrent Sets," *Journal of the London Mathematical Society* 2(33): 89–100 (1986).

[Billingsley 1965] Billingsley, P., *Ergodic Theory and Information,* John Wiley and Sons, New York (1965).

[Blanchard 1984] Blanchard, P., "Complex Analytic Dynamics On The Riemann Sphere," *Bulletin of the American Mathematical Society 11*: 88–144 (1984).

[Brolin] Brolin, H., "Invariant Sets under Iteration of Rational Functions," *Arkiv För Matematik,* Band 6 nr 6: 103–144.

[Brown 1976] Brown, J. R., *Ergodic Theory and Topological Dynamics,* Academic Press, New York (1976).

[Curry 1983] Curry, J., Garnett, L., and Sullivan, D., "On the Iteration of Rational Functions: Computer Experiments with Newton's Method," *Communications in Mathematical Physics 91*: 267–277 (1983).

[Demko 1985] Demko, S., Hodges, L., and Naylor, B., "Construction of Fractal Objects with Iterated Function Systems," *Computer Graphics 19*(3): 271–278 (July 1985).

[Devaney 1986] Devaney, R., *An Introduction to Chaotic Dynamical Systems,* Addison-Wesley, New York, 1986.

[Dewdney 1987] Dewdney, A. K., "Beauty and Profundity: The Mandelbrot Set and a Flock of Its Cousins Called Julia," *Scientific American 257*: 140–146 (November 1987).

[Diaconis 1986] Diaconis, P. M. and Shahshahani, M., "Products of Random Matrices and Computer Image Generation," *Contemporary Mathematics, 50*: 173–182 (1986).

[Douady 1982] Douady, A., and Hubbard, J., *Comptes Rendus (Paris) 294*: 123–126 (1982).

[Douady 1985] Douady, A. and Hubbard, J. H., "On the Dynamics of Polynomial-like Mappings," *Annales Scientifiques de l'Ecole Normale Supérieure,* 4ᵉ Série, *18*: 287–343 (1985).

[Dunford 1966] Dunford, N., and Schwartz, J. T., *Linear Operators Part I: General Theory,* 3rd edition, John Wiley and Sons, New York, 1966.

[Eisen 1969] Eisen, M., *Introduction to Mathematical Probability Theory,* Prentice Hall, Englewood Cliffs, NJ, 1969.

[Elton 1987] Elton, J., "An Ergodic Theorem for Iterated Maps," *Journal of Ergodic Theory and Dynamical Systems,* 7: 481–488 (1987).

[Elton 1988] Elton, J., "A Simultaneously Contractive Remetrization Theorem for Iterated Function Systems," Georgia Institute of Technology Preprint, 1988.

[Fatou 1919–20] Fatou, P., "Sur les Equations Fonctionelles," *Bulletin Societé Math. France 47*: 161–271 (1919); *48*: 33-94, 208–314 (1920).

[Federer 1969] Federer, H., *Geometric Measure Theory,* Springer-Verlag, New York, 1969.

[Feigenbaum 1979] Feigenbaum, M. J., "The Universal Metric Properties of Nonlinear Transformations," *Journal of Statistical Physics 21*: 669–706 (1979).

[Fournier 1982] Fournier, A., Fussell, D., and Carpenter, L., "Computer Rendering of Stochastic Models," *Communications of the AMC 25*(6) (June 1982).

[Gilbert 1982] Gilbert, W. J., "Fractal Geometry Derived from Complex Bases," *The Mathematical Intelligencer 4*: 78–86 (1982).

[Gleick 1987] Gleick, J., *Chaos: Making a New Science,* Viking Press, New York, 1987.

[Halmos 1974] Halmos, P. R., *Measure Theory,* Springer-Verlag, New York, 1974.

[Hardin 1985] Hardin, D. P., "Hyperbolic Iterated Function Systems and Applications," Ph.D. Thesis, Georgia Institute of Technology, 1985.

[Hata 1985] Hata, M., "On the Structure of Self-Similar Sets," *Japan Journal of Applied Mathematics 2*(2): 381–414 (December 1985).

[Hutchinson 1981] Hutchinson, J., "Fractals and Self-Similarity," *Indiana University Journal of Mathematics 30*: 713–747 (1981).

[Julia 1918] Julia, G., "Memoire sur l'Itération des Fonctions Rationelles," *Journal des Mathematiques Pures et Appliquées 4*: 47–245 (1918).

[Kasriel 1971] Kasriel, R. H., *Undergraduate Topology,* Saunders, Philadelphia, 1971.

[Kawaguchi 1982] Kawaguchi, Y., "A Morphological Study of the Form of Nature," *Computer Graphics 16*(3), (July 1982). SIGGRAPH 1982 Proceedings.

[Lasota 1973] Lasota, A., and Yorke, J. A., "On the Existence of Invariant Measures for Piecewise Monotonic Transformations," *Transactions of the American Mathematical Society 186*: 481–488 (1973).

[Mandelbrot 1982] Mandelbrot, B., *The Fractal Geometry of Nature,* W.H. Freeman and Co., San Francisco, 1982.

[Massopust 1986] Massopust, P., Ph.D. Thesis, Georgia Institute of Technology, 1986.

[May 1976] May, R. B., "Simple Mathematical Models with Very Complicated Dynamics," *Nature 261*: 459–467 (1976).

[Mendelson 1963] Mendelson, B., *Introduction to Topology,* Blackie & Son Limited, London, 1963.

[Miller 1986] Miller, G. S. P., "The Definition and Rendering of Terrain Maps," *Computer Graphics 20*(4), (August 1986).

[Oppenheimer 1986] Oppenheimer, P. E., "Real Time Design and Animation of Fractal Plants and Trees," *Computer Graphics 20*(4), (August 1986).

[Peitgen 1986] Peitgen, H.-O., and Richter, P. H., *The Beauty of Fractals,* Springer-Verlag, Berlin, New York, 1986.

[Rudin 1964] Rudin, W., *Principles of Mathematical Analysis,* 2nd Edition, McGraw-Hill, New York, 1964.

[Rudin 1966] Rudin, W., *Real and Complex Analysis,* McGraw-Hill, New York, 1966.

[Reuter 1987] Reuter, L., "Rendering and Magnification of Fractals Using Iterated Function Systems," Ph.D. Thesis, Georgia Institute of Technology, (December 1987).

[Scia 1987] Scia 1987, see Dewdney 1987.

[Scott 1917] Scott, D. H., *An Introduction to Structural Botany: Part I, Flowering Plants,* A. & C. Black, Ltd., London, 1917.

[Sinai 1976] Sinai, Ya. G., *Introduction to Ergodic Theory,* Princeton University Press, Princeton, 1976.

[Smith 1984] Smith, A. R., "Plants, Fractals, and Formal Languages," *Computer Graphics 18*(3): 1–10 (July 1984). SIGGRAPH 1984 Proceedings.

[Strahle 1991] Turbulent Combustion Data Analysis Using Fractals, *AIAA Journal 29* (3) (1991).

[Sullivan 1982] Sullivan, D., "Quasi-Conformal Homeomorphisms and Dynamics,"I, II, and III. Preprints, Institut des Hautes Etudes Scientifiques, Bures-sur-Yvette, France, 1982.

[Vrscay 1986] Vrscay, E. R., "Julia Sets and Mandelbrot-like Sets Associated with Higher Order Schröder Iteration Functions: A Computer-Assisted Study," *Mathematics of Computation 46*: 151–169 (1986).

[Withers 1987] Withers, W. D., "Calculating Derivatives with Respect to Parameters of Average Values in Iterated Function Systems," *Physica D 28*: 206–214 (1987).

Selected Answers

Answers to Chapter II, section 1

II.1.8 The Riemann sphere is not a vector space because of the action of the 'Point at Infinity'. To be a vector space, each point must have an additive inverse, that is, another point which summed to it yields the additive identity element 0. Since there is no such inverse for ∞, the space is not a vector space. Note that the usual check $\alpha(x + y) = \alpha x + \alpha y$ does hold however.

The examples in 1.7 are not vector spaces for the simple reason that they are not closed under algebraic operations. That is, multiply any element in **X** by a large enough scalar, and it will not be in the space. The example in 1.6 has the same problem. If one defines multiplication by scalars in codespace in a reasonable way, namely multiply the distances by multiplying the codes and 'carrying' one will find a sequence for which there are no codes. (Multiply $x = (N - 1)(N - 1) \cdots$ by 2).

II.1.9 By definition $\{x\} \subset \mathbf{X}$ if every element in $\{x\}$ is in **X**. The only element in $\{x\}$ is x which by assumption is in **X**. Hence $\{x\} \subset \mathbf{X}$.

Answers to Chapter II, section 2

II.2.1 We use extensively the inequality $|x| + |y| \geq |x + y|$. It is proven by separating it into three cases: a) $x, y \geq 0$. In this case $|x| = x, |y| = y$, and since the sum of two positive numbers is positive,$|x + y| = x + y = |x| + |y|$. b) $x, y \geq 0$. In this case $|x| = -x, |y| = -y$. Since the sum of two negative numbers is negative, $|x + y| = -x - y = |x| + |y|$. c) $x \geq 0, y \leq 0$. By changing the roles of x, y this proves the other possibility as well. $|x| = x, |y| = -y$. $|x + y| = |x - (-y)| = ||x| - |y||$. This latter number is either equal to $|x| - |y|$ or its negative, and both are less than or equal to $|x| + |y|$ with equality only if one or the other are zero. Hence $|x + y| \leq |x| + |y|$. We proceed to verify the metric space hypotheses:

(a)

(i) $d(x, y) = |x - y| = |-(x - y)| = |y - x| = d(y, x)$.
(ii) $x - y \neq 0$ unless $x = y$. It is not infinite if x, y are not infinite. Since taking the absolute value does not alter either of these properties $0 < d(x, y) < \infty$.
(iii) $d(x, x) = |x - x| = |0| = 0$.
(iv) $d(x, y) = |x - y| = |x - z + z - y| \leq |x - z| + |z - y|$ by means of the proof above.

(b) (i), (ii), (iii) can be verified by multiplying everything in part a) by 2. (iv) $d(x, y) = 2|x - y| = 2|x - z + z - y| \leq 2(|x - z| + |z - y| = 2|x - z| + 2|z - y|$.
(c) All properties can be done by a) substituting x^3, y^3, z^3 everywhere in the above arguments.

II.2.3 $d(x, y) = |xy|$ is not a metric for $\mathbb{R}$. Choose $x = 0$ and $y \neq 0$. Then $x \neq y$, but $d(x, y) = |0 \cdot y| = 0$, which violates property (i).

II.2.4 On the punctured plane $\mathbb{R}^2 \setminus \mathcal{O}$ define $d(x, y) = |r_1 - r_2| + \theta$. Then for two points x, y, we have

(1) $d(x, y) = |r_1 - r_2 + \theta| = |r_2 - r_1| + \theta = d(y, x)$ since θ is still the smallest angle subtended by the two rays.
(2) If $x \neq y$ and $|r_1 - r_2| = 0$ then $\theta \neq 0$ so $d(x, y) > 0$ if $x \neq y$. $\theta \leq \pi$ so that for $d(x, y) = \infty$ we must have either r_1 or r_2 infinite. But $r_1 = x_1^2 + x_2^2$ and $r_2 = y_1^2 + y_2^2$ so if either is infinite so is either x or y. Hence $0 < d(x, y) < \infty$ if $x \neq y$.

(3) $d(x, x) = |r - r| + 0 = 0 + 0 = 0$.

(4) $d(x, z) + d(z, y) = |r_1 - r_3| + |r_3 - r_2| + \theta_1 + \theta_2$. $\theta_1 + \theta_2 > \theta$ if the ray to z is not between the other two, and $\theta_1 + \theta_2 = \theta$ otherwise, so $d(x, z) + d(z, y) \geq |r_1 - r_3| + |r_3 - r_2| + \theta \geq |r_1 - r_3 + r_3 - r_2| + \theta = |r_1 - r_2| + \theta = d(x, y)$ by problem 1 above.

II.2.5 If x, y are not equal, then there is an x_i, y_i pair which do not agree. Hence $|x_i - y_i|/(N + 1)^i \neq 0$. Since all the terms in the sum are positive or zero, there is no cancelling, and therefore $d(x, y) \neq 0$. We can switch the order in each term because of the absolute value signs, consequently $d(x, y) = d(y, x)$. The series converges since the numerators are less than $2N$ so the series is dominated by

$$2N \cdot \sum_{i=1}^{\infty} \frac{1}{(N + 1)^i}$$

which converges for all $N < \infty$. Because the series converges, the proof of all the metric requirements, except the above, follow from problem on each term, and therefore for the whole series.

II.2.6 The easiest way to show that this equation for a distance measure satisfies metric properties turns out to be that we show that it really is a Euclidean metric, in the usual sense. Let x be an element of Σ the code space on N symbols. Then the series

$$\sum_{i=1}^{\infty} \frac{x_i}{(N + 1)^i}$$

converges to a number in $[0, 1]$. In fact it is the $(N + 1)$-ary expansion of a real number in this range, having the property that it has no digits equal to N, because the code space symbols range from $0, \ldots, N - 1$. For instance, when $N = 9$ we have

$$\sum_{i=1}^{\infty} \frac{x_i}{10^i} = \frac{x_1}{10} + \frac{x_2}{100} + \cdots = .x_1 x_2 x_3 \ldots$$

Then the mapping $\Sigma \to [0, 1]$ given by

$$x \mapsto \tilde{x} = \sum_{i=1}^{\infty} \frac{x_i}{(N + 1)^i}$$

is a metric equivalence, with constants c_1, c_2 in the definition equal to 1. Using this map, let $(x, y), (u, v)$ be two points in the space of shifts on N symbols. By the above equivalence, we have corresponding points $(\tilde{x}, \tilde{y}), (\tilde{u}, \tilde{v})$ in the square $[0, 1] \times [0, 1] \subset \mathbb{R}^2$. Since the series converge, we have

$$\tilde{x} - \tilde{u} = \sum_{i=1}^{\infty} \frac{x_i}{(N + 1)^i} - \sum_{i=1}^{\infty} \frac{u_i}{(N + 1)^i} = \sum_{i=1}^{\infty} \frac{(x_i - u_i)}{(N + 1)^i}$$

The metric in the statement of the problem can therefore be written as

$$d((x, y), (u, v)) = \sqrt{(\tilde{x} - \tilde{u})^2 + (\tilde{y} - \tilde{v})^2}$$

which is Euclidean distance in $\mathbb{R}^2$, and hence a metric.

II.2.7 In body space, the shortest distance between two points is the length of the path through the body. The shortest distance from a point to itself is then 0. The shortest distance between two points is the length of the path which is also the shortest path in reverse order. The length of a path between two distinct points cannot be zero, and a path of infinite length inside the body must not be the shortest path between two points. Finally the shortest path between two points in the body is less than the path which takes a detour to a third point unless that point is on the shortest path already. Hence $d(x, y)$ is a metric.

II.2.8 For a function d which is not a metric, see for example exercise 2.3. For the annulus to look like a cylinder, we must have the distance around any concentric path look the same. This implies that we need the distance around all concentric paths equal. Use the angle subtended between two points to do this. To look like height, we need a linear radial distance, use the radius. If we want to look like the surface of a cylinder with Euclidean metric (unrolled), we thus use $\sqrt{(r_1 - r_2)^2 + \theta^2}$. The proof that this is a metric follows exercise 2.4.

II.2.9 That a distance measure formed by great circle distance is a metric is true on any two-sphere. We verify the conditions as follows:

(i) The plane defined by the center C of the sphere and any two distinct points x, y on its surface cuts the sphere on the great circle path between the two points (that is its definition), so there is one and only one great circle distance between the two points. This distance is $r\theta$ where θ is the interior angle xCy. Hence $d(x, y) = 0$ if and only if $\theta = 0$ that is $x = y$.

(ii) Because we are dealing with interior angles, the angle xCy and the angle yCx are equal hence $d(x, y) = d(y, x) > 0$. The maximum value of an interior angle is π, hence $d(x, y) \leq r\pi < \infty$.

(iii) To prove the triangle inequality, we note first that if x, y, z lie on the same great circle, then if z is interior to x, y, $d(x, z) + d(z, y) = d(x, y)$, whereas if it is exterior then since an exterior angle is larger than an interior one, $d(x, z) + d(z, y) > d(x, y)$. Consequently, we deal only with z not on the great circle defined by x, y. For this case, the $\mathbb{R}^3$ distance between the points x and z is given by $\sqrt{r_1^2 + r_2^2 - 2r_1 r_2 \cos\theta}$ by the law of cosines. Without loss of generality, $r_1 = r_2 = 1$ and the square root is monotonic increasing, so that this distance is proportional to $1 - \cos\theta$ and hence decreases for decreasing θ. Consider the plane triangle in $\mathbb{R}^3$ defined by x, y, z. By moving z towards the line $\overline{xy}$, we decrease the distances between x and z and between z and y (we are moving down the perpendicular from z). Since the above law of cosines formula insists that these distances can only decrease if the corresponding angles xCz and yCz decrease, the distances on the sphere are also strictly decreasing. This implies that $d(x, z) + d(y, z) > d(x, y)$ in this case. Hence the triangle inequality holds for great circle distance on any sphere. When this sphere is interpreted as $\hat{\mathbb{C}}$, the great circle between ∞ and any other point is along the projection of vertical lines in the complex plane. Hence since $1 + i$ is above the real axis, its distance to infinity is less than that of zero, which is $\pi/2$.

II.2.10 Suppose that we have

$$c_1 d_1(x, y) \leq d_2(x, y) \leq c_2 d_1(x, y)$$

where c_1, c_2 are positive real numbers. Using the first inequality, we have

$$d_1(x, y) \leq \frac{1}{c_1} d_2(x, y)$$

and the second gives

$$\frac{1}{c_2} d_2(x, y) \leq d_1(x, y)$$

Let $e_1 = \frac{1}{c_2}$ and $e_2 = \frac{1}{c_1}$, and combine the above two inequalities, we get

$$e_1 d_2(x, y) \leq d_1(x, y) \leq e_2 d_2(x, y)$$

as desired.

II.2.11 Yes in both cases. Let $d_1(x, y)$ be Manhattan metric, $d_2(x, y)$ be Euclidean, and use, for instance $c_1 = 1/2, c_2 = 2$. Since the Manhattan metric is at most $\sqrt{2}$ times the Euclidean metric (when they lie on a line at a 45° angle) these constants will do fine.

II.2.12 Two metrics are uniformly equivalent if and only if there are constants c_1, c_2 such that

$$c_1 d_1(x, y) \leq d_2(x, y) \leq c_2 d_1(x, y)$$

for every pair x, y in the space. To see that the metric $d(x, y) = |r_1 - r_2| + \theta$ is not equivalent to the Euclidean metric on the annulus, choose d_1 to be the Euclidean metric, and d_2 to be this one. Let x, y be on the real line $x < 0$ and $y > 0$. Then $d_2(x, y) > \theta = \pi$. Regardless of the value of c_2 by moving x, y towards the origin we can insure that $d_1(x, y)$ is less than any ϵ by letting $x = (-\epsilon/3, 0)$ and $y = (\epsilon/3, 0)$. Hence $c_2 d_1(x, y) < c_2\epsilon$. Then if $\epsilon < c_2/\pi$ the inequality fails. Hence the two metrics are not equivalent.

II.2.13 Use $h(x) = x - 1$. Then $\tilde{d}_1(h(x), h(y)) = 2 \cdot |(x-1), (y-1)| = 2 \cdot |x - y|$. This metric is equivalent to $d_1(x, y)$ by using $c_1 = 1/4$, $c_2 = 1$.

II.2.14 To show that ($\blacksquare$,Manhattan) and ($\blacksquare$,Euclidean) are equivalent metric spaces, use $h(x) = x$. It is easiest to see the equivalence by making the substitution $x' = 0$, $y' = y - x = y_1 - x_1$, $y_2 - x_2$. Then the Euclidean distance $d(x, y) = d(\mathcal{O}, y') = \sqrt{y_1'^2 + y_2'^2} = r$. The Manhattan distance in the same notation is $|r \cos \theta + r \sin \theta|$. Maximizing the last expression with respect to θ yields $\theta = \pi/2$ or $\cos \theta = \sin \theta = \sqrt{2}/2$. So the maximum difference in size between the two metrics is $d_M = \sqrt{2} d_E$. Then

$$\frac{1}{\sqrt{2}} d_M(x, y) \le d_E(x, y) \le d_M(x, y).$$

II.2.15 Using $h(z) = h(z_1 + iz_2) = (z_1, z_2)$, we have $\tilde{d}(x, y) = d_E(x, y)$ on $\mathbb{R}^2$. The above problem gives the metric equivalence for this with the Manhattan metric.

II.2.16 Suppose they were equivalent. Then there are constants c_1, c_2 such that

$$c_1 d_1(x, y) \le d_2(x, y) \le c_2 d_1(x, y)$$

Let $x = 1$ and $y = \frac{1}{2c_2 + 1}$. Then $d_2(x, y) = 2c_2 d_1(x, y)$, a contradiction. Hence they are not equivalent.

II.2.18 If two metric spaces are equivalent, there is a one-to-one, onto map $h : X_1 \to X_2$ such that

$$c_1 d_1(x, y) \le d_2(h(x), h(y)) \le c_2 d_1(x, y)$$

The second inequality implies that if

$$d_1(x, y) < \delta = \epsilon/c_2 \text{ then } d_2(h(x), h(y)) < c_2 \delta = \epsilon.$$

Since h is invertible, we may write the first inequality as

$$d_1(h^{-1}(x), h^{-1}(y)) \le \frac{1}{c_1} d_2(x, y)$$

for points $x, y \in X_2$. Then $d_1(h^{-1}(x), h^{-1}(y)) < \epsilon$ whenever $d_2(x, y) < \delta = c_1 \epsilon$. Hence h and its inverse are continuous, so $h(x)$ is a homeomorphism.

II.2.19 Intuitively, these metrics behave similar to base 10 decimals. In fact, if we look at Σ_{10} the symbols are 0–9, and by choosing $k = N = 10$, we get decimal expressions (with some ambiguity). To show that these metrics are topologically identical, we must demonstrate that the identity map $\iota(x) = x$ is a homeomorphism. This is equivalent in epsilon-delta format to showing that we need to show that for every $x, \epsilon > 0$ there is a δ such that for every y in Σ_N the two statements

$$d_{k_1}(x, y) < \delta \Rightarrow d_{k_2}(x, y) < \epsilon$$

and,

$$d_{k_2}(x, y) < \delta \Rightarrow d_{k_1}(x, y) < \epsilon$$

are true.

We can simplify matters by breaking the proof into cases and reducing it to a single statement to prove as follows:

Case 1: $k_1 = k_2$. In this case there is nothing to prove, since a metric is always equivalent to itself.

Case 2: $k_1 < k_2$. In this case, by comparing terms of the two series, since $k_2^i > k_1^i$, $d_{k_2}(x, y) \le d_{k_1}(x, y)$ for any x, y with equality holding only for $x = y$. Consequently, we may choose $\delta = \epsilon$ in the first statement:

$$d_{k_1}(x, y) < \epsilon \Rightarrow d_{k_2}(x, y) \le d_{k_1} < \epsilon.$$

and we need only prove the second statement, namely

$$d_{k_2}(x, y) < \delta \Rightarrow d_{k_1}(x, y) < \epsilon$$

Case 3: This case requires only the proof of the first statement for the same reasons as case 2. There is no loss in just assuming case 2, and reversing the subscripts for case 3.

Consequently, all we need to do is show that if $k_1 < k_2$,

$$d_{k_2}(x, y) < \delta_1 \Rightarrow d_{k_1}(x, y) < \epsilon$$

and then take $\delta = \min(\delta_1, \epsilon)$.

Looking at the analogy between decimals and Σ_N, we can get the distance $d_k(x, y)$ arbitrarily small if $x_i = y_i$ up to some big number i_0. This would correspond in decimal arithmetic to the first i_0 decimal places being equal to zero in the distance formula. Writing it without the analogy, since the series is less than

$$C = N \sum_{i=1}^{\infty} \frac{1}{k^i}$$

then

$$\sum_{i=i_0}^{\infty} \frac{|x_i - y_i|}{k^i} < \sum_{i=i_0}^{\infty} \frac{N}{k^i} = \frac{C}{k^i}$$

To insure this is less than, say α we only need to choose i_0 large enough. This is nice, we can get a set we can describe by just using the digits that put us inside a given $B_\epsilon(x)$. What we don't know is whether it contains a ball inside it (For large enough N it is a ball, but we don't know that). On the other hand, we can look at the smallest possible distance we could get if any of the first i_0 "digits" were not zero. This would be the distance between an x and a y with *only* the i_0 digits different, namely

$$|x_{i_0} - y_{i_0}| = 1$$

and $x_i = y_i$ whenever $i \neq i_0$. This distance is

$$d_k(x, y) = \frac{1}{k^{i_0}}$$

So any ball $B_\delta(x)$ with $\delta < 1/k^{i_0}$ is guaranteed to have no elements which disagree on the first i_0 places.

We now proceed as follows: Choose i_0 such that the set

$$B = \{y : x_i = y_i \text{ if } i \leq i_0\}$$

is inside of the d_{k_1} ball $B_\epsilon(x)$. Now choose δ_0 such that the d_{k_2} ball $B_{\delta_0}(x)$. Since all elements of this d_{k_2} ball have at least i_0 "digits" of agreement with x, everything in the ball is in the d_{k_1} ball of radius ϵ. Therefore,

$$d_{k_2}(x, y) < \delta_0 \Rightarrow d_{k_1}(x, y) < \epsilon$$

which was all we decided we needed to prove.

Answers to Chapter II, section 3

II.3.1 By Definition 3.3, if $\{x_n\}$ is a Cauchy sequence in (X, d) a complete metric space, then it converges to a point $x \in X$. This implies that given an $\epsilon > 0$ there is an N such that if $n > N$ then $x_n \in B(x, \epsilon)$. Since there are infinitely many integers greater than any integer N, $x_n \in B(x, \epsilon)$ for infinitely many integers n.

II.3.2 We will use the property of $(\mathbb{R}, \text{Euclidean})$ that every set bounded above has a Least Upper Bound (LUB). This implies that a monotone increasing sequence $\{l_n\}$ which is bounded above has a limit. Since the LUB cannot be less than l_n for any n (the sequence is increasing) and there is no minimum distance to all the l_n (since it is the LUB) it follows that for any $\epsilon > 0$ there is an N such that $x_n \in B(LUB, \epsilon)$ for $n > N$. We use this property to prove completeness of $(\mathbb{R}, d)$ as follows: Let $\epsilon_i = 1/2^i$ and N_i be for each i a number such that $N_i > N_{i-1}$ and $d(x_n, x_m) < \epsilon_i$ whenever $n, m > N_i$. It follows that there is an interval of length ϵ_i such that all the x_n are in this interval if $n > N_i$. Let l_i be the greatest lower bound of the set $\{x_n : n > N_i\}$ (which exists by subtracting everything from a large number and using the LUB principle). The sequence $\{l_i\}$ is then monotone increasing, and is bounded above by the LUB of the same set of x_n which we call L_i. Then this sequence has a limit which we will call x. We claim that x is the limit of the Cauchy sequence as well. Given $\epsilon > 0$, $B(x, \epsilon) \supset [l_i, L_i]$ for some i, hence if $n > N_i$, $x_n \in B(x, \epsilon)$ which means x is the limit of $\{x_n\}$ and the sequence converges. Hence $(\mathbb{R}, \text{Euclidean})$ is complete.

II.3.3 We show that $\mathbb{R}^2$ is complete by using the fact that $\mathbb{R}$ is complete. Let $\{(x_n, y_n)\}$ be a sequence in $\mathbb{R}^2$ which converges to (x, y). Then $\{x_n\} \to x$, $\{y_n\} \to y$. The converse is also true in $\mathbb{R}^n$, since given $\epsilon > 0$ we can find a small rectangle $B(x, \epsilon_x) \times B(x, \epsilon_y) \subset B((x, y), \epsilon)$. In turn, we can find N_x, N_y such that $x_n \in B(x, \epsilon_x)$ if $n > N_x$ and $y_n \in B(y, \epsilon_y)$ if $n > N_y$. Choose $N = \max(N_x, N_y)$. Then for $n > N$, $(x_n, y_n) \in B((x, y), \epsilon)$ since it is in the small rectangle inside. With this in mind, let $\{(x_n, y_n)\}$ be a Cauchy sequence. Then because $\sqrt{(x_n - x_m)^2 + (y_n - y_m)^2} < \epsilon$ implies $|x_n - x_m| < \epsilon$, $|y_n - y_m| < \epsilon$, $\{x_n\}$ and $\{y_n\}$ are each Cauchy sequences in $\mathbb{R}$. By problem 2, these converge, so $\{(x_n, y_n)\}$ converges. Hence $\mathbb{R}^2$ is complete.

II.3.4 There are two ways of approaching this. We can either repeat the arguments in the answers to exercises 3.2 and 3.3 for ■, or we can peek ahead to the definition of a closed set. If (X, d) is a complete metric space, the only way a Cauchy sequence in a subset can fail to converge in the subset is if it converges to a limit which is not in that subset, because viewed as a Cauchy sequence in X, it has a limit. The limit of this sequence is thus a limit point of the subset, and hence if the set is closed, it is in the set. So any closed subset of a complete metric space is complete. ■ is a closed subset of $\mathbb{R}^2$ so it is complete by problem 3.

II.3.5 The sphere can be thought of as an embedded closed subset of $\mathbb{R}^3$. The spherical metric is equivalent to the $\mathbb{R}^3$ Euclidean metric. To see this, note that the Euclidean metric, used on the sphere gives to each pair of points x, y the length of the chord between them. This length, for a sphere of radius 1 is $\sqrt{2 - 2\cos\theta}$ (by the law of cosines). Using the constants $2/\pi$ and $1/\pi$ this is a metric equivalence to the distance measure of great circles. If $\{x_i\}$ is a Cauchy sequence on the sphere, it is in $\mathbb{R}^3$ as well by this equivalence, consequently it has some $(\mathbb{R}^3)$ limit x. Since the sphere is a closed subset of $\mathbb{R}^3$, it contains its limit points, and therefore contains x. Hence the sequence has a limit, and the sphere is complete.

II.3.6 Let $\{x_\sigma\}$ be a Cauchy sequence on the codespace Σ on N symbols. Then given $\epsilon > 0$ there is an M such that if $\sigma_1, \sigma_2 > M$ then $d(x_{\sigma_1}, x_{\sigma_2}) < \epsilon$. In particular, if we choose $\epsilon = \frac{1}{(N+1)^i}$ then since any of the first i terms must either be 0 or contribute more than this value, $x_{\sigma_1}, x_{\sigma_2}$ must agree on the first i terms. Choose x such that for each i, the first i terms of x agree with all those elements of $\{x_\sigma\}$ with $\sigma > M_i$, where M_i is the M corresponding to the above choice of ϵ. Then $x \in \Sigma$ and for each $\epsilon > 0$ there is an M such that $d(x, x_\sigma) < \epsilon$ if $\sigma > M$. In other words x is the limit of $\{x_\sigma\}$. Hence codespace is complete.

II.3.7 Let $\{f_i\}$ be a Cauchy sequence on $(C[0, 1], D)$. Then for each $s \in [0, 1]$, the sequence $\{f_i(s)\}$ is Cauchy in $\mathbb{R}$, since given ϵ and N guaranteed by the assumptions, $d(f_i(s), f_j(s))$ in the Euclidean metric is certainly less than the maximum value over s. We will anticipate a little and call the limit of each such sequence for each s $f(s)$. As defined, $f(s)$ is clearly a function on $[0, 1]$, we must show that it is continuous. Given $\epsilon > 0$ there is a $\delta > 0$ such that if $d(s_1, s_2) < \delta$ then for any i, $D(f_i(s_1), f_i(s_2)) < \epsilon/3$. This property, called equicontinuity, is true for bounded functions on a closed interval, such as we have. Then there is also an N such that if $i > N$ $d(f_i(s_1), f(s_1)) < \epsilon/3$ and $d(f_i(s_2), f(s_2)) < \epsilon/3$ since the sequence converges. Consequently, for our choice of δ,

$$d(f(s_1), f(s_2)) < d(f_i(s_1), f(s_1)) + d(f_i(s_1), f_i(s_2)) + d(f_i(s_2), f(s_2)) < \epsilon$$

by the triangle inequality, and hence $f(s)$ is continuous. Then $f \in (C[0, 1], D)$ and the space is complete.

II.3.8 Let (X_1, d_1) and (X_2, d_2) be equivalent metric spaces. Then there is a homeomorphism $h : X_1 \to X_2$ such that $\tilde{d}(x, y) = d_2(h(x), h(y))$ is equivalent to $d_1(x, y)$ on X_1. Let $\{x_n\}$ be a Cauchy sequence in X_2, and let $\epsilon > 0$, and N large enough that $d_2(x_n, x_m) < \epsilon$ whenever $n, m > N$. There exists a sequence $\{y_n\}$ in X_1 such that $h(y_n) = x_n \forall n$, and by the metric equivalence, we have that

$$d_1(y_n, y_m) < c_2\tilde{d}(y_n, y_m) = c_2 d_n(x_n, x_m) < c_2\epsilon$$

Consequently, by specifying $\delta = c_2\epsilon$, for any $\epsilon > 0$ there is an N such that $d_2(x_n, x_m) < \delta$ which forces $d_1(y_n, y_m) < \epsilon \forall n, m > N$, in other words $\{y_n\}$ is a Cauchy sequence in X_1. This has a limit y because X_1 is complete. Since for all $\delta > 0$ there is an N such that $d_1(y, y_n) < \delta$ for $n > N$, then since $\tilde{(d)}(y, y_n) < (1/c_1)d_1(y, y_n) < (1/c_1)\delta = \epsilon$, it follows that

$\tilde{d}(y, y_n) = d_2(h(y), x_n)$, $\{x_n\} \to h(y) = x$ and the sequence converges in $\mathbf{X}_2$. Hence $\mathbf{X}_2$ is complete. In the process, if two spaces are metrically equivalent, then a Cauchy sequence in one corresponds to a Cauchy sequence in the other under the metric equivalence.

II.3.9 Given two points $x = (x_1, x_2)$, $y = (y_1, y_2)$ in (■,Manhattan), any path which consists of straight horizontal and vertical segments, which are monotonic in each coordinate have the same length. Since the sum of these lengths is equal to the distance between the two points, they are all shortest paths.

II.3.10 Suppose $\{x_n\} \to x$ in the space $\mathbf{X}$. Then given $\epsilon/2 > 0$ there exists an N such that $d(x, x_n) < \epsilon/2$ if $n > N$. Choose two such points x_n, x_m then by the triangle inequality

$$d(x_n, x_m) \leq d(x_n, x) + d(x, x_m) < \epsilon$$

So given ϵ, if $n, m > N$, $d(x_n, x_m) < \epsilon$, in other words, the sequence is Cauchy.

II.3.11 Suppose $\{x_n\} \to x$ and $\{x_n\} \to y$. Then given $\epsilon/2 > 0$ there is an N_1 such that $d(x_n, x) < \epsilon/2$, and an N_2 such that $d(x_n, y) < \epsilon/2$. Choose $N = max(N_1, N_2)$. Then for $n > N$

$$d(x, y) \leq d(x, x_n) + d(x_n, y) < \epsilon$$

Consequently, for any $\epsilon > 0$, $d(x, y) < \epsilon$. In other words, $d(x, y) = 0$, or $x = y$.

II.3.12 Problem is misstated. 0 is the limit regardless, in the second space the sequence does not converge because 0 is not in the space.

II.3.13 Define the sequence $\{x_n\}$ by $x_n = a \forall n$. This sequence is clearly Cauchy, $d(x_n, x_m) = 0 < \epsilon$ for any $\epsilon > 0$ and any n, m. It also converges since $d(a, x_n) = 0 < \epsilon$ for all n and any $\epsilon > 0$. The space contains no limit points since a limit point is defined as being the limit of a sequence but not a member of the sequence. Since this is the only sequence in the space, the space contains no limit points and therefore it contains all its limit points (there are none) but is not perfect since a is not a limit point.

II.3.14 $\{x_n = n\}$ does not converge in ($\mathbb{R}$,Euclidean). Suppose it did. Then by Theorem 3.1 this is a Cauchy sequence. But for any $n \neq m$ $d(x_n, x_m) \geq 1$ and thus is not less than any given $\epsilon > 0$ for any N. For ($\hat{\mathbb{C}}$,spherical), the distance to the Point at Infinity is calculated by looking at the tangent of the angle subtended by the real line and the projection through the sphere. The radius of the sphere is 1 so this tangent is $1/n$. Given $\epsilon > 0$ we may choose N such that $1/N < \epsilon$, and therefore if $n > N$ $d(\infty, x_n) = 1/n < \epsilon$. Hence the sequence converges in ($\hat{\mathbb{C}}$,spherical).

II.3.15 Let x be a limit point in $\mathbf{X}_1$. Then there is a sequence $\{x_n\} \to x$, $x \neq x_n \forall n$, in $\mathbf{X}_1$ by definition of a limit point. Since h provides a metric equivalence, given $\epsilon > 0$ there is a $\delta > 0$ such that if $n > N$, $d_1(x, x_n) < \delta = (1/c_1)\epsilon$ implies that $d_2(h(x), h(x_n)) < \epsilon$. Hence since h is onto $\{h(x_n)\} \to h(x) \in \mathbf{X}_2$. Since h is 1–1 and onto, $h(x_n) \neq h(x) \forall n$. Hence $h(x)$ is a limit point in $\mathbf{X}_2$.

II.3.16 (■,Euclidean) is defined as being those points whose x and y coordinates both lie in $[0, 1]$. The points in this set are all greater than 1 in the y coordinate, hence are not in ■. The set, taken in $\mathbb{R}^2$ has two limit points in ■ (it does not converge as a sequence). These are at $(1, 1)$ and $(-1, 1)$.

II.3.17 Any point of the set S is the limit of a sequence which converges because eventually all the terms are the same. This type of sequence is always possible (see exercise 3.13), however such convergences do not qualify the point as a limit point of the set. The only such point would be 0, since this is the only Cauchy sequence converging to point not in the sequence. Since this point is not in the space $(0, 1]$, the set of limit points of S is empty and therefore a subset of S. Hence S is closed in $(0, 1]$,Euclidean).

II.3.18 Let $S = [0, 1] \subset (\mathbb{R},$Euclidean). Let $x \in S$, and let $a = min(x, 1 - x)$. If $a = 0$, set $a = 1$. In other words, a is the least distance to an end of the interval. Then the sequence $\{x_n = x + a/2^n\}_{n=1}^{\infty}$ lies entirely in S. Given $\epsilon > 0$ there is an N such that $\epsilon > a/2^N$. This implies that for $n > N$, $d(x, x_n) = d(x, x - a/2^n) = a/2^n < \epsilon$, in other words this sequence in $[0, 1]$ converges to x, hence x is a limit point. Since x was any point in the set, the set consists entirely of limit points. It is closed, and therefore perfect.

II.3.19 By reference to exercise 3.17, this set contains its only limit point, 0. It is not perfect since for any n there is no

Cauchy sequence (and hence no convergent sequence converging to $1/n$ which does not contain $1/n$. Hence $1/n$ is not a limit point, and the set is not perfect. It is closed, since it contains it's limit point.

II.3.20 That Σ contains its limit points is true because it is the whole space and is therefore closed. We need only verify that each point in Σ is the limit of a sequence $\{y_n\}$ which does not include it. This is done as follows: Let $x = x_1 x_2 \ldots \in \Sigma$ and suppose we have N symbols. For each n let y_n be the point which agrees with x on all symbols except the nth, and is equal to the next greater symbol at the nth place unless that symbol in x is $N - 1$ in which case use $N - 2$. Then for each n we have

$$d(x, x_n) = \sum_{i=1}^{\infty} \frac{|x_i - y_{n_i}|}{(N+1)^i} = \frac{|x_n - y_{nn}|}{(N+1)^n} = \frac{1}{(N+1)^n}$$

because there is only one non-zero term in the sum. This sequence is then convergent–given $\epsilon > 0$ there is an M such that for $n > M$ $d(x, y_n) = 1/(N+1)^n < \epsilon$. No term in the sequence is equal to x because it disagrees at the nth term. Since x is arbitrary, every point in Σ is a limit point and Σ is a perfect set.

II.3.21 If (S, d) is complete, then given a Cauchy sequence in S, it converges to a point in S. Since every limit point is the limit of a convergent sequence in S, and hence a Cauchy sequence in S, S contains its limit points and is therefore closed. Suppose S is closed, and let $\{x_n\}$ be a Cauchy sequence in S. Then $\{x_n\}$ is a Cauchy sequence in $\mathbf{X}$ as well and there is a point $x \in \mathbf{X}$ such that $\{x_n\} \to x$, because $\mathbf{X}$ is complete. Hence x is a limit point for S. Since S is closed, $x \in S$.

Answers to Chapter II, section 4

II.4.1 In order for $\mathbf{X}$ to be closed, it must contain all its limit points. Thus let $\{x_n\}$ be a *convergent* sequence in $\mathbf{X}$. By definition, there is an $x \in \mathbf{X}$ such that $\{x_n\} \to x$. Hence $\mathbf{X}$ is closed. A Cauchy sequence need not converge however. Let $\mathbf{X} \subset Y$, such that Y is complete. Then if $\{x_n\} \to y \notin \mathbf{X}$, then this sequence does not converge in $\mathbf{X}$, and y is not a limit point in $\mathbf{X}$. The sequence is nevertheless Cauchy, and since it does not converge, $\mathbf{X}$ is not a complete metric space. An example is $\mathbf{X} = (0, 1)$.

II.4.2 See the answer to exercise 3.21 (Chapter II).

II.4.3 By exercise II.3.15, if x is a limit point of the set S in $\mathbf{X}_1$ then $\theta(x)$ is a limit point of the set $\theta(S) \subset \mathbf{X}_2$. Suppose there is an element in $\theta(S)$ which is a limit point, namely that there exists a sequence $\{(\theta(s))_n\} \to s \in \theta(S)$. Then given $\epsilon > 0$ there is an N such that for $n > N$,

$$d_1(\theta^{-1}(\theta(s)_n), \theta^{-1}(\theta(s))) \le (1/c_1) d_2(\theta(s)_n, \theta(s)) < \epsilon$$

Letting $\theta^{-1}(\theta(s)_n) = s_n$ we see that $\theta(s)$ is the image of a limit point in S. Since S is closed, the image of s is in $\theta(S)$, thus $\theta(S)$ contains its limit points, and is therefore closed.

II.4.4 Given $x \in \mathbf{X}$, by definition $B(x, \epsilon) = \{y in \mathbf{X} : d(x, y) < \epsilon\}$ and hence is a subset of $\mathbf{X}$ for any ϵ. Hence $\mathbf{X}$ is open.

II.4.6 Let S be a bounded subset of $(\mathbb{R}^N, \text{Euclidean})$. By definition, there is a point $a \in S$ and a number R, such that $d(s, a) < R \forall s \in S$. Explicitly,

$$\sqrt{\sum_{n=1}^{N} (s_n - a_n)^2} < R$$

The terms in the sum are all non-negative, it therefore follows that $(s_n - a_n)^2 < R^2$ or that $|s_n - a_n| < R$, for each entry. This means that S is contained in the box $[a_1 - R, a_1 + R] \times [a_2 - R, a_2 + R] \times \cdots \times [a_N - R, a_N + R]$. To provide a finite ϵ-net for S, we subdivide each interval into M intervals such that $N/M < \epsilon$. Then the product of the endpoints of each interval is a point in a finite ϵ-net of this N-cube, its intersection with S is also finite and each point of S is less than ϵ away from one of the above corners. Hence S is totally bounded. To show that $\mathbb{R}^N$ is complete, it suffices to note that $\mathbb{R}$ is complete (exercise II.3.2) and that a sequence is Cauchy or convergent if and only if it is Cauchy or convergent on each of its coordinates. Finally, since $\mathbb{R}^N$ is complete, and S is totally bounded if it is bounded, it follows that if S is closed and bounded it is compact.

II.4.7 Let $(\mathbf{X}, d)$ be a metric space, $f : \mathbf{X} \to \mathbf{X}$ be continuous, and A be a compact non-empty subset of $\mathbf{X}$. It follows immediately that if $a \in A$, then $f(a) \in f(A)$ by definition, so $f(A)$ is non-empty. Let $\{b_n\}$ be a sequence of points in $f(A)$. Then each b_n is the image of at least one point in A. Let $\{a_n\}$ be any sequence in A such that $f(a_n) = b_n$ for each n. Because A is compact, there is a subsequence $\{a_{n_i}\} \subset \{a_n\}$ which converges to some point $a \in A$. The sequence $\{f(a_{n_i})\}$ is a subsequence of $\{b_n\}$, we must show that it converges to a point in $f(A)$. In fact, because f is continuous, given $\epsilon > 0$, there is a $\delta > 0$ such that if $d(a_{n_i}, a) < \delta$ then $d(f(a_{n_i}), f(a) < \epsilon$. Since this subsequence converges in A, given $\delta > 0$ there is an N such that $i > N$ implies $d(a_{n_i}, a) < \delta$ and therefore that $d(f(a_{n_i}), f(a)) < \epsilon$. This subsequence therefore converges in $f(A)$, and letting $b_{n_i} = f(a_{n_i})$ we have found a convergent subsequence for the sequence $\{b_n\} \subset f(A)$. Hence $f(A)$ is a compact non-empty subset of $\mathbf{X}$.

II.4.8 Let $S \subset (\mathbf{X}_1, d_1)$ be open. Then it follows that by example 2.4.5, $\mathbf{X}_1 \setminus S$ is closed. By exercise II.4.3, $h(\mathbf{X}_1 \setminus S)$ is closed in $\mathbf{X}_2$. Let $t \in h(\mathbf{X}_1 \setminus S)$. Then it follows that since h is invertible, $h^{-1}(t) \notin S$, hence $t \in \mathbf{X}_2 \setminus h(S)$. Hence $h(\mathbf{X}_1 \setminus S) = \mathbf{X}_2 \setminus h(S)$. Hence $\mathbf{X}_2 \setminus h(S)$ is closed which means that $h(S)$ is open in $\mathbf{X}_2$. Note that the proof depends on the invertibility of h. It does *not* hold even for arbitrary continuous functions.

II.4.10 Given $x \in (0, 1)$ we have $0 < x < 1$ so choose $\epsilon = \min(x, 1 - x)$. Then the set $B(x, \epsilon) = \{y \in \mathbf{X} : d(x, y) < \epsilon\}$ is a subset of $(0, 1)$. Hence $(0, 1)$ is open in $\mathbf{X}$. Let $x \in \{2\}$. This is a singleton set, so $x = 2$. Choose $\epsilon < 1$. Then the set $B(x, \epsilon) = \{y \in \mathbf{X} : d(x, y) < 1\}$ contains only the point 2 and $d(x, 2) = 0 < \epsilon$. Hence $\{2\}$ is open in $\mathbf{X}$. The two sets are both closed: $(0, 1) = \mathbf{X} \setminus \{2\}$ is the complement of an open set, by 2.4.5 it is closed. By the same token $\{2\} = \mathbf{X} \setminus (0, 1)$ so it is the complement of an open set and is closed. $\{2\}$ is compact, since given any sequence in $\{2\}$ all entries are 2, hence the sequence converges, and has a convergent subsequence (any subsequence). $(0, 1)$ is not compact, choose the sequence $\{x_n = 1/n\}$. This converges to $0 \notin (0, 1)$, so any subsequence must likewise converge to 0. Hence it has no convergent subsequence.

II.4.11 Let $S \subset (\mathbf{X}, d)$. By definition, its boundary ∂S is the set of points s such that any open ball $B(s, \epsilon)$ contains points in S and in $\mathbf{X} \setminus S$. But $S = \mathbf{X} \setminus (\mathbf{X} \setminus S)$, since it is the set of points which are in $\mathbf{X}$ which are not in $Xsp \setminus S$. Consequenctly if any open ball $B(s, \epsilon)$ around a point s contains points in both sets it is in the boundary $\partial(\mathbf{X} \setminus S)$, and the two boundaries are the same. $\mathbf{X}$ is open (exercise II.4.4) hence $\emptyset$ is closed, and contains it's boundary. Since this implies the boundary of $\mathbf{X}$ is contained in the empty set, $\partial \mathbf{X} = \emptyset$.

II.4.12 Let $(\mathbf{X}_1, d_1)$ and $(\mathbf{X}_2, d_2)$ be equivalent metric spaces and let $h : \mathbf{X}_1 \to \mathbf{X}_2$ be the metric equivalence. Let $S \subset \mathbf{X}_1$, and let $s \in \partial S$. If $h(s) \notin \partial h(S)$ then there is an open ball $B(h(s), \epsilon)$ which is either a subset of $h(S)$ or of $\mathbf{X}_2 \setminus h(S)$. If it is a subset of $h(S)$, then it is the image of an open set in S (see problem II.4.8), and by definition there is an open ball around s in S, which contradicts the statement $s \in \partial S$. If it is a subset of $\mathbf{X}_2 \setminus h(S)$, then there are no points in $B(h(s), \epsilon)$ which are the image of points in S. Hence $B(h(s), \epsilon)$ is the image of an open set containing s which is contained in $\mathbf{X}_1 \setminus S$. Then there is an open ball within this open set, which is centered at s, and contains no points in S. Again this contradicts $s \in \partial S$. Consequently, since this is true for all s in ∂S, $h(\partial S) \subset \partial h(S)$. If we do this again for the transformation h^{-1} with the boundary of $h(S)$ we get that the inverse image of the boundary of $h(S)$ is contained in ∂S. Then the two boundaries are images of each other, and $h(\partial S) = \partial h(S)$.

II.4.13 $S \subset (\mathbb{R}, \text{Euclidean})$ is the set of rational points in the real line. We want to show that the boundary of S is $\mathbb{R}$ itself. Let $x \in \mathbb{R}$, and let $\epsilon > 0$ be given. Suppose a, b are two integers such that $b = a + 1$ and $a < x < b$. Now let N be chosen such that $1/2^N < \epsilon$. Since the numbers $a + k/2^N$ for $k \leq 2^N$ subdivide the interval $[a, b]$ into 2^N equal parts of width $1/2^N < \epsilon$ it follows that any interval which is greater than ϵ in width contains at least one of them. But $B(x, \epsilon) = [x - \epsilon, x + \epsilon]$ is of width 2ϵ, hence there is a rational number in $B(x, \epsilon)$. The number $\sqrt{2}$ is not a rational number: Let $p/q = \sqrt{2}$ and suppose the fraction p/q is in lowest terms. Squaring both sides yields $p^2 = 2q^2$ implying that p is even. Hence p^2 is divisible by 4. Hence q is even, a contradiction. If c is irrational, so is $p/(cq)$ for any rational p/q. If not, by simple algebra we may write c in fractional form. Now take all the numbers in $[a, b]$ of the form $a + k/(\sqrt{2}2^N)$. Once again these are evenly spaced and any interval of width less than ϵ contains at least one of them. Hence there is a number in $B(x, \epsilon)$ which is in $\mathbb{R} \setminus S$. Hence $x \in \partial S$. Since x was arbitrary, $\partial S = \mathbb{R}$.

II.4.14 The only point in $\hat{\mathbb{C}}$ for which every ϵ-ball contains points in $\mathbb{C}$ and points in $\hat{\mathbb{C}} \setminus \mathbb{C} = \{\infty\}$ is ∞. Hence the boundary of $\mathbb{C}$ in $\hat{\mathbb{C}}$ is $\{\infty\}$.

II.4.15 Let $S \subset \mathbf{X}$ be closed. Suppose that there is a point in $\mathbf{X}$ such that for every $\epsilon > 0$ $B(x, \epsilon)$ contains points in S and points in $\mathbf{X} \setminus S$. Define a sequence $\{x_n\}$ by $x_n \in B(x, 1/n)$, $x_n \neq x$. If this cannot be done, then $\partial S = \emptyset$ and the statement is true. Then the above sequence makes x a limit point of S and hence $x \in S$. Hence $\partial S \subset S$.

II.4.16 Suppose S is open. Let $s \in \partial S$. Then for every $\epsilon > 0$ there is a point $x \in B(s, \epsilon)$ such that $x \notin S$. Then there is no $\epsilon > 0$ such that $B(s, \epsilon) \subset S$. Since S is open $s \notin S$. Hence $\partial S \cap S = \emptyset$.

II.4.17 By definition $S^0 = S \setminus \partial S$. Since $S \cap \partial S = \emptyset$, $S \setminus \partial S = S$. Hence $S^0 = S$. Conversely, suppose $S^0 = S$ Then $\partial S = \emptyset$. Consequently for any $s \in S$ there is an $\epsilon > 0$ for which $B(s, \epsilon)$ contains no points in $\mathbf{X} \setminus S$. Then $B(s, \epsilon) \subset S$, and S is open.

II.4.18 Since S is closed, by exercise II.4.16, $\partial S \subset S$. By definition $S^0 = S \setminus \partial S$. We have $S \setminus \partial S \cup \partial S = S$.

II.4.19 By exercise II.4.12, being the boundary of a set is invariant under metric equivalence. Thus if $s \in S^0$, then $s \in S$ and $s \notin \partial S$. Since $h(s) \in h(S)$, and $h(s) \in \partial h(S) = h(\partial S)$ implies that $s \in \partial S$ which contradicts hypothesis, $h(s) \in (h(S) \setminus \partial h(S))$ which means $h(s) \in (h(S))^0$.

II.4.20 $\mathbf{X} \setminus \partial S$ is $S \setminus \partial S \cup (\mathbf{X} \setminus S) \setminus \partial S$. Since $\partial S = \partial (Xsp \setminus S)$ (by exercise II.4.11) $\mathbf{X} \setminus \partial S = (\mathbf{X} \setminus S)^0 \cup S^0$, which are disjoint sets, and by II.4.17, are open. Note that one or the other or both may be empty, however. For the examples, (a) the boundary of the set $S = \{(x, y) \in \mathbb{R}^2 : x^2 + y^2 < 1\}$ is the set $\partial S = \{(x, y) \in \mathbb{R}^2 : x^2 + y^2 = 1\}$, since every ϵ-ball around a point in this set contains $(x, y) \in S$ and as well (x, y) such that $x^2 + y^2 > 1$. This is a circle of unit radius and divides the plane into the open disk interior, and the plane with a circular hole of radius one in it. (b) $S = \mathbb{R}^2$ the whole space, so $\partial S = \emptyset$. The two sets $\mathbb{R}^2$ and $\emptyset$ are both open as shown in II.4.1 and II.4.4.

II.4.21 By problem II.4.20, $\mathbf{X} \setminus \partial S$ is the union of two disjoint open sets. The union of open sets is open, hence ∂S is closed.

II.4.22 We start this problem with an elementary fact about compact sets. Let C be a compact set. Then if $A \subset C$ is a closed subset, let C_n be an open cover for A. We can extend this cover to an open cover of C, by adding ϵ-balls in $C \setminus A$ (which is open) we can form an open cover for C. There is therefore a finite subcover $\{C_1, \ldots C_N\}$. Since by our choice, all of the C_n which cover A are from the original open cover, and they are a subset of this finite subcover, there is a finite subcover for A. It follows that any closed subset of a compact set is compact.

We now turn to the problem at hand. S is a subset of a compact metric space. By problem II.4.21, ∂S is closed, and also a subset of the same compact metric space. Therefore ∂S is compact. The assertion is best proven by proving the following statement which is true for arbitrary subsets S (closed, open, otherwise). Let S be a subset of a metric space $(\mathbf{X}, d)$. Then $A \subset S$ is open in the space (S, d) if and only if we can write $A = S \cap O$ for some open subset of $\mathbf{X}$. If A is open in (S, d) then for each $a \in A$ and each $\epsilon > 0$

$$B_S(a, \epsilon) = \{y \in S : d(a, y) < \epsilon\} = \{y \in \mathbf{X} : d(a, y) < \epsilon\} \cap S = B_\mathbf{X}(a, \epsilon) \cap S$$

so $A = S \cap O$ where O is the union of such balls in $\mathbf{X}$. Conversely, if $A = S \cap O$ for some open set $O \in \mathbf{X}$ then given $a \in A$, there is an $\epsilon > 0$ such that $B_\mathbf{X}(a, \epsilon) \subset O$, and

$$B_S(a, \epsilon) = \{y \in S : d(x, y) < \epsilon\} = \{y \in \mathbf{X} : d(x, y) < \epsilon\} \cap S = B_\mathbf{X}(a, \epsilon) \cap S$$

The latter intersection is in $A = S \cap O$, so A is open.

The assertion that the interior of the land is open in Z despite the fact that it includes points in the "border" of Z, is true by the above statement. The fact that it appears to include border points comes about because these points are points in $\partial Z \subset \mathbf{X}$, whereas when Z is itself the metric space, the border of the whole space is the empty set.

Answers to Chapter II, section 5

II.5.1 The properties of connectedness preserved by metric equivalence: (i) Connectedness. Let $(\mathbf{X}_1, d_1)$ and $(\mathbf{X}_2, d_2)$ be equivalent metric spaces and $h : \mathbf{X}_1 \to \mathbf{X}_2$ be the invertible function giving this equivalence. It was proven last section (problem II.4.8) that if S is an open set, then $h(S)$ is open, and (problem II.4.3) if S is a closed set, then $h(S)$ is closed. Because h^{-1} also establishes an equivalence, the same is true about closed and open sets in $\mathbf{X}_2$ under h^{-1}. Let $\mathbf{X}_1$ be connected. Suppose $S \subset \mathbf{X}_2$ is both closed and open. Then $h^{-1}(S)$ is closed because S is closed, and open because S is open. Since $\mathbf{X}_1$ is connected, $h^{-1}(S)$ is either $\mathbf{X}_1$ or $\emptyset$. In the first case, because h is onto, $S = \mathbf{X}_2$. In the second $S = \emptyset$. Hence $\mathbf{X}_2$ is connected.

(ii) By studying the definition of metric equivalence, we see that given $\epsilon > 0$ there is a $\delta = (1/c_2)\epsilon$ such that $d_2(h(x), h(y)) < \epsilon$ whenever $d_1(x, y) < \delta$. Consequently h is a continuous function. The composition of continuous functions is continuous. Suppose X_1 is pathwise connected. Given two points $x, y \in X_2$ because h is onto, there are a pair of points $u, v \in X_1$ such that $h(u) = x, h(v) = y$. Because X_1 is pathwise connected, there exists a continuous function $f : [0, 1] \to X_1$ taking $f(0) = u$, $f(1) = v$. But then there is a continuous function $h(f(y))$ taking a point $y \in [0, 1]$ to X_2 such that $h(f(0)) = h(u) = x$ and $h(f(1)) = h(v) = y$. Consequently, X_2 is pathwise connected.

(iii) Suppose X_1 is disconnected. Then there exist a set $A \neq X_1$ and $A \neq \emptyset$ which is both open and closed in X_1. As in the proof of i), since this set is open, $h(A)$ is open in X_2, since the set is non-empty, $h(A)$ is non-empty, and since $A \neq X_1$ and h is onto, $h(A) \neq X_2$. Hence X_2 is also disconnected.

(iv) Suppose X_1 is pathwise disconnected. Then there exist points $x, y \in X_1$ such that there is no continuous function mapping $[0, 1]$ to X_1 with $f(0) = x$, $f(1) = y$. Suppose there were a continuous function $g : [0, 1] \to X_2$ such that $g(0) = h(x)$, $g(1) = h(y)$. Then, because h^{-1} is continuous, the function $h^{-1}(g(w))$ taking the interval to X_1 would take $h^{-1}(g(0)) = x, h^{-1}(g(1)) = y$, a contradiction. Hence X_2 is pathwise disconnected.

(v) Suppose X_1 is simply connected, and, because h is continuous and onto, let $f_1, f_2 : [0, 1] \to X_2$ be two paths connecting points $h(x), h(y)$ in X_2. Then there exist paths $h^{-1}(f_1([0, 1])), h^{-1}(f_2([0, 1]))$ connecting $x, y \in X_1$. There is then a continuous deformation $g(s, t)$ transforming $h^{-1}(f_1([0, 1]))$ to $h^{-1}(f_2([0, 1]))$. Then $h(g([0, 1] \times [0, 1])) \to X_2$ is a continuous function, and

$$h(g(s, 0)) = h(h^{-1}(f_1(s))) = f_1(s),$$

$$h(g(s, 1)) = h(h^{-1}(f_2(s))) = f_2(s),$$

while

$$h(g(0, t)) = h(h^{-1}(h(x)) = h(x), \text{ and}$$

$$h(g(1, t)) = h(h^{-1}(h(y))) = h(y),$$

hence X_2 is simply connected.

(vi) Finally, assume the converse, that X_1 is multiply connected. Then there exist two points $x, y \in X_1$ and two paths $f_1, f_2 : [0, 1] \to X_1$ with no continuous deformation between them. We again transfer everything to X_2. The points $h(x), h(y)$ in X_2 have the two paths $h(f_1([0, 1])), h(f_2([0, 1]))$ connecting them. If there were a continuous deformation $g : [0, 1] \times [0, 1] \to X_2$ deforming one path to the other, then $h^{-1}(g([0, 1] \times [0, 1])) \to X_1$ would provide a deformation in X_1. Hence there is no such deformation, and X_2 is multiply connected.

II.5.2 The metric space (■,Euclidean) is simply connected: Choose x, y and two functions from $[0, 1] \to$ ■ such that $f_0(0) = f_1(0) = x$ and $f_0(1) = f_1(1) = y$. Let g in the definition be given by $g(s, t) = t \cdot f_0(s) + (1 - t) \cdot f_1(s)$, where $a(x, y) = (ax, ay)$. This satisfies the four demands of the definition, we must only check that for all values of (s, t) it remains within the square. It would be maximized in, say, the x coordinate by setting $f_0(s) = f_1(s) = (1, y)$ in which case for any t the function would equal $t + 1 - t = (1, y)$. Since this is clearly in the square, there is no problem there. The other cases are identical. Hence ■ is simply connected.

II.5.3 To show that $X = ((0, 1) \cup \{2\}$, Euclidean) is disconnected, we examine the subset $\{2\}$. This set is closed since points are closed in metric spaces. It is also open: The ball $B_\epsilon(2) \subset \{2\}$ for $\epsilon = 1/2$ since by definition this ball is the set of all points in X which are closer than $1/2$ to the point 2 of which there is only the point itself. Consequently, for every point in the set $\{2\} \subset X$, there is an open ball around that point in the subset, hence the subset is open. Since this set is both open and closed, and not empty or equal to X, the set is disconnected.

II.5.4 Given the codespace metric we have used, for the space Σ on the N symbols $0, \ldots, (N - 1)$, there is a metric equivalence to the subset of $[0, 1]$ written in base $N + 1$ of real numbers with no $N + 1$-ary digits equal to N. The equivalence is generated by the mapping $x_1 x_2 \ldots \mapsto .x_1 x_2 \ldots$. The map has the same distance formula in both cases, and is clearly 1–1 and onto. The set of such real numbers is a totally disconnected subset of $([0, 1]$, Euclidean): Given any closed subset of the real line, either that subset is a single point, or an interval (possibly $[a, \infty)$). If not, we may write the subset as the union of two closed disjoint subsets, and there is a distance between them say c. Then for any point in each subinterval, an ϵ-ball around that point of radius $c/3$ contains only other points from the same subinterval, hence each subinterval is open as well as closed and the set is disconnected. Since the set given above contains no intervals of with numbers with the digit N it is clearly not an interval, and it clearly contains more than one point. Hence codespace is disconnected. To see that it is totally disconnected, suppose there were an interval enclosed in the equivalent subspace of $[0, 1]$. Then if $a = .x_1x_2\ldots$ and $b = .y_1y_2\ldots$ suppose they agree to $x_i \leq y_i$. Then the point $.x_1x_2\ldots(z_i + 1 = N)\ldots$ is in the interval $[a, b]$ and not in the set. Hence there are no intervals in the set, and it is totally disconnected, by our metric equivalence, codespace is totally disconnected.

II.5.5 Put the annulus in $\mathbb{R}^2$ centered at the origin, take x, y on either side of the origin, and take f_0 to be a path which goes around the upper half of the annulus, f_1 a path which goes around the lower half. Suppose the function $g(s, t)$ exists as in the definition. Then for each point $f_0(s)$, $s \neq x$, y we have $f_0(s)|_y > 0$ and the corresponding value for $f_1(s)$ is negative. Holding s fixed the intermediate value theorem says that there is a value of t such that $g(s, t)|_y = 0$. Since we may take the annulus as thin as we please, this says that every point must pass through either x or y as it is deformed from one path to the other. If some points pass through x and some through y there is a value of t such that $g(s, t)$ is not connected, hence is not a path. Hence the annulus is multiply connected.

II.5.6 Choose the set of nested subsets of $\mathbb{R}^2$ given by $S_i = \mathbb{R}^2 B_i$ where $B_i = \{(x, y) : -1 \leq x \leq 1, y \leq i\}$. These sets are nested, they are all connected, and the intersection is the plane minus a strip around the y-axis 2 units wide, which is disconnected.

II.5.8 The body space is multiply connected, it has a hole corresponding to the entire gastro-intestinal tract, if not others (e.g. pierced ears).

Answers to Chapter II, section 6

II.6.1 If x and y are elements of $\mathcal{H}(\mathbf{X})$ then by definition they are both non-empty and both compact. Their union is therefore non-empty. To see that it is also compact, let $\mathcal{C}$ be a covering of $x \cup y$ by open sets. Then there is a subset of $\mathcal{C}$ which covers x, say $\mathcal{C}_x$ and another $\mathcal{C}_y$ which covers y. Because x is compact, there is a finite subset of $\mathcal{C}_x$ which covers x, likewise for y. The collection of sets formed by taking all the sets in the finite subcover of x and all of those in the finite subcover y forms a subcover of $x \cup y$, and since its cardinality is the sum of two finite integers, is also finite. Hence $\mathcal{C}$ admits a finite subcover of $x \cup y$ which is therefore compact. Since $x \cup y$ is compact and non-empty, it is in $\mathcal{H}(\mathbf{X})$. To see that the intersection of two elements in $\mathcal{H}(\mathbf{X})$ need not be in $\mathcal{H}(\mathbf{X})$, take x and y disjoint. Then $x \cap y$ is empty and therefore not in $\mathcal{H}(\mathbf{X})$.

II.6.2 The elements of $\mathcal{H}(\mathbf{X})$ are compact non-empty subsets of $\mathbf{X}$, while a compact non-empty subset of $\mathbf{X}$ is only a single point in $\mathcal{H}(\mathbf{X})$. An infinite collection of compact non-empty subsets of $\mathbf{X}$ is a valid subset of $\mathcal{H}(\mathbf{X})$, while it's union may or may not be a compact non-empty subset of $\mathbf{X}$.

II.6.3 Let $B, C \in \mathcal{H}(\mathbf{X})$ and $B \subset C$. Let $x \in \mathbf{X}$. Then

$$d(x, C) = \min_{y \in C}(d(x, y))$$
$$= \min(\min_{y \in B}(d(x, y)), \min_{y \in (C \setminus B)}(d(x, y)))$$
$$\leq \min_{y \in B}(d(x, y)) = d(x, B)$$

II.6.4 The distance from $(1, 1)$ to the closed disk of radius $\frac{1}{2}$ centered at $(\frac{1}{2}, 0)$ is given by the distance to the point on the boundary of the disk closest to $(1, 1)$, since this point is not in the disk. The closest point on the boundary is on a radius of the disk, by the triangle inequality. This means the distance is the distance from $(1, 1)$ to $(\frac{1}{2}, 0)$ minus the distance along this chord which is inside the disk, namely $\frac{1}{2}$. The whole distance is $\sqrt{1 + (1 - 1/2)^2} = \sqrt{5/4}$ so the distance $d(x, B) = (\sqrt{5} - 1)/2$.

II.6.5 In this case, the closed disk of radius $\frac{1}{2}$ looks like a square with corners at $(0, 0)$, $(0, 1)$, $(1/2, 1/2)$, $(1/2, -1, 2)$ since this is the set of points equidistant from $(1/2, 0)$ at a distance $1/2$. In the Manhattan metric, the distance to any point on the right upper face of this square is $1/2 + 1/2 = 1$. Since the distance to any other face is larger, $d(x, B) = 1$.

II.6.6 The elements of the set B are less than 3 only if n is odd. The sequence $\{\frac{-n}{n^2+1}\}$ increases monotonically towards zero, consequently it has its least value at $1/2$. The least value in B then is the value $x_1 = 3 - 1/2$ or $5/2$. The distance to the point $x = 1/2$ is then $d(x, B) = 2$.

II.6.7 It is easiest to provide a counterexample. Let $B \subset A$ be a proper subset, such that there is an $x \in A$ with $d(x, B) = 1$. (For instance two disks centered at the origin one with radius 1 one with radius 2). Then $d(A, B) \geq 1$. Since $B \subset A$ the minimum distance from any point in B to A is to the point itself, namely 0. Hence $d(B, A) = 0$. The two are not equal, so this is not a metric on $\mathcal{H}(\mathbf{X})$.

II.6.11 Let $A, B \in \mathcal{H}(\mathbf{X})$, $A \neq B$. Suppose $d(A, B) = 0$. Then for every $x \in A$, $d(x, B) = 0$, since we take the maximum over these to form $d(A, B)$. This can only occur if $\min_{y \in B}(d(x, y)) = 0$, for every point $x \in A$, which means that for each one, there is a $y \in B$ with $y = x$. In other words $A \subset B$. By the same argument, if $d(B, A)$ is also 0, then $B \subset A$. If $A \subset B$ and $B \subset A$ then $A = B$. This is not the case so one or the other distance cannot be 0. The converse of the above is that if $A \subset B$, then for each point in A,

$$d(x, B) = \min_{y \in B}(d(x, y)) = d(x, x) = 0$$

so that the maximum over these is 0. Hence if $A \subset B$ then $d(A, B) = 0$.

II.6.12 Using the solution to problem II.6.3, for each $x \in A$ we have $d(x, C) \leq d(x, B)$. Suppose $d(A, B) < d(A, C)$. Then there is a point $x \in A$ such that $d(x, C) > d(y, B)$ for all $y \in A$. This is a contradiction, choosing $y = x$. Hence $d(A, C) \leq d(A, B)$.

II.6.14 We write out the expressions on either side of the inequality.

$$d(A, B) = \max_{x \in A}(\min_{y \in B} d(x, y))$$

On the other side, we have

$$d(A, C) + d(C, B) = \max_{x \in A}(\min_{z \in C}(d(x, z))) + \max_{w \in C}(\min_{y \in B}(d(w, y)))$$

Since the maximum over $w \in C$ is greater than or equal to the value if w is replaced by z, we can write

$$d(A, C) + d(C, B) \geq \max_{x \in A}(\min_{z \in C}(d(x, z)) + \min_{y \in B}(d(z, y)))$$

For any choice of $x \in A$ and $y \in B$ the sum is greater than the distance $d(x, y)$ by the triangle inequality, including the case where z has been chosen to minimize $d(x, z)$ and where y has been chosen to minimize $d(z, y)$. Hence for the choice of y the sum of the minima above is greater than $d(x, y)$ for that choice of y and hence greater than the distance $d(x, y)$ which is the minimum chosen to minimize $d(x, y)$ directly. We may then write

$$d(A, C) + d(C, B) \geq \max_{x \in A}(\min_{y \in B}(d(x, y))) = d(A, B)$$

For the other case, whether or not $d(A, B) \leq d(C, A) + d(C, B)$, we use the case $C \subset B \subset A$. Then the right hand side is identically zero, while if $A \neq B$, the left hand side is not. Hence the expression is not true in general.

II.6.16 We make extensive use of II.6.12. To this end, we have

$$h(A \cup B, C \cup D) = d(A \cup B, C \cup D) \vee d(C \cup D, A \cup B).$$

If $h(A \cup B, C \cup D) = d(A \cup B, C \cup D)$ then either $h(A \cup B, C \cup D) = d(A, C \cup D)$ or $h(A \cup B, C \cup D) = d(B, C \cup D)$. Choose the former. then by II.6.12,

$$h(A \cup B, C \cup D) \leq d(A, C) \leq h(A, C) \leq h(A, C) \vee h(B, D).$$

If we choose the latter,

$$h(A \cup B, C \cup D) \leq d(B, D) \leq h(B, D) \leq h(A, C) \vee h(B, D)$$

If $h(A \cup B, C \cup D) = d(C \cup D, A \cup B)$ then either $h(A \cup B, C \cup D) = d(C, A \cup B)$ or $h(A \cup B, C \cup D) = d(D, A \cup B)$.

Choose the former. then by II.6.12,

$$h(A \cup B, C \cup D) \le d(C, A) \le h(A, C) \le h(A, C) \vee h(B, D).$$

If we choose the latter,

$$h(A \cup B, C \cup D) \le d(D, B) \le h(B, D) \le h(A, C) \vee h(B, D).$$

Since this is all the possibilities, the result follows.

Answers to Chapter II, section 7

II.7.3 The reasonable assumptions we could make are that each image is a set of points characterized by spatial coordinates and color coordinates. The spatial coordinates for a photograph are in two dimensions and form a filled square, they are thus compact. If we assume that the colors are a closed and totally bounded region, say grey scales from 1–100, then the product space of pictures is compact. We assume some continuity in time, this is backed up by physical assumptions, the tree will probably not do something totally disconnected from its motion in the interval before. Finally, we assume that the tree waves at finite speed. This implies that the change between photographs in picture space will decrease as the time interval shortens. This should yield a Cauchy sequence given that the distance between two photographs is the Hausdorff metric, applied to successive pictures as sets of points with (x,y,color) coordinates, and, say a Euclidean metric on this 3-space. A looks like a photograph of the tree at 1 second.

II.7.4 In the space of non-empty compact subsets of the Sierpinski triangle, an example of an infinite set is (among others) the set of all singleton sets $\{x\}$ with $x \in \Delta$. A Cauchy sequence in this set is given by successive righthand corners of triangles with lefthand (lower) corners at the lefthand lower corner of Δ. This sequence converges to the lefthand lower corner of Δ, the convergence implies that it is Cauchy.

II.7.6 If $\mathbf{X}$ is complete, then so is $\mathcal{H}(\mathbf{X})$. Since this is the case if $\mathbf{X}$ is compact, it remains to show that $\mathcal{H}(\mathbf{X})$ is totally bounded. Let ϵ be given, and let $\{y_1, \ldots, y_k\}$ be an ϵ-net for $\mathbf{X}$. The singleton sets $\{y_1\}, \ldots, \{y_k\}$ are elements of $\mathcal{H}(\mathbf{X})$ since such sets are compact in a metric space. Let $A \in \mathcal{H}(\mathbf{X})$. The distance $h(\{y_i\}, A)$ is given by

$$d(y_i, A) \vee d(A, y_i)$$

where by definition, the first distance is the distance from y_i to the closest point in A, and the second is given by the maximum over points in A of the shortest distance from one of those points to y_i that is the maximum distance of a point to y_i so it is clear that this set of singleton sets does not in itself form an ϵ-net of $\mathcal{H}(\mathbf{X})$. However, since the collection is finite, its power set is also finite (the collection of all possible sets made from the points y_i. Take that union of singletons such that the distance from each of the singletons to A is less than ϵ and call this set Y. This certainly exists, since each element of A is closer than ϵ to some y_i. Then the first element of the above formula is less than ϵ. So is the second. For every element of A the distance to one of the elements in $\{y_i\}$ is less than ϵ consequently $d(a, Y) < \epsilon \forall a \in A$. Hence $\mathcal{H}(\mathbf{X})$ is totally bounded and hence compact.

Answers to Chapter III, section 1

III.1.1 For all $x \in X$ we may write

$$f^m \circ f^n(x) = \overbrace{f(f(\cdots f(f^n(x))\cdots))}^{m\text{copies}}$$
$$= \underbrace{\overbrace{f(f(\cdots f}^{m\text{copies}} \, \overbrace{(f(f(\cdots f}^{n\text{copies}}(x)\cdots))\cdots)))}_{m+n\text{copies}}$$
$$= f^{m+n}(x).$$

III.1.2 Let $f : \mathbb{R} \to \mathbb{R}$ with $f(x) = 2x$. Let $f(x_1) = f(x_2)$. Then $2x_1 = 2x_2$ which implies that $x_1 = x_2$ so f is one-to-one. Let $y \in \mathbb{R}$. Then for any y, $1/2y \in \mathbb{R}$, so that $f(1/2y) = 2(1/2y) = y$. Then f is onto. Since f is both one-to-one and onto, it is invertible. Each iteration multiplies the result of the previous iteration by 2. Hence we have

$$f^n(x) = \underbrace{2(2(2 \cdots (2x) \cdots))}_{n \text{copies}} = 2^n x.$$

Similarly,

$$f^{-n}(x) = \underbrace{\frac{1}{2}(\frac{1}{2}(\frac{1}{2} \cdots (\frac{1}{2}x) \cdots))}_{n \text{copies}} = 2^{-n} x.$$

Thus for any integer, we have $f^n(x) = 2^n x$.

III.1.3 Let $f : [0, 1] \rightarrow [0, 1]$ with $f(x) = 1/2x$. f is one-to-one since $1/2x_1 = 1/2x_2 \Rightarrow x_1 = x_2$. But f is not onto, since $f(x) = 1 \Rightarrow x = 2 \notin [0, 1]$. Hence f is not invertible.

III.1.4 Let $f : [0, 1] \rightarrow [0, 1]$ with $f(x) = 4x(1 - x)$. Taking the first derivative, we have $f'(x) = 4 - 8x$, which has a zero at $x = 1/2$ which is a maximum. The value of f there is 1. The minimum value must be at one or the other or both endpoints. The function takes the value 0 at both endpoints. Hence the function is onto. It is not one-to-one however, for example $f(1/4) = f(3/4) = 3/4$.

III.1.5 The first thing to show is that $f(x) \in C$ if $x \in C$. We demonstrate this by noting that if $x \in C$ then x is the endpoint of some closed interval of the form either $[(2^n3^m)/(3^k), (2^n3^m + 1)/(3^k)]$ or $[(3^n - 1)/(3^k), (3^n)/(3^k)]$. Hence f raises the power of the denominator by 1 and is still of this form. So let $f(x) \in C$. f is one-to-one since $x_1/3 = x_2/3 \Rightarrow x_1 = x_2$. It is not onto, since $1 \in C$ and $x/3 = 1 \Rightarrow x = 3 \notin C$. To find another transformation, note that if $x \in C$ then $1 - x \in C$. So $f(x) = (1 - x)/3$ is another such transformation.

III.1.6 Let $f : \mathbb{R}^2 \rightarrow \mathbb{R}^2$ with $f(x_1, x_2) = (2x_1, x_2^2 + x_1)$. f is not invertible because it is not one-to-one. Let $y = f(x) = f(x_1, x_2)$. Then $y = f(x_1, -x_2)$ as well, since x_2 appears in the expression for f only in squared form.

III.1.8 An affine transformation applied to an interval takes it to another interval, possibly reflecting it (transposing the endpoints in interval notation), hence

$$f([1, 2]) = a[1, 2] + b = \begin{cases} [a + b, 2a + b] & \text{if } a > 0, \text{ and} \\ [2a + b, a + b] & \text{if } a < 0. \end{cases}$$

To map into the interval we must have

$$1 \leq a + b \leq 2, \qquad 1 \leq 2a + b \leq 2$$

The new length of the interval will be a, so that $0 \leq a \leq 1$, with $a = 0$ mapping the interval to a single point.

If $f, g : [1, 2] \rightarrow [1, 2]$ then $f([1, 2]) \subset [1, 2]$, hence

$$g \circ f([1, 2]) \subset g([1, 2]) \subset [1, 2]$$

$$f \circ g([1, 2]) \subset f([1, 2]) \subset [1, 2]$$

so that the compositions also map $[1, 2]$ into itself.

We now want the conditions such that

$$f \circ g(\mathbf{X}) \cup g \circ f(\mathbf{X}) = \mathbf{X}$$

Let $f(x) = ax + b$, and $g(x) = cx + d$. We make the simplifying assumption that $f \circ g(\mathbf{X})$ maps the interval to the lower end of itself, and $g \circ f(\mathbf{X})$ to the upper end, and the two overlap across the middle at at least one point. We write out the compositions explicitly:

$$f \circ g(x) = a(cx + d) + b = acx + ad + b;$$

$$g \circ f(x) = c(ax + b) + d = acx + bc + d.$$

The contraction of the interval is the same in each case, and is equal to ac. In order for the two maps to overlap, then, the length of each interval must be at least $1/2$. We therefore have $1/2 \leq |ac| \leq 1$. There are two cases: If $ac > 0$ then $f \circ g$

must map the interval to $[1, 1 + ac]$ and $g \circ f$ must map it to $[2 - ac, 2]$. We have

$$ac + ad + b = 1$$
$$2ac + bc + d = 2$$
$$ad + b = 1 - ac$$
$$bc + d = 2 - 2ac$$

If $ac < 0$ then $f \circ g$ must take $1 \mapsto 1 - ac$ and $2 \mapsto 1$. Similarly $g \circ f$ takes $2 \mapsto 2 + ac$ and $1 \mapsto 2$. This gives

$$2ac + ad + b = 1$$
$$ac + bc + d = 2$$
$$ad + b = 1 - 2ac$$
$$bc + d = 2 - ac$$

Given two values for a, c satisfying $1/2 \le |ac| \le 1$, these equations allow calculation of b, d such that the above condition is satisfied.

III.1.10 $I_n = f^n(I_0)$. Consequently, we have $f(I_n) = f(f^n(I_0)) = f^{n+1}(I_0) = I_{n+1}$. We are given that $I = \cup_{n=0}^{\infty} I_n$ is a finite connected interval, hence

$$f(\cup_{n=0}^{\infty} I_n) = \cup_{n=0}^{\infty} I_{n+1} = \cup_{n=1}^{\infty} I_n = I \setminus I_0$$

Then the length $\ell = \text{length}(I_0) + \text{length}(f(I))$. The length of $f(I)$ is the amount that it was contracted, hence is equal to $a\ell$. This is equal to the length of $I \setminus I_0 = \ell - b$. This gives $a\ell = \ell - b$ or $\ell = b/(1 - a)$. This yields

$$\sum_{n=1}^{\infty} b \cdot a^n = \frac{b}{(1 - a)}$$

III.1.11 Let $f, g : \mathbb{R} \to \mathbb{R}$ be polynomial transformations. Then if

$$f(x) = a_0 + a_1 x + \cdots + a_N x^N \; ; \; g(x) = b_0 + b_1 + \cdots + b_M x^M$$

we have

$$f \circ g(x) = a_0 + a_1 g(x) + a_2 (g(x))^2 + \cdots + a_N (g(x))^N$$

If we expand $g(x)$ in the last term in the expression, the term with the higest power of x is $a_N (b_M x^M)^N$. Since by definition, $a_N, b_M \neq 0$, $a_N b_M^N \neq 0$ so this is a polynomial transformation of degree $M \cdot N$. Using this, the degree of $f^m(x)$ is N^m.

III.1.12 For $n > 1$ a polynomial function $f : \mathbb{R} \to \mathbb{R}$ of degree n is generally not invertible. A quick example is any polynomial having $a_i = 0$ for all odd i. More precisely, suppose that $f(x)$ is invertible. Then there is exactly one number b such that

$$a_0 + a_1 b + \cdots a_N b^N = 0$$

because the function must be one-to-one. This polynomial must then factor into the two terms

$$(x - b)(c_0 + c_1 x + \cdots + c_{n-1} x^{n-1}) = a_0 + \cdots + a_N x^N$$

The polynomial $c_0 + c_1 x + \cdots + c_{N-1} x^{N-1}$ either has no zeroes or it has only one at $x = b$. Assume not. Then by adding or subtracting from c_0, we can force it to have a zero, thus creating a new function with two roots. Such a function is then not one-to-one and therefore not invertible.

III.1.13 For large enough x, $|x^N| \gg |x^{N-1}| \gg \cdots$. So for large $|x|$, $f(x)$ looks like $a_N x^N$ which, for $|x| > 1$ is expanding, since $|x_1^n - x_2^n| > |x_1 - x_2|$. If I is an interval $\{x : |x - a| \le b\}$ for fixed $a, b \in \mathbb{R}$, then if $M > 0$, $\ell(f(I))/f(I) > M$ for $b > \beta$. Let $I = [a - b, a + b]$. Construct the polynomials and subtract.

III.1.15 Parabolae opening up or down will work. For example $f(x) = a(x - 1)^2$ where $a \le 2$.

III.1.16 The function is continuous. $f_\lambda(0) = 0$ for all λ. It is maximized by setting the derivative equal to zero, as follows:

$$f_\lambda(x) = 2\lambda x - \lambda x^2 = \lambda(2x - x^2)$$
$$f_\lambda'(x) = \lambda(2 - 2x)$$

so that $f_\lambda'(x) = 0$ when $x = 1$. This is a maximum, since the second derivative is negative, and has the value $f_\lambda(1) = \lambda$. Hence if $\lambda \in [0, 2]$ then $f_\lambda(x) : [0, 2] \to [0, 2]$. The fold occurs at $(1, \lambda)$.

III.1.17 To be a fold point of $f(x)$, x must be a local maximum or minimum. If $f(x)$ is differentiable, the condition for this is $f'(x) = 0$ and $f''(x) \neq 0$. Polynomial transformations are differentiable. If $f''(x) = 0$ we have an inflection for some functions, (for example $f(x) = x^3$) so no fold exists at x.

III.1.21 To show that the Möbius transformation is invertible, we show that it is one-to-one and onto, as follows:
one-to-one:
Let $f(x_1) = f(x_2)$. Then we have

$$\frac{ax_1 + b}{cx_1 + d} = \frac{ax_2 + b}{cx_2 + d}$$

which yields

$$acx_1x_2 + bcx_2 + adx_1 + bd = acx_1x_2 + bcx_1 + adx_2 + bd$$

or, $bcx_2 + adx_1 = bcx_1 + adx_2$. That is

$$(ad - bc)x_1 = (ad - bc)x_2$$

Since the 'determinant' $ad - bc$ is assumed to be non-zero, $x_1 = x_2$ and the function is one-to-one.
onto:
Let $y \in \mathbb{R}$, we need to find x such that

$$y = \frac{ax + b}{cx + d}$$

We have

$$cxy + dy = ax + b$$
$$cyx - ax = b - dy$$
$$(cy - a)x = b - dy$$
$$x = \frac{-dy + b}{cy - a}$$

Since $ad - bc \neq 0$ we have $(-a)(-d) - bc \neq 0$ so this is another Möbius transformation, and f is onto.

III.1.22 We want to show that if f_1, f_2 are Möbius transformations then $f_1 \circ f_2$ is. Writing

$$f_1 = \frac{a_1 x + b_1}{c_1 x + d_1} \; ; f_2 = \frac{a_2 x + b_2}{c_2 x + d_2}$$

we make the necessary substitutions:

$$\frac{a_1 \frac{a_2 x + b_2}{c_2 x + d_2} + b_1}{c_1 \frac{a_2 x + b_2}{c_2 x + d_2} + d_1} = \frac{a_1(a_2 x + b_2) + b_1(c_2 x + d_2)}{c_1(a_2 x + b_2) + d_1(c_2 x + d_2)}$$

$$= \frac{(a_1 a_2 + b_1 c_2)x + (a_1 b_2 + b_1 d_2)}{(c_1 a_2 + d_1 c_2)x + (b_2 c_1 + d_1 d_2)}$$

which has the form of a Möbius transformation. We need to check:

$$(a_1a_2 + b_1c_2)(b_2c_1 + d_1d_2) - (a_1b_2 + b_1d_2)(c_1a_2 + d_1c_2)$$
$$= (a_1a_2b_2c_1 + a_1a_2d_1d_2 + b_1c_2b_2c_1 + b_1c_2d_1d_2)$$
$$\quad - (a_1b_2c_1a_2 + a_1b_2d_1c_2 + b_1d_2c_1a_2 + b_1d_2d_1c_2)$$
$$= (a_1a_2d_1d_2 + b_1c_1b_2c_2) - (a_1d_1b_2c_2 + a_2d_2b_1c_1)$$
$$= b_2c_2(b_1c_1 - a_1d_1) + a_2d_2(a_1d_1 - b_1c_1)$$
$$= (a_2d_2 - b_2c_2)(a_1d_1 - b_1c_1) \neq 0$$

because f_1, f_2 are Möbius transformations by assumption.

III.1.23 The action of the transformation $f(z) = 1/z$ on the Riemann sphere is to flip it over. It takes $1 \mapsto 1$, $-1 \mapsto -1$ so there is no rotation, and $0 \mapsto \infty$, $\infty \mapsto 0$ which inverts the two poles.

III.1.24 When $f(\infty) = \infty$ then $a/c \to \infty$, since $a < \infty$ we have $c = 0$. Conversly, if $c = 0$ then $f(\infty) = \infty$ by definition. Hence if $\infty \mapsto \infty$ then the transformation looks like

$$f(x) = \frac{ax + b}{d} = \frac{a}{d}x + \frac{b}{d}$$

which is an affine transformation.

III.1.25 $f : \mathbb{R} \to \mathbb{R}$ takes $1 \mapsto 2$, $2 \mapsto 0$ and $0 \mapsto \infty$. So we have $d = 0$ and the function looks like

$$f(x) = \frac{ax + b}{cx}$$

$1 \mapsto 2$ implies $(a + b)/c = 2$. $2 \mapsto 0$ implies $2a + b = 0$ or $b = -2a$. This yields $-a/c = 2$ so that $f(\infty) = 2$. We can only determine the transformation up to a constant (that is, multiplying top and bottom by the same constant yields another identical transformation) so set $c = 1$. This gives $a = -2$, $b = 4$ so the Möbius transformation

$$f(x) = \frac{-2x + 4}{x}$$

will do it.

Answers to Chapter III, section 2

III.2.1 Let the affine transformation be denoted, as usual, by

$$f\begin{pmatrix} x \\ y \end{pmatrix} = \begin{pmatrix} a & b \\ c & d \end{pmatrix}\begin{pmatrix} x \\ y \end{pmatrix} + \begin{pmatrix} e \\ f \end{pmatrix}$$

Since $(0, 0)$ is mapped to $(4, 5)$, we have $e = 4$, and $f = 5$. From the relation $(0, 1) \mapsto (-1, 2)$ we have

$$\begin{pmatrix} -1 \\ 2 \end{pmatrix} = \begin{pmatrix} a & b \\ c & d \end{pmatrix}\begin{pmatrix} 0 \\ 1 \end{pmatrix} + \begin{pmatrix} 4 \\ 5 \end{pmatrix}$$

or,

$$\begin{pmatrix} -1 \\ 2 \end{pmatrix} = \begin{pmatrix} b \\ d \end{pmatrix} + \begin{pmatrix} 4 \\ 5 \end{pmatrix}$$

which gives $b = -5$ and $d = -3$. Finally, using $(1, 0) \mapsto (3, 0)$, we have

$$\begin{pmatrix} 3 \\ 0 \end{pmatrix} = \begin{pmatrix} a & -5 \\ c & -3 \end{pmatrix}\begin{pmatrix} 1 \\ 0 \end{pmatrix} + \begin{pmatrix} 4 \\ 5 \end{pmatrix}$$

or,

$$\begin{pmatrix} 3 \\ 0 \end{pmatrix} = \begin{pmatrix} a \\ c \end{pmatrix} + \begin{pmatrix} 4 \\ 5 \end{pmatrix}$$

which gives $a = -1$ and $c = -5$. The function then is

$$f\begin{pmatrix} x \\ y \end{pmatrix} = \begin{pmatrix} -1 & -5 \\ -5 & -3 \end{pmatrix} \begin{pmatrix} x \\ y \end{pmatrix} + \begin{pmatrix} 4 \\ 5 \end{pmatrix}.$$

III.2.2 The affine transformation will be invertible if the matrix A is invertible, since if $y = Ax + t$, then $A^{-1}(y - t)$ yields x again if A^{-1} exists. Let A and A^{-1} be denoted by

$$A = \begin{pmatrix} a & b \\ c & d \end{pmatrix}$$

$$A^{-1} = \begin{pmatrix} e & f \\ g & h \end{pmatrix}$$

Then we have, by definition,

$$\begin{pmatrix} e & f \\ g & h \end{pmatrix} \begin{pmatrix} a & b \\ c & d \end{pmatrix} = \begin{pmatrix} 1 & 0 \\ 0 & 1 \end{pmatrix}$$

or,

$$ea + fc = 1$$
$$eb + fd = 0$$
$$ga + hc = 0$$
$$gb + hd = 1.$$

Solving the first two for e, f,

$$e = \frac{-fd}{b}$$

$$-fad + fbc = b$$

or

$$f = \frac{-b}{\det A} \; ; e = \frac{d}{\det A}$$

Solving the second two for g, h,

$$g = \frac{-hc}{a}$$

$$-hbc + had = a$$

or

$$g = \frac{-c}{\det A} \; ; h = \frac{a}{\det A}$$

Putting these together, we get

$$A^{-1} = \frac{1}{\det A} \begin{pmatrix} d & -b \\ -c & a \end{pmatrix}$$

Consequently, A^{-1} exists if and only if $\det A \neq 0$. If it does, then if $(x_1, x_2) \mapsto (y_1, y_2)$ we have

$$\begin{pmatrix} x_1 \\ x_2 \end{pmatrix} = \frac{1}{\det A} \begin{pmatrix} d & -b \\ -c & a \end{pmatrix} \begin{pmatrix} y_1 - e \\ y_2 - f \end{pmatrix}$$

III.2.3 Let $f_1, f_2 : \mathbb{R}^2 \to \mathbb{R}^2$ with

$$f_1 \begin{pmatrix} x \\ y \end{pmatrix} = \begin{pmatrix} a & b \\ c & d \end{pmatrix} \begin{pmatrix} x \\ y \end{pmatrix} + \begin{pmatrix} e \\ f \end{pmatrix}$$

and

$$f_2 \begin{pmatrix} x \\ y \end{pmatrix} = \begin{pmatrix} a' & b' \\ c' & d' \end{pmatrix} \begin{pmatrix} x \\ y \end{pmatrix} + \begin{pmatrix} e' \\ f' \end{pmatrix}.$$

Then we have

$$f_3 = f_1 \circ f_2 = \begin{pmatrix} a & b \\ c & d \end{pmatrix} \left(\begin{pmatrix} a' & b' \\ c' & d' \end{pmatrix} \begin{pmatrix} x \\ y \end{pmatrix} + \begin{pmatrix} e' \\ f' \end{pmatrix} \right) + \begin{pmatrix} e \\ f \end{pmatrix}$$

$$= \begin{pmatrix} aa' + bc' & ab' + bd' \\ ca' + dc' & cb' + dd' \end{pmatrix} \begin{pmatrix} x \\ y \end{pmatrix} + \begin{pmatrix} ae' + bf' + e \\ ce' + df' + f \end{pmatrix}$$

Letting

$$a'' = aa' + bc'$$
$$b'' = ab' + bd'$$
$$c'' = ca' + dc'$$
$$d'' = cb' + dd'$$
$$e'' = ae' + bf' + e$$
$$f'' = ce' + df' + f$$

then,

$$f_3 \begin{pmatrix} x \\ y \end{pmatrix} = \begin{pmatrix} a'' & b'' \\ c'' & d'' \end{pmatrix} \begin{pmatrix} x \\ y \end{pmatrix} + \begin{pmatrix} e'' \\ f'' \end{pmatrix}$$

which is affine, as desired.

III.2.4 Let

$$A = \begin{pmatrix} a & b \\ c & d \end{pmatrix}$$

$$B = \begin{pmatrix} e & f \\ g & h \end{pmatrix}$$

Then we have

$$AB = \begin{pmatrix} a & b \\ c & d \end{pmatrix} \begin{pmatrix} e & f \\ g & h \end{pmatrix} = \begin{pmatrix} ae + bg & af + bh \\ ce + dg & cf + dh \end{pmatrix}$$

The products subtracted to form the determinant $\det(AB)$ are given by

$$(ae + bg)(cf + dh) = aecf + aedh + bgcf + bgdh$$
$$(af + bh)(ce + dg) = aecf + afdg + bhce + bgdh$$

Subtracting these yields

$$\det(AB) = aedh + bgcf - afdg - bhce$$
$$= ad(eh - fg) - bc(eh - fg)$$
$$= (ad - bc)(eh - fg)$$
$$= \det A \cdot \det B$$

as desired.

III.2.5 The easiest way to visualize this is to break the desired transformation into two transformations, one which takes the triangle to the line between $(0, 0)$ and $(1, 1)$ and one which takes this line to the line between $(2, 2)$ and $(1, 1)$. The second transformation is an inversion and translation:

$$\begin{pmatrix} 2 \\ 2 \end{pmatrix} = \begin{pmatrix} a & b \\ c & d \end{pmatrix} \begin{pmatrix} 0 \\ 0 \end{pmatrix} + \begin{pmatrix} e \\ f \end{pmatrix}$$

which gives $e = f = 2$ and

$$\begin{pmatrix} 1 \\ 1 \end{pmatrix} = \begin{pmatrix} a & b \\ c & d \end{pmatrix} \begin{pmatrix} 1 \\ 1 \end{pmatrix} + \begin{pmatrix} 2 \\ 2 \end{pmatrix}$$

which gives $a + b - 2 = 1$ and $c + d - 2 = 1$, which by letting $b = c = 0$ gives $a = -1$ and $d = -1$. The first transformation takes the triangle to the line between $(0, 0)$ and $(1, 1)$. $(0, 0) \mapsto (0, 0)$ so that $e = f = 0$. We have

$$\begin{pmatrix} 1 \\ 1 \end{pmatrix} = \begin{pmatrix} a & b \\ c & d \end{pmatrix} \begin{pmatrix} 0 \\ 1 \end{pmatrix}$$

so that $b = d = 1$, and

$$\begin{pmatrix} 1 \\ 1 \end{pmatrix} = \begin{pmatrix} a & b \\ c & d \end{pmatrix} \begin{pmatrix} 1 \\ 0 \end{pmatrix}$$

so that $a = c = 1$. Multiplying the two together, we have

$$\begin{pmatrix} -1 & 0 \\ 0 & -1 \end{pmatrix} \begin{pmatrix} 1 & 1 \\ 1 & 1 \end{pmatrix} \begin{pmatrix} x_1 \\ x_2 \end{pmatrix} + \begin{pmatrix} 2 \\ 2 \end{pmatrix} = \begin{pmatrix} -1 & -1 \\ -1 & -1 \end{pmatrix} \begin{pmatrix} x_1 \\ x_2 \end{pmatrix} + \begin{pmatrix} 2 \\ 2 \end{pmatrix}$$

To put this in scaling ratios and rotation angles, let

$$r_1 + r_2 = \sqrt{2}$$

Then the needed rotation angles are

$$\cos\theta_1 = \sin\theta 1 = \cos\theta_2 = \sin\theta_2 = \frac{-1}{\sqrt{2}}$$

which gives $\theta_1 = \theta_2 = \frac{5\pi}{4}$.

III.2.6 We start with the hint. Let

$$w = \begin{pmatrix} a & b \\ c & d \end{pmatrix} \begin{pmatrix} x_1 \\ x_2 \end{pmatrix} + \begin{pmatrix} e \\ f \end{pmatrix}$$

Since $\begin{pmatrix} e \\ f \end{pmatrix}$ is a translation term, and translation does not affect the area of any transformed polygon, without loss of generality $e = f = 0$. As well, it does not change anything to assume that one vertex of the triangle is at the origin. A triangle is then determined by specifying two points (x_1, x_2) and (y_1, y_2) such that $(y_1, y_2) \neq a \cdot (x_1, x_2)$ (which makes them non-collinear), and neither is the origin (which makes less than three distinct vertices, hence no area). First, show that the area of the parallelogram in figure 2.6 is given by

$$\text{area} = |x_1 y_2 - x_2 y_1|$$

(this is the length of the vector in a vector (cross) product). To see this, rotate the parallelogram until (y_1, y_2) becomes $(y_1', 0)$. Then if the new coordinates for (x_1, x_2) are given by (x_1', x_2') the area of the parallelogram is the base times the height, or $y_1' x_2'$. The rotation is $-\theta$ with

$$\cos\theta = \frac{y_1}{\sqrt{y_1^2 + y_2^2}} \; ; \; \sin\theta = \frac{y_2}{\sqrt{y_1^2 + y_2^2}}$$

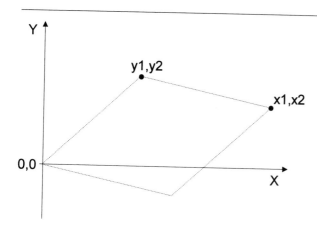

Figure 2.6. Getting the area of a triangle from a parallelogram

We have:

$$\begin{pmatrix} \cos\theta & \sin\theta \\ -\sin\theta & \cos\theta \end{pmatrix}\begin{pmatrix} x_1 \\ x_2 \end{pmatrix} = \begin{pmatrix} x_1\cos\theta + x_2\sin\theta \\ x_2\cos\theta - x_1\sin\theta \end{pmatrix}$$

$$\begin{pmatrix} \cos\theta & \sin\theta \\ -\sin\theta & \cos\theta \end{pmatrix}\begin{pmatrix} y_1 \\ y_2 \end{pmatrix} = \begin{pmatrix} y_1\cos\theta + y_2\sin\theta \\ y_2\cos\theta - y_1\sin\theta \end{pmatrix}$$

so that

$$y_1'x_2' = (y_1\cos\theta + y_2\sin\theta)(x_2\cos\theta - x_1\sin\theta)$$

$$(y_1^2 + y_2^2)y_1'x_2' = y_1x_2y_1^2 - y_1x_1y_1y_2 + y_2x_2y_1x_2 - x_1y_2y_2^2$$

$$= y_1^2(x_2y_1 - x_1y_2) + y_2^2(x_2y_1 - x_1y_2)$$

yielding $|y_1'x_1'| = |x_1y_2 - y_1x_2|$ as desired. The area of the desired triangle is then $\frac{1}{2}|x_1y_2 - y_2x_1|$. Apply the transformation

$$w\begin{pmatrix} x_1 \\ x_2 \end{pmatrix} = \begin{pmatrix} a & b \\ c & d \end{pmatrix}\begin{pmatrix} x_1 \\ x_2 \end{pmatrix}$$

to each of the two vectors. Then the new area is derived by

$$x_1' = ax_1 + bx_2; \qquad y_1' = ay_1 + by_2$$
$$x_2' = cx_1 + dx_2; \qquad y_2' = cy_1 + dy_2$$

yielding

$$\frac{1}{2}|acx_1y_1 + adx_1y_2 + bcx_2y_1 + bdx_2y_2 - acx_1y_1 - bcx_1y_2 - adx_2y_1 - bcx_2y_2|$$

$$= \frac{1}{2}|(ad - bc)x_1y_2 - (ad - bc)x_2y_1|$$

$$= \frac{1}{2}|\det A||x_1y_2 - x_2y_1|$$

A polygon (and many other surfaces) can be broken into triangles on each of which this formula applies, which finishes the proof. To see that $\det A < 0$ corresponds to reflection, write

$$\det\left(\begin{pmatrix} -1 & 0 \\ 0 & 1 \end{pmatrix}\begin{pmatrix} a & b \\ c & d \end{pmatrix}\right) = -\det\begin{pmatrix} a & b \\ c & d \end{pmatrix}$$

Note: The vector/cross/wedge product used to do the proof this way can be generalized to be applicable on other types of surfaces and hypersurfaces. The generalized version preserves the form $|\det A|\cdot$("area") to such surfaces and volumes.

III.2.7 Let $w(x) = Ax + t$ be an $\mathbb{R}^2$ similitude. Then w has one of the forms:

$$w\begin{pmatrix} x_1 \\ x_2 \end{pmatrix} = \begin{pmatrix} r\cos\theta & -r\sin\theta \\ r\sin\theta & r\cos\theta \end{pmatrix}\begin{pmatrix} x_1 \\ x_2 \end{pmatrix} + \begin{pmatrix} e \\ f \end{pmatrix}$$

$$w\begin{pmatrix} x_1 \\ x_2 \end{pmatrix} = \begin{pmatrix} r\cos\theta & r\sin\theta \\ r\sin\theta & -r\cos\theta \end{pmatrix}\begin{pmatrix} x_1 \\ x_2 \end{pmatrix} + \begin{pmatrix} e \\ f \end{pmatrix}$$

The first can be rewritten as

$$w\begin{pmatrix} x_1 \\ x_2 \end{pmatrix} = r\begin{pmatrix} \cos\theta & -\sin\theta \\ \sin\theta & \cos\theta \end{pmatrix}\begin{pmatrix} x_1 \\ x_2 \end{pmatrix} + \begin{pmatrix} e \\ f \end{pmatrix} = rR_\theta\begin{pmatrix} x_1 \\ x_2 \end{pmatrix} + \begin{pmatrix} e \\ f \end{pmatrix}$$

By making the substitution $\phi = -\theta$ the transformation

$$RR_\theta = \begin{pmatrix} 1 & 0 \\ 0 & -1 \end{pmatrix}\begin{pmatrix} \cos\theta & -\sin\theta \\ \sin\theta & \cos\theta \end{pmatrix}$$

$$= \begin{pmatrix} \cos\theta & -\sin\theta \\ -\sin\theta & -\cos\theta \end{pmatrix}$$

$$= \begin{pmatrix} \cos\phi & -\sin\phi \\ \sin\phi & \cos\phi \end{pmatrix}$$

so with this substitution, the second form for a similitude becomes

$$w\begin{pmatrix} x_1 \\ x_2 \end{pmatrix} = rRR_\phi\begin{pmatrix} x_1 \\ x_2 \end{pmatrix} + \begin{pmatrix} e \\ f \end{pmatrix}$$

which has the proper form.

III.2.9 There are many possible ways to write a similitude $w(x) = Ax + t$ which will take the Sierpinski triangle to a subset of itself, and we give the simplest one. To take the triangle to the subtriangle with corner $(0, 0)$ we note that by taking

$$(0, 0) \mapsto (0, 0)$$

$$(1, 0) \mapsto (\frac{1}{2}, 0)$$

$$(\frac{1}{2}, 1) \mapsto (\frac{1}{4}, \frac{1}{2})$$

we have completely determined w. The first transformation implies that

$$t = \begin{pmatrix} e \\ f \end{pmatrix} = \begin{pmatrix} 0 \\ 0 \end{pmatrix}$$

For the other two,

$$\begin{pmatrix} a & b \\ c & d \end{pmatrix}\begin{pmatrix} 1 \\ 0 \end{pmatrix} = \begin{pmatrix} a \\ c \end{pmatrix} = \begin{pmatrix} \frac{1}{2} \\ 0 \end{pmatrix}$$

so that $a = \frac{1}{2}$ and $c = 0$.

$$\begin{pmatrix} \frac{1}{2} & b \\ 0 & d \end{pmatrix}\begin{pmatrix} \frac{1}{2} \\ 1 \end{pmatrix} = \begin{pmatrix} \frac{1}{4} + b \\ d \end{pmatrix} = \begin{pmatrix} \frac{1}{4} \\ \frac{1}{2} \end{pmatrix}$$

so that $b = 0$ and $d = \frac{1}{2}$. Then w looks like

$$w\begin{pmatrix} x_1 \\ x_2 \end{pmatrix} = \begin{pmatrix} \frac{1}{2} & 0 \\ 0 & \frac{1}{2} \end{pmatrix}\begin{pmatrix} x_1 \\ x_2 \end{pmatrix}$$

III.2.10 The simplest way to show this decomposition is to multiply it out.

$$\begin{pmatrix} r_1 & 0 \\ 0 & r_2 \end{pmatrix} \begin{pmatrix} \cos\theta & -\sin\theta \\ \sin\theta & \cos\theta \end{pmatrix} \begin{pmatrix} r_3 & 0 \\ 0 & r_4 \end{pmatrix}$$

$$= \begin{pmatrix} r_1\cos\theta & -r_1\sin\theta \\ r_2\sin\theta & r_2\cos\theta \end{pmatrix} \begin{pmatrix} r_3 & 0 \\ 0 & r_4 \end{pmatrix}$$

$$= \begin{pmatrix} r_1 r_3\cos\theta & -r_1 r_4\sin\theta \\ r_2 r_3\sin\theta & r_2 r_4\cos\theta \end{pmatrix}$$

This is a general affine transformation because the four terms $r_1 r_3$, $r_1 r_4$, $r_2 r_3$, and $r_2 r_4$ are *linearly independent*. This means that none of these can be written as a linear combination (sum with constant coefficients) of the other three, hence any arbitrary value can be written this way. It is more important to understand what is going on geometrically, which is done by looking at the action of this transformation on the coordinate axes. The transformation by r_3, r_4 rescales the coordinates, that is stretches or shrinks each coordinate and possibly reflects one or both. The rotation matrix rotates the coordinates by some angle θ, leaving the rescaled coordinates off axis. The action of the second rescaling has two functions. It further stretches or shrinks the rotated axes, but because they are no longer lined up with horizontal or vertical, it also changes the *angle* between them. This has the effect of skewing the new map of the plane. We are left with a transformation which translates, rotates, rescales, reflects and skews. This is the most general affine transformation, and requires all three parts of the decomposition to accomplish it.

III.2.12 We multiply the two matrices to get the identity matrix:

$$\begin{pmatrix} a & b \\ c & d \end{pmatrix} \begin{pmatrix} \frac{d}{\det A} & \frac{-b}{\det A} \\ \frac{-c}{\det A} & \frac{a}{\det A} \end{pmatrix} = \begin{pmatrix} \frac{ad-bc}{\det A} & \frac{ba-ab}{\det A} \\ \frac{cd-dc}{\det A} & \frac{-bc+ad}{\det A} \end{pmatrix}$$

$$= \begin{pmatrix} \frac{\det A}{\det A} & 0 \\ 0 & \frac{\det A}{\det A} \end{pmatrix}$$

$$= \begin{pmatrix} 1 & 0 \\ 0 & 1 \end{pmatrix}$$

which proves the statement.

III.2.13 Let

$$B = \begin{pmatrix} a & b \\ c & d \end{pmatrix}$$

$$A = \begin{pmatrix} e & f \\ g & h \end{pmatrix}$$

By exercise III.2.12 we have

$$B^{-1} = \frac{1}{\det B} \begin{pmatrix} d & -b \\ -c & a \end{pmatrix}$$

$$= \begin{pmatrix} \frac{d}{\det B} & \frac{-b}{\det B} \\ \frac{-c}{\det B} & \frac{a}{\det B} \end{pmatrix}$$

Then we have

$$BAB^{-1} = \begin{pmatrix} a & b \\ c & d \end{pmatrix} \begin{pmatrix} e & f \\ g & h \end{pmatrix} \begin{pmatrix} \frac{d}{\det B} & \frac{-b}{\det B} \\ \frac{-c}{\det B} & \frac{a}{\det B} \end{pmatrix}$$

$$= \begin{pmatrix} a & b \\ c & d \end{pmatrix} \begin{pmatrix} \frac{ed-fc}{\det B} & \frac{af-eb}{\det B} \\ \frac{gd-hc}{\det B} & \frac{ah-gb}{\det B} \end{pmatrix}$$

$$= \begin{pmatrix} \frac{aed-afc+bgd-bhc}{\det B} & \frac{a^2-aeb+bah-gb^2}{\det B} \\ \frac{ced-fc^2+gd^2-dhc}{\det B} & \frac{caf-ceb+dah-dgb}{\det B} \end{pmatrix}$$

which gives

$$\text{tr}(BAB^{-1}) = \frac{1}{\det B}(aed - afc + bgd - bhc + caf - ceb + dah - dgb)$$

$$= \frac{1}{\det B}(ade - bce + adh - bch)$$

$$= \frac{\det B}{\det B}(e + h) = \text{tr}\, A$$

The other equation is somewhat simpler:

$$\det(BAB^{-1}) = \det B \det A \det B^{-1} = \det B \det B^{-1} \det A = \det(BB^{-1}) \det A$$

Now $BB^{-1} = I$ the identity matrix, and $\det I = 1$, so

$$\det(BAB^{-1}) = \det(BB^{-1}) \det A = \det A.$$

III.2.14 We simplify the problem by noting that if q is any scalar,

$$\frac{|A(qx)|}{|qx|} = \left(\frac{(aqx_1 + bqx_2)^2 + (cqx_1 + dqx_2)^2}{q^2(x_1^2 + x_2^2)} \right)^{\frac{1}{2}}$$

$$= \left(\frac{q^2(ax_1 + bx_2)^2 + q^2(cx_1 + dx_2)^2}{q^2(x_1^2 + x_2^2)} \right)^{\frac{1}{2}}$$

$$= \frac{|Ax|}{|x|}$$

consequently, it suffices to maximize over x such that $|x| = 1$, that is, over the unit circle. For the Euclidean metric then, we let $x_1 = \cos\theta$ and $x_2 = \sin\theta$. Then recognizing that the derivative will be zero only if the derivative of the quantity inside the square root sign is zero, we maximize

$$0 = \frac{d}{d\theta}((a\cos\theta + b\sin\theta)^2 + (c\cos\theta + d\sin\theta)^2)$$

$$= 2(a\cos\theta + b\sin\theta)(-a\sin\theta + b\cos\theta)$$

$$\quad + 2(c\cos\theta + d\sin\theta)(-c\sin\theta + d\cos\theta)$$

$$= -a^2\sin\theta\cos\theta + ab\cos^2\theta - ab\sin^2\theta + b^2\sin\theta\cos\theta$$

$$\quad - c^2\sin\theta\cos\theta + cd\cos^2\theta - cd\sin^2\theta + d^2\sin\theta\cos\theta$$

$$= \frac{b^2 - a^2}{2}\sin 2\theta + ab\cos 2\theta + \frac{d^2 + c^2}{2}\sin 2\theta + cd\cos 2\theta$$

$$0 = 2(ab + cd)\cos 2\theta + ((b^2 + d^2) - (a^2 + c^2))\sin 2\theta, \text{ or,}$$

$$\tan 2\theta = \frac{((a^2 + c^2) - (b^2 + d^2))}{2(ab + cd)}$$

This last equation has (in general) four solutions. Two are minima, two are maxima, and the norm exists. The unit circle has been transformed to an ellipse, the semi-major axis of which is the norm of A.

For the Manhattan metric, we again can restrict to the unit circle which in this metric is the set of points (x_1, x_2) such that $|x_1 + x_2| = 1$, and is a diamond centered at the origin, with vertices on the axes at $(\pm 1, 0)$ and $(0, \pm 1)$. By looking at the Euclidean case, we can see that the transformed version of this figure under A will contain two maxima and two minima. From exercise II.2.9, this transformation is a composition of a coordinate rescaling, a rotation, and another coordinate rescaling (and possibly a reflection). The result is again a parallelogram centered at the origin. The triangle inequality will now tell us that the maximum is at one pair of vertices. These are the images of the original vertices, hence one of these two images is the maximum. We have

$$A(x_1, x_2) = (ax_1 + bx_2, cx_1 + dx_2)$$

at the vertices we have $|A| = max((a + c), (b + d))$.

Answers to Chapter III, section 3

III.3.1 (i) The most general Möbius transformation such that $f(\infty) = \infty$ is of the form $f(z) = az + b$. To show this we need to show that if $f(\infty) = \infty$ then $c = 0$. Suppose not. Then for $c \neq 0$ we have $f(\infty) = a/c \neq \infty$ because a is finite. We thus have

$$f(z) = \frac{az + b}{d} = \frac{a\overline{d}z + b\overline{d}}{|d|^2}$$

Define $a' = a\overline{d}/|d|^2$ and $b' = b\overline{d}/|d|^2$. then $f(z) = a'z + b'$ as desired.

(ii) Suppose $f : \mathbb{R}^2 \to \mathbb{R}^2$ is a two dimensional similitude without reflection. The we may write f as

$$f\begin{pmatrix} x \\ y \end{pmatrix} = \begin{pmatrix} \cos\theta & -\sin\theta \\ \sin\theta & \cos\theta \end{pmatrix}\begin{pmatrix} x \\ y \end{pmatrix} + \begin{pmatrix} e \\ f \end{pmatrix}$$

Let $z = x + iy$, $b = e + if$ and $a = a_1 + ia_2$. We have $az = (a_1 x - a_2 y) + i(a_1 y + a_2 x)$ which we can write using real coordinates and matrix notation as

$$\begin{pmatrix} a_1 & -a_2 \\ a_2 & a_1 \end{pmatrix} = A\begin{pmatrix} x \\ y \end{pmatrix}$$

Letting $\det A = r^2$ ($\det A$ is positive since we presumed no reflection) and $\theta = \cos^{-1}(a_1/r)$ we have our result. Since we divided out $\det A$ the determinant is 1 and with $a = d$ and $b = -c$, the new determinant must read $\cos^2\theta + \sin^2\theta = 1$, as a check that the new values match up.

III.3.2 We wish to prove that the operation $f(z) = \frac{1}{z}$ can be expressed as mapping the complex plane with infinity, to a sphere of radius 1, with equator at $|z| = 1$, rotating the sphere by π around the real axis (the axis between -1 and 1), and mapping the result back to the plane. We first establish that for at least some points this is correct.

(i) The results of this mapping on the points $\mathcal{O}$ and ∞ would be to exchange them. The map $f(z) = \frac{1}{z}$ does indeed exchange these two points.

(ii) The result on the equator is to flip it about the real line, and map it back to itself. Thus every point on the equator $|z| = 1$ is mapped to the equator. If we were to look down from above, the result of the rotation would be to exchange the halves of the equator above and below the real line, that is, each point on $|z| = 1$ should go to it's complex conjugate. We have

$$f(z) = \frac{1}{z} = \frac{\overline{z}}{|z|^2}$$

For points on the equator, the denominator of the last expression is 1 and each point does indeed move to it's conjugate. We note in passing that having specified the action on these many points, as well as that we are using a Möbius transformation is in fact enough to prove the assertion, but we will perform the explicit map.

(iii) We need a coordinate system to represent points on the sphere. Since the map used to generate the Riemann sphere is to map each point to the point on the surface where a line connecting the north pole to the plane point intersects the sphere, we choose the angle ϕ between this line and the diameter of the sphere connecting the poles (interior angle). This gives us a "latitude" coordinate, for a longitude, we take the polar coordinate angle of the plane point, with respect to the positive real line. Expressing the point z in the plane as $re^{i\theta}$, this amounts to choosing θ. We need the coordinate transformation for ϕ. The line from the north pole to the point z forms the hypotenuse of a right triangle, the side adjacent is the radius of the sphere to the north pole, and the side opposite is the plane vector to z. Consequently, we have

$$r = |z| = \tan \phi$$

We may thus express the coordinate transformation as

$$z = z_1 + iz_2 = \tan \phi \cos \theta + i \tan \phi \sin \theta$$

or more simply as

$$z = \tan \phi e^{i\theta}.$$

How do these coordinates change when we rotate the sphere? As with the equator, the θ coordinate becomes the negative angle with the positive real line, that is $\theta \mapsto -\theta$. For ϕ we need some plane geometry. First, we note that if we have a diameter of a circle in the plane, and a point not on that diameter, the triangle formed by the chords between the endpoints and the third point is a right triangle. To see this, add the radius to the third point to divide the triangle into two isoceles triangles, and sum up the angles at the center of the circle. Second, we take the same planar circle, and draw the lines connecting a point on the perpendicular bisector of the diameter to the endpoints. If we now draw a line from one of the ends of the diameter to the point at which the line from the other end to the bisector intersects the circle, the triangle so formed is similar to the two triangles formed by the lines previously drawn to the bisector. Consequently, given that the tangent of the angle from the north pole to a point on the sphere is $\tan \phi$, the tangent of the angle formed by the *south* pole is $\cot \phi$.

When we rotate the Riemann sphere around the real line, we have exactly this situation, namely that the angle ϕ of the original point on the sphere is now the angle measured from the south pole. Consequently, the tangent of the new angle to the north pole is $\cot \phi$. We now look at the action of $f(z)$ on $z = \tan \phi e^{i\theta}$. We have

$$f(z) = \frac{1}{z} = \frac{1}{\tan \phi e^{i\theta}} = \cot \phi e^{i(-\theta)}$$

so the spherical coordinates have taken $\theta \mapsto -\theta$ and $\tan \phi \mapsto \cot \phi$ as desired.

III.3.3 If $f(z)$ is not a similitude then, by problem 1, $c \neq 0$. We may thus let

$$e = \frac{a}{c}$$

$$f = -\frac{ad - bc}{c^2}$$

$$g = \frac{d}{c}.$$

This writes $f(z)$ in the proper form as we now check:

$$e + \frac{f}{z + g} = \frac{a}{c} - \frac{\left(\frac{ad-bc}{c^2}\right)}{z + \frac{d}{c}}$$

$$= \frac{az + \frac{d}{c}a}{cz + d} - \frac{\left(\frac{ad-bc}{c}\right)}{cz + d}$$

$$= \frac{az + b}{cz + d}$$

$$= f(z)$$

as desired.

III.3.6 We wish to show that a Möbius transformation $f : \hat{\mathbb{C}} \to \hat{\mathbb{C}}$ is always invertible. We divide the problem into it's two cases: Let

$$f(z) = \frac{az + b}{cz + d}$$

(i) Suppose $c = 0$. Then the Möbius transformation can be rewritten as $f(z) = a'z + b'$. Let $w = f(z)$. Then the inverse is given by

$$z = \frac{w - b}{a}$$

(ii) Suppose $c \neq 0$. Then the transformation can be rewritten in the form

$$f(z) = e + \frac{f}{z + g}$$

Let $w = f(z)$. Then the inverse may be written

$$z = g + \frac{f}{w - e}$$

III.3.7 See answer to exercise 1.22 in Chapter III.

III.3.8 We wish to transform the real line to the unit circle in $\hat{\mathbb{C}}$. This means taking $z = z_1 + iz_2$ such that $z_2 = 0$ to $|z| = 1$. Looking at the Riemann sphere, this can be accomplished by rotating the north and south poles to the equator. Specifically, we take $\infty \mapsto i$ and $0 \mapsto -i$. Since we need one more point to define the transformation, we may do this by insisting that we rotate about the axis between 1 and -1 by taking $1 \mapsto 1$. We have

$$\infty \mapsto i \Rightarrow \frac{a}{c} = i$$

$$0 \mapsto -i \Rightarrow \frac{b}{d} = -i$$

$$1 \mapsto 1 \Rightarrow \frac{a + b}{c + d} = 1$$

From this we get

$$a = ic$$
$$b = -id$$
$$\frac{a + b}{c + d} = 1$$
$$a + b = c + d$$
$$ic - id = c + d$$
$$d = -c \left(\frac{1 - i}{1 + i} \right)$$
$$= -c \left(\frac{1 - 2i + i^2}{1 - i^2} \right)$$
$$d = ic$$

Choose $a = 1$. Then we have

$$f(z) = \frac{z + i}{-iz + 1} = \frac{iz - 1}{z + i}$$

III.3.9 We do out the first few terms of $f^n(z)$ to see what we are dealing with:

$$f^0(z) = z$$

$$f^1(z) = \frac{1}{1+z}$$

$$f^2(z) = \frac{1}{1 + \frac{1}{1+z}}$$

$$f^n(z) = \frac{1}{1 + \cfrac{1}{1 + \ddots \frac{1}{z}}}$$

This object is known as a *continued fraction* and is usually written as

$$[a_0, a_1, \ldots, a_k] = a_0 + \cfrac{1}{a_1 + \cfrac{1}{a_2 + \cfrac{1}{\ddots \frac{1}{a_k}}}}$$

There is a whole body of mathematics to deal with these interesting creatures, they can be used to represent numbers, functions, traces of matrices, and other things. We need only the following relations: Let $p_0 = a_0$, $p_1 = a_1 a_0 + 1$, $p_k = a_k p_{k-1} + p_{k-2}$ and $q_0 = 1$, $q_1 = a_1$, $q_k = a_k q_{k-1} + q_{k-2}$. Then the following relation holds:

$$[a_0, a_1, \ldots, a_k] = \frac{p_k}{q_k}$$

For our function, we have

$$f^n(z) = [0, 1, 1, \ldots, 1, z]$$

with n one's implied. If we look at the growth of the terms p_i they represent (for terms before the nth term, the terms in a *Fibonacci series*. That is, let $F_0 = 1$, $F_1 = 1$, $F_n = F_{n-1} + F_{n-2}$. The first few terms look like

$$1, 1, 2, 3, 5, 8, 13, \ldots$$

Since $p_0 = 0$, $p_1 = 1$, $p_2 = 1$, $p_3 = 2$, and so forth, we may write $p_k = F_{k-1}$, for $k < n$. Similarly, we may write $q_k = F_k$ for $k < n$. Then $f^n(z)$ may be written, using the above relations,

$$f^n(z) = \frac{z F_{n-2} + F_{n-3}}{z F_{n-1} + F_{n-2}}$$

For the inverse transformations, we look at

$$w = \frac{1}{1+z} \Rightarrow z = -1 + \frac{1}{w}$$

Writing out the first few iterations, we have

$$f^{-1}(w) = -1 + \frac{1}{w}$$

$$f^{-2}(w) = -1 + \frac{1}{-1 + \frac{1}{w}}$$

Again we have the form of a continued fraction, this time

$$f^{-n}(w) = [-1, -1, -1, \ldots, -1, w]$$

Using exactly the same kind of analysis as before, we have

$$
\begin{aligned}
p_0 &= -1 &;&& q_0 &= 1 \\
p_1 &= 2 &;&& q_1 &= -1 \\
p_2 &= -3 &;&& q_2 &= 2 \\
&\vdots &;&& &\vdots \\
p_k &= \mp F_{k+1} &;&& q_k &= \pm F_k
\end{aligned}
$$

which gives

$$
f^{-n}(w) = -\frac{w F_n + F_{n-1}}{w F_{n-1} + F_{n-2}}
$$

III.3.10 We want the geometrical interpretation of the map

$$
f(z) = i + \frac{1}{z - i}
$$

To get this we first note that by rewriting

$$
f(z) = i + \frac{1}{z - i} = \frac{iz + 1 + 1}{z - i} = \frac{iz + 2}{z - 1}
$$

we have $f(\infty) = a/c = i$, $f(i) = \infty$. This suggests we first translate the plane to bring the point i to the origin. For this we have

$$
\theta(z) = z - i
$$
$$
\theta^{-1}(z) = z + i
$$
$$
\theta \circ f \circ \theta^{-1}(z) = \left(i + \frac{1}{z + i - i}\right) - i = \frac{1}{z}
$$

Consequently we can describe the mapping as translating the plane up by one unit, mapping the plane to the Riemann sphere, rotating the Riemann sphere about the real line, and mapping back. The net effect is a translation and an inversion.

Answers to Chapter III, section 4

III.4.2 To be correct inverse functions for $f(z)$ we must verify that $f(f^{-1}(z)) = z$. For points in the upper half plane, we have

$$
\begin{aligned}
&f(a(x_1, x_2) + ib(x_1, x_2)) \\
&= (a^2 - b^2 - 1) + 2iab \\
&= \frac{\sqrt{(1 + x_1)^2 + x_2^2} + 1 + x_1}{2} - \frac{\sqrt{(1 + x_1)^2 + x_2^2} - 1 - x_1}{2} - \frac{2}{2} \\
&\quad + 2i\sqrt{\frac{\sqrt{(1 + x_1)^2 + x_2^2} + 1 + x_1}{2}}\sqrt{\frac{\sqrt{(1 + x_1)^2 + x_2^2} - 1 - x_1}{2}} \\
&= x_1 + 2i\sqrt{\frac{(1 + x_1)^2 + x_2^2 - (1 + x_1)^2}{4}} \\
&= x_1 + ix_2 = z
\end{aligned}
$$

so that w_1 does indeed invert $f(z)$ there. w_2 differs from w_1 in the above derivation only in the imaginary part which replaces $2iab$ with $2i(-a)(-b)$ and hence also inverts $f(z)$. Finally, for the lower half of the plane, we notice that we are in need of a negative sign preceding the imaginary term, since the positive square root is indicated. This is provided by replacing a

by $-a$. This has no effect on the real part, because a appears there only as an a^2 term, completing the test for the inverse of $f(z)$.

III.4.3 Let $f(z) = z^2 + 1$. The critical points are those values of z such that $f'(z) = 0$, that is, the first derivative vanishes. $f'(z) = 2z = 0$ when $z = 0$ so this is the critical point of $f(z)$. The critical values of $f(z)$ are by definition the range values at which $f'(z)$ vanishes, that is $f(z)$ such that z is a critical point. This occurs at $f(0) = 1$ so 1 is a critical point for $f(z)$.

III.4.5 We determine the local approximate affine transformations for these functions by expanding about z_0 in a Taylor (Laurent) series for $f(z)$, and restricting to the first two coefficients a_0 and a_1.

(a) $f(z) = z^2$ and $z_0 = 1$. $a_0 = f(1) = 1$. For the linear coefficient, we have

$$a_1 = \frac{f'(1)}{1!} = 2$$

Letting $w(z)$ be the linear approximation, we have

$$w(z) = 1 + 2(z - 1) = 2z - 1$$

(b) $f(z) = \frac{1}{z}$ and $z_0 = 1 + i$. We have

$$a_0 = f(z_0) = \frac{1}{1 + i}$$

$$a_1 = f'(z_0) = \frac{-1}{z_0^2} = \frac{-1}{(1 + i)^2}$$

Using the same notation as in (a), we have

$$w(z) = \frac{1}{1 + i} + \frac{-1}{(1 + i)^2}(z - (1 + i))$$

$$= \frac{1}{1 + i} + \frac{-z}{(1 + i)^2} + \frac{1 + i}{(1 + i)^2}$$

$$= \frac{2}{1 + i} - \frac{z}{2i}$$

$$= \frac{i}{2}z + \frac{2}{1 + i}$$

(c) $f(z) = (z - 1)^3$ and $z_0 = 1 - i$. We have

$$a_0 = f(z_0) = ((1 - i) - 1)^3 = (-i)^3 = i$$

$$a_1 = f'(z_0) = 3(z_0 - 1)^2 = 3(-i)^2 = -3$$

Putting these together gives

$$w(z) = i + -3(z - (1 - i))$$

$$= i + 3 - 3i - 3z$$

$$= -3z + 3 - 2i$$

Answers to Chapter III, section 5

III.5.3 Let x_f be the fixed point of $f(x) = Ax + t$. Let $\theta(x) = x - x_f$. Then $\theta^{-1}(x') = x' + x_f$, and we have

$$\theta \circ \theta^{-1}(x') = A(x' + x_f) + t - x_f$$

$$= Ax' + Ax_f + t - x_f$$

$$= Ax' + x_f - x_f$$

$$= Ax' = f'(x')$$

as desired.

III.5.4 Using the coordinate change notation $\theta : \mathbf{X} \to \mathbf{X}$ we have $\theta(x) = 2x - 1$, which yields $\theta^{-1}(x') = (x' + 1)/2$. We get the transformed f' in the new space by

$$f'(x') = \theta \circ f \circ \theta^{-1}(x') : \mathbf{X}' \to \mathbf{X}'$$

which we get from

$$f \circ \theta^{-1}(x') = \left(\left(\frac{x' + 1}{2} \right) - 1 \right)^2 + 1 = \left(\frac{x' - 1}{2} \right)^2 + 1$$

giving us,

$$\theta \circ f \circ \theta^{-1}(x') = 2 \left(\left(\frac{x' - 1}{2} \right)^2 + 1 \right) - 1 = \frac{1}{2}(x' - 1)^2 + 1$$

III.5.5 We first find the fixed points of the Möbius transformation

$$f(z) = \frac{(z + 2)}{(4 - z)} = z$$

Multiplying through we get

$$z + 2 = 4z - z^2$$

or

$$z^2 - 3z + 2 = 0$$

which factors into

$$(z - 1)(z - 2) = 0$$

There are two fixed points, at $z = 1$ and $z = 2$.

Make the coordinate transformation which takes $1 \mapsto 0$ and $2 \mapsto \infty$. The transformation

$$\theta(z) = \frac{z - 1}{z - 2}$$

will work. We need the inverse of this:

$$w = \frac{z - 1}{z - 2}$$
$$(z - 2)w = z - 1$$
$$zw - 2w = z - 1$$
$$wz - z = 2w - 1$$
$$z = \theta^{-1}(w) = \frac{2w - 1}{w - 1}$$

We make the composition $f'(w) = \theta \circ f \circ \theta^{-1}(w)$ as follows:

$$f \circ \theta^{-1}(w) = \frac{\frac{2w-1}{w-1} + 2}{4 - \frac{2w-1}{w-1}} = \frac{2w - 1 + 2w - 2}{4w - 4 - 2w + 1} = \frac{4w - 3}{2w - 3}$$

so that,

$$\theta \circ f \circ \theta^{-1}(w) = \frac{\frac{4w-3}{2w-3} - 1}{\frac{4w-3}{2w-3} - 2} = \frac{4w - 3 - 2w + 3}{4w - 3 - 4w + 6} = \frac{2}{3}w$$

This map shrinks the sphere by a factor of $2/3$. The whole process can be written by expanding the real line from 1 to 2 to the the whole positive real line, then shrinking the resultant Riemann sphere, then reducing the positive real line to the line between 1 and 2 again. The net result is shrinking the sphere in a non-uniform manner.

III.5.6 Let $W(x) = Ax + t$ which we express as usual. The fixed point of this transformation is

$$x_f = \begin{pmatrix} x_1 \\ x_2 \end{pmatrix} = \begin{pmatrix} a & b \\ c & d \end{pmatrix} \begin{pmatrix} x_1 \\ x_2 \end{pmatrix} + \begin{pmatrix} e \\ f \end{pmatrix}$$

which we rewrite as

$$x_{f_1} = ax_{f_1} + bx_{f_2} + e \; ; \; x_{f_2} = cx_{f_1} + dx_{f_2} + f$$

We solve this as a system of linear equations:

$$0 = (a - 1)x_{f_1} + bx_{f_2} + e$$
$$0 = cx_{f_1} + (d - 1)x_{f_2} + f$$

which gives

$$0 = c(a - 1)x_{f_1} + bcx_{f_2} + ce$$
$$0 = c(a - 1)x_{f_1} + (a - 1)(d - 1)x_{f_2} + e$$
$$0 = ((a - 1)(d - 1) - bc)x_{f_2} + (a - 1)f - ce$$
$$0 = (\det(A - I))x_{f_2} + (a - 1)f - ce$$

yielding

$$x_{f_2} = \frac{ce - (a - 1)f}{\det(A - I)}$$

A nearly identical derivation gives

$$x_{f_1} = \frac{bf - (d - 1)e}{\det(A - I)}$$

Notice that these equations do indeed become invalid if $\det(A - I) = 0$, when the "fixed point" tends to infinity. To transform coordinates so that x_f is the origin, choose the transformation $\theta(x) = x - x_f$. The inverse of this, letting $w = \theta(x)$ is just $x = w + x_f$. The whole coordinate transformation is then

$$\theta \circ f \circ \theta^{-1}(w) = (A(w + x_f) + t) - x_f$$
$$= Aw + Ax_f + t - x_f$$
$$= Aw + (Ax_f + t) - x_f$$
$$= Aw + x_f - x_f$$
$$= Aw$$

So that the effect, in the transformed system is just the action of the linear transformation Ax.

III.5.7 By exercise III.2.13 we know that

$$\det D = \lambda_1 \lambda_2 = \det A$$

and

$$\operatorname{tr} D = \lambda_1 + \lambda_2 = \operatorname{tr} A$$

Then by multiplying $(\lambda - \lambda_1)(\lambda - \lambda_2)$ each λ_i is a solution to the equation

$$\lambda^2 - (\lambda_1 + \lambda_2)\lambda + (\lambda_1 \lambda_2) = 0$$

or

$$\lambda^2 - \text{tr } A\lambda + \det A = 0$$

Now

$$\det \begin{pmatrix} e - \lambda & f \\ g & h - \lambda \end{pmatrix} = (e - \lambda)(h - \lambda) - fg$$

$$= eh - e\lambda - h\lambda + \lambda^2 - fg$$

$$= \lambda^2 - (e + h)\lambda + eh - fg$$

$$= \lambda^2 - \text{tr } A\lambda + \det A$$

as desired. The above equation is called the *eigenvalue equation* for A, the numbers λ_1, λ_2 are called *eigenvalues*, and have the property that there exist lines in $\mathbb{R}^2$ such that for any v on those lines

$$Av = \lambda v.$$

The lines are called the *eigenvectors* of the matrix A.

III.5.8 To make the transformation $h(z) = \frac{1}{z}$ for analyzing the behavior at the point at infinity, we first note that $h^{-1}(w) = \frac{1}{w}$ has the same functional form. Composing, we have

$$\theta \circ f \circ \theta^{-1} = \frac{1}{7(\frac{1}{w}) + 1} = \frac{w}{7 + w}$$

The point at 0 is of interest, and is now a fixed point. Locally we may expand

$$\frac{w}{7 + w} \approx a_0 + a_1 w$$

where, $a_0 = 0$ and a_1 is given by the Taylor expansion

$$a_1 = \frac{d}{dw}\left(\frac{w}{7 + w}\right)(0) = \left.\frac{(7 + w) - w}{(7 + w)^2}\right|_{w=0} = \frac{7}{49}$$

so that near zero, this looks like $f'(w) = (1/7)w$ which is a contraction.

III.5.9 We have

$$f_\mu(x) = x^2 - \mu; \qquad g_\lambda(x) = \lambda x(1 - x)$$

and we want to reexpress f as $f_{\mu(\lambda)}$. f is a parabola opening up, centered at the y-axis with vertex at $-\mu$, g is a parabola centered at $x = 1/2$ opening down, with zeroes at 0, 1. Consequently, we need to reflect f, and stretch it so that it's zeroes are one unit apart, and shift it to the right by 1/2. First, the transformation $\theta(x) = -x$ will reflect the parabola as follows:

$$\theta \circ f \circ \theta^{-1}(x) = \theta \circ f(-x) = -f(-x)$$

which for any value of μ is

$$-f(-x) = -((-x)^2 - \mu) = -x^2 + \mu$$

a parabola pointing down with vertex μ centered on the y-axis. The transformation which will move the vertex to $x = 1/2$ is given by $\theta(x) = x - 1/2$. Combining these yields $\theta(x) = -x - 1/2$, or,

$$\theta \circ f \circ \theta^{-1}(x) = -x^2 + x - 1/4 + \mu - 1/2 = -x^2 + x + (\mu - 3/4)$$

We now need to adjust this so as to produce a multiplication by λ and remove the constant term. We use the transformation $\theta(x) = x/\lambda$. Once again composing, we now have

$$\theta(x) = \frac{-x}{\lambda} - \frac{1}{2}$$

so that

$$\theta^{-1}(x) = -\lambda x + \frac{\lambda}{2}$$

We have

$$\theta \circ f_\mu \circ \theta^{-1}(x') = \theta \circ f_\mu(-\lambda x' + \frac{\lambda}{2})$$

$$= \theta(\lambda^2(\frac{1}{4} - x' + x'^2) - \mu)$$

$$= -\lambda(\frac{1}{4} - x' + x'^2) - \frac{\mu}{\lambda} - \frac{1}{2}$$

$$= -\lambda x'^2 + \lambda x' - (\frac{\lambda}{4} + \frac{1}{2} + \frac{\mu}{\lambda})$$

This is identical to the expression for $g_\lambda(x')$ except for the constant term, we now find $\mu(\lambda)$ by setting this term to zero and solving for μ.

$$0 = \frac{\lambda}{4} + \frac{1}{2} + \frac{\mu}{\lambda}$$

$$\frac{-\mu}{\lambda} = \frac{\lambda + 2}{4}$$

$$\mu = \frac{-1}{4}\lambda^2 + \frac{-1}{2}\lambda$$

We have

$$\theta(x) = \frac{1}{2} - \frac{x}{\lambda} \quad ; \quad \mu(\lambda) = \frac{-1}{4}\lambda^2 + \frac{-1}{2}\lambda$$

III.5.10 We want the behavior of $g(x) = x^2 - 1/2$ near the fixed points. We have

$$x_f = x_f^2 - \frac{1}{2}$$

$$x_f^2 - x_f - \frac{1}{2} = 0$$

$$x_f = \frac{1 \pm \sqrt{3}}{2}$$

and we show the behavior around these points using two methods.
Analysis by moving the fixed points to the origin:
  Let

$$\theta(x) = x - x_{f_1} = x - (\frac{1 + \sqrt{3}}{2})$$

$$\theta^{-1}(x) = x + x_{f_1} = x + (\frac{1 + \sqrt{3}}{2})$$

Then we have

$$g \circ \theta^{-1}(x') = x'^2 + (1 + \sqrt{3})x' + \frac{1 + 2\sqrt{3} + 3}{4} - \frac{1}{2}$$

$$= x'^2(1 + \sqrt{3})x' + \frac{1 + \sqrt{3}}{2}$$

$$g'(x') = \theta \circ g \circ \theta^{-1}(x') = x'^2 + (1 + \sqrt{3})x'$$

which has a fixed point at the origin. Very close to the origin, since $x'^2 \ll x'$, we have

$$g'(x') \approx (1 + \sqrt{3})x'$$

which is an expansion. For the other fixed point, take

$$\theta(x) = x - x_{f_2} = x - \left(\frac{1 - \sqrt{3}}{2}\right)$$

$$\theta^{-1}(x) = x + x_{f_2} = x + \left(\frac{1 - \sqrt{3}}{2}\right)$$

which gives us

$$g \circ \theta^{-1}(x') = x'^2 + (1 - \sqrt{3})x' + \frac{1 - 2\sqrt{3} + 3}{4} - \frac{1}{2}$$

$$= x'^2(1 - \sqrt{3})x' + \frac{1 - \sqrt{3}}{2}$$

$$g'(x') = \theta \circ g \circ \theta^{-1}(x') = x'^2 + (1 - \sqrt{3})x'$$

Again looking at behavior near the origin, we have

$$g'(x') \approx (1 - \sqrt{3})x'$$

Since the coefficient is less than zero, there is a reflection, and since $|1 - \sqrt{3}| < 1$, it is composed with a contraction. Analysis by expanding $g(x)$ in a Taylor series:

The derivatives needed to expand $g(x)$ in Taylor series up to the linear term, at the fixed points are

$$g^{(0)}(x) = x^2 - \frac{1}{2}$$

$$g^{(1)}(x) = 2x$$

At the first fixed point, x_{f_1} we have

$$a_0 = g^{(0)}(x)\big|_{\frac{1+\sqrt{3}}{2}} = \frac{1 + \sqrt{3}}{2}$$

$$a_1 = g^{(1)}(x)\big|_{\frac{1+\sqrt{3}}{2}} = 1 + \sqrt{3}$$

and the expansion to the linear term looks like

$$\frac{1 + \sqrt{3}}{2} + (1 + \sqrt{3})\left(x - \frac{1 + \sqrt{3}}{2}\right) = (1 + \sqrt{3})x + \left(\frac{1 + \sqrt{3}}{2} - 2 - \sqrt{3}\right)$$

which is an expansion. For the fixed point x_{f_2} we have

$$a_0 = g^{(0)}(x)\big|_{\frac{1-\sqrt{3}}{2}} = \frac{1 - \sqrt{3}}{2}$$

$$a_1 = g^{(1)}(x)\big|_{\frac{1-\sqrt{3}}{2}} = 1 - \sqrt{3}$$

and the expansion to the linear term looks like

$$\frac{1 - \sqrt{3}}{2} + (1 - \sqrt{3})\left(x - \frac{1 - \sqrt{3}}{2}\right) = (1 - \sqrt{3})x + \left(\frac{1 - \sqrt{3}}{2} - 2 = \sqrt{3}\right)$$

which is a reflection and contraction as before. In general, it depends on the nature of the function which method is easier.

If it can be transformed by inspection, transforming the fixed points to the origin is preferable. In general, the Taylor series expansion method is more straight forward and simpler.

III.5.11 In exercise III.5.7 we showed that the entries of the diagonal matrix D satisfy the eigenvalue equation

$$\lambda^2 - \text{tr } A\lambda + \det A = 0$$

Using the quadratic equation, the solutions to this equation are

$$\lambda_i = \frac{\text{tr } A \pm \sqrt{(\text{tr } A)^2 - 4\det A}}{2}$$

and the condition stated in the problem is that the discriminant of this equation be greater than zero, in order to have two real solutions to the equation. If the discriminant is less than zero, then the solutions are complex conjugates of the form $\lambda_k = x \pm iy = re^{\pm i\theta}$. The matrix defines a transformation which contains a rotation by θ and therefore cannot be put into diagonal form, although it can be written as a rescaling and a rotation.

Turning to the problem of finding the matrix B, we separate out for a moment the normalizing factor of $1/\det B$ for the inverse of B, which will turn out not to matter, and write

$$\det B \cdot BAB^{-1} = \det B \cdot D = \det B \cdot \begin{pmatrix} \lambda_1 & 0 \\ 0 & \lambda_2 \end{pmatrix}$$

Letting

$$B = \begin{pmatrix} a & b \\ c & d \end{pmatrix}$$

we have

$$\det B \cdot BAB^{-1} = \begin{pmatrix} aed - afc + bgd - bhc & a^2 - aeb + bah - gb^2 \\ ced - fc^2 + gd^2 - dhc & caf - ceb + dah - dgb \end{pmatrix}$$

which yields

$$ade - afc + bdg - bch = \det B \cdot \lambda_1$$
$$adh + afc - bdg - bce = \det B \cdot \lambda_2$$
$$a^2 f - b^2 g - abe + abh = 0$$
$$d^2 g - c^2 f + cde - cdh = 0$$

The first two equations when subtracted or added, merely return to us the result of exercise III.2.13, that the trace and determinant are invariant under this kind of coordinate change. Since only these are dependent on the det B term in front, we can safely ignore it in what follows, which is why it was carried separately. Rewriting the last two equations as

$$a^2 f - ab(e - h) - b^2 g = 0$$
$$c^2 f - cd(e - h) - d^2 g = 0$$

we may solve either one, solving the last one for c, we have

$$c = \frac{d(e - h) \pm \sqrt{d^2(e - h)^2 + 4fd^2 g}}{2f}$$

$$= \frac{d}{f}(\frac{(e - h) \pm \sqrt{(e - h)^2 + 4fg}}{2})$$

$$= \frac{d}{f}(\frac{\text{tr } A \pm \sqrt{(e - h)^2 + 4fg}}{2} - h)$$

Examining the discriminant, we have

$$(e - h)^2 + 4fg = e^2 + h^2 - 2eh + 4fg = e^2 + h^2 + 2fg - 2 \det A$$
$$= e^2 + h^2 + 2 \det A + 2fg - 4 \det A = e^2 + h^2 + 2eh - 4 \det A$$
$$= (e + h)^2 - 4 \det A = (\text{tr } A)^2 - 4 \det A$$

which is the discriminant of the eigenvalue equation. Putting this in, we have

$$c = \frac{d}{f}(\frac{\text{tr } A \pm \sqrt{(\text{tr } A)^2 - 4 \det A}}{2} - h) = \frac{d}{f}(\lambda_i - h)$$

or

$$c(\frac{f}{\lambda_i - h}) = d \; ; \; a(\frac{f}{\lambda_i - h}) = d.$$

Since $\det B \neq 0$ the columns are independent, so we must choose a different λ_i for each of the above equations. Only the ratios are important here, so we choose $a = c = 1$, and have

$$B = (\frac{(\lambda_1 - h)(\lambda_2 - h)}{f(\lambda_1 - \lambda_2)}) \begin{pmatrix} 1 & \frac{f}{\lambda_1 - h} \\ 1 & \frac{f}{\lambda_2 - h} \end{pmatrix}$$

In closing, we note that the *eigenvectors* of A are the solutions to the pair of equations

$$ex_1 + fx_2 = \lambda x_1$$
$$gx_1 + hx_2 = \lambda x_2$$

which appears to yield the two solutions

$$\frac{(\lambda - e)}{f} x_1 = x_2 \; ; \; \frac{g}{(\lambda - h)} x_1 = x_2$$

These are actually identical, since by the eigenvalue equation

$$\frac{\lambda - e}{f} = \frac{g}{\lambda - h}$$

To solve for the eigenvectors, simply let $x_1 = 1$ and use one of the corresponding values for x_2. Hence the rows of the matrix B found above are precisely the eigenvectors of A. This is true in general, (for $N \times N$ matrices the eigenvalue equation will have N real roots if it is diagonalizable), and depending on the problem presents another way to find a useful transformation.

III.5.12 To solve this problem, we make use of the methods developed in exercise III.5.11. We have

$$f(x) = \begin{pmatrix} 1 & 2 \\ 2 & 3 \end{pmatrix} \begin{pmatrix} x_1 \\ x_2 \end{pmatrix} + \begin{pmatrix} 1 \\ 1 \end{pmatrix}$$

Solving for the fixed point, we may rewrite this as

$$f(x) = \begin{pmatrix} 1 & 2 \\ 2 & 3 \end{pmatrix} \begin{pmatrix} x_1 \\ x_2 + 1/2 \end{pmatrix} + \begin{pmatrix} 0 \\ -1/2 \end{pmatrix}$$

so there is a necessary translation by $(0, -1/2)$ to bring the fixed point to the origin. To diagonalize the matrix we have

$$A = \begin{pmatrix} 1 & 2 \\ 2 & 3 \end{pmatrix}$$

so that $\text{tr } A = 4$ and $\det A = -1$. This gives us the eigenvalue equation

$$\lambda^2 - 4\lambda - 1 = 0$$

yielding

$$\lambda_i = 2 \pm \sqrt{5}$$

We may now write down immediately the final form of the transformation, we want B and the above translation to produce

$$f'(x') = \begin{pmatrix} 2 + \sqrt{5} & 0 \\ 0 & 2 - \sqrt{5} \end{pmatrix}$$

which is indeed a coordinate rescaling. A matrix for B is then given in exercise III.5.11 by

$$B = \begin{pmatrix} 1 & \frac{c}{\lambda_1 - d} \\ 1 & \frac{c}{\lambda_2 - d} \end{pmatrix}$$
$$= \begin{pmatrix} 1 & \frac{2}{\lambda_1 - 3} \\ 1 & \frac{2}{\lambda_2 - 3} \end{pmatrix}$$

or to write out the whole expression for the coordinate transformation $\theta(x)$,

$$\theta(x) = \begin{pmatrix} 1 & \frac{2}{\sqrt{5}-1} \\ 1 & \frac{-2}{\sqrt{5}+1} \end{pmatrix} \begin{pmatrix} x_1 \\ x_2 \end{pmatrix} + \begin{pmatrix} 0 \\ \frac{1}{2} \end{pmatrix}$$

III.5.13 Let f be a transformation $f: \mathbf{X} \to \mathbf{X}$. We want to show that $\{f^n\}_{n=0}^\infty$ is a semigroup, which requires showing that $f^n \circ f^m: \mathbf{X} \to \mathbf{X}$ is in this set of transformations. But

$$f^n \circ f^m(x) = f^n(f^m(x)) = f^{n+m}(x)$$

so the set of iterates of f is indeed a semigroup.

III.5.14 Let $T: \Sigma \to \Sigma$ be a left shift transformation on code space, i.e.

$$x = x_1 x_2 x_3 \dots$$
$$T(x) = x_2 x_3 x_4 \dots$$

Since each application of T shifts the sequence to the left and drops the initial term, we have

$$T^n(x) = x_{n+1} x_{n+2} x_{n+3} \dots$$

The fixed points of $T^3(x) = x_4 x_5 \dots$ must therefore obey

$$x_{f_1} x_{f_2} x_{f_3} \dots = x_{f_4} x_{f_5} x_{f_6} \dots$$

Listing out the first few equalities, we have

$$x_1 = x_4$$
$$x_2 = x_5$$
$$x_3 = x_6$$
$$x_4 = x_7$$
$$\vdots$$

which yields

$$x_1 = x_{3n+1}$$
$$x_2 = x_{3n+2}$$
$$x_3 = x_{3n+3}$$

In other words, any fixed point of $T^3(x)$ must repeat the first three symbols indefinitely. Explicitly, the fixed points of $T^3(x)$ are

$$\overline{000}, \overline{001}, \overline{010}, \overline{011}, \overline{100}, \overline{101}, \overline{110}, \overline{111}.$$

III.5.15 We need to show that the Möbius transformations on $\hat{\mathbb{R}}$ form a group under composition. To check closure, we let

$$w_1(x) = \frac{ax + b}{cx + d} \; ; \; w_2(x) = \frac{ex + f}{gx + h}$$

The composition is then

$$
\begin{aligned}
w_1 \circ w_2(x) &= \frac{a(\frac{ex+f}{gx+h}) + b}{c(\frac{ex+f}{gx+h}) + d} \\
&= \frac{aex + af + bgx + bh}{cex + cf + dgx + dh} \\
&= \frac{(ae + bg)x + (af + bh)}{(ce + dg)x + (cf + dh)}
\end{aligned}
$$

which has the proper form. We check

$$
\begin{aligned}
(ae + bg)(cf + dh) &- (ce + dg)(af + bh) \\
&= aecf + adeh + bcgf + bdgh - aecf - bceh - adgf - bdgh \\
&= (ad - bc)eh - (ad - bc)gf \\
&= (ad - bc)(eh - gf) \neq 0
\end{aligned}
$$

so the Möbius transformations on $\hat{\mathbb{R}}$ form a semigroup. They are invertible as follows: Let

$$w = \frac{ax + b}{cx + d}$$

We have

$$
\begin{aligned}
wcx + wd &= ax + b \\
wcx - ax &= b - wd \\
x &= \frac{-dw + b}{cw - a} = \frac{dw - b}{-cw + a}
\end{aligned}
$$

which has the proper form, with

$$da - (-b)(-c) = ad - bc \neq 0$$

so there is an inverse for each element, and they form a group.

III.5.16 This problem is identical in algebraic manipulations to exercise 5.12, with the replacement of $a, b, c, d, z, w \in \hat{\mathbb{C}}$ instead of in $\hat{\mathbb{R}}$.

III.5.17 Let $f(x) = Ax + t$, and $g(x) = Bx + u$ be invertible affine transformations on $\mathbb{R}^2$. Then if $w = f(x)$, we have

$$w - t = Ax \; ; \; A^{-1}(w - t) = x = A^{-1}w - A^{-1}t = f^{-1}(w).$$

Similarly, $g^{-1}(w) = B^{-1}w - B^{-1}u$.
We have

$$f \circ g(x) = A(Bx + u) + t = ABx + (Au + t)$$

which is an affine transformation. We need to check whether it is invertible both because it needs to be in our set of transformations, and to prove group structure. Let

$$
\begin{aligned}
w &= ABx + (Au + t) \\
w - (Au + t) &= ABx \\
A^{-1}(w - (Au + t)) &= Bx \\
B^{-1}A^{-1}(w - (Au + t)) &= x
\end{aligned}
$$

Rewriting the left hand side, we have

$$f^{-1}(w) = B^{-1}A^{-1}w - B^{-1}A^{-1}(Au + t) = B^{-1}A^{-1}w - (B^{-1}u + B^{-1}A^{-1}t),$$

which is affine. The inverse matrices B^{-1}, A^{-1} must exist because by assumption f, g are invertible. Hence the set of invertible affine transformations is a group under composition.

III.5.18 Suppose $f(\Delta) \subset \Delta$. This implies $f^n(\Delta) \subset \Delta$, as follows: It is certainly true for $n = 1$. Suppose for every $0 < i < k$, the statement holds. Since by hypothesis $f^{k-1}(\Delta) \subset \Delta$, and $f(A) \subset \Delta$ if $A \subset \Delta$, we have $f(f^{k-1}(\Delta)) \subset \Delta$. But $f(f^{k-1}(\Delta)) = f^k(\Delta)$ so the claim is true by induction. To show that this implies that $\{f^n\}$ is a semigroup, recall that $f^n \circ f^m = f^{n+m}$ which is in the collection. Hence $\{f^n\}$ is a semigroup of transformations.

III.5.19 Let $w(x) = Ax + t$ where

$$A = \begin{pmatrix} a & 0 \\ b & c \end{pmatrix}$$

and $ac \neq 0$. This transformation is invertible with $x = A^{-1}(w - t)$, as long as A^{-1} exists. This is the case since $\det A = ac \neq 0$. To show that the set of transformations of this form are a group under composition, we need to show that if B, A are of this form then BA is as well, and, if A is of this form then A^{-1} is as well.

Let

$$A = \begin{pmatrix} a & 0 \\ b & c \end{pmatrix} \quad \text{and} \quad B = \begin{pmatrix} e & 0 \\ f & g \end{pmatrix}$$

Then

$$BA = \begin{pmatrix} e & 0 \\ f & g \end{pmatrix} \begin{pmatrix} a & 0 \\ b & c \end{pmatrix} = \begin{pmatrix} ae & 0 \\ af + bg & cg \end{pmatrix}$$

which is also of the proper form. We have

$$ae \cdot cg = (ac)(eg) \neq 0$$

so the transformation is in the set. Furthermore,

$$A^{-1} = \frac{1}{\det A} \begin{pmatrix} c & 0 \\ -b & a \end{pmatrix}$$

which is of the correct form. $ca = ac$, so the inverse is in the set. Hence the set of all such transformations $w(x)$ on $\mathbb{R}^2$ forms a group.

III.5.20 Given $f(z) = Az^2 + Bz + C$, we want $\theta(z) = az + b$ such that

$$\theta \circ f \circ \theta^{-1}(z') = (z')^2 + \tilde{C}$$

Since

$$\theta^{-1}(z') = \frac{1}{a}(z' - b)$$

we have

$$f \circ \theta^{-1}(z') = \frac{A}{a^2}(z'^2 - 2bz' + b^2) + \frac{B}{a}(z' - b) + C$$

$$= \frac{A}{a^2}z'^2 + (B - \frac{2bA}{a^2})z' + (\frac{Ab^2}{a^2} - \frac{Bb}{a} + C)$$

Composing $\theta(z)$ with this yields

$$\theta \circ f \circ \theta^{-1}(z) = \frac{A}{a}z'^2 + (B - 2\frac{bA}{a})z' + (\frac{Ab^2}{a} - Bb + Ca + b)$$

We must have $A/a = 1$ and $B - 2(bA/a) = 0$ for the desired transformation, meaning $a = A$ and $b = B/2$. Using these to simplify the constant term we have

$$\theta(z) = Az + \frac{B}{2}; \qquad \tilde{C} = CA + \frac{B}{2} - \frac{B^2}{4}$$

Answers to Chapter III, section 6

III.6.1

(a) $\sup(-\infty, 3) = 3$; $\inf(-\infty, 3) = -\infty$.
(b) $\sup(\mathcal{C}) = 1$; $\inf(\mathcal{C}) = 0$.
(c) $\sup\{1, 2, 3, 4, \ldots\} = +\infty$; $\inf\{1, 2, 3, 4, \ldots\} = 1$.
(d) $\sup(\mathbb{R}^+) = +\infty$; $\inf(\mathbb{R}^+) = 0$.

III.6.2 Let $f : \mathbf{X} \to \mathbf{X}$ be a contraction mapping on $(\mathbf{X}, d)$, and let $s_0 = \inf\{s \in \mathbb{R} : s$ is a contractivity factor for $f\}$. Then s_0 is a contractivity factor for f.

Proof: Clearly $0 \leq s < 1$ since if s is any contractivity factor for f, then $0 \leq s < 1$ so that $0 \leq s_0 \leq s < 1$. As well, the infimum is taken over a non-empty set since there is at least some s which is a contractivity factor for f by definition. Consequently, we need only show that s_0 satisfies

$$d(f(x), f(y)) \leq s_0 \cdot d(x, y) \; \forall x, y \in \mathbf{X}.$$

Suppose not. Then there is a pair $x, y \in \mathbf{X}$ such that

$$d(f(x), f(y)) > s_0 \cdot d(x, y).$$

But this says that

$$d(f(x), f(y)) - s_0 \cdot d(x, y) > 0$$

and since

$$d(f(x), f(y)) - d(x, y) < 0$$

by the intermediate value theorem, there is an s' with $1 > s' > s_0$ such that

$$d(f(x), f(y)) - s' \cdot d(x, y) = 0$$

Hence s' is less than all valid contractivity factors, and there is an interval (s_0, s') which contains no values which are valid contractivity factors for f. This is a contradiction of the assumption that s_0 is the infimum of the contractivity factors for f.

III.6.3 Let $f : \mathbf{X} \to \mathbf{X}$ and $g : \mathbf{X} \to \mathbf{X}$ be contractions with contractivity factors s, t respectively. Then the contractivity factor of $f \circ g$ is st.

Proof: For all $x, y \in \mathbf{X}$ we have

$$d(g(x), g(y)) \leq t \cdot d(x, y) \Rightarrow s \cdot d(g(x), g(y)) \leq st \cdot d(x, y)$$

But we also have

$$d(f(g(x)), f(g(y))) \leq s \cdot d(g(x), g(y)) \leq st \cdot d(x, y)$$

which by definition makes st a contractivity factor for $f(g(x)) = f \circ g(x)$ as desired.

III.6.4 Let $w(x) = Ax + t : \mathbb{R}^2 \to \mathbb{R}^2$ with fixed point x_f and $\det(A - I) \neq 0$. Let $h(x) = x - x_f = x'$. Then $h^{-1}(x') = x' + x_f$. Represent

$$w(x) = \begin{pmatrix} a & b \\ c & d \end{pmatrix} \begin{pmatrix} x_1 \\ x_2 \end{pmatrix} + \begin{pmatrix} e \\ f \end{pmatrix}$$

and

$$x_f = \begin{pmatrix} x_{f_1} \\ x_{f_2} \end{pmatrix}$$

Then we have

$$w \circ h^{-1}(x') = \begin{pmatrix} a & b \\ c & d \end{pmatrix} \begin{pmatrix} x_1' + x_{f_1} \\ x_2' + x_{f_2} \end{pmatrix} + \begin{pmatrix} e \\ f \end{pmatrix}$$

$$= \begin{pmatrix} ax_1' + ax_{f_1} + bx_2' + bx_{f_2} + e \\ cx_1' + cx_{f_1} + dx_2' + dx_{f_2} + f \end{pmatrix}$$

$$= \begin{pmatrix} ax_1' + bx_2' + x_{f_1} \\ cx_1' + dx_2' + x_{f_2} \end{pmatrix}$$

because x_f is a fixed point. Then

$$h \circ w \circ h^{-1}(x') = \begin{pmatrix} ax_1' + bx_2' + x_{f_1} \\ cx_1' + dx_2' + x_{f_2} \end{pmatrix} - \begin{pmatrix} x_{f_1} \\ x_{f_2} \end{pmatrix}$$

$$= \begin{pmatrix} ax_1' + bx_2' \\ cx_1' + dx_2' \end{pmatrix}$$

$$= Ax' = w'(x')$$

Inverting this we have

$$h^{-1} \circ w \circ h(x) = h^{-1} \circ w'(x - x_f) = h^{-1}(A(x - x_f)) = A(x - x_f) + x_f$$

Finally, we check that $A(x - x_f) + x_f = Ax + t$. Since

$$\begin{pmatrix} x_{f_1} \\ x_{f_2} \end{pmatrix} = \begin{pmatrix} ax_{f_1} + bx_{f_2} + e \\ cx_{f_1} + dx_{f_2} + f \end{pmatrix}$$

we have

$$\begin{pmatrix} ax_1 + bx_2 - (ax_{f_1} + bx_{f_2}) + (ax_{f_1} + bx_{f_2} + e) \\ cx_1 + dx_2 - (cx_{f_1} + dx_{f_2}) + (cx_{f_1} + dx_{f_2} + f) \end{pmatrix}$$

$$= \begin{pmatrix} ax_1 + bx_2 + e \\ cx_1 + dx_2 + f \end{pmatrix}$$

$$= \begin{pmatrix} a & b \\ c & d \end{pmatrix} \begin{pmatrix} x_1 \\ x_2 \end{pmatrix} + \begin{pmatrix} e \\ f \end{pmatrix}$$

$$= Ax + t$$

To prove the next claim we have

$$w(x) = w^1(x) = A^1(x - x_f) + x_f$$

and

$$w^2(x) = A(w(x) - x_f) + x_f$$

$$= A((A(x - x_f) + x_f) - x_f) + x_f$$

$$= A(A(x - x_f)) + x_f$$

$$= A^2(x - x_f) + x_f$$

Suppose that

$$w^n(x) = A^n(x - x_f) + x_f$$

Then we have

$$w^{n+1}(x) = A((A^n(x - x_f) + x_f) - x_f) + x_f$$
$$= A(A^n(x - x_f)) + x_f$$
$$= A^{n+1}(x - x_f) + x_f$$

so the result follows by induction on n.

We now describe the conditions on A under which the map is contractive for the Euclidean metric. We want $d(w(x), w(y)) \leq s \cdot d(x, y)$ for some $0 \leq s < 1$. This means,

$$d((A(x - x_f) + x_f), (A(y - x_f) + x_f)) \leq s \cdot d(x, y)$$

or

$$(ax_1 - ax_{f_1} + bx_2 - bx_{f_2} + x_{f_1} - ay_1 + ax_{f_1} - by_2 + bx_{f_2} - x_{f_1})^2$$
$$+ (cx_1 - cx_{f_1} + dx_2 - dx_{f_2} + x_{f_2} - cy_1 + cx_{f_1} - dy_2 + dx_{f_2} - x_{f_2})^2$$
$$= (ax_1 - ay_1 + bx_2 - by_2)^2 + (cx_1 - cy_1 + dx_2 - dy_2)^2$$
$$= (A(x - y)_1)^2 + (A(x - y)_2)^2$$

Using the notation of exercise III.2.14 we need

$$|A(x - y)| \leq s \cdot |x - y|$$

or

$$\frac{|A(x - y)|}{|x - y|} \leq s \forall x, y \in X$$

This is certainly true if we maximize over $|x - y|$ so we have as a condition that $|A| \leq s < 1$ for $w(x)$ to be contractive. The proof is identical with the Manhattan metric substituted and the appropriate changes made.

III.6.6 Let $f : \mathbb{R} \to \mathbb{R}$ be given by $f(x) = 1/2x + 1/2$. Then f is a contraction mapping as follows:

$$d(f(x), f(y)) = d(\frac{1}{2}x + \frac{1}{2}, \frac{1}{2}y + \frac{1}{2})$$
$$= |\frac{1}{2}x + \frac{1}{2} - (\frac{1}{2}y + \frac{1}{2})|$$
$$= |\frac{1}{2}x - \frac{1}{2}y|$$
$$= \frac{1}{2}|x - y|$$
$$\leq \frac{1}{2}d(x, y)$$

so that $s = \frac{1}{2}$ is a contractivity factor. To find the fixed point, we write

$$x_f = \frac{1}{2}x_f + \frac{1}{2}$$
$$\frac{1}{2}x_f = \frac{1}{2}$$

or $x_f = 1$. Rewriting in the form $f(x) = A(x - x_f) + x_f$, we have

$$f(x) = \frac{1}{2}(x-1) + 1$$

$$f^n(x) = \frac{1}{2^n}(x-1) + 1$$

$$\lim_{n \to \infty} f^n(x) = \lim_{n \to \infty} \frac{1}{2^n}(x-1) + 1$$

$$= 1 = x_f$$

With $x = 0$, $f^n(x) = 1 - 1/(2^n)$.

Writing the original,

$$f^2(x) = \frac{1}{2}(\frac{1}{2}x + \frac{1}{2}) + \frac{1}{2} = \frac{1}{2} + \frac{1}{4} + \frac{1}{4}x$$

If

$$f^n(x) = \frac{1}{2}(f^{n-1}(x)) + \frac{1}{2} = \sum_{i=1}^{n} \frac{1}{2^i} + \frac{1}{2^n}x$$

then

$$f^{n+1} = \frac{1}{2}(\sum_{i=1}^{n} \frac{1}{2^i} + \frac{1}{2^n}x) + \frac{1}{2}$$

$$= (\sum_{i=1}^{n} \frac{1}{2^{i+1}} + \frac{1}{2^{n+1}}x) + \frac{1}{2}$$

$$= \sum_{i=2}^{n+1} \frac{1}{2^i} + \frac{1}{2^{n+1}}x + \frac{1}{2}$$

$$= \sum_{i=1}^{n+1} \frac{1}{2^i} + \frac{1}{2^{n+1}}x$$

so the formula for $f^n(x)$ follows by induction. In the limit,

$$x_f = \sum_{n=1}^{\infty} \frac{1}{2^n} = 1$$

independent of x.

III.6.7 Let (X, d) be compact, and $f : X \to X$ be a contraction with contractivity factor s and fixed point x_f. Then $f(X) \subset X$, $f(X) \neq X$.

Proof: (X, d) is compact so it is bounded. In particular, since the function $g(x) = d(x_f, x)$ is a continuous function from (X, d) to $\mathbb{R}$, it has a maximum M, and a point $x_M \in X$ such that $d(x_f, x_M) = M$. Then for all $x \in X$,

$$d(f(x), f(x_f)) = d(f(x), x_f)$$

$$\leq d(f(x_M), x_f)$$

$$\leq s \cdot M < M$$

hence there is no x such that $f(x) = x_M$, and $f(X) \neq X$.

III.6.8 The set of contraction mappings on (X, d) are a semigroup under composition. To see this, we only need to know that if f, g are contractions then so is $f \circ g$. By problem 3, if the contractivity factors for f, g are s, t respectively, then so is $f \circ g$ with contractivity factor st.

III.6.9 Let $w(x) = Ax + t$ where

$$A = \begin{pmatrix} \frac{1}{2} \cos 120° & -\frac{1}{2} \sin 120° \\ \frac{1}{2} \sin 120° & \frac{1}{2} \cos 120° \end{pmatrix}$$

and $t = \begin{pmatrix} 1/2 \\ 0 \end{pmatrix}$. The easiest way to see that w takes the equilateral Sierpinski triangle to a subset of itself, and is a contraction is to break the transformation apart. The equilateral Sierpinski triangle is expressed by the following IFS:

$$w_1 \begin{pmatrix} x_1 \\ x_2 \end{pmatrix} = \begin{pmatrix} \frac{1}{2} & 0 \\ 0 & \frac{1}{2} \end{pmatrix} \begin{pmatrix} x_1 \\ x_2 \end{pmatrix}$$

$$w_2 \begin{pmatrix} x_1 \\ x_2 \end{pmatrix} = \begin{pmatrix} \frac{1}{2} & 0 \\ 0 & \frac{1}{2} \end{pmatrix} \begin{pmatrix} x_1 \\ x_2 \end{pmatrix} + \begin{pmatrix} \frac{1}{4} \\ \frac{\sqrt{3}}{4} \end{pmatrix}$$

$$w_2 \begin{pmatrix} x_1 \\ x_2 \end{pmatrix} = \begin{pmatrix} \frac{1}{2} & 0 \\ 0 & \frac{1}{2} \end{pmatrix} \begin{pmatrix} x_1 \\ x_2 \end{pmatrix} + \begin{pmatrix} \frac{1}{2} \\ 0 \end{pmatrix}$$

Each transformation shrinks the triangle towards one of its vertices by a factor of 2, the vertices are $(0, 0)$, $(1/2, \sqrt{3}/2)$, and $(1, 0)$ respectively. Looking at w we may rewrite it as

$$w \begin{pmatrix} x_1 \\ x_2 \end{pmatrix} = \frac{1}{2} \begin{pmatrix} \cos 120° & -\sin 120° \\ \sin 120° & \cos 120° \end{pmatrix} \begin{pmatrix} x_1 \\ x_2 \end{pmatrix} + \begin{pmatrix} \frac{1}{2} \\ 0 \end{pmatrix}$$

$$= \begin{pmatrix} \frac{1}{2} & 0 \\ 0 & \frac{1}{2} \end{pmatrix} \begin{pmatrix} \cos 120° & -\sin 120° \\ \sin 120° & \cos 120° \end{pmatrix} \begin{pmatrix} x_1 \\ x_2 \end{pmatrix} + \begin{pmatrix} \frac{1}{2} \\ 0 \end{pmatrix}$$

$$= \begin{pmatrix} \frac{1}{2} & 0 \\ 0 & \frac{1}{2} \end{pmatrix} (R_\theta \begin{pmatrix} x_1 \\ x_2 \end{pmatrix}) + \begin{pmatrix} \frac{1}{2} \\ 0 \end{pmatrix}$$

where R_θ is a rotation by $\theta = 120°$. For the whole attractor A, the map therefore rotates A by $120°$ about its left lower vertex (O), by an angle equal to the exterior angle of an equilateral triangle, depositing it as an equilateral Sierpinski triangle on the left side of the y-axis with vertices $(-1, 0)$, $(0, 0)$, and $(-1/2, \sqrt{3}/2)$. The next action is a contraction, since it is identical to w_1, and creates, setwise, a mirror image of $w_1(A)$ on the left side of the y-axis. Finally, there is a translation to the right by $1/2$ bringing the image to a half-sized equilateral triangle with vertices $(0, 0)$, $(1/2, 0)$, and $(1/4, \sqrt{3}/4)$. In other words, it lies exactly (setwise) on the image $w_1(A)$. So the map does indeed contract the triangle, and take it to a subset of itself. The fixed point of w can be calculated by

$$\begin{pmatrix} x_1 \\ x_2 \end{pmatrix} = \begin{pmatrix} \frac{1}{2} \cos 120° & -\frac{1}{2} \sin 120° \\ \frac{1}{2} \sin 120° & \frac{1}{2} \cos 120° \end{pmatrix} \begin{pmatrix} x_1 \\ x_2 \end{pmatrix} + \begin{pmatrix} 1/2 \\ 0 \end{pmatrix}$$

$$= \begin{pmatrix} \frac{-1}{4} & \frac{-\sqrt{3}}{4} \\ \frac{\sqrt{3}}{4} & \frac{-1}{4} \end{pmatrix} \begin{pmatrix} x_1 \\ x_2 \end{pmatrix} + \begin{pmatrix} 1/2 \\ 0 \end{pmatrix}$$

which yields

$$\frac{5}{4}x_1 + \frac{\sqrt{3}}{4}x_2 - \frac{1}{2} = 0$$

$$\frac{\sqrt{3}}{4}x_1 - \frac{5}{4}x_2 = 0$$

$$\frac{\sqrt{3}}{5}x_1 = x_2$$

$$x_1 = \frac{10}{31}$$

$$x_2 = \frac{2\sqrt{3}}{31}$$

III.6.10 Let Σ be a code space on the two symbols $\{0, 1\}$ and $f(x) = 1x = 1x_1x_2 \ldots$. Then using the metric

$$d(x, y) = \sum_{n=1}^{\infty} \frac{|x_i - y_i|}{3^i}$$

we have

$$d(f(x), f(y)) = \sum_{n=1}^{\infty} \frac{|x_i - y_i|}{3^{i+1}} + \frac{|1 - 1|}{3}$$

$$= \sum_{n=1}^{\infty} \frac{|x_i - y_i|}{3^{i+1}}$$

$$= \frac{1}{3} \sum_{n=1}^{\infty} \frac{|x_i - y_i|}{3^i}$$

$$= \frac{1}{3} d(x, y)$$

so that f is a contraction mapping with contractivity factor $1/3$ and fixed point $\bar{1}$.

III.6.11 Let $(\mathbf{X}, d)$ be a compact metric space, and $f : \mathbf{X} \to \mathbf{X}$ be a contraction mapping. Then $\{f^n(\mathbf{X})$ is a Cauchy sequence in $(\mathcal{H}(\mathbf{X}), h)$ and $\lim_{n \to \infty} f^n(\mathbf{X}) = \{x_f\}$ where x_f is the fixed point of $f(x)$.

Proof: We first verify that $\{f(\mathbf{X})\}$ is in fact a sequence in $\mathcal{H}(\mathbf{X})$. Since f is a contraction mapping, f is continuous, and the continuous image of a compact set is compact. Hence $f(\mathbf{X})$ is compact. By induction, $f^n(\mathbf{X})$ is compact. Furthermore, since $x_f \in \mathbf{X}$ and $f^n(x_f) = x_f \in f^n(\mathbf{X})$, $f^n(\mathbf{X})$ is non-empty. Hence, $f^n(\mathbf{X}) \in \mathcal{H}(\mathbf{X})$ for all n, so this is indeed a sequence in $\mathcal{H}(\mathbf{X})$. We next verify that the sequence is Cauchy. If so, since $\mathbf{X}$ is compact, $\mathbf{X}$ is complete, and therefore so is $\mathcal{H}(\mathbf{X})$. Hence if the sequence is Cauchy, it has a limit in $\mathcal{H}(\mathbf{X})$. Let s be a contractivity factor for $f(x)$, and let $\epsilon > 0$ be given. Since $s < 1$, we may choose N such that for any $n > N$, $s^n < \delta$ where δ is a number of choice less than 1. Let n, m be chosen $n < m$, and $n, m > N$. Since $m > n$, we have $f^m(\mathbf{X}) \subset f^n(\mathbf{X})$, so that

$$h(f^m(\mathbf{X}), f^n(\mathbf{X})) = \max_{x \in f^n(\mathbf{X})} d(x, f^m(\mathbf{X}))$$

But this must be smaller than the diameter of $f^n(\mathbf{X})$ defined by

$$\operatorname{diam}(f^n(\mathbf{X})) = \max_{x, y \in f^n(\mathbf{X})} d(x, y)$$

We have

$$\operatorname{diam}(f^n(\mathbf{X})) \le s^n \operatorname{diam}(\mathbf{X})$$

since s is a contractivity factor for f. By choosing N such that $s^N \operatorname{diam}(\mathbf{X}) < \epsilon$, we guarantee that

$$h(f^m(\mathbf{X}), f^n(\mathbf{X})) = \max_{x \in f^n(\mathbf{X})} d(x, f^m(\mathbf{X})) \le s^N \operatorname{diam}(\mathbf{X}) < \epsilon$$

because all the distances considered for the maximum are less than $\operatorname{diam}(f^n(\mathbf{X}))$. Hence the sequence is Cauchy and has a limit in $\mathcal{H}(\mathbf{X})$.

We now show that that limit is x_f. We know that $x_f \in \lim_{n \to \infty} f^n(\mathbf{X})$. We need only show that

$$\lim_{n \to \infty} \operatorname{diam}(f^n(\mathbf{X})) = 0$$

But this limit is bounded by

$$\lim_{n \to \infty} \operatorname{diam}(f^n(\mathbf{X})) \le \lim_{n \to \infty} s^n \operatorname{diam}(\mathbf{X}) = 0$$

Hence, if $y \in \lim_{n \to \infty} f^n(\mathbf{X})$ then $d(y, x_f) = 0$, but d is a metric, so this implies $y = x_f$. Hence

$$\lim_{n \to \infty} f^n(\mathbf{X}) = \{x_f\}$$

as desired.

III.6.13 We are asked to show that if $A = \begin{pmatrix} a & b \\ c & d \end{pmatrix}$ is strictly positive definite (that is $a, b, c, d > 0$) then the transformation $f(\theta)$ defined on the first quadrant by

$$A \begin{pmatrix} \cos \theta \\ \sin \theta \end{pmatrix} = \lambda \begin{pmatrix} \cos f(\theta) \\ \sin f(\theta) \end{pmatrix}$$

is a contraction mapping, where λ is some positive number. Since we may scale λ to fit the matrix A, we first check the case where $\det A = 0$. This happens if the rows or columns of A are multiples of each other, and we have $f(\theta) = \theta_f$ a constant, so in this case the function f converges on the first application. Otherwise, by rescaling λ this will work for any strictly positive definite A so long as it works for a strictly positive definite A such that $\det A = 1$. We need to have real eigenvalues, so that $\operatorname{tr} A = a + d \ge 2$. We have $ad - bc = 1$ or $bc = ad - 1$ and $bc > 0$ so that $ad \ge 1$. Suppose $ad = 1$. Then $d = 1/a$ and we have a discriminant

$$a^2 + 2ad + d^2 - 4 = a^2 + \frac{1}{a^2} - 2$$

The expression

$$a^2 + \frac{1}{a^2}$$

has derivative

$$\frac{d}{da}(a^2 + \frac{1}{a^2}) = 2a - \frac{2}{a^3}$$

showing zeroes when $a^4 = 1$ which since $a > 0$ occurs when $a = 1$. This is a minimum, and yields a discriminant of 0, but because the determinant is 1, this would yield $bc = 0$, which it can not be, so that we do indeed have such a λ, and one choice is greater than zero.

To show that f is a contraction mapping, we must show that

$$d(f(\theta_1), f(\theta_2)) \le s \cdot d(\theta_1, \theta_2).$$

The sine of an angle is an increasing function in the first quadrant, so writing

$$\sin(f(\theta_1) - f(\theta_2)) = \sin f(\theta_1) \cos f(\theta_2) - \cos f(\theta_1) \sin f(\theta_2)$$

$$= (\frac{c \cos\theta_1 + d \sin\theta_1}{\lambda})(\frac{a \cos\theta_2 + b \sin\theta_2}{\lambda}) - (\frac{a \cos\theta_1 + b \sin\theta_1}{\lambda})(\frac{c \cos\theta_2 + d \sin\theta_2}{\lambda})$$

$$= \frac{1}{\lambda^2}(ac \cos\theta_1 \cos\theta_2 + bc \cos\theta_1 \sin\theta_2 + ad \sin\theta_1 \cos\theta_2 + bd \sin\theta_2 \sin\theta_1)$$

$$- \frac{1}{\lambda^2}(ac \cos\theta_1 \cos\theta_2 + ad \cos\theta_1 \sin\theta_2 + bc \sin\theta_1 \cos\theta_2 + bd \sin\theta_1 \sin\theta_2)$$

$$= \frac{1}{\lambda^2}(ad(\sin\theta_1 \cos\theta_2 - \cos\theta_1 \sin\theta_2) - bc(\sin\theta_1 \cos\theta_2 - \cos\theta_1 \sin\theta_2)$$

$$= \frac{\det A}{\lambda^2}(\sin\theta_1 \cos\theta_2 - \cos\theta_1 \sin\theta_2)$$

$$= \frac{\det A}{\lambda^2} \sin(\theta_1 - \theta_2)$$

so that, to get a contraction mapping, we need

$$\frac{\sin(f(\theta_1) - f(\theta_2))}{\sin(\theta_1 - \theta_2)} = \frac{\det A}{\lambda^2} = s < 1$$

To test this, recall that assuming $\det A = 1$, we have tr $A > 2$ and

$$\lambda = \frac{\text{tr } A}{2} \pm \frac{\sqrt{(\text{tr } A)^2 - 4}}{2}$$

indicating $\lambda > 1$. So f is a contraction, implying that $\{f^n(\theta)\} \to \theta_f$ as desired. The fixed angle is in fact the angle of the eigenvector, yielding

$$\theta_f = \arctan(\frac{\lambda - a}{b})$$

Answers to Chapter III, section 7

III.7.1 We have $w_1(x) = 1/3x$ and $w_2(x) = 1/3x + 2/3$. We compute the contractivity factor for the IFS.

$$d(w_1(x), w_1(y)) = |\frac{1}{3}x - \frac{1}{3}y| = \frac{1}{3}|x - y| = \frac{1}{3}d(x, y)$$

$$s_1 = \frac{1}{3}$$

$$d(w_2(x), w_2(y)) = |(\frac{1}{3}x + \frac{2}{3}) - (\frac{1}{3}y + \frac{2}{3})| = \frac{1}{3}|x - y| = \frac{1}{3}d(x, y)$$

$$s_2 = \frac{1}{3}$$

$$s = max(s_1, s_2) = max(\frac{1}{3}, \frac{1}{3}) = \frac{1}{3}$$

We prove that the attractor of this IFS is the Cantor set by proving by induction that the action of $W(B)$ is to remove the middle third of each interval in B. To this end, our hypothesis is: Let $[a, b]$ be a component of B_k with $k < n$. Then

$$[a, a + \frac{1}{3}(b - a)] \cup [b - \frac{1}{3}(b - a), b] \subset B_{k+1}$$

and

$$(a + \frac{1}{3}(b - a), b - \frac{1}{3}(b - a)) \cap B_{k+1} = \emptyset$$

We establish that this is true for the first case: Let $B_0 = [0, 1]$. Then

$$B_1 = W(B_0) = w_1(B_0) \cup w_2(B_0) = [0, \frac{1}{3}] \cup [\frac{2}{3}, 1]$$

Now suppose that $[a, b] \subset B_n$ is a component of B_n. Then there is a component $[a', b'] \subset B_{n-1}$ whose image is $[a, b]$ under some w_i ($i = 1, 2$). By hypothesis, the component $[a', b']$ has had its middle third removed in B_n. Then $[a, b]$ is replaced in B_{n+1} by the set

$$w_i([a' + \frac{1}{3}(b' - a')] \cup [b' - \frac{1}{3}(b' - a')]) = [a + \frac{1}{3}(b - a)] \cup [b - \frac{1}{3}(b - a)]$$

since under w_i the length $b' - a' \mapsto b - a$ and clearly the two points $a', b' \mapsto a, b$ respectively. Hence this operation is a removal of middle thirds, and the induction goes through. Then the attractor of the IFS is the classical Cantor set.

III.7.2 We have $w_1(x) = s_1 x$ and $w_2(x) = (1 - s_1)x + s_1$. Let $B_0 = [0, 1]$. Then

$$B_1 = W(B_0)$$
$$= w_1([0, 1]) \cup w_2([0, 1])$$
$$= [0, s_1] \cup [s_1, 1 - s_1 + s_1]$$
$$= [0, s_1] \cup [s_1, 1] = [0, 1] = B_0$$

Hence $B_0 = [0, 1]$ is the attractor of the IFS.

III.7.3 Let $w_1(x) = \frac{1}{3}x$ and $w_2(x) = \frac{1}{2}x + \frac{1}{2}$. Let $B_0 = [0, 1]$. Then

$$B_1 = W(B_0)$$
$$= w_1([0, 1]) \cup w_2([0, 1])$$
$$= [0, \frac{1}{3}] \cup [\frac{1}{2}, 1]$$

We have, at least on the first iteration, an operation which removes an interval of length $\frac{1}{6}$ from an interval, with this removed interval being just to the left of the center of the interval. We test this by applying the induction hypothesis as in problem 1. Assume that for each $k < n$, any component interval in B_k has a $\frac{1}{6}$ slice taken out to the left of the halfway mark, specifically, $[a, b]$ goes to

$$[a, a + \frac{1}{3}(b - a)] \cup [b - \frac{1}{2}(b - a), b]$$

and

$$(a + \frac{1}{3}(b - a), b - \frac{1}{2}(b - a)) \cap B_{k+1} = \emptyset$$

Then if $[a, b]$ is a component of B_n, then it is the image under one of the maps, say w_i of some interval $[a', b'] \subset B_{n-1}$. In B_n this interval has been replaced by

$$[a', a' + \frac{1}{3}(b' - a')] \cup [b' - \frac{1}{2}(b' - a'), b']$$

the image of which under w_i is

$$[a, a + \frac{1}{3}(b - a)] \cup [b - \frac{1}{2}(b - a), b]$$

which says that that B_{n+1} is derived from B_n by taking out the $\frac{1}{6}$ slice. A is a Cantor set, with uncountably many points, and no intervals. To see this last, if $[a, b]$ were a component interval of A, then since $W(A) = A$, this operation does not change $[a, b]$ but the above proof says it must lose a slice, hence $[a, b] \subset A \Rightarrow a = b$.

III.7.5 Let $w_1(x) = ax + b$ and $w_2(x) = cx + d$ as in the statement of the problem. The attractor A of the IFS is compact, hence it is bounded, and acheives its bounds. Then there is a closed interval $[a, b] \subset \mathbb{R}$ such that $a = \min\{x : x \in A\}$ and

$b = \max\{x : x \in A\}$. Suppose

$$w_1([a, b]) \cap w_2([a, b]) \neq \emptyset$$

Then since $[a, b]$ is connected, and the w_i are continuous, $w_1([a, b])$ and $w_2([a, b])$ are connected, their intersection is non-empty, hence $w_1([a, b]) \cup w_2([a, b])$ is connected. Furthermore, there are points $a_1, b_1 \in A \subset [a, b]$ whose image in one or the other map is a, b respectively. If there were any points in one of the images which were outside the interval $[a, b]$, then the image of either a or b under one of the maps would lie outside of $[a, b]$ and therefore not be in A which it is. Hence this image, $w_1([a, b]) \cup w_2([a, b])$ is mapped to itself exactly by W and is therefore the attractor A itself. Thus A is connected if $w_1([a, b]) \cap w_2([a, b]) \neq \emptyset$. Suppose $w_1([a, b]) \cap w_2([a, b]) = \emptyset$. Then it follows since $A \subset [a, b]$ that $w_1(A) \cap w_2(A) = \emptyset$, as well. Let $[a', b']$ be the longest connected component of A. Since $w_1(A) \cap w_2(A) = \emptyset$, there is an $i \in \{1, 2\}$ and an interval $[a'', b'']$ (connected component of A) such that

$$w_i([a'', b'']) = [a', b']$$

Let $s_i < 1$ be a contractivity factor for w_i. Then

$$d(a', b') \leq s_i d(a'', b'')$$

which says that either $[a'', b'']$ is longer than $[a', b']$ or the length of each is zero. The first contradicts the statement that $[a', b']$ is the longest connected component of A, consequently $a' = b'$, and the only connected components of A are points. Hence A is totally disconnected.

III.7.6 No. The reason for this is as follows: Because each map must have a contractivity factor of less than one, each map must take the proposed attractor A, that is the union of two disjoint intervals, to a subset of either one interval or the other. The image of each map will be the union of two disjoint intervals itself. Consequently, let w_1 take A into the first interval, and w_2 take A into the second. This situation must arise (possibly by renumbering the maps. It is possible to generate a single connected interval of either of these two images with w_3, constructing one of the intervals of A. It is not possible to cover the gap in the other, since the image of w_3 must lie entirely in one interval or the other. (Note that the attractor could be constructed from four maps).

III.7.7 Let $\{\mathbb{R}^2 : w_1, w_2\}$ be an IFS with w_i given by

$$w_1 \begin{pmatrix} x \\ y \end{pmatrix} = \begin{pmatrix} \frac{1}{2} & 0 \\ 0 & \frac{1}{2} \end{pmatrix} \begin{pmatrix} x \\ y \end{pmatrix} + \begin{pmatrix} \frac{1}{2} \\ \frac{1}{2} \end{pmatrix}$$

$$w_2 \begin{pmatrix} x \\ y \end{pmatrix} = \begin{pmatrix} \frac{1}{2} & 0 \\ 0 & \frac{1}{2} \end{pmatrix} \begin{pmatrix} x \\ y \end{pmatrix}$$

Let the attractor of the IFS be A. Then $(x, y) \in A$ implies that $x = y$, and $x, y \in [0, 1]$.

Proof: Let $B = \{(x, y) \in \mathbb{R}^2 : x = y, x, y \in [0, 1]\}$ Then

$$w_1 \begin{pmatrix} x \\ y \end{pmatrix} = \begin{pmatrix} \frac{1}{2}x + \frac{1}{2} \\ \frac{1}{2}y + \frac{1}{2} \end{pmatrix}$$

$$w_2 \begin{pmatrix} x \\ y \end{pmatrix} = \begin{pmatrix} \frac{1}{2}x \\ \frac{1}{2}y \end{pmatrix}$$

Since by supposition, $x, y \leq 1$, $1/2x, 1/2y \leq 1/2$ so adding $1/2$ to each remains in $[0, 1]$. Furthermore, if $x = y$ then $1/2x = 1/2y$ and $1/2x + 1/2 = 1/2y + 1/2$ so the image of B is contained in B. The image of B under w_1 is $\{(x, y) \in B : x, y \in [1/2, 1]\}$ exactly. This is because $0 \mapsto 1/2$ in each coordinate, and $1 \mapsto 1$ in each coordinate. The image of B under w_2 is $\{(x, y) \in B : x, y \in [0, 1/2]\}$ exactly as well. Hence the image $W(B) = B$, and since the attractor is unique, $B = A$.

III.7.8 We show that the attractor is the point $\{1\}$, together with the images $w_2^n(0)$. That these points are in the attractor of the IFS follows from the fact that the two fixed points $0, 1$ must be in the attractor, and therefore all forward images of them must be as well. Let

$$A = \{x \in \mathbb{R} : x = 1 \text{ or } x = w_2^n(0); n = 0, \ldots \infty\}$$

Let $x \in A$. Then $w_1(x) = 0 = w_2^0(0)$, if $x = 1$ then $w_2(x) = 1$, and otherwise there is a k such that $w_2^k(0) = x$ so that $w_2(x) = w_2^{k+1}(0)$. Hence $W(A) \subset A$. Furthermore if $x = 0$ then there is an element of A (for instance 0) such that $x = w_1(0)$, if $x = 1$ then $x = w_2(1)$, and otherwise if $x = w_2^k(0)$ then $x = w_2(w_2^{k-1}(0))$. Consequently W takes A onto itself as well. Therefore $W(A) = A$. A is a countable sequence of increasing points by definition. We want the geometric series for x_n. Writing $w_2(x) = 2/3(x - 1) + 1$ we have $w_2^n(x) = (2/3)^n(x - 1) + 1$. Substituting $x = 0$, we have $x_n = 1 - (2/3)^n$. To put this in the form of a geometric series,

$$x_n - x_{n-1} = 1 - (\frac{2}{3})^n - 1 + (\frac{2}{3})^{n-1}$$

$$= (1 - \frac{2}{3})(\frac{2}{3})^{n-1}$$

$$= \frac{1}{2} \cdot (\frac{2}{3})^n$$

We may thus write the terms x_n as the partial sums of the geometric series

$$\frac{1}{2} \sum_{n=1}^{\infty} (\frac{2}{3})^n$$

Answers to Chapter III, section 8

There are no answers to section 8, all the exercises are programming experiments.

Answers to Chapter III, section 9

III.9.2 We assume that $w_0(B) = C$ for any $B \in \mathcal{H}(\mathbf{X})$. We have

$$w_1 \begin{pmatrix} x \\ y \end{pmatrix} = \begin{pmatrix} 0.75 & 0 \\ 0 & 0.75 \end{pmatrix} \begin{pmatrix} x \\ y \end{pmatrix} + \begin{pmatrix} 0.25 \\ 0 \end{pmatrix}$$

To show that $\{\mathbb{R}^2 : w_0, w_1\}$ we need only establish the contractivity factor for w_1. We may rewrite w_1 using multiplication by a scalar as

$$w_1 \begin{pmatrix} x \\ y \end{pmatrix} = 0.75 \begin{pmatrix} x \\ y \end{pmatrix} + \begin{pmatrix} 0.25 \\ 0 \end{pmatrix}$$

$$d(w_1(x), w_1(y)) = \sqrt{(.75x_1 + .25 - .75y_1 - .25)^2 + (.75x_2 - .75y_2)^2}$$

$$= \sqrt{(.75^2((x_1 - y_1)^2 + (x_2 - y_2)^2))}$$

$$= .75\sqrt{(x_1 - y_1)^2 + (x_2 - y_2)^2}$$

$$= .75d(x, y)$$

so the contractivity factor for w_1 is 0.75, the map is contractive and this is a hyperbolic IFS with condensation. Let $A_0 = C$. We calculate $w_1^n(A_0)$ by using the $A(x - x_f) + x_f$ form of the transformation. First the fixed point.

$$x_{f_1} = .75x_{f_1} + .25 \Rightarrow .25x_{f_1} = .25 \Rightarrow x_{f_1} = 1$$

$$x_{f_2} = .75x_{f_2} \Rightarrow x_{f_2} = 0$$

Rewriting w_1^n as

$$w_1 \begin{pmatrix} x \\ y \end{pmatrix} = (0.75)^n \begin{pmatrix} x - 1 \\ y \end{pmatrix} + \begin{pmatrix} 1 \\ 0 \end{pmatrix}$$

we see that $w_1^n(A_0)$ is C shrunk by $(0.75)^n$ and moved to $1 - (0.75)^n$. In other words, as the figure shows, the trees get smaller, closer together, and closer to 1. By example 1,

$$A = \bigcup_{n=0}^{\infty} W^n(A_0)$$

with

$$A_1 = C \cup w_1(C)$$
$$A_2 = C \cup w_1(C \cup w_1(C)) = C \cup w_1(C) \cup w_1^2(C)$$

$$\vdots$$

Suppose that for $k \le n$ we have

$$A_k = \bigcup_{i=0}^{k} w_1^i(C)$$

then

$$A_{n+1} = C \cup w_1(A_n)$$
$$= C \cup w_1 \left(\bigcup_{i=0}^{n} w_1^i(C) \right)$$
$$= C \cup \bigcup_{i=1}^{n+1} w_1^i(C)$$
$$= \bigcup_{i=0}^{n+1} w_1^i(C)$$

Counting the number of transformations of C for A_n produces $n + 1$ trees as desired. Let C be the trunk of the tree. That is $w_0(B) = C$ for any $B \in \mathcal{H}(\mathbf{X})$. We define w_1, w_2 so as to form the two branches off the trunk, as follows:

$$w_1 \begin{pmatrix} x \\ y \end{pmatrix} = \begin{pmatrix} r\cos\theta & -r\sin\theta \\ r\sin\theta & r\cos\theta \end{pmatrix} + \begin{pmatrix} 0 \\ 1 \end{pmatrix}$$
$$w_2 \begin{pmatrix} x \\ y \end{pmatrix} = \begin{pmatrix} r\cos\theta & r\sin\theta \\ -r\sin\theta & r\cos\theta \end{pmatrix} + \begin{pmatrix} 0 \\ 1 \end{pmatrix}$$

In other words, w_1 rotates the trunk by θ, shrinks it by r and moves its base to $(0, 1)$. w_2 rotates the trunk by $-\theta$, shrinks it by r and moves its base to $(0, 1)$. Successive applications of either of these transformations rotates the trunk by $\pm n\theta$, shrinks it by r^n, and moves the base to the top of the last piece. The figure becomes disconnected if the translation moves the piece above the top of the last trunk, it becomes not simply connected if r is close to 1 and θ is large.

III.9.8 Without the 'cls' instruction we get effectively the union of W applied to the initial set. By example 1, this is the attractor of an IFS with condensation, the condensation set being the initial set chosen.

Answers to Chapter III, section 10

III.10.10 We make reference to the general form of the affine transformation expressed in terms of the fixed point derived in exercise III.6.4. That is, for a contractive affine map, we have

$$f(x) = Ax + t = A(x - x_f) + x_f$$
$$f^n(x) = A^n(x - x_f) + x_f$$

which we regard as the *canonical* form of a contraction mapping. We have, in this case

$$w_i(z) = s_i + (1 - s_i)a_i = s_i(z - a_i) + a_i$$

Regarding multiplication by a complex number as an $\mathbb{R}^2$ transformation, namely

$$s_i z = s_{i_1} z_1 + i s_{i_1} z_2 + i s_{i_2} z_1 - s_{i_2} z_2$$

$$= (s_{i_1} z_1 - s_{i_2} z_2) + i(s_{i_1} z_2 + s_{i_2} z_1)$$

$$\sim \begin{pmatrix} s_{i_1} & -s_{i_2} \\ s_{i_2} & s_{i_1} \end{pmatrix} \begin{pmatrix} z_1 \\ z_2 \end{pmatrix}$$

this transformation is equivalent to the $\mathbb{R}^2$ transformation

$$w_i \begin{pmatrix} x \\ y \end{pmatrix} = \begin{pmatrix} s_{i_1} & -s_{i_2} \\ s_{i_2} & s_{i_1} \end{pmatrix} \begin{pmatrix} x - a_{i_1} \\ y - a_{i_2} \end{pmatrix} + \begin{pmatrix} a_{i_1} \\ a_{i_2} \end{pmatrix}$$

from which we can read off the fixed point as a_i. Note that s_i is a rotation and scaling, making the whole operation a similitude.

Answers to Chapter III, section 11

III.11.3 In order that the attractor of an IFS be two lines joined at the origin, and variable with a parameter p, we may pick three maps as follows: Let $w_1(x, y)$ shrink the angle towards the origin with no rotation. This creates the angle again, with smaller clock hands. To get the hands back to the same length, create two maps which map the whole attractor to the tip of each hand, that is to a line at the same angle as each hand, with a fixed point at the end of the hand. Consequently, choose w_1 to be

$$w_1(x, y) = \begin{pmatrix} .9 & 0 \\ 0 & .9 \end{pmatrix}$$

which shrinks the clock face slightly. To collapse the attractor to a line of slope θ, we need

$$w(x, y) = \begin{pmatrix} a & b \\ c & d \end{pmatrix} \begin{pmatrix} x \\ y \end{pmatrix} = \begin{pmatrix} ax + by \\ cx + dy \end{pmatrix}$$

such that

$$\frac{cx + dy}{ax + by} = \tan\theta \quad \forall\, x, y$$

This is most easily accomplished by starting with the relations

$$\frac{c}{a} = \frac{d}{b} = \tan\theta$$

for which one solution is the matrix

$$A = \begin{pmatrix} \cos\theta & \cos\theta \\ \sin\theta & \sin\theta \end{pmatrix}$$

This matrix has det $A = 0$, and generates the proper slope. Now suppose we want the length of the new line to be less than 1. The clock hands can be any set of lengths, we will take the minute hand to be length 1, and the hour hand to be length $1/2$. Choosing the minute hand itself, we have

$$A \begin{pmatrix} \cos\theta \\ \sin\theta \end{pmatrix} = \begin{pmatrix} \cos^2\theta + \sin\theta\cos\theta \\ \sin^2\theta + \sin\theta\cos\theta \end{pmatrix}$$

so that the right hand side is of length less than two. Add to this the possibility of the whole other hand, which must be less than one, and we have to divide this matrix by 3 to assure that it fits within the given hand. So let

$$A = \begin{pmatrix} \frac{1}{3}\cos\theta & \frac{1}{3}\cos\theta \\ \frac{1}{3}\sin\theta & \frac{1}{3}\sin\theta \end{pmatrix}$$

Now we arrange to put the fixed point at the end of the hand.

$$\begin{pmatrix} \cos\theta \\ \sin\theta \end{pmatrix} = A \begin{pmatrix} \cos\theta \\ \sin\theta \end{pmatrix} + \begin{pmatrix} e \\ f \end{pmatrix}$$

or

$$e = \cos\theta - \frac{\cos^2\theta}{3} - \frac{\sin\theta\cos\theta}{3}$$

$$f = \sin\theta - \frac{\sin^2\theta}{3} - \frac{\sin\theta\cos\theta}{3}$$

We now let $\theta_h = \pi/2 - (\pi/6)p$ be the hour angle, which increases clockwise from $\pi/2$ so that it reaches 2π when $p = 12$. We let $\theta_m = \pi/2 - (2/\pi)p$ which travels the same direction but reaches 2π every hour. Having arranged that

$$w_m(x, y) = A|_{\theta=\theta_m} \begin{pmatrix} x \\ y \end{pmatrix} + \begin{pmatrix} e_m \\ f_m \end{pmatrix}$$

will be the map to the minute hand. The map to the hour hand is similar, but we must shorten the hand by a factor of 2. So,

$$w_h(x, y) = A_h \begin{pmatrix} x \\ y \end{pmatrix} + \begin{pmatrix} e_h \\ f_h \end{pmatrix}$$

where $A_h = \frac{1}{2}A$ and

$$e = \cos\theta_h - \frac{\cos^2\theta}{6} - \frac{\sin\theta_h\cos\theta_h}{6}$$

$$f = \sin\theta_h - \frac{\sin^2\theta}{6} - \frac{\sin\theta_h\cos\theta_h}{6}$$

Answers to Chapter IV, section 1

IV.1.1 The addresses are $a = 2121\overline{122}$, $b = 33\overline{3}$, and $c = 12231\overline{222}$. the height, and half the width, and then, depending on the map, moves it to the appropriate corner of the square. To shrink the square by one half on each side, we define the map

$$w_1 \begin{pmatrix} x \\ y \end{pmatrix} = \begin{pmatrix} \frac{1}{2} & 0 \\ 0 & \frac{1}{2} \end{pmatrix} \begin{pmatrix} x \\ y \end{pmatrix} = A \begin{pmatrix} x \\ y \end{pmatrix}$$

and assign the symbol 0 to an iteration of this map. To get the quarter labelled 1, we must do the same shrink, and move the lower left corner over by $1/2$ in the x direction, which gives the map

$$w_2 \begin{pmatrix} x \\ y \end{pmatrix} = A \begin{pmatrix} x \\ y \end{pmatrix} + \begin{pmatrix} \frac{1}{2} \\ 0 \end{pmatrix}$$

An iteration by w_2 corresponds to an address symbol of 1. The third map must move the lower left corner to $(1/2, 1/2)$, consequently we define

$$w_3 \begin{pmatrix} x \\ y \end{pmatrix} = A \begin{pmatrix} x \\ y \end{pmatrix} + \begin{pmatrix} \frac{1}{2} \\ \frac{1}{2} \end{pmatrix}$$

An iteration by w_3 corresponds to the address symbol 2. Finally, the last map moves the lower left corner up by 1/2, giving

$$w_3 \begin{pmatrix} x \\ y \end{pmatrix} = A \begin{pmatrix} x \\ y \end{pmatrix} + \begin{pmatrix} 0 \\ \frac{1}{2} \end{pmatrix}$$

and an iteration by this map gets the address label 3.

IV.1.4 The easiest way to see this problem is to look for a subset of the IFS maps which have as their attractor one of the sides of the Sierpinski triangle. Consider the IFS $\{\mathbb{R}^2; w_1, w_2\}$ giving rise only to addresses which contain no 3's. If a point is on the x-axis, neither of the two IFS

$$w_1 \begin{pmatrix} x \\ y \end{pmatrix} = \begin{pmatrix} \frac{1}{2} & 0 \\ 0 & \frac{1}{2} \end{pmatrix} \begin{pmatrix} x \\ y \end{pmatrix}$$

$$w_2 \begin{pmatrix} x \\ y \end{pmatrix} = \begin{pmatrix} \frac{1}{2} & 0 \\ 0 & \frac{1}{2} \end{pmatrix} + \begin{pmatrix} \frac{1}{2} \\ 0 \end{pmatrix}$$

Either of these two will return a point on the x-axis to the x-axis, and their attractor is the line between $1\bar{1}$ and $2\bar{2}$. On the other hand, any point with a y component not equal to zero will on a single iteration have a new y value equal to half the original, which is still non-zero. Hence this edge of the Sierpinski triangle consists of all points whose addresses contain only 1 and 2.

We may use identical arguments to verify that the points along the left edge of the triangle have addresses containing only 1's and 3's. The final edge contains all addresses containing only 2's and 3's.

IV.1.6 One scheme is to use the Cartesian product of the two codespaces. Let the IFS with condensation be $\{X; w_0, w_1, \ldots, w_N\}$ and the IFS whose attractor A is used for the condensation set $\{X; v_1, v_2, \ldots, v_M\}$. The attractor B for the IFS with condensation is then the union of the successive iterates

$$B = \bigcup_{n=0}^{\infty} W^n(A)$$

Any point on B may be described by giving the sequence of iterates done on A to get to the particular image on B of A, combined with the address on the image, which corresponds to an address on the attractor A. Consequently, we have two code space strings, possibly on two code spaces, which give the whole address. Any address can occur.

IV.1.8 This is part of the proof of Theorem 1 of section 4.5. Let $\{X; w_1, \ldots, w_N\}$ be the IFS in question, let its contractivity be s, and let its attractor be A. Consider the images of A for two points which are near by in code space. That is let some small $\epsilon > 0$ be given, and let $x = x_1 x_2 x_3 \ldots$ and $y = y_1 y_2 y_3 \ldots$ be two code space points $d(x, y) < \epsilon$. The points in A having these addresses are contained in the images

$$\tilde{x} \in w_{x_1} \circ w_{x_2} \circ \cdots w_{x_n}(A)$$
$$\tilde{y} \in w_{y_1} \circ w_{y_2} \circ \cdots w_{y_n}(A)$$

Now, for ϵ small, we may find a number n such that

$$\frac{1}{(N+1)^n} < \epsilon$$

which means that $x_i = y_i$ for all $i < n$. Then both points $\tilde{x}$, $\tilde{y}$ are contained in

$$w_{x_1} \circ w_{x_2} \circ \cdots w_{x_n}(A)$$

since the contractivity is s this image must be smaller than $s^n \cdot \text{diam}(A)$. Hence the closer the addresses in code space, the closer the points having those addresses are on the attractor.

IV.1.12 The scheme which is used in Figure IV.99 is the following: Write the numbers with the same denominators in horizontal lists beginning with the denominator 1, and putting each successive denominator under the list before. Starting at the upper right hand corner, go down if we have just arrived at the left edge, go right if we have just arrived at the top edge, go up and to the right if we are leaving the left edge, or have just gone right and up, go down and to the left if we are leaving the top edge or have just gone down and left. This winds us through the rationals as shown below:

$$\begin{array}{cccccc}
\frac{1}{1} & \frac{2}{1} & \rightarrow & \frac{3}{1} & \frac{4}{1} & \rightarrow & \frac{5}{1} & \cdots \\
\downarrow & \nearrow & & \swarrow & \nearrow & & \swarrow & \cdots \\
\frac{1}{2} & \frac{2}{2} & & \frac{3}{2} & \frac{4}{2} & & \frac{5}{2} & \cdots \\
& \swarrow & \nearrow & & \swarrow & & & \\
\frac{1}{3} & \frac{2}{3} & & \frac{3}{3} & \frac{4}{3} & & \frac{5}{3} & \cdots \\
\downarrow & \nearrow & & \swarrow & & & & \\
\frac{1}{4} & \frac{2}{4} & & \frac{3}{4} & \frac{4}{4} & & \frac{5}{4} & \cdots \\
\vdots & \vdots & & \vdots & \vdots & & \vdots &
\end{array}$$

There are two things to note: The scheme works regardless of the entries as long as every rational is in some horizontal list, with the lists ordered in a countable manner, and, all redundant terms are not in lowest terms. Since any horizontal list becomes 'shorter' but still infinite if such terms are removed, we may use the same scheme if we agree to write down only those numbers in lowest terms in each horizontal list. To get the negative rationals, put the negative list corresponding to each positive one in below it. That is, the second row is numbers with denominator -1 the fourth with -2 and so forth. This method of converting the rationals to a linear list indexed by the counting numbers was invented by Cantor.

IV.1.13 The triangles in the Sierpinski triangle can be labelled by size: There is 1 triangle of size 1, 3 of size $\frac{1}{3}$ and in general 3^n triangles of size $\frac{1}{3^n}$ since on each iteration of $W(A)$ we generate 3 times as many triangles of $\frac{1}{3}$ the size. Suppose we label each of the triangles with size $\frac{1}{3^n}$ with the labels

$$3^n + 1, 3^n + 2, \ldots, 3^n + 3^n$$

Since this last term is $2 \cdot 3^n < 3^{n+1}$, each is a unique label, there is a label for each triangle, and the collection of labels is a proper subset of the positive integers. The number of triangles is therefore countable.

Answers to Chapter IV, section 2

IV.2.1 The IFS $\{\mathbb{R}; \frac{1}{2}x, \frac{1}{2}x + \frac{1}{2}\}$ is just touching. Let A be the attractor, and let $\mathcal{O}$ be the set

$$\mathcal{O} = (0, \frac{1}{2}) \cap A \cup (\frac{1}{2}, 1) \cap A.$$

Then we have

$$w_1(\mathcal{O}) = (0, \frac{1}{4}) \cup (\frac{1}{4}, \frac{1}{2})$$
$$w_2(\mathcal{O}) = (\frac{1}{2}, \frac{3}{4}) \cup (\frac{3}{4}, 1)$$

which are disjoint. By construction

$$\bigcup_{i=1}^{n} w_i(\mathcal{O}) \subset \mathcal{O}$$

Furthermore, $\mathcal{O}$ is open because it is the finite union of open intervals. The IFS is not totally disconnected because it has the attractor

$$A = [0, 1] = [0, \frac{1}{2}] \cup [\frac{1}{2}, 1] = w_1(A) \cup w_2(A)$$

and $w_1(1) = w_2(0)$.

IV.2.2 The attractor of the IFS $\{\mathbb{R}; \frac{1}{2}x, \frac{3}{4}x + \frac{1}{4}\}$ is $[0, 1]$ as follows:

$$w_1([0, 1]) = [0, \frac{1}{2}]$$

$$w_2([0, 1]) = [\frac{1}{4}, 1]$$

$$w_1([0, 1]) \cup w_2([0, 1]) = [0, 1]$$

Since there is a unique attractor with the property that $W(A) = A$ this is it. Furthermore, there is an open set $(\frac{1}{4}, \frac{1}{2})$ interior to the overlap between the images, that is

$$(\frac{1}{4}, \frac{1}{2}) \subset w_1(A) \cap w_2(A) = [\frac{1}{4}, \frac{1}{2}]$$

so the IFS is overlapping.

IV.2.3 The first thing to show is that the attractor of this IFS is $[0, 1]$, as follows: For any affine transformation $f(x) = ax + b$ on $\mathbb{R}$, an interval transforms to either

$$[c, d] \mapsto \begin{cases} [ac + b, ad + b] & \text{if } a > 0 \\ [ad + b, ac + b] & \text{if } a < 0 \end{cases}$$

For each w_n here, $a = \frac{1}{10} > 0$, so we have,

$$w_n([0, 1]) = [\frac{n-1}{10}, \frac{n}{10}]$$

Thus the union of these is

$$W([0, 1]) = \bigcup_{n=1}^{10} w_n([0, 1]) = [0, \frac{1}{10}] \cup \cdots \cup [\frac{9}{10}, 1] = [0, 1]$$

Since the attractor A is unique, this is it. The IFS is clearly not totally disconnected because $w_n([0, 1]) \cap w_{n+1}([0, 1]) = \{\frac{n}{10}\}$. However, let

$$\mathcal{O} = \bigcup_{n=1}^{10} (\frac{n-1}{10}, \frac{n}{10}) \subset A$$

Then

$$w_n(\mathcal{O}) \subset w_n((0, 1)) = (\frac{n-1}{10}, \frac{n}{10}) \subset \mathcal{O}$$

and for any $i \neq j$

$$w_i(\mathcal{O}) \cap w_j(\mathcal{O}) \subset w_i((0, 1)) \cap w_i((0, 1)) = (\frac{i-1}{10}, \frac{i}{10}) \cap (\frac{j-1}{10}, \frac{j}{10}) = \emptyset$$

So the attractor is just-touching.

IV.2.4 Our IFS is $\{[0, 1]; w_1 = \frac{1}{3}x, w_2(x) = \frac{1}{3}x + \frac{2}{3}\}$. The attractor A is a subset of $[0, 1]$, and $w_1([0, 1]) = [0, \frac{1}{3}]$ and $w_2([0, 1]) = [\frac{2}{3}, 1]$ so that $w_1([0, 1]) \cap w_2([0, 1]) = \emptyset$ Then

$$w_1(A) \cap w_2(A) \subset w_1([0, 1]) \cap w_2([0, 1]) = \emptyset$$

so the IFS is totally disconnected.

IV.2.6 We have $w_1(x) = \frac{1}{2}$ and $w_2(x) = \frac{1}{2}$. The attractor is clearly $A = \{\frac{1}{2}\}$. The images of the two maps intersect, so the IFS is not totally disconnected. Furthermore, since the entire attractor is in the intersection there is clearly no open set $\mathcal{O}$ satisfying the conditions for a just-touching IFS. Consequently the IFS is overlapping, by default.

IV.2.8 The smallest number with no non-zero entries in its decimal expansion is $.\bar{1}$. Any other choice, with non-zero entries, must have at least one digit which is greater than 1. Reducing this digit to 1 produces a smaller choice. Hence reducing all of them to 1 produces the smallest.

IV.2.10 Let $\{X; w_1, w_2, \ldots, w_N\}$ be our IFS with attractor A. We first prove $I \Longleftrightarrow II$. Let x be a periodic point. Then by definition there exists a finite sequence of transformations $w_{\sigma_1}, \ldots, w_{\sigma_n}$ such that

$$w_{\sigma_1} \circ \cdots \circ w_{\sigma_n}(x) = x$$

If we define $f(x)$ to be this transformation, we have

$$f(x) = f^m(x) = x = (w_{\sigma_1} \circ \cdots \circ w_{\sigma_n}) \circ \cdots \circ (w_{\sigma_1} \circ \cdots \circ w_{\sigma_n})(x)$$

consequently x has the address $\overline{\sigma_1 \cdots \sigma_n}$, which is a periodic address. Now suppose that x has the periodic address $\overline{\sigma_1 \cdots \sigma_n}$. Then x is given by

$$\phi(\overline{\sigma_1 \cdots \sigma_n}) = x = \phi(\sigma_1 \cdots \sigma_n \overline{\sigma_1 \cdots \sigma_n})$$
$$= w_{\sigma_1} \circ \cdots \circ w_{\sigma_n} \phi(\overline{\sigma_1 \cdots \sigma_n}) = w_{\sigma_1} \circ \cdots \circ w_{\sigma_n}(x)$$

so x is periodic. We next prove $I \Longleftrightarrow III$. The semigroup of transformations generated by $\{w_1, \ldots w_N\}$ includes all finite compositions of the maps for the IFS. In particular since $w_{\sigma_1} \circ \cdots \circ w_{\sigma_n}$ is a finite composition of these maps, it is an element of the semigroup. It's fixed point is x. Now suppose that x is a fixed point of an element of the semigroup of transformations generated by these maps. Then by definition of the semigroup, there is an element of the form $w_{\sigma_1} \circ \cdots \circ w_{\sigma_n}$ with the property that

$$w_{\sigma_1} \circ \cdots \circ w_{\sigma_n}(x) = x$$

Consequently x is periodic.

IV.2.11 We wish to show that any periodic point of the IFS $\{[0, 1] : \frac{1}{2}x, \frac{1}{2}x + \frac{1}{2}\}$ is of the form $p/(2^n - 1)$. This is certainly the case for the two fixed points 0, 1. Choose any point x in $[0, 1]$. Then we claim that after any sequence of k operations, we have

$$w_{\sigma_1} \circ \cdots \circ w_{\sigma_k}(x) = \frac{1}{2^k}x + C$$

for some number C. This is true since each iteration multiplies x by $\frac{1}{2}$ and multiplies any constant term by the same factor, and possibly adds another part to the constant term. If x is k periodic then we have

$$x = \frac{1}{2^k}x + C \quad \text{or} \quad x = \frac{C2^k}{(2^k - 1)}$$

Now C is composed by taking the constant term from the last iteration and multiplying it by $\frac{1}{2}$ and either adding a half or adding 0. Thus the binary expansion of C contains at each digit a 1 if $\frac{1}{2}$ was added on that iteration, and a 0 if it was not. Hence C has a binary expansion of length k, which, multiplied by 2^k gives any possible number from 0 to $2^k - 1$, as desired.

IV.2.14 The IFS $\{\mathbb{R}; w_1(x) = 0, w_2(x) = \frac{1}{2}x + \frac{1}{2}\}$ is an IFS with condensation, the condensation set is $\{0\}$ and consequently the attractor A is given by

$$A = \bigcup_{n=0}^{\infty} w_2^n(0) \cup \{1\}$$

Any point $p \in A$ is periodic since there is some k for which $p = w_2^k(0)$ unless $p = 1$ which is a fixed point and therefore periodic. But this means that

$$p = w_2^k(w_1(p)) = w_2^k(p) \circ w_1(p)$$

so p is $k + 1$ periodic. a totally disconnected IFS is invertible, it follows that the cardinality of A is the same as that of its related code space. But code space on more than one symbol is uncountable by Theorem 1.1 in Chapter IV. Hence the attractor is too.

IV.2.16 Any uncountable attractor which is overlapping has uncountably many points with multiple addresses. This is because it contains an open set of points with multiple addresses which maps to an open set in code space by ϕ^{-1}. Any open set with non-empty interior in code space on more than one symbol is uncountable.

IV.2.17 One example of such an IFS is any overlapping IFS of two maps on the interval $[0, 1]$ such that the end points of the overlapping region $w_1(A) \cap w_2(A)$ are 3-periodic. This causes some of the sets of 4 symbols in *any* given position to map to the same point on the attractor, allowing uncountably many addresses for some points.

IV.2.18 The set of interval end points in C is equal to exactly twice the number of intervals used in the derivation. Following the proof of the countable number of triangles in the Sierpinski triangle, (exercise 1.13 in Chapter IV), we have at each subdivision, 2^n intervals added, and we number them $3^n + 1, 3^n + 2, \ldots, 3^n + 2^n < 3^{n+1}$. Since this numbering scheme produces an integer in order for each interval without repeating, the intervals have been put into correspondence with the positive integers. Then their end points can too, since we may give all the left-hand end points the negative integer corresponding to the right hand end point labelled as above. Since the integers are countable so are the endpoints.

Answers to Chapter IV, section 3

IV.3.10 We first show that the point $12\overline{12}$ is a periodic point of period 2.

$$f(12\overline{12}) = 2\overline{12} = 21\overline{21}$$
$$f^2(12\overline{12}) = f(21\overline{21}) = 1\overline{21} = 12\overline{12}$$

so this is a periodic point of period two, and the other element of the cycle is $21\overline{21}$. Next we show that it is repulsive. Let $x_f = 12\overline{12}$, and $y \in \Sigma$. Then

$$d(f^2(x_f), f^2(y)) = d(x_f, f^2(y))$$
$$= \sum_{i=3}^{\infty} \frac{|x_i - y_i|}{3^{(i-2)}} = 3 \cdot 3 \cdot \sum_{i=3}^{\infty} \frac{|x_i - y_i|}{3^i}$$
$$= 9 \cdot d(x_f, y) - (3|1 - y_1| + |2 - y_2|)$$

If we choose $\frac{1}{9} > \epsilon > 0$ then for $y \in B_\epsilon(x_f)$ we have $y_1 = 1$ and $y_2 = 2$, so for any such ϵ

$$d(x_f, f^2(y)) \geq 9 \cdot d(x_f, y)$$

and the point is repulsive, the cycle is therefore repulsive too.

IV.3.11 The dynamical system $\{[0, 1]; 2x(1 - x)\}$ has an attractive fixed point at $x_f = \frac{1}{2}$. $f(0) = 0$ so that 0 is also a fixed point of the system. To examine the behavior near 0, we first notice that if $x \neq 0$ then as $x \to 0$ we have $(1 - x) \to 1$. Consequently, as we approach 0 the function approaches $f(x) \to 2x$. Since for any $x > 0$ we have $d(2x, 0) = 2d(x, 0) > d(x, 0)$ it is easy to find an ϵ such that the function is repulsive on an ϵ-ball around 0. For example, let $\epsilon = .1$. Then $d(0, f(x)) \geq 1.8d(0, x)$ for any $x \in B_\epsilon(0)$.

IV.3.14 We want the behavior of the dynamical system $\{\mathbb{R} \cup \{\infty\}; f(x) = \lambda x\}$ in four cases:

(i) $\lambda = 0$ gives us $f(x) = 0$ which makes 0 the only fixed point, and attractive.
(ii) $0 < |\lambda| < 1$. ∞ is a repulsive fixed point since $|f(x)| < |x|$, likewise, 0 is an attractive fixed point.
(iii) $\lambda = -1$. 0 is a neutral fixed point, as is ∞. Every other point is period 2, and neutral.
(iv) $1 < \lambda < \infty$. Since $|f(x)| > x$ for any $x \neq 0, \infty$, 0 is a repulsive fixed point, and ∞ an attractive one.

IV.3.17 We have

$$f(x) = \begin{cases} x + 1 & \text{if } x \leq 0 \\ -x + 1 & \text{if } x \geq 0 \end{cases}$$

If $x = 0$ then $f(x) = 1$ and $f^2(x) = 0$ so that $\{0, 1\}$ is a cycle of period 2. For any other x split x into its integer part and fractional part, that is

$$x = n + \tilde{x} \qquad |\tilde{x}| < 1, n \in \mathbb{Z}$$

Then if $x < 0$, $f^n(x) = \tilde{x}$ which is between -1 and zero, and $f^{n+1}(x) = 1 + \tilde{x}$ which is between 0 and 1. Applying f to this number gives $f(1 + \tilde{x}) = -\tilde{x}$, which is again greater than zero, and

$$f^2(1 + \tilde{x}) = f(-\tilde{x}) = 1 + \tilde{x}$$

which is a cycle of period 2. Consequently for any $x < 0$, x is eventually periodic of period 2. For $x > 0$, we have

$$f(x) = f(n + \tilde{x}) = -(n - 1) - \tilde{x}$$

so that redefining $n := -(n - 1)$ and $\tilde{x} := -\tilde{x}$ gives us the case just examined, and these points are eventually periodic of period 2 as well.

IV.3.18 If $\{\mathbf{X}; w_1, \ldots, w_N\}$ is a hyperbolic IFS, we may assume a contractivity $s < 1$. By Theorem 7.1 in Chapter III, the dynamical system $\{\mathcal{H}(\mathbf{X}); W\}$ possesses a unique fixed point $A \in \mathcal{H}(\mathbf{X})$. Also by Theorem 7.1, for any $B, C \in \mathcal{H}(\mathbf{X})$ we have $h(W(B), W(C)) \geq s \cdot h(B, C)$. Consequently,

$$d(W(A), W(B)) = d(A, W(B)) \geq s \cdot d(A, B)$$

and the fixed point A is globally attractive (globally attractive means ϵ may be chosen any size).

IV.3.20 We need a small lemma to proceed. If a space is compact, then for any $\epsilon > 0$, there is an ϵ-net $\{y_i\}_{i=1}^{n(\epsilon)}$ of points such that every point in the space is less than ϵ from some element of the net. We need to produce a sequence of such nets.

   Lemma: *Given $\epsilon_1 > \epsilon_2$ and an ϵ_1-net $\{y_i\}_{i=1}^{n(\epsilon_1)}$ there is a minimal ϵ_2-net $\{z_i\}_{i=1}^{n(\epsilon_2)}$ such that $\{y_i\} \subset \{z_i\}$.*

   Proof: There is always *some* ϵ_2-net which has the ϵ_1-net as a subset, since given any ϵ_2-net $\{z_i\}_{i=1}^{n(\epsilon_2)}$, the finite collection of points

$$\{y_i\}_{i=1}^{n(\epsilon_1)} \cup \{z_i\}_{i=1}^{n(\epsilon_2)}$$

is an ϵ_2-net containing the ϵ_1-net. Now taking the collection of all ϵ_2-nets containing this particular ϵ_1-net, there is an $n(\epsilon_2) \geq n(\epsilon_1)$ which is the minimum size of any of the nets in the collection. Let $\{z_i\}_{i=1}^{n(\epsilon_2)}$ be one of these.

   Now suppose that our code space has the metric

$$d(x, y) = \sum_{i=1}^{\infty} \frac{|x_i - y_i|}{k^i}$$

with some $k > 1$, perhaps $k = N + 1$. Let $\epsilon_i = \frac{1}{k^i}$, and, with the help of the above lemma, construct a *nested* sequence of ϵ_i-nets,

$$\{y_{1_i}\}_{i=1}^{n(\epsilon_1)} \subset \{y_{2_i}\}_{i=1}^{n(\epsilon_2)} \subset \{y_{3_i}\}_{i=1}^{n(\epsilon_3)} \subset \cdots$$

We build a sequence of points $\{x_j\}_{j=1}^{\infty}$ by letting

$$x_j = y_{1_1} \cdots y_{1_{j+1}} y_{2_1} \cdots y_{2_{j+1}} \cdots y_{n(\epsilon_j)_1} \cdots y_{n(\epsilon_j)_{j+1}}$$

By construction, x_j visits each ϵ_j ball around each element of the ϵ_j net by repeated applications of the shift transformation T. But code space is compact, and therefore the sequence $\{x_j\}_{j=1}^{\infty}$ must have a convergent subsequence, which converges to a point x which visits every neighborhood in code space. Hence x has a dense orbit under the shift transformation.

IV.3.21 Let A, B be two open sets in code space. Choose an $\epsilon > 0$ and points $\sigma_A \in A$, and $\sigma_B \in B$, such that $B_\epsilon(\sigma_A) \subset A$. Then if the metric for Σ is

$$d(x, y) = \sum_{i=1}^{\infty} \frac{|x_i - y_i|}{(N + 1)^i}$$

there is an M such that for any $\sigma \in \Sigma$ such that

$$\sigma_1 = \sigma_{A_1}, \sigma_2 = \sigma_{A_2}, \cdots \sigma_M = \sigma_{A_M}$$

we have $\sigma \in B_\epsilon(\sigma_A) \subset A$. Then we only need to choose

$$\sigma = \sigma_{A_1} \cdots \sigma_{A_M} \sigma_B$$

which is in A, but $T^M(\sigma) = \sigma_B \in B$. Since by choosing the sequence of σ's which agree on $M, M+1, M+2, \ldots$ first symbols with σ_A, followed by σ_B, as above, we get that for all $m > M$ there is an element of A such that $T^m(\sigma) \in B$, T is mixing on code space.

Answers to Chapter IV, section 4

All the problems are examples or figure exercises.

Answers to Chapter IV, section 5

IV.5.1 Let $\{X_1; f_1\}$ and $\{X_2; f_2\}$ be equivalent dynamical systems with the equivalence provided by the homeomorphism $\theta : X_1 \to X_2$. Let $\{x_1, x_2, \ldots, x_p\}$ be a cycle of period p in X_1. That is $f_1^p(x_i) = x_i$, and

$$f_1(x_i) = \begin{cases} x_{i+1} & \text{if } i \neq p \\ x_1 & \text{if } i = p \end{cases}$$

Then

$$f_2(\theta(x_i)) = \theta \circ f_1 \circ \theta^{-1} \circ \theta(x_i) = \theta(f_1(x_i))$$

and

$$f_2^p(x_i) = \theta \circ f_1 \circ \theta^{-1} \circ \theta \circ \cdots \circ f_1 \circ \theta^{-1} \circ \theta(x_i) = \theta(x_i)$$

so that $\{\theta(x_1), \ldots \theta(x_p)\}$ is a cycle of period p. To show that the statement holds in reverse, define the homeomorphism $\theta_2 : X_2 \to X_1$ by $\theta_2 = \theta^{-1}$, and repeat the above argument.

The second statement, regarding attractive fixed points, is highly dependent on the restricted definition of the term given in the text, and, since it is not preserved by homeomorphism must be dependent on the metric. Choose two copies of the real line, $(\mathbb{R}, d_1)$ and $(\mathbb{R}, d_2)$ where $d_1(x, y)$ is the usual metric, and $d_2(x, y)$ is given by

$$d_2(x, y) = \begin{cases} |3x - 3y| & \text{if } x \leq 0 \\ |x - y| & \text{if } x \geq 0 \end{cases}$$

The identity map is then a homeomorphism between these spaces. Now choose the function $f(x) = -\frac{1}{2}x$ for both systems. In $(\mathbb{R}, d_1)$, 0 is an attractive fixed point. For $(\mathbb{R}, d_2)$, no matter what the size of the ϵ ball is around 0, it will include points on the positive side that, because the ball is shrunken on the negative side (for example, $B_1(0) = (-1/3, 1)$), fail to land back in the ball on one iteration. Hence this is not an attractive fixed point, even though the two systems are equivalent.

IV.5.2 An eventually periodic orbit is has the orbit $\{x_1, x_2, \ldots, x_n, x_{n+1}, \ldots x_{n+p} = x_{n+1}\}$, for some n and some period p. As in exercise IV.5.1, we write

$$f_1(x_i) = \begin{cases} x_{i+1} & \text{if } i \neq n+p \\ x_{n+1} & \text{if } i = n+p \end{cases}$$

The argument is then identical to that in IV.5.1.

IV.5.3 We modify the example in IV.5.1. Let $(\mathbb{R}, d_1)$ and $(\mathbb{R}, d_2)$ be as in the solution to problem IV.5.1. Let the homeomorphism $\theta : \mathbb{R} \to \mathbb{R}$ be $\theta(x) = x$. Then if $f_1(x) = -2x$, under our homeomorphism we have $f_2(x) = -2x$. However, in $(\mathbb{R}, d_1)$ 0 is a repulsive fixed point, but given any $\epsilon > 0$ a point $x < 0$ such that $x \in B_\epsilon(0)$ moves under 1 iteration to a point which is closer to 0 under the d_2 metric than it started.

IV.5.4 θ is a homeomorphism, which means that θ and θ^{-1} are continuous. This implies that $\theta^{-1}(\mathcal{O})$ is open if $\mathcal{O}$ is open and $(\theta^{-1})^{-1}(\mathcal{O}) = \theta(\mathcal{O})$ is open if $\mathcal{O}$ is open. Consequently, if there is an open neighborhood of x_f in $\mathbf{X}_1$, then $\theta(\mathcal{O})$ is open in $\mathbf{X}_2$ and $\theta(x_f) \in \theta(\mathcal{O})$ implies that there is a neighborhood $\mathcal{C}$ of $\theta(x_f)$ in $\mathbf{X}_2$ such that $\mathcal{C} \subset \theta(\mathcal{O})$. Suppose that for any $y \in \mathcal{O}$ we have

$$\lim_{n \to \infty} f_1^n(y) = x_f.$$

Then since θ is continuous

$$\lim_{n \to \infty} f_2^n(\theta(y)) = \lim_{n \to \infty} \theta(f_1^n(y)) = \theta(x_f)$$

for all $\theta(y) \in \theta(\mathcal{O})$, in particular for all elements of $\mathcal{C}$. Then $x_{2_f} = \theta(x_f)$ is an attractive fixed point of f_2. *Note that this is the standard definition of an attractive fixed point. We will use the more restrictive version given in the text, however.*

IV.5.5 The solution above to IV.5.4 shows that this definition of attractive fixed points is preserved under equivalence and conjugacy. (and in fact does not require that the spaces be metric spaces). All that is needed is to give a similar definition of a repulsive fixed point. For invertible maps, this is done by saying that x_f is a repulsive fixed point if x_f is an attractive fixed point of the map f^{-1}, that is, there is a neighborhood $\mathcal{O}$ of x_f such that for every $y \in \mathcal{O}$, $\lim_{n \to \infty} f^{-n}(y) = x_f$. For non-invertible maps, this is extended by saying that there is a neighborhood $\mathcal{O}$ of x_f such that for every $y \in \mathcal{O}$ there is an n such that $f^n(y) \notin \mathcal{O}$. This is likewise preserved under equivalence and conjugacy, a moment's reflection says that if this is true for $\mathcal{O}$, then it is true for any neighborhood $\mathcal{O}_1 \subset \mathcal{O}$.

IV.5.7 We want to show that if θ, a homeomorphism establishing the equivalence of $\{\mathbb{R}; f\}$ and $\{\mathbb{R}; g\}$ is a diffeomorphism, then the definition of an attractive fixed point given in the text is preserved under the equivalence. Let x_f be an attractive fixed point of $\{\mathbb{R}; f\}$. The definition here establishes that a fixed point is attractive if and only if it is a contraction on some ϵ-ball around x_f. We let $\theta(x_f) = x_g$ be the corresponding fixed point of $\{\mathbb{R}; g\}$, and we will show that g is also locally a contraction. Choose $\epsilon > 0$ such that f is a contraction on $B_\epsilon(x_f)$. Then for $y \in B_\epsilon(x_f)$, there is an s, $0 \leq s < 1$ such that $|f(x_f) - f(y)| \leq s|x_f - y|$. Since $\theta'(x_f) \neq 0$, there is an ϵ small enough (the derivative is continuous by assumption) such that for any $y \in B_\epsilon(x_f)$, the following expression is well-defined (i.e. none of the fractions are equal to zero in the denominator) for $y \neq x_f$:

$$\left| \frac{\theta(f(x_f)) - \theta(f(y))}{f(x_f) - f(y)} \right| \left| \frac{x_f - y}{\theta(x_f) - \theta(y)} \right|$$

In the limit as $y \to x_f$, the two fractions on the right approach, respectively, $|\theta'(x_f)| |\frac{1}{\theta'(x_f)}| = 1$. But since f is a contraction close enough to x_f, we have

$$\left| \frac{x_f - y}{f(x_f) - f(y)} \right| \geq \frac{1}{s} > 1$$

which implies that

$$\left| \frac{\theta(f(x_f)) - \theta(f(y))}{\theta(x_f) - \theta(y)} \right| \leq s$$

Then for $\epsilon > 0$ close enough to $\theta(x_f)$, we have

$$|\theta(f(x_f)) - \theta(f(y))| \leq s|\theta(x_f) - \theta(y)|$$

so that $|g(x_g) - g(z)| \leq s|x_g - z|$ for z close to x_g, and the fixed point is attractive because it is a local contraction.

IV.5.8 Assume $\{\mathbb{R}; f\}$ is a dynamical system such that f is differentiable for all $x \in \mathbb{R}$. A web diagram for a given initial condition x_0 starts at the point $y = x_0$, and successive points are then taken from $(x_n, x_n) \mapsto (x_n, x_{n+1}) \mapsto (x_{n+1}, x_{n+1})$ where $x_{n+1} = f(x_n)$, and the points of the form (x_n, x_{n+1}) are on the graph of the function $f(x)$. It follows that if a point is fixed, $x_n = x_{n+1} = f(x_n)$ so that this is a point of intersection of the graph of $f(x)$ with the line (x, x) or $y = x$.

To prove the second assertion, we note that for our definition of attractive fixed points we have an $\epsilon > 0$ such that for any

$y \in B_\epsilon(x_f)$, we have a constant $0 \leq s < 1$ such that $|f(y) - f(x_f)| \leq s|y - x_f|$. Then for $y \neq x_f$ we may write

$$\frac{|f(y) - f(x_f)|}{|y - x_f|} = \left|\frac{f(y) - f(x_f)}{y - x_f}\right| \leq s$$

Since this is true for all $y \in B_\epsilon(x_f)$, it is certainly true as $y \to x_f$, yielding

$$\lim_{y \to x_f}\left|\frac{f(y) - f(x_f)}{y - x_f}\right| = |f'(x_f)| \leq s < 1$$

To go the other way, if the derivative is continuous in a neighborhood of x_f, and $|f'(x_f)| < 1$, it follows that for some $\delta > 0$ there is a neighborhood $B_\delta(x_f)$ such that for any $y \in B_\delta(x_f)$, $|f'(y)| < 1$. We may then find an $0 < \epsilon < \delta$ and a number $s < 1$ such that $|f'(y)| < s$ for all $y \in B_\epsilon$. Then the map is a local contraction since

$$\lim_{y \to y_0}\left|\frac{f(y) - f(y_0)}{y - y_0}\right| \leq s$$

implies that close enough to y_0, we have

$$|f(y) - f(y_0)| \leq s|y - y_0|$$

in particular, $|f(y) - f(x_f)| \leq s|y - x_f|$, as desired.

Finally, to generalize to cycles, let $x_1, x_2, \ldots, x_p$ be a cycle of period p, and suppose that the cycle is attractive. By definition, this implies that x_i is an attractive fixed point of the dynamical system $\{\mathbb{R}; f^p\}$, for some x_i in the cycle. By the result above, we have $|(f^p)'(x_i)| < 1$ for this point. This says that

$$|(f^p)'(x_i)| = |f'(x_i) \cdots f'(x_p)f'(x_1) \cdots f'(x_{i-1})| < 1$$

Notice if this condition holds at x_i it holds at any x_j in the cycle since the expression is the same product. If this product is less than 1, it follows that for at least one of the elements in the product $|f'(x_j)| < 1$, but it can also be the case that some of them do not satisfy this condition, only their product must.

IV.5.9 Since we would like our IFS to have an associated shift dynamical system equivalent to

$$f(x) = \begin{cases} 1 - 2x & x \in [0, \frac{1}{2}] \\ 2x - 1 & x \in [\frac{1}{2}, 1] \end{cases}$$

we look at an IFS which is the inverse of $f(x)$ on the regions in question. Consider the IFS

$$\{[0, 1]; \frac{1}{2}x + \frac{1}{2}, -\frac{1}{2}x + \frac{1}{2}\}$$

The attractor of this system is $[0, 1]$ as follows:

$$W([0, 1]) = \frac{1}{2}([0, 1]) + \frac{1}{2} \cup -\frac{1}{2}([0, 1]) + \frac{1}{2}$$

$$= [0, \frac{1}{2}] + \frac{1}{2} \cup [-\frac{1}{2}, 0] + \frac{1}{2}$$

$$= [\frac{1}{2}, 1] \cup [0, \frac{1}{2}] = [0, 1]$$

That this IFS is just-touching results from the intersection above containing the point $\frac{1}{2}$, and the fact that $\mathcal{O} = (0, \frac{1}{2}) \cup (\frac{1}{2}, 1)$ suffices for the open set. The points of multiple addresses are the images of this point. Defining the associated shift dynamical system, we have

$$S(a) = \begin{cases} w_1^{-1}(a) & \text{if } a \in (\frac{1}{2}, 1] \\ w_2^{-1}(a) & \text{if } a \in [0, \frac{1}{2}) \end{cases}$$

The point of intersection yields

$$w_1^{-1}(\frac{1}{2}) = 2(\frac{1}{2} - \frac{1}{2}) = 0$$

$$w_2^{-1}(\frac{1}{2}) = -2(\frac{1}{2} - \frac{1}{2}) = 0$$

so we define

$$S(\frac{1}{2}) = w_1^{-1}(\frac{1}{2}) = w_2^{-1}(\frac{1}{2}) = 0$$

and there is no conflict. This explicitly gives $S(a)$ as

$$S(a) = \begin{cases} 2a - 1 & a \in [\frac{1}{2}, 1] \\ 1 - 2a & a \in [0, \frac{1}{2}] \end{cases}$$

which is clearly equivalent to $f(x)$ with the homeomorphism being the identity on $[0, 1]$.

IV.5.10 When $p = 0$ we have

$$w_1 \begin{pmatrix} x \\ y \end{pmatrix} = \begin{pmatrix} \frac{1}{4} & 0 \\ 0 & \frac{1}{4} \end{pmatrix} \begin{pmatrix} x \\ y \end{pmatrix}$$

$$w_1 \begin{pmatrix} x \\ y \end{pmatrix} = \begin{pmatrix} \frac{1}{4} & 0 \\ 0 & \frac{1}{4} \end{pmatrix} \begin{pmatrix} x \\ y \end{pmatrix} + \begin{pmatrix} \frac{3}{8} \\ 0 \end{pmatrix}$$

$$w_1 \begin{pmatrix} x \\ y \end{pmatrix} = \begin{pmatrix} \frac{1}{4} & 0 \\ 0 & \frac{1}{4} \end{pmatrix} \begin{pmatrix} x \\ y \end{pmatrix} + \begin{pmatrix} \frac{3}{4} \\ 0 \end{pmatrix}$$

If we let $y = 0$, then

$$w_1(x, 0) = (\frac{1}{4}x, 0)$$

$$w_2(x, 0) = (\frac{1}{4}x + \frac{3}{8}, 0)$$

$$w_1(x, 0) = (\frac{1}{4}x + \frac{1}{4}, 0)$$

so that for any point on the x-axis, $w_i(x, y)$ is also on the x-axis. We now consider the image of the set $[0, 1]$. This maps to three subintervals, each of length $\frac{1}{4}$ and arranged with their left endpoints at 0, $\frac{3}{8}$ and $\frac{1}{4}$, respectively. The set formed by repeated iteration removes two subintervals from each interval formed by the previous iteration, and the attractor must be $\lim_{n \to \infty} W^n([0, 1], 0)$. Thus for $p = 0$, the attractor is a Cantor set.

For $p = 1$ we have

$$w_1 \begin{pmatrix} x \\ y \end{pmatrix} = \begin{pmatrix} \frac{1}{2} & 0 \\ 0 & \frac{1}{2} \end{pmatrix} \begin{pmatrix} x \\ y \end{pmatrix}$$

$$w_1 \begin{pmatrix} x \\ y \end{pmatrix} = \begin{pmatrix} \frac{1}{2} & 0 \\ 0 & \frac{1}{2} \end{pmatrix} \begin{pmatrix} x \\ y \end{pmatrix} + \begin{pmatrix} \frac{1}{2} \\ \frac{1}{2} \end{pmatrix}$$

$$w_1 \begin{pmatrix} x \\ y \end{pmatrix} = \begin{pmatrix} \frac{1}{2} & 0 \\ 0 & \frac{1}{2} \end{pmatrix} \begin{pmatrix} x \\ y \end{pmatrix} \begin{pmatrix} \frac{1}{2} \\ 0 \end{pmatrix}$$

which is a just-touching Sierpinski triangle.

We next prove that for fixed $\sigma \in \Sigma$, the map $\phi(p) \to A(p)$ is continuous in p, for $p \in [0, 1]$. We may write the three transformations in matrix notation as $w_i(x) = Ax + t_i$, where

$$A = \begin{pmatrix} \frac{1+p}{4} & 0 \\ 0 & \frac{1+p}{4} \end{pmatrix}$$

$$t_i = \begin{cases} (0,0) & i = 1 \\ (\frac{3+p}{8}, \frac{p}{2}) & i = 2 \\ (\frac{3-p}{4}, 0) & i = 3 \end{cases}$$

so that we have

$$w_{\sigma_1} \circ w_{\sigma_2}(x) = A^2 x + A t_{\sigma_2} + t_{\sigma_1}$$

in general, we have

$$w_{\sigma_1} \circ w_{\sigma_2} \circ \cdots \circ w_{\sigma_n}(x)$$
$$= A^n x + A^{n-1} t_{\sigma_n} + A^{n-2} t_{\sigma_{n-1}} + \cdots + A t_{\sigma_2} + t_{\sigma_1}$$

so that

$$\phi(p) = \lim_{n \to \infty} A^n x + \sum_{i=0}^{n} A^i t_{\sigma_{i+1}} = \sum_{i=0}^{\infty} A^i t_{\sigma_{i+1}}$$

Writing this out for fixed σ, and letting

$$t_{\sigma_i} = \begin{pmatrix} e_{\sigma_i} \\ f_{\sigma_i} \end{pmatrix}$$

we have

$$\phi_\sigma(p) = \begin{pmatrix} \sum_{i=0}^{\infty} (\frac{1+p}{4})^i e_{\sigma_{i+1}} \\ \sum_{i=0}^{\infty} (\frac{1+p}{4})^i f_{\sigma_{i+1}} \end{pmatrix}$$

which are continuous functions of p with σ fixed (they are, for fixed σ, differentiable in p, being a form of Taylor series expansion).

That the set $J(x) = \{y \in A(0) : \theta(y) = x\}$ where $\theta(x) = \phi(1)(\phi(0)^{-1}(x)$ is the set of y whose paths meet at x is true by construction.

Answers to Chapter IV, section 6

IV.6.3 The three points in question have addresses $1\bar{3} = 3\bar{1}$, $1\bar{2} = 2\bar{1}$, and $2\bar{3} = 3\bar{2}$. Defining $S(a) = w_1^{-1}(a)$ forces the point a to become $1\bar{2}$, which is eventually periodic since $T(1\bar{2}) = \bar{2}$, a fixed point. Similarly, if we define $S(a) = w_2^{-1}(a)$ we force the point a to become $2\bar{1}$ which since $T(2\bar{1}) = \bar{1}$ is also eventually periodic. The points b, c are analyzed similarly, being forced to either of the two addresses, which after one iteration land on the fixed points $\bar{2}$ or $\bar{3}$.

IV.6.4 If we create an IFS for the equilateral Sierpinski triangle in which we rotate the subtriangle corresponding to $w_2(\triangle)$ by 240° and that corresponding to $w_3(\triangle)$ by 120° we get a just-touching Sierpinski triangle in which the multiple address points correspond to equating the addresses

$$1\bar{3} = 3\bar{3} \qquad 3\bar{1} = 2\bar{1} \qquad 1\bar{2} = 2\bar{2}$$

Since these points, regardless of which map is used in the equation $S(a) = w_i^{-1}(a)$ map to the points $\bar{3}$, $\bar{1}$, and $\bar{2}$, on one application of the map, $S(a)$ may be chosen without conflict. Specifically, the IFS in question is

$$w_1 \begin{pmatrix} x \\ y \end{pmatrix} = \begin{pmatrix} \frac{1}{2} & 0 \\ 0 & \frac{1}{2} \end{pmatrix} \begin{pmatrix} x \\ y \end{pmatrix}$$

$$w_2 \begin{pmatrix} x \\ y \end{pmatrix} = \begin{pmatrix} \frac{1}{2}\cos 240° & -\frac{1}{2}\sin 240° \\ \frac{1}{2}\sin 240° & \frac{1}{2}\cos 240° \end{pmatrix} \begin{pmatrix} x \\ y \end{pmatrix} + \begin{pmatrix} \frac{1}{2} \\ 0 \end{pmatrix}$$

$$w_3 \begin{pmatrix} x \\ y \end{pmatrix} = \begin{pmatrix} \frac{1}{2}\cos 120° & -\frac{1}{2}\sin 120° \\ \frac{1}{2}\sin 120° & \frac{1}{2}\cos 120° \end{pmatrix} \begin{pmatrix} x \\ y \end{pmatrix} \begin{pmatrix} \frac{1}{4} \\ \frac{\sqrt{3}}{4} \end{pmatrix}$$

IV.6.6 The IFS $\{\Sigma; t_1, \ldots t_N\}$ where $t_n = n\sigma$ has by definition an associated shift dynamical system defined by $S(\sigma) = t_{\sigma_1}^{-1}(\sigma)$ where $\sigma = \sigma_1\sigma_2\cdots$. Then $S(\sigma) = \sigma_2\sigma_3\cdots$, which is exactly $T(\sigma)$. To see whether two such dynamical systems can be equivalent for different values of N, we use exercise IV.5.1, namely if two dynamical systems are equivalent, then a cycle of period p maps to a cycle of period p. Given $N_1 > N_2$, we have for either that any of the cycles of period 1 are given by $\bar{n}$ where $0 \leq n \leq N_i$. These are distinct, and if θ were the homeomorphism establishing the equivalence, would map to distinct fixed points. But this system for Σ_{N_2} has N_2 such points, which is less than N_1. Thus

$$\theta : \Sigma_{N_1} \to \Sigma_{N_2}$$

either does not map fixed points to fixed points or is not one-to-one, in other words there is no such equivalence.

IV.6.14 First we prove, as claimed in the text, that

$$\tilde{A} = \{(\phi(\sigma), \sigma) : \sigma \in \Sigma\}$$

is the attractor of the lifted IFS. To do this we need to confirm hyperbolicity, in order to assert the existence of a unique fixed point in $\mathcal{H}(X \times \Sigma)$. Let the metric on X be denoted by d_X and that on Σ be denoted d_Σ. To make things simple, we use a Manhattan metric on the product space, namely,

$$d((x_1, \sigma_1), (x_2, \sigma_2)) = d_X(x_1, x_2) + d_\Sigma(\sigma_1, \sigma_2)$$

We have, by definition,

$$\tilde{w}_i((x, \sigma)) = (w_i(x), i\sigma)$$

and we must show that this is a contraction. w_i are the maps of the original IFS, and we may assume that that system has a contractivity of s. Then

$$d(\tilde{w}_i(x_1, \sigma_1), \tilde{w}_i(x_2, \sigma_2)) = d_X(w_i(x_1), w_i(x_2)) + \frac{1}{N+1}\sum_{j=1}^{\infty} \frac{|\sigma_{1_j} - \sigma_{2_j}|}{(N+1)^j}$$

$$\leq s d_X(x_1, x_2) + \frac{1}{N+1} d_\Sigma(\sigma_1, \sigma_2)$$

Let $m = \max(s, \frac{1}{N+1})$. Then $0 < m < 1$ and

$$d(\tilde{w}_i(x_1, \sigma_1), \tilde{w}_i(x_2, \sigma_2)) \leq m d((x_1, \sigma_1), (x_2, \sigma_2))$$

so each map $\tilde{w}_i$ is contractive, and the IFS is hyperbolic. Consequently since X and Σ are complete, and therefore so is $X \times \Sigma$, we have the attractor $\tilde{A}$ is the unique fixed point in $\mathcal{H}(X \times \Sigma)$ of the map

$$\tilde{W}(B) = \bigcup_{i=1}^{N} \tilde{w}_i(B)$$

for B a compact, non-empty subset of $X \times \Sigma$. We now verify that the attractor $\tilde{A}$ is as given in the text. For any element of the set $\tilde{A} = \{(\phi(\sigma), \sigma) : \sigma \in \Sigma\}$ and some $\tilde{w}_i$ we have

$$\tilde{w}_i(\tilde{A}) = \{(w_i(\phi(\sigma)), i\sigma) : \sigma \in \Sigma\}$$

but $i\sigma = \phi^{-1}(w_i(\phi(\sigma)))$ so the image $\tilde{W}(\tilde{A})$ is in $\tilde{A}$. Let $a \in \tilde{A}$. Then a is given by

$$a = (w_{\sigma_1}(T(\sigma)), \sigma_1 T(\sigma))$$

so a is in the image of an element of $\tilde{A}$ under $\tilde{w}_{\sigma_1}$. Hence $\tilde{W}(\tilde{A}) = \tilde{A}$ and this is the unique attractor of the lifted IFS. Finally, to complete Lemma 1, we have

$$w_1(\tilde{A}) = \{a \in \tilde{A} : a = (x, 1\sigma)\}$$
$$w_2(\tilde{A}) = \{a \in \tilde{A} : a = (x, 2\sigma)\}$$

These sets are points with mutually exclusive sets of second coordinates, so their intersection is empty. Hence $\tilde{A}$ is totally disconnected.

IV.6.15 Let $\{x_n\}$ be any orbit of the associated shift dynamical system $\{A; S\}$. Then by definition, $S(x_n) = x_{n+1} = w_i^{-1}(x_n)$ for some w_i. Thus, regardless of the rule for assigning $S(a)$ on overlapping regions, there is a rule of assignment which assigns either w_1^{-1} or w_2^{-1} to each application of S to the orbit. Let τ be the address such that if $S(x_n) = w_i^{-1}(x_{n+1})$ then $\tau_n = i$. Let

$$\sigma_n = \tau_n \tau_{n+1} \cdots$$

Then $\phi(\sigma_n) = x_n$, $\sigma_{n+1} = T(\sigma_n)$, so that the orbit $\{(\phi(\sigma_n), \sigma_n)\}$ of the shift dynamical system $\{\tilde{A}; \tilde{S}\}$ has the desired property.

IV.6.16 By definition, the lift is $(\sigma_1, \sigma_2) \mapsto (T\sigma_1, \sigma_1\sigma_2)$ which looks like the shift automorphism: it takes the symbol off the first element and puts it on the second. The suitably defined inverse would then be to take a symbol off the second element and put it back on the first. This defines a system on $\Sigma \times \Sigma$, as the space of shifts, which is identical to the shift automorphism.

Answers to Chapter IV, section 7

IV.7.2 $\tilde{x}_0 = 21\overline{2}$ consequently the true orbit of $\tilde{x}_0$ is given by

$$\tilde{x}_0 = 21\overline{2} \qquad T(\tilde{x}_0) = 1\overline{2} \qquad T^2(\tilde{x}_0) = \overline{2}$$
$$T^n(\tilde{x}_0) = \overline{2} \qquad \text{for } n \geq 2$$

By contrast, we have $\tilde{x}_n = 1\overline{2}$ after the first iteration, so that $d(\tilde{x}_n, T^n(\tilde{x}_0)) = 1/3$ after the second iteration, under the code space metric for two symbols. We wish to find a shadowing orbit. Taking $\theta = 1/3$ and $s = 1/3$ we need

$$d(\tilde{x}_n, x_n) \leq \frac{s\theta}{1-s} = \frac{1}{6}$$

Any real orbit which has the symbol 2 in it will eventually be separated from $1\overline{2}$ by at least $1/3$ so the only possible choice for a shadowing orbit is $\overline{1}$. Computing the distance yields

$$d(1\overline{2}, \overline{1}) = \frac{1}{3}\sum_{n=1}^{\infty}\frac{1}{3^n} = \frac{1}{1-\frac{1}{3}} - 1 = \frac{1}{6}$$

which satisfies the shadowing theorem.

Answers to Chapter IV, section 8

IV.8.1 It suffices to show that the positive rationals are dense in the positive reals, since the function $\theta(x) = -x$ will show that the negative rationals are then dense in the negative reals. To this end let $r \in \mathbb{R}$, and let its decimal expansion be $.r_1r_2r_3\ldots$. Then the sequence of rationals

$$.r_1\overline{0}, .r_1r_2\overline{0}, .r_1r_2r_3\overline{0}, \ldots$$

converges to r. Hence every number $r \in \mathbb{R}$ is a limit point of the rationals, and since the closure of a set contains its limit points, the closure of the rationals is the reals. Hence the rationals are dense in $\mathbb{R}$.

IV.8.2 The sequence in question is an enumeration of the rationals in $[0, 1]$, i.e. $\mathbb{Q} \cap [0, 1]$. Since $\mathbb{Q}$ is dense in $\mathbb{R}$, $\mathbb{Q} \cap [0, 1]$ is dense in $[0, 1]$.

IV.8.3 The function $f(x) = 2x \bmod 1$ can be viewed as follows: On the circle of radius $1/2\pi$, that is, of circumference 1, define $x(\theta)$ to be the distance along the circumference in the counterclockwise direction from the positive x-axis. Then taking the usual equation of angles $\theta = \theta + 2\pi = \theta \bmod 2\pi$, the function $\theta \mapsto 2\theta$ gives us our function on $[0, 1]$. If θ is not a rational number times 2π, it cannot be return to itself under this map, since $2^n\theta = \theta$ implies that $2^n\theta = \theta + m\pi$ for some integer m. Hence any irrational number times 2π for θ results in a dense orbit on the circle.

IV.8.4 The dynamical system $\{[0, \infty); f(x) = 2x\}$ is sensitive to initial conditions as follows: Choose $\delta = 1$, and for any $\epsilon > 0$ let $y \neq x$, and $y \in B_\epsilon(x)$. Then since $d(f(x), f(y)) = 2d(x, y)$, we have $d(f^n(x), f^n(y)) = 2^n d(x, y)$. Choose n, given y such that $2^n d(x, y) > 1$. In particular, for any $n > N$ where $2^N > 1$, $d(f^n(x), f^n(y)) > \delta$.

The dynamical system $\{[0, \infty); f(x) = .5x\}$ is not sensitive. In this case, we have $d(f(x), f(y)) = 0.5d(x, y)$ so that for any $\epsilon > 0$ and *any* δ, there is an N such that for any $y \in B_\epsilon(x)$ $d(f^n(x), f^n(y)) = (0.5)^n d(x, y)$ is less than δ.

IV.8.5 Let Σ be the code space on two symbols, and let T be the shift operator. Then $\{\Sigma; T\}$ is sensitive to initial conditions: Choose $\delta < \frac{1}{3}$. Then if $x, y \in \Sigma$ and $x \neq y$, there is an n such that $x_n \neq y_n$. Then since

$$T^n(x) = T^n(x_1 x_2 \ldots) = x_n x_{n+1} \ldots$$
$$T^n(y) = T^n(y_1 y_2 \ldots) = y_n y_{n+1} \ldots$$

it follows that

$$d(T^n(x), T^n(y)) \geq \frac{|x_n - y_n|}{3} > \delta$$

Hence whatever ϵ is chosen, we may find a y and an n satisfying sensitivity to initial conditions.

$\{\Sigma; T\}$ is transitive as well: Let $\mathcal{U}, \mathcal{V} \subset \Sigma$ be open sets. Then we can choose $y \in \mathcal{V}$ and $\delta > 0$ such that $B_\delta(y) \subset \mathcal{V}$, and $x \in \mathcal{U}$ and $\epsilon > 0$ such that $B_\epsilon(x) \subset \mathcal{U}$. Choose n such that $\frac{1}{3^n} < \epsilon$. Then any $v \in \Sigma$ such that

$$v = x_1 x_2 \ldots x_n v_1 v_2 \ldots \in B_\epsilon(x) \subset \mathcal{U}$$

Similarly, choose M such that for any $m > M$,

$$u = y_1 y_2 \ldots y_m u_1 u_2 \ldots \in B_\delta(y) \subset \mathcal{V}$$

Then any $\sigma \in \Sigma$ of the form

$$\sigma = y_1 y_2 \ldots y_m x_1 x_2 \ldots x_n \sigma_1 \sigma_2 \ldots$$

is in $\mathcal{V}$. $T^n(\sigma) = x_1 x_2 \ldots x_n \sigma_1 \sigma_2 \ldots$, which is in $\mathcal{U}$. Hence $T^n(\mathcal{V}) \cap \mathcal{U} \neq \emptyset$, as desired.

IV.8.6 Suppose that $\{X; f\}$ and $\{Y; g\}$ are equivalent under the homeomorphism $\theta : X \to Y$. Let $\mathcal{U}, \mathcal{V} \subset Y$ be open sets. Then because θ is continuous, $\theta^{-1}(\mathcal{U}), \theta^{-1}(\mathcal{V})$ are open sets in X. Since $\{X; f\}$ is transitive, there is an n such that

$$\theta^{-1}(\mathcal{U}) \cap f^n \circ \theta^{-1}(\mathcal{V}) \neq \emptyset$$

Let x be a member of this non-empty intersection. Then $\theta(x) \in \mathcal{U}$ and $x \in f^n \circ \theta^{-1}(\mathcal{V})$ implies

$$\theta \circ \theta^{-1}(\mathcal{U}) \cap \theta \circ f^n \circ \theta^{-1}(\mathcal{V}) = \mathcal{U} \cap g^n(\mathcal{V}) \neq \emptyset$$

so that $\{Y; g\}$ is also transitive. Reversing the roles of θ and θ^{-1}, which we may do because θ^{-1} is also continuous, yields the converse.

Answers to Chapter V, section 1

V.1.3 Let (X, d) be a metric space, $A \subset X$ with $A = \{a, b, c\}$. We wish to calculate the fractal dimension of A. Let $r = \min_{x, y \in A}(d(x, y))$. Then for any $\epsilon < r/2$, we have $\mathcal{N}(A, \epsilon) = 3$. Hence the fractal dimension is

$$D(A) = \lim_{\epsilon \to 0} \frac{\ln 3}{\ln(1/\epsilon)} = 0.$$

V.1.6 Using the Box Counting Theorem to cover the classical Cantor set $\mathcal{C}$ with boxes of size $1/3^n$, we will get

$$D = \lim_{n \to \infty} \left(\frac{\ln \mathcal{N}(\mathcal{C})}{\ln 3^n} \right).$$

We need to calculate $\mathcal{N}_n(\mathcal{C})$. It takes $1 = 2^0$ box when $n = 0$, subdividing and deleting the middle third gives $2 = 2^1$ boxes when $n = 1$. On each subdivision and deletion of a set of middle thirds, we divide each interval covered by a box of size $1/3^n$ in thirds and retain two of these thirds, so that it will take 2^{n+1} boxes of size $1/3^{n+1}$ to cover the set at the next step. Hence $\mathcal{N}_n(\mathcal{C}) = 2^n$, and we have

$$D(\mathcal{C} = lim_{n \to \infty} \frac{\ln(2^n)}{\ln 3^n} = \lim_{n \to \infty} \frac{n \ln 2}{n \ln 3} = \frac{\ln 2}{\ln 3}.$$

V.1.7 Use boxes of side length $1/2^n$. For each quartering of the figure, and its subdivisions, we divide each box into 4 parts, and one part contains none of the fractal subset points. Consequently, on each subdivision, the number of boxes increases threefold. There is one box of side length $1/2^0$ hence

$$D(\text{Figure}) = \lim_{n \to \infty} \frac{\ln 3^n}{\ln 2^n} = \frac{\ln 3}{\ln 2}.$$

V.1.8 One way to choose the boxes is to take side lengths $1/2^n$, with the first subdivision running horizontally from just below the bottom of the caption, the vertical running vertically from the lefthand 'tip' of the bottom piece. Do it with a ruler to assure yourself that this divides the figure into three congruent pieces, and leaves the fourth of the four squares empty. Following the lead of the previous problem, the dimension is $\frac{\ln 3}{\ln 2}$.

V.1.9 We wish to show that using a Cantor set constructed by removing middle thirds on the interval $[0, 3]$ has the same fractal dimension as the usual Cantor set. We start with 2^0 intervals of length $1/3^{0-1}$. The rest of the procedure is the same, except that we now have

$$D(\text{Big Cantor}) = \lim_{n \to \infty} \frac{\ln(2^n)}{\ln(3^{n-1})} = \lim_{n \to \infty} \frac{n \ln 2}{(n-1) \ln 3} = \frac{\ln 2}{\ln 3} \lim_{n \to \infty} \frac{n}{n-1} = \frac{\ln 2}{\ln 3}.$$

V.1.10 Let $A \subset \mathbb{R}^2$ be a compact non-empty subset. A has fractal dimension D_1 in the Euclidean metric, and D_2 in the Manhattan metric. The metrics are equivalent, consequently there is a transformation $\theta : \mathbb{R}^2 \to \mathbb{R}^2$ which forms the equivalence. By Theorem 1.2 in Chapter V then, we have $D_1(A) = D_2(\theta(A))$ or $D_1 = D_2$.

V.1.11 Let $B = \begin{pmatrix} 1/2 & 0 \\ 0 & 1/2 \end{pmatrix}$, and define two IFS

$$B \begin{pmatrix} x \\ y \end{pmatrix} + \begin{pmatrix} 1 \\ 0 \end{pmatrix}, B \begin{pmatrix} x \\ y \end{pmatrix}, B \begin{pmatrix} x \\ y \end{pmatrix} + \begin{pmatrix} 0 \\ 1 \end{pmatrix}$$

with attractor A_1, and

$$B \begin{pmatrix} x \\ y \end{pmatrix} + \begin{pmatrix} 2 \\ 0 \end{pmatrix}, B \begin{pmatrix} x \\ y \end{pmatrix}, B \begin{pmatrix} x \\ y \end{pmatrix} \begin{pmatrix} 1 \\ 1 \end{pmatrix}$$

with attractor A_2.

These are two Sierpinski triangles, one with right angle at the origin, and hypotenuse between $(1, 0)$ and $(0, 1)$, the other with right angle at $(1, 1)$ and hypotenuse between the origin and $(0, 2)$. We rotate one into the other, expanding by a factor of two, to generate an equivalence between the two systems. This requires an invertible affine map. One which works is

$$\theta \begin{pmatrix} x \\ y \end{pmatrix} = \begin{pmatrix} -1 & -1 \\ 1 & -1 \end{pmatrix} \begin{pmatrix} x \\ y \end{pmatrix} + \begin{pmatrix} 1 \\ 1 \end{pmatrix}.$$

The two attractors are metrically equivalent, and therefore have the same fractal dimension.

Answers to Chapter V, section 2

V.2.3 Let B be given by

$$B = \begin{pmatrix} 1/2 & 0 \\ 0 & 1/2 \end{pmatrix}$$

Then we can use the following IFS to generate ∎:

$$w_1 = Bx$$

$$w_2 = Bx + \begin{pmatrix} 0 \\ 1/2 \end{pmatrix}$$

$$w_3 = Bx + \begin{pmatrix} 1/2 \\ 0 \end{pmatrix}$$

$$w_2 = Bx + \begin{pmatrix} 1/2 \\ 1/2 \end{pmatrix}$$

Each map has a contractivity factor of $1/2$, so the fractal dimension by Theorem 2.3 in Chapter V is given by

$$1/2^D + 1/2^D + 1/2^D + 1/2^D = 1$$

This yields $1/2^D = 1/4$ or $D = 2$ as desired.

V.2.4 We get the fractal dimension of the Cantor set from the IFS

$$\{[0, 1]; \ w_1(x) = \frac{1}{3}x; \ w_2(x) = \frac{1}{3}x + \frac{2}{3}\}.$$

Each map has contractivity $1/3$ so we have

$$1/3^D + 1/3^D = 1 \text{ or } 1/3^D = 1/2.$$

Then $-D \ln 3 = -\ln 2$ or $D = \frac{\ln 2}{\ln 3}$. to the equation $3(1/2^D) + (2/5)^D = 1$. Solved numerically, this gives $D \approx 1.874$.

V.2.5 The attractor contains 3 images of itself scaled by $1/2$, with two of them rotated by $\pi/2$ and shifted. Assuming that the attractor is centered at the origin, and extends two units in either x-direction and unit in either y direction, the IFS is given by

$$w_1 = B \begin{pmatrix} x \\ y \end{pmatrix}$$

$$w_2 = B \begin{pmatrix} x \\ y \end{pmatrix} + \begin{pmatrix} 3/2 \\ -3/4 \end{pmatrix}$$

$$w_2 = B \begin{pmatrix} x \\ y \end{pmatrix} + \begin{pmatrix} -3/2 \\ 3/4 \end{pmatrix}$$

where B is given by

$$\begin{matrix} 0 & -1/2 \\ 1/2 & 0 \end{matrix}$$

The scaling factors are all $1/2$, leading to the equation $3/2^D = 1$, or $D = \frac{\ln 3}{\ln 2}$.

V.2.8 This attractor has 6 maps, 3 at $1/2$ scale, 3 at $1/4$ scale. It is overlapping, we need an upper bound which we get from Theorem 3.1 in Chapter V. We have,

$$3 \cdot (1/2^D) + 3 \cdot (1/4^D) = 1$$

Rewriting this as

$$3 \cdot (1/2)^{2D} + 3 \cdot (1/2)^D - 1 = 0$$

yields $1/2^D = \pm(\sqrt{21}/6) - 1/2$ or

$$D = \left| \frac{\ln(\sqrt{21}/6 - 1/2)}{\ln 2} \right|$$

V.2.9 Let $w_i : \mathbb{R}^7 \to \mathbb{R}^7$ for $i = 1, 2$, be an IFS with scaling factors s_1, s_2 respectively.

That the new maps are similitudes follows from the general fact that the composition of similitudes is a similitude. To see this write the two similitudes as

$$r_1 R_1 R_{\theta_1} \circ r_2 R_2 R_{\theta_2}(x)$$

Since reflections are diagonal, they commute with the other matrices, the scalars commute as well. The composition of rotations is a rotation, the composition of reflections is a reflection, multiplying scalars together is a scalar. We are left with

$$r_1 r_2 R_1 R_2 R_{\theta_1} R_{\theta_2}$$

which is a similitude.

The new contraction factors are

$$s_{v_1} = s_1^2 s_{v_2} = s_1 s_2 s_{v_3} = s_2 s_1 s_{v_4} = s_2^2$$

Using the original system in Theorem 3.1 in Chapter V, we have

$$|s_1|^D + |s_2|^D = 1$$

The new system yields

$$|s_1^2|^D + 2|s_1 s_2|^D + |s_2^2|^D = (|s_1|^D + |s_2|^D)^2 = 1$$

The calculation of dimension remains unchanged, since the first equation is a solution of the second.

Answers to Chapter V, section 3

The exercises in Section 3 are all experimental in nature.

Answers to Chapter V, section 4

V.4.1 The fact that $\mathcal{M}(A, p, \epsilon)$ is a non-increasing function of ϵ is proved by looking at the sums over which the infimum is taken. Suppose $\epsilon_1 < \epsilon_2$ and that $\mathcal{M}(A, p, \epsilon_1) < \mathcal{M}(A, p, \epsilon_2)$. Since any sequence $\{A_i\}$ which qualifies for ϵ_1 is a sequence of sets with diameters less than ϵ_2, it follows that by taking the same sequence in the computation of $\mathcal{M}(A, p, \epsilon_2)$, the infimum must be less than or equal to the sum generated by this sequence. But this is a contradiction.

V.4.2 Let $A = \{a, b, c, d, e, f, g\}$ be a subset of $(\mathbb{R}^2, \text{Euclidean})$. Let $2r$ be the minimum distance between points in A. By choosing $\epsilon < r$, we must, for any ϵ cover the set with at least 7 sets $A_1, \ldots A_7$. We can also cover A exactly with 7 sets of any epsilon of this size. Consequently, the infimum is generated for any p by the limit of the finite sum as $\epsilon \to 0$ which is 0 for any $p > 0$. When $p = 0$ the sum is exactly 7.

V.4.3 Let A be a bounded countably infinite subset of $\mathbb{R}^2$, under the Euclidean metric. Let $\rho : \mathbb{Z}^+ \to A$ be a rule of assignment which takes each positive integer i to an element $a_i \in A$, where ρ is 1–1 and onto. Such a function exists by definition of countability. Then the sequence of subsets $A_i = \{a_i\}$ of A is in $\mathcal{A}$, since their union is the set A. But the diameter of any singleton set in $\mathbb{R}^2$ is zero, so that for any $p > 0$

$$\sum_{i=1}^{\infty} (\text{diam}(\{a_i\}))^p = 0$$

Hence for any $p > 0$, $\mathcal{M}(A, p, \epsilon) = 0$ for any $\epsilon > 0$, consequently, $\mathcal{M}(A, p) = 0$. On the other hand, if $p = 0$, then for any non-empty subset of A, $(\text{diam}(A_i))^p = 1$. For each $n > 1$, there is an $\epsilon > 0$ such that there are more than n disjoint subsets of A, hence $\mathcal{M}(A, 0) = \infty$.

V.4.4 For $p = 0$, given $\epsilon < \frac{1}{3^n}$, there are at least 2^n subsets in any sequence of subsets of diameter less than ϵ which cover $\mathcal{C}$. To see this, the endpoints of 2^n disjoint intervals of diameter $\frac{1}{3^n}$ are elements of $\mathcal{C}$, and can not lie in the same subset. Hence as $\epsilon \to 0$, $\mathcal{M}(\mathcal{C}, 0, \epsilon) \to \infty$. Since $\mathcal{M}(\mathcal{C}, 0)$ is the supremum of these numbers, it is also infinite. For $p = 1$, given $\epsilon > 0$, we may choose the smallest n such that $\frac{1}{3^n} < \epsilon$, and cover $\mathcal{C}$ with exactly 2^n subsets with diameter $< \epsilon$, by taking the intersection of $\mathcal{C}$ with subintervals of $[0, 1]$. The infimum for any $\epsilon > 0$ is the value of the sum as $n \to \infty$, which is

$$\lim_{n \to \infty} \frac{2^n}{3^n} = 0$$

Since $\mathcal{M}(\mathcal{C}, p, \epsilon) = 0$ for any $\epsilon > 0$ we have $\mathcal{M}(\mathcal{C}, 1) = 0$.

V.4.5 The limit scheme forces a minimization of the number of sets by the diameter of sets. Suppose we attempt to find $\mathcal{M}(A, p)$ such that this number is less than infinity, but non-zero (this is not always possible). For the Sierpinski triangle, we may proceed by using a sequence of ϵ values that are the diameters of the subtriangles. That this is sufficient is guaranteed by problem 1. For any p then, the values calculated for exactly the sequence of subtriangles leads to the infimum desired. We know that calculating on the subtriangles, we can minimally cover $\triangle$ with 3^n triangles of diameter $d \cdot 2^{-n}$ where d is the diameter of $\triangle$. Then we have, for each n, p a value of $d \cdot 3^n \cdot 2^{-np}$ as a candidate for the infimum. Notice that as $n \to \infty$ for p small, this is an increasing sequence, consequently our infimum at fixed n, p is the value at n. For p large, the sequence converges to 0, so that we have one infimum for all ϵ values. These p values correspond to infinite or zero values for $\mathcal{M}(A, p, \epsilon)$. The value at which we will get a finite nonzero number is when, independent of n, we have $2^{np} = 3^n$. This value is $p = \log_2 3 = (\ln 3)/(\ln 2)$. At this value, the terms in the sequence are identically d, the diameter of the Sierpinski triangle.

V.4.6 As the following proof will indicate (exercise V.4.7), to get strict inequality one must have a situation where it takes more sets which are box shaped to cover A than if another, better shape is chosen.

V.4.7 We have by definition,

$$D(A) = \lim_{\epsilon \to 0} \frac{\mathcal{N}_\epsilon}{\frac{1}{\epsilon}}$$

for the fractal dimension, where the diameter of each box is ϵ times some fixed constant. Fixing $\epsilon > 0$ and rearranging this formula, we have

$$D(A) \ln\left(\frac{1}{\epsilon}\right) = \ln \mathcal{N}_\epsilon$$

$$\frac{1}{\epsilon^{D(A)}} = \mathcal{N}_\epsilon$$

$$1 = \mathcal{N}_\epsilon \epsilon^{D(A)} = \sum_{i=1}^{\mathcal{N}} \epsilon^{D(A)}$$

Since for A_i chosen such that $A_i = B_i \cap A$, that is the intersection of each box with the set A, we have $\text{diam}(A_i) \leq B_i$, it follows that for this choice of ϵ, $\mathcal{M}(A, D(A), \epsilon) < \infty$, since it must be less than or equal to this choice of A_i. Since $D(A)$ converges, this must then be the case for every choice of ϵ, and we have that

$$\lim_{\epsilon \to 0} \sum_{i=1}^{\infty} (\text{diam}(A_i))^p \leq 1$$

for $p = D(A)$. If this sum is zero, then $D_H(A) \leq D(A)$, if it is not, then $D_H(A) = D(A)$. Notice that if $\mathcal{M}(A, D(A)) = 0$, we cannot automatically conclude that these two are unequal, if it is non-zero then they are equal.

V.4.10 Let $A_1, A_2 \in (\mathbb{R}^m, \text{Euclidean})$, be two metrically equivalent sets, and let $\theta : A_1 \to A_2$ be a metric equivalence. Then for any $B \subset A_1$, we have $\text{diam}(\theta(B)) \leq C_1 \text{diam}(B)$ and $\text{diam}(B) \leq C_2 \text{diam}(\theta(B))$ for some constants C_1, C_2. Consequently, if $\mathcal{M}(A_1, p) = \infty$, then for any choice of subsets, and $\epsilon > 0$, we must have $\sum_{i=1}^{\infty} (\text{diam}(B_i))^p \to \infty$. Thus for any choice of subsets on A_2 this sum for the images of B_i is equal to $\sum_{i=1}^{\infty} (C_2 \text{diam}(B_i))^p$ which is therefore also infinite.

Hence $\mathcal{M}(A_2, p) = \infty$. Conversely, if for some p, we have $\mathcal{M}(A_1, p) = 0$, then the infimum over ϵ values for all choices of subsets of A_1 is zero. Consequently, for any $\delta > 0$, we may find a set of subsets satisfying $\text{diam}(C_i) < \epsilon$ for which $\mathcal{M}(A, p, \epsilon) < \delta$ for every epsilon by choosing a sequence of subsets whose sum is less than δ / C_1 and using the metric equivalence to generate the sequence in A_2. Consequently $\mathcal{M}(A_2, p) = 0$ as well. Obviously by using θ^{-1}, and reversing the role of the constants, we can do the same, given a value on A_2. Since the two sets agree on infinite and zero values for $\mathcal{M}(A, p)$, then the point at which the two values change are the same. Hence $D_H(A_1) = D_H(A_2)$.

V.4.11 Proving that a metric equivalence generates the same Hausdorff-Besicovitch dimension, is identical to exercise V.4.10. The second statement, that the Hausdorff D-dimensional measure $\mathcal{M}(A, D_H(A))$ is not preserved may be shown by example: Looking at example V.4.8, we may take a Sierpinski triangle with vertices at $(0, 0)$, $(2, 0)$, and $(0, 2)$. Then for each step n in the example, we cover the triangle with 3^n closed disks of diameter $2\sqrt{2} \cdot \frac{1}{2^n}$, and the example proceeds as in the text, except that

$$\mathcal{M}(\Delta, D_H(\Delta)) = (2\sqrt{2})^{\ln 3 / \ln 2}$$

Consequently we may take the original Sierpinski triangle, with the metric $d_2(x, y) = 2 \cdot d(x, y)$, or twice the Euclidean metric, and obtain a different value for the Hausdorff p-dimensional measure.

V.4.12 They should be (inches)$^{1.391}$.

Answers to Chapter VI, section 1

VI.1.2 The Taylor expansion for $\sin(x)$ near 0 is given by

$$\sin(x) = \sum_{n=0}^{\infty} \frac{(-1)^n x^{(2n+1)}}{(2n + 1)!} = x - \frac{1}{3!}x^3 + \frac{1}{5!}x^5 + \cdots$$

giving a linear approximation of $\ell(x) = x$. Let $\epsilon > 0$. Then a linear coordinate transformation from $[0, \epsilon] \times [0, \epsilon]$ to $[0, 1] \times [0, 1]$ is given by $\theta((x, y)) = (\frac{x}{\epsilon}, \frac{y}{\epsilon})$. With the change of coordinates we still get $\ell'(x') = x'$, and the graph of this function, L is $(x, x) \in [0, 1] \times [0, 1]$. Transforming the function $y = f(x) = \sin(x)$, gives us

$$y' = f'(x') = \theta \circ f \circ \theta^{-1}(x') = \frac{\sin(\epsilon x)}{\epsilon}$$

so the graph G is the set $(x, \frac{\sin(\epsilon x)}{\epsilon}) \in [0, 1] \times [0, 1]$.

The Manhattan metric distance between a point $g = (x, \frac{\sin(\epsilon x)}{\epsilon})$ on G and a point $l = (y, y)$ on L is given by

$$d(g, l) = |x - y| + |\frac{\sin(\epsilon x)}{\epsilon} - y|$$

The shape of an ϵ ball in Manhattan metric is a square with vertical and horizontal corners. So to get the point on L closest to a given point $g \in G$, we look for the intersection of the smallest isosceles right triangle with right angle at g intersecting L, which since the slope of L is 1, intersects from (x, x) to $(\frac{\sin(\epsilon x)}{\epsilon}, \frac{\sin(\epsilon x)}{\epsilon})$ at the same Manhattan distance. Hence for any $g \in G$ the Manhattan distance to L is

$$d(g, L) = x - \frac{\sin(\epsilon x)}{\epsilon}$$

The derivative of $\frac{\sin(\epsilon x)}{\epsilon}$ is decreasing, and this function is always less than x in this domain, hence the distance from G to L is given by the maximum, reached when $x = 1$, or

$$d(G, L) = 1 - \frac{\sin(\epsilon)}{\epsilon}$$

Going the other way, we base an isosceles right triangle with the right angle at (y, y) and look for the smallest such triangle intersecting G. Once again, since the slope of G is decreasing, this will touch when we take the first coordinates equal, hence

$$d(\ell, G) = y - \frac{\sin(\epsilon y)}{\epsilon}$$

and this is maximized at $(1, 1)$. Hence

$$d(L, G) = 1 - \frac{\sin(\epsilon)}{\epsilon} = d(L, G) = h(L, G)$$

Noting that this is indeed decreasing as $\epsilon \to 0$ since

$$\lim_{\epsilon \to 0} \frac{\sin(\epsilon)}{\epsilon} = 1$$

we calculate

$$h(L, G) = .01 \Rightarrow .99\epsilon = \sin(\epsilon)$$

or ϵ less than approximately .1.

Answers to Chapter VI, section 2

VI.2.2 Let

$$f(x) = F_{i-1} + \frac{x - x_{i-1}}{x_i - x_{i-1}}(F_i - F_{i-1})$$

The first step is to show that if $(x, y) \in G$ then so is $w_i(x, y)$. Let $(x, y) \in G$, that is $(x, y) = (x, f(x))$. Then (x, y) goes to

$$\begin{pmatrix} a_i x \\ c_i x \end{pmatrix} + \begin{pmatrix} e_i \\ f_i \end{pmatrix}$$

or

$$\begin{pmatrix} \left(\frac{x_i - x_{i-1}}{x_N - x_0}\right) x + \frac{x_N x_{i-1} - x_0 x_i}{x_N - x_0} \\ \left(\frac{F_i - F_{i-1}}{x_N - x_0}\right) x + \frac{x_N F_{i-1} - x_0 F_i}{x_N - x_0} \end{pmatrix}$$

To compute $f(x)$ for the transformed x we do

$$\frac{x_i - x_{i-1}}{x_N - x_0}x + \frac{x_N x_{i-1} - x_0 x_i}{x_N - x_0} - x_i$$

$$= \frac{x_i - x_{i-1}}{x_N - x_0}x + \frac{x_N x_{i-1} - x_0 x_i}{x_N - x_0} - \frac{x_{i-1} x_N - x_{i-1} x_0}{x_N - x_0}$$

$$= \frac{x_i - x_{i-1}}{x_N - x_0} + \frac{x_{i-1} x_0 - x_0 x_i}{x_N - x_0}$$

$$= (x - x_0)\left(\frac{x_i - x_{i-1}}{x_N - x_0}\right)$$

Dividing by $x_i - x_{i-1}$ yields

$$\frac{x - x_0}{x_N - x_0}$$

and multiplying by $(F_i - F_{i-1})$ and adding F_{i-1} yields

$$\frac{(F_i - F_{i-1})(x - x_0)}{x_N - x_0} + \frac{F_{i-1} x_N - F_{i-1} x_0}{x_N - x_0}$$

$$= \left(\frac{F_i - F_{i-1}}{x_N - x_0}\right) x + \frac{F_{i-1} x_N - F_i x_0}{x_N - x_0} = c_i x + f_i$$

hence

$$\bigcup_{i=1}^{N} w_i(G) \subset G$$

To show equality it suffices to show that the endpoints of the graph map to the endpoints of the ith interval under w_i hence showing that the map is onto. We substitute x_0, x_N for x above to get

$$a_i x_N + e_i = \left(\frac{x_i - x_{i-1}}{x_N - x_0}\right) x_N + \frac{x_N x_{i-1} - x_0 x_i}{x_N - x_0} = x_i$$

$$a_i x_0 + e_i = \left(\frac{x_i - x_{i-1}}{x_N - x_0}\right) x_0 + \frac{x_N x_{i-1} - x_0 x_i}{x_N - x_0} = x_{i-1}$$

so that the union of the w_i is onto. Hence

$$\bigcup_{i=1}^{N} w_i(G) = G$$

as desired.

VI.2.5 For this exercise we are asked to find an IFS whose attractor is a parabola passing through the points $(0, 0)$, $(1, 1)$, and $(2, 4)$. There is only one such parabola of the form $y = Ax^2 + Bx + C$, which we will find, and its equation is $f(x) = x^2$. The IFS is of the form

$$w_1\begin{pmatrix} x \\ y \end{pmatrix} = \begin{pmatrix} a_1 & 0 \\ c_1 & d_1 \end{pmatrix} \begin{pmatrix} x \\ y \end{pmatrix} + \begin{pmatrix} e_1 \\ f_1 \end{pmatrix}$$

$$w_2\begin{pmatrix} x \\ y \end{pmatrix} = \begin{pmatrix} a_2 & 0 \\ c_2 & d_2 \end{pmatrix} \begin{pmatrix} x \\ y \end{pmatrix} + \begin{pmatrix} e_2 \\ f_2 \end{pmatrix}$$

which we can insure passes through the interpolation points by using the equations for the various terms given in the text. To insure that this IFS has the parabola $y = f(x)$ as its attractor, we substitute a point from this graph into the equation to yield

$$w_1\begin{pmatrix} x \\ f(x) \end{pmatrix} = \begin{pmatrix} a_1 x + e_1 \\ c_1 x + d_1 f(x) + f_1 \end{pmatrix}$$

$$w_2\begin{pmatrix} x \\ f(x) \end{pmatrix} = \begin{pmatrix} a_2 x + e_2 \\ c_2 x + d_2 f(x) + f_2 \end{pmatrix}$$

so that in both cases,

$$(a_i x + e_i)^2 = c_i x + d_i x^2 + f_i$$

We calculate

$$a_1 = \frac{1-0}{2-0} = \frac{1}{2} \qquad\qquad e_1 = \frac{0-0}{2-0} = 0$$

$$a_2 = \frac{2-1}{2-0} = \frac{1}{2} \qquad\qquad e_2 = \frac{2-0}{2-0} = 1$$

$$c_1 = \frac{1-0}{2-0} - d_1\frac{4-0}{2-0} = \frac{1}{2} - 2d_1 \quad f_1 = \frac{0-0}{2-0} - d_1\frac{0-0}{2-0} = 0$$

$$c_2 = \frac{4-1}{2-0} - d_2\frac{4-0}{2-0} = \frac{3}{2} - 2d_2 \quad f_2 = \frac{2-0}{2-0} - d_2\frac{0-0}{2-0} = 1$$

yielding

$$(\frac{1}{2}x)^2 = (\frac{1}{2} - 2d_1)x + d_1 x^2$$

$$(\frac{1}{2}x + 1)^2 = (1 - 2d_2)x + d_2 x^2 + 1$$

These are solved by equation of coefficients:

$$\frac{1}{4} = d_1$$

$$\frac{1}{2} - 2d_1 = 0$$

$$d_1 = \frac{1}{4}$$

and

$$\frac{1}{4}x^2 + x + 1 = \frac{3}{2}x - 2d_2x + d_2x^2 + 1$$

$$\frac{1}{4} = d_2$$

$$1 = \frac{3}{2} - 2d_2$$

This leaves $c_1 = 0$, and $c_2 = 1$. We have

$$w_1 \begin{pmatrix} x \\ y \end{pmatrix} = \begin{pmatrix} \frac{1}{2} & 0 \\ 0 & \frac{1}{4} \end{pmatrix} \begin{pmatrix} x \\ y \end{pmatrix}$$

$$w_2 \begin{pmatrix} x \\ y \end{pmatrix} = \begin{pmatrix} \frac{1}{2} & 0 \\ 1 & \frac{1}{4} \end{pmatrix} + \begin{pmatrix} 1 \\ 1 \end{pmatrix}$$

VI.2.10 By exercise II.2.11, the Euclidean and Manhattan metrics on $\mathbb{R}^2$ are equivalent, so it suffices to show that the metric in Theorem 2.1 in Chapter VI is equivalent to the Manhattan metric on $\mathbb{R}^2$. We do this by looking at the three possible cases:

If $\theta = 1$ the two metrics, d from Theorem 2.1 in Chapter VI and d_M the Manhattan metric, are identical and therefore equivalent.

If $\theta > 1$, then since $\theta|y_1 - y_2| \geq |y_1 - y_2|$ we have $d((x_1, y_1), (x_2, y_2)) \geq d_M((x_1, y_1), (x_2, y_2))$, and since $\frac{1}{\theta}|x_1 - x_2| \leq |x_1 - x_2|$ we have

$$\frac{1}{\theta}d((x_1, y_1), (x_2, y_2)) \leq d_M((x_1, y_1), (x_2, y_2)) \leq d((x_1, y_1), (x_2, y_2))$$

and the metrics are equivalent.

Finally, if $\theta < 1$ then we have, by the same argument

$$\theta d_M((x_1, y_1), (x_2, y_2)) \leq d((x_1, y_1), (x_2, y_2)) \leq d_M((x_1, y_1), (x_2, y_2))$$

so they are equivalent in this case as well.

VI.2.13 If we look at *any* IFS of the form $\{\mathbb{R}^2; w_1(x, y)\}$, then if this IFS is hyperbolic, by definition w_1 is a contraction mapping, and the attractor of the IFS is a single point. But for $N = 1$, the curve we are attempting to interpolate has two interpolation points. Hence the attempt will always fail. Another way to look at it is that by exercise VI.2.5, these interpolations can always be arranged to be quadratic, a quadratic interpolation always requires at least 3 values.

VI.2.14 By definition, we have

$$(Tf)(x) = c_n \ell_n^{-1}(x) + d_n f(\ell_n^{-1}(x)) + f_n$$

where $\ell_n(x) = a_n x + e_n$, on each subinterval $[x_{n-1}, x_n]$. We need to show that for $f \in \mathcal{F}$, Tf is an interpolation function. It is shown in the text that $Tf \in \mathcal{F}$ and hence does the proper job at the endpoints. What is necessary is to show that Tf takes each function f to a function which agrees with f at the internal interpolation points. By definition, ℓ_n^{-1} takes x_{n-1} to x_0 and x_n to x_N. Because $f \in \mathcal{F}$, f takes x_0 to F_0, and x_N to F_N, and we have

$$(Tf)(x_{n-1}) = c_n x_0 + d_n F_0 + f_n$$
$$(Tf)(x_n) = c_n x_N + d_n F_N + f_n$$

into which we substitute the definitions of c_n, d_n, and f_n, to get

$$(Tf)(x_{n-1}) = \frac{1}{x_N - x_0}(F_n x_0 - F_{n-1} x_0 - d_n F_N x_0 + d_n F_0 x_0$$
$$= quad + d_n F_0 x_N - d_n F_0 x_0 + x_N F_{n-1} - x_0 F_n - d_n x_N F_0 + d_n x_0 F_N)$$
$$= F_{n-1}$$

$$(Tf)(x_n) = \frac{1}{x_N - x_0}(F_n x_N - F_{n-1} x_N - d_n F_N x_N + d_n F_0 x_N$$
$$+ d_n F_N x_N - d_n F_N x_0 + x_N F_{n-1} - x_0 F_n - d_n x_N F_0 + d_n x_0 F_N)$$
$$= F_n$$

as desired. Since Tf interpolates the data for any $f \in \mathcal{F}$ if f is a fixed point of the transformation, then by definition, the graph of f is the attractor of the prescribed IFS, and f is an interpolation function to begin with.

VI.2.15 We wish to construct an interpolation using quadratic scaling, in which the $w_n(x, y)$ take the form

$$w_n(x, y) = ((a_n x + e_n), (c_n x + d_n y + g_n y^2 + f_n))$$

As before, we need to take the data pairs (x_0, F_0) to (x_{n-1}, F_{n-1}) and (x_N, F_N) to (x_n, F_n). Since the transformation in x, is the same, we may define a_n, e_n as before, namely

$$a_n = \frac{x_n - x_{n-1}}{x_N - x_0} \qquad \text{and} \qquad e_n = \frac{x_{n-1} x_N - x_n x_0}{x_N - x_0}$$

This will insure that the first coordinates of the data pairs are mapped properly. For the second coordinates, we must satisfy the equations

$$F_{n-1} = c_n x_0 + d_n F_0 + g_n F_0^2 + f_n$$
$$F_n = c_n x_N + d_n F_N + g_n F_N^2 + f_n$$

Notice that c_n and f_n must coincide with the linear case when g_n is zero. This gives

$$c_n = \frac{F_n - F_{n-1}}{x_N - x_0} - d_n \frac{F_N - F_0}{x_N - x_0} - g_n X_c$$

where we need to fill in for X_g, and

$$f_n = \frac{x_N F_{n-1} - x_0 F_n}{x_N - x_0} - d_n \frac{x_N F_0 - x_0 F_N}{x_N - x_0} - g_n X_f$$

There are no quadratic terms in the equations to be satisfied, consequently, these must cancel at x_0, x_N, giving us

$$g_n F_0^2 = g_n X_c x_0 + g_n X_f$$
$$g_n F_N^2 = g_n X_c x_N + g_n X_f$$

Dividing out g_n, and subtracting these yields

$$F_N^2 - F_0^2 = (x_N - x_0) X_c$$
$$X_c = \frac{F_N^2 - F_0^2}{x_N - x_0}$$

Putting this back into the first equation after dividing out g_n, yields

$$X_f = F_0^2 - \frac{F_N^2 - F_0^2}{x_N - x_0} x_0$$

$$= \frac{x_N F_0^2 - x_0 F_0^2 - F_N^2 x_0 + F_0^2 x_0}{x_N - x_0}$$

$$= \frac{x_N F_0^2 - x_0 F_N^2}{x_N - x_0}$$

so that we have

$$c_n = \frac{F_n - F_{n-1}}{x_N - x_0} - d_n \frac{F_N - F_0}{x_N - x_0} - g_n \frac{F_N^2 - F_0^2}{x_N - x_0}$$

$$f_n = \frac{x_N F_{n-1} - x_0 F_n}{x_N - x_0} - d_n \frac{x_N F_0 - x_0 F_N}{x_N - x_0} - g_n \frac{x_N F_0^2 - x_0 F_N^2}{x_N - x_0}$$

to complete the coefficients for the quadratic case.

VI.2.16 Consider the integral

$$I = \int_{x_0}^{x_N} f(x) dx$$

which here is well defined, since $f(x)$ is continuous and of bounded variation. We assume that f is the fixed point of the fractal interpolation transformation, i.e. $Tf = f$, where $(\mathcal{F}, d)$ is the metric space of functions which match the endpoint data, and are continuous. Then we have

$$I = \int_{x_0}^{x_N} (Tf)(x) dx = \sum_{n=1}^{N} \int_{x_{n-1}}^{x_n} (Tf)(x) dx$$

$$= \sum_{n=1}^{N} \int_{x_0}^{x_N} (c_n x + d_n f(x) + f_n) d(a_n x + e_n) = \alpha I + \beta = \frac{\beta}{1 - \alpha}$$

where

$$\alpha = \sum_{n=1}^{N} a_n d_n \qquad \text{and} \qquad \beta = \sum_{n=1}^{N} a_n \int_{x_0}^{x_N} (c_n x + f_n) dx$$

We need to show that under the standard assumptions, $|\alpha| < 1$, we have

$$|\alpha| = |\sum_{n=1}^{N} a_n d_n| \le \sum_{n=1}^{N} |a_n d_n|$$

and

$$\sum_{n=1}^{N} a_n = 1 \qquad \text{and} \qquad |a_n| < 1$$

The x_n are ordered, consequently $a_n x_0 + e_n < a_n x_N + e_n$ which implies $a_n > 0$, hence $|a_n d_n| = a_n |d_n|$, and therefore

$$|\alpha| \le \sum_{n=1}^{N} a_n |d_n|$$

which implies $|\alpha| < 1$. To show that

$$\int_{x_0}^{x_N} f_0(x) dx = \beta$$

we merely note that β is independent of d_n, and the result follows from above.

VI.2.17 We want the integral

$$I_1 = \int_{x_0}^{x_N} x f(x) dx$$

under the same assumptions as in exercise VI.2.16. Once again since $Tf = f$, we have

$$I_1 = \int_{x_0}^{x_N} x(Tf)(x) dx = \sum_{n=1}^{N} \int_{x_{n-1}}^{x_n} x(Tf)(x) dx$$

so we transform variables.

$$(Tf)(a_n x + e_n) = c_n x + d_n f(x) + f_n$$

and we replace x with $\frac{x - e_n}{a_n}$ to write the same integral as

$$\sum_{n=1}^{N} \int_{x_0}^{x_N} \frac{(x - e_n)}{a_n} (c_n x + d_n f(x) + f_n) d(a_n x + e_n)$$

$$= \sum_{n=1}^{N} \int_{x_0}^{x_N} \frac{x}{a_n} (c_n x + d_n f(x) + f_n) d(a_n x + e_n)$$

$$- \sum_{n=1}^{N} \int_{x_0}^{x_N} \frac{e_n}{a_n} (c_n x + d_n f(x) + f_n) d(a_n x + e_n)$$

The first integral can be written as

$$\int_{x_0}^{x_N} \frac{x}{a_n} (c_n x + f_n) d(a_n x + e_n) + \int_{x_0}^{x_N} \frac{d_n}{a_n} x f(x) d(a_n x + e_n)$$

$$= \int_{x_0}^{x_N} (c_n x^2 + f_n x) dx + d_n \int_{x_0}^{x_N} x f(x) dx$$

$$= \int_{x_0}^{x_N} (c_n x^2 + f_n x) dx + d_n I_1$$

The other half looks like

$$e_n \int_{x_0}^{x_N} (c_n x + f_n) dx + a_n e_n \int_{x_0}^{x_N} f(x) dx$$

which using the notation I_0 for the integral I in exercise VI.2.16 yields

$$e_n \int_{x_0}^{x_N} (c_n x + f_n) dx + d_n e_n I_0$$

We now let

$$\gamma = \sum_{n=1}^{N} \int_{x_0}^{x_N} (c_n x^2 + f_n x)\,dx$$

$$\delta = \sum_{n=1}^{N} e_n \int_{x_0}^{x_N} (c_n x + f_n)\,dx$$

$$\epsilon = \sum_{n=1}^{N} d_n$$

$$\phi = \sum_{n=1}^{N} d_n e_n$$

Then we have

$$I_1 = \gamma + \epsilon I_1 - (\phi I_0 + \delta)$$
$$= \frac{\gamma - (\phi I_0 + \delta)}{1 - \epsilon}$$
$$= \frac{\gamma - \delta - (\phi \beta/(1 - \alpha))}{1 - \epsilon}$$

where we have used the notation of VI.2.16 for the integral I_0. We therefore must restrict the d_n such that $\epsilon \neq 1$. Once again, γ, δ are independent of d_n, and we arrive at the integral $I_1 = \gamma - \delta$ in the case that $d_n = 0$ for all n.

Answers to Chapter VI, section 3

The exercises in Section 3 are all experimental in nature.

Answers to Chapter VI, section 4

VI.4.1 Define the metric on $\mathbb{R} \times \mathbf{Y}$ by

$$d((x_1, y_1), (x_2, y_2)) = |x_1 - x_2| + \theta d_Y(y_1, y_2)$$

where $a = \max\{a_i : i = 1, \dots, N\}$.

$$d_Y(M_n(a, y), M_n(b, y)) \leq c|a - b|$$
$$d_Y(M_n(x, a), M_n(b, y)) \leq s d_Y(a, b)$$

and $\theta = \frac{1-a}{2c}$. $c, s > 0$, and under the usual assumptions about the ordering of x_i, $1 > a > 0$. We wish to show that under these assumptions, the IFS is hyperbolic with respect to the metric d. Under each map, $(x, y) \mapsto (L_n(x), M_n(x, y))$ so that

$$d(w_n(x_1, y_1), w_n(x_2, y_2)) = d((L_n(x_1), M_n(x_1, y_1)), (L_n(x_2), M_n(x_2, y_2)))$$
$$= |L_n(x_1) - L_n(x_2)| + \theta d_Y(M_n(x_1, y_1), M_n(x_2, y_2))$$
$$= |a_n x_1 - a_n x_2| + \theta d_Y(M_n(x_1, y_1), M_n(x_2, y_2))$$
$$= |a_n||x_1 - x_2| + \theta d_Y(M_n(x_1, y_1), M_n(x_2, y_2))$$
$$\leq a|x_1 - x_2| + \theta d_Y(M_n(x_1, y_1), M_n(x_2, y_2))$$

The second half of this expression can be decomposed using the triangle inequality to read

$$d_Y(M_n(x_1, y_1), M_n(x_2, y_2))$$
$$\leq d_Y(M_n(x_1, y_1), M_n(x_2, y_1)) + d_Y(M_n(x_2, y_1), M_n(x_2, y_2))$$
$$\leq c|x_1 - x_2| + s d_Y(y_1, y_2)$$

Putting this together with the statements above, we have

$$d(w_n(x_1, y_1), w_n(x_2, y_2)) \le a|x_1 - x_2| + \theta c|x_1 - x_2| + s\theta d_Y(y_1, y_2)$$

$$\le \frac{2a + 1 - a}{2}|x_1 - x_2| + s\theta d_Y(y_1, y_2)$$

$$\le \frac{1 + a}{2}|x_1 - x_2| + s\theta d_Y(y_1, y_2)$$

Since $0 < a < 1$, we have $1 + a < 2$ so let $s' = \max(\frac{1+a}{2}, s) < 1$ then

$$d(w_n(x_1, y_1), w_n(x_2, y_2)) \le s'|x_1 - x_2| + s'\theta d_Y(y_1, y_2) = s'd((x_1, y_1), (x_2, y_2))$$

so the IFS is hyperbolic as desired.

VI.4.2 By Theorem 4.1 in Chapter VI, the attractor exists. We devise the operator $T : \mathcal{F} \to \mathcal{F}$ as in Theorem 2.2 in Chapter VI to be

$$(Tf)(x) = M_n(f(L_n^{-1}(x))) \quad x \in [x_{n-1}, x_n]$$

where $\mathcal{F}$ is the set of continuous functions $f : [x_0, x_N] \to Y$. L_n is $a_n x + e_n$ which is continuous with continuous inverse, f is assumed continuous, and M_n is again linear, so that Tf is a continuous function on each of the $[x_{n-1}, x_n]$. We must assure that T takes $\mathcal{F}$ into itself. We have

$$L_1(x_0) = a_1 x_0 + e_1 = \frac{x_1 - x_0}{x_N - x_0}x_0 + \frac{x_N x_0 - x_0 x_1}{x_N - x_0} = x_0$$

$$L_N(x_N) = a_N x_N + e_N = \frac{x_N - x_{N-1}}{x_N - x_0}x_N + \frac{x_N x_{N-1} - x_N x_0}{x_N - x_0} = x_N$$

$$L_n(x_{n-1}) = a_n x_{n-1} + e_n = \frac{x_n - x_{n-1}}{x_N - x_0}x_{n-1} + \frac{x_N x_{n-1} - x_0 x_{n-1}}{x_N - x_0} = x_{n-1}$$

$$L_n(x_n) = a_n x_n + e_n = \frac{x_n - x_{n-1}}{x_N - x_0}x_n + \frac{x_N x_{n-1} - x_n x_0}{x_N - x_0} = x_n$$

so the endpoints of each segment conincide with the next and last, and x_0, x_N map to themselves. Hence these points are fixed under L_n and therefore under L_n^{-1}. By definition,

$$(Tf)(x_0) = M_n(x_0, F_0) = (x_{n-1}, F_{n-1})$$
$$(Tf)(x_N) = M_n(x_N, F_N) = (x_n, F_n)$$

hence $T : \mathcal{F} \to \mathcal{F}$ as desired. We now show that T is a contraction mapping. Let $f, g \in \mathcal{F}$. Then we have

$$(Tf)(x) = M_n(L_n^{-1}, f(L_n^{-1}(x)))$$
$$(Tg)(x) = M_n(L_n^{-1}, g(L_n^{-1}(x)))$$

With the given definition of a metric, and the restrictions imposed on M_n we have

$$d(Tf, Tg) = \max_{x \in [x_{n-1}, x_n]} d_Y(M_n(L_n^{-1}(x), f(L_n^{-1}(x))), M_n(L_n^{-1}(x), g(L_n^{-1}(x))))$$

$$\le \max_{x \in [x_{n-1}, x_n]} sd_Y(f(L_n^{-1}(x)), g(L_n^{-1}(x)))$$

$$\le d(f, g)$$

as desired.

VI.4.8 We want to show we can generate polynomial graphs with affine hidden variable interpolation. Using the proof given in exercise VI.4.2, we have $(Tf)(x) = M_n(L_n^{-1}(x), f(L_n^{-1}(x)))$. Consequently, the attractor of the IFS is given by

$$f(x) = M_n(L_n^{-1}(x), f(L_n^{-1}(x)))$$

where we let $f(x) = (y(x), z(x))$. We are interested only in fitting the first component, the second component may be chosen for ease of manipulation of the equations. We use the notation of the book, namely

$$L_n(x) = a_n x + e_n$$

$$M_n \begin{pmatrix} x \\ y \\ z \end{pmatrix} = A_n \begin{pmatrix} y \\ z \end{pmatrix} + \begin{pmatrix} c_n x + f_n \\ k_n x + g_n \end{pmatrix}$$

$$A_n = \begin{pmatrix} d_n & h_n \\ l_n & m_n \end{pmatrix}$$

From this we get

$$\begin{pmatrix} y(x) \\ z(x) \end{pmatrix} = A_n \begin{pmatrix} y(L_n^{-1}(x)) \\ z(L_n^{-1}(x)) \end{pmatrix} + \begin{pmatrix} c_n(L_n^{-1}(x)) + f_n \\ k_n(L_n^{-1}(x)) + g_n \end{pmatrix}$$

from which we extract the equation for $y(x)$ to get

$$y(x) = d_n y(L_n^{-1}(x)) + h_n z(L_n^{-1}(x)) + f_n + c_n L_n^{-1}(x)$$

Newton proved that it requires $N + 1$ interpolation points to generate an Nth degree polynomial, so that we assume that we have been given these points, and that they fit $y(x)$, and likewise assume

$$y(x) = \alpha_N x^N + \alpha_{N-1} x^{N-1} + \cdots + \alpha_0$$

Since L_n is linear, so is L_n^{-1}, so the degree of $y(L_n^{-1}(x))$ is also N, and we let (for notational convenience)

$$y(L_n^{-1}(x)) = \beta_{N n} x^N + \beta_{N-1 n} x^{N-1} + \cdots \beta_{0 n}$$

Then we have

$$\alpha_N x^N + \cdots + \alpha_0 = d_n y(L_n^{-1}(x)) + h_n x(L_n^{-1}(x)) + f_n + c_n L_n^{-1}(x)$$
$$= d_n \beta_{N n} x^N + \cdots d_n \beta_{0 n} + h_n z(L_n^{-1}(x)) + f_n + c_n L_n^{-1}(x)$$

which gives

$$(\alpha_N - d_n \beta_{N n}) x^N + \cdots (\alpha_0 - d_n \beta_{0 n}) = h_n z(L_n^{-1}(x)) + f_n + c_n L_n^{-1}(x).$$

Now d_n, as in the non-hidden variable case, is independent of the fitting of the points (x_i, F_i), hence on each subinterval, we may choose

$$d_n = \frac{\beta_{N n}}{\alpha_N}$$

Then the left hand side is a polynomial of degree $N - 1$ so we may choose h_n, and let

$$z(x) = \gamma_{N-1} x^{N-1} + \cdots + \gamma_0$$

in such a way so that if

$$\delta_{N-1 n} x^{N-1} + \cdots + \delta_{0 n} = z(L_n^{-1}(x))$$

then we have

$$h_n \delta_{i_n} = \alpha_i - d_n \beta_{i_n}$$

for $i \geq 2$ and

$$h_n \delta_{1 n} x + h_n \delta_{0 n} + c_n L_n^{-1}(x)) + f_n = (\alpha_1 - d_n \beta_{1 n}) x + (\alpha_0 - d_n \beta_{0 n})$$

by equation of coefficients. Thus we may generate a hidden variable fractal interpolation function for a polynomial of degree N.

Answers to Chapter VI, section 5

VI.5.1 From Theorem 5.1 in Chapter VI, using three maps, we may choose

$$w_1 \begin{pmatrix} x \\ y \\ z \end{pmatrix} = \begin{pmatrix} \frac{1}{3} & 0 & 0 \\ 0 & a_1 & b_1 \\ 0 & c_1 & d_1 \end{pmatrix} \begin{pmatrix} x \\ y \\ z \end{pmatrix} + \begin{pmatrix} 0 \\ e_1 \\ f_1 \end{pmatrix}$$

$$w_2 \begin{pmatrix} x \\ y \\ z \end{pmatrix} = \begin{pmatrix} \frac{1}{3} & 0 & 0 \\ 0 & a_2 & b_2 \\ 0 & c_2 & d_2 \end{pmatrix} \begin{pmatrix} x \\ y \\ z \end{pmatrix} + \begin{pmatrix} \frac{1}{2} \\ e_2 \\ f_2 \end{pmatrix}$$

$$w_3 \begin{pmatrix} x \\ y \\ z \end{pmatrix} = \begin{pmatrix} \frac{1}{3} & 0 & 0 \\ 0 & a_3 & b_3 \\ 0 & c_3 & d_3 \end{pmatrix} \begin{pmatrix} x \\ y \\ z \end{pmatrix} + \begin{pmatrix} \frac{2}{3} \\ e_3 \\ f_3 \end{pmatrix}$$

For the Sierpinski triangle, the appropriate values are given by $a_i = d_i = \frac{1}{2}$, $b_i = c_i = 0$, and $e_1 = f_1 = 0$, $e_2 = \frac{1}{2}$, $f_2 = 0$ and $e_3 = 0$, $f_3 = \frac{1}{2}$.

Answers to Chapter VII, section 1

VII.1.1 Let $\{\mathbb{R}^2, f\}$ be given by

$$f(x, y) = \begin{cases} (2x, 2y - 1) & \text{if } y > 1/2, \\ (2x - 1, 2y) & \text{if } x > 1/2, \\ (2x, 2y) & \text{otherwise.} \end{cases}$$

If $(x, y) > 0$ and $(x, y) \notin S$, then eventually $y > 1/2$ since y is doubled if $x > 1/2$ or if neither $x, y > 1/2$. Once this happens we have

$$f^n(x, y) = (2^n x, 2^n y - 2^{n-1})$$

Given $y > 1/2$, we have

$$\lim_{n \to \infty} 2^n y - 2^{n-1} = \lim_{n \to \infty} (2y - 1) 2^{n-1} = \infty$$

and $x \to \infty$, as well. Hence these points escape to infinity. If $y < 0$, $x > 0$, then eventually $2^n x > 1/2$, if $x < 1/2$. Once this happens we have

$$\lim_{n \to \infty} 2^n x - 2^{n-1} = \lim_{n \to \infty} (2x - 1) 2^{n-1} = \infty$$

for any point whose x coordinate is larger than $1/2$. The y coordinate remains negative, and $2^n y \to -\infty$. Hence these points also escape, in this case in the direction of $(\infty, -\infty)$. For points $y < 0$, $x < 0$, each coordinate remains negative indefinitely and these points diverge as $2^n x \to -\infty$ and $2^n y \to -\infty$, so these points escape to $(-\infty, -\infty)$.

VII.1.3 The condition $x^2 + y^2 > R$ determines the color based on the iteration number. This is passage past a circle of radius $\sqrt{R}$. Since if $f(x, y) = (2x, 2y)$ we have $f^n(x, y) = (2^n x, 2^n y)$, and

$$(2^n x)^2 + (2^n y)^2 = 2^{2n}(x^2 + y^2) > R \Rightarrow x^2 + y^2 > \frac{R}{2^{2n}}$$

we have concentric circles of radius $\frac{\sqrt{R}}{2^n}$.

VII.1.6 These two functions,

$$w_1(z) = re^{i\theta} z - 1 \quad w_2(z) = re^{i\theta} z + 1$$

take a disk of radius R, to a disk of radius rR, rotate it by θ, and move its center to either ± 1. Consequently, if we can find a disk which is mapped into itself by both maps, we know that the attractor of the IFS is interior to the disk (see exercise VII.1.8). We thus look for R as a function of r such that this is the case, and restrict it such that its two images do not intersect. This will force the attractor of the IFS to be totally disconnected.

We make one further simplification: If the rotation were 0, we could examine the transformation on the real line, since these transformations take circles to circles, this will correctly predict the values of R. This reduces the problem to looking at $w_i(x) = -rx \pm 1$. We look for a cross section of the ball $[a, b]$ such that the images

$$w_1([a, b]) = [-rb - 1, -ra - 1] \quad w_2([a, b]) = [-rb + 1, -ra + 1]$$

both lie within $[a, b]$. These two intervals are of identical length and orientation, differing only by a shift of two along the real line. Consequently, our conditions are

$$a \leq -rb - 1 \quad b \geq -ra + 1$$

Solving at equality yields

$$a = -rb - 1$$
$$b = -ra + 1$$
$$a = r^2 a - r - 1$$
$$(1 - r^2)a = -r - 1$$
$$a = -\frac{1 + r}{1 - r^2} = -\frac{1}{1 - r}$$

Symmetry now dictates that $b = \frac{1}{1-r}$, so that a disk of radius $R = 1/(1 - r)$ maps inside itself under each map of the original IFS. We now want the condition on r such that these images of the disk $B_R(\mathcal{O})$ do not intersect. This happens when a disk of radius rR, centered about 1 does not intersect the y-axis, namely that

$$r \cdot \frac{1}{1 - r} < 1$$
$$r < 1 - r$$
$$2r < 1$$
$$r < 1/2$$

Consequently, for the original maps, as long as $r < 1/2$, the maps

$$w_1(A) \cap w_2(A) \subset w_1(B_R(\mathcal{O})) \cap w_2(B_R(\mathcal{O})) = \emptyset$$

and the IFS is totally disconnected. We must now define $f(z)$ such that on the attractor, we do this with three maps (letting $z = z_1 + iz_2$),

$$f(z) = \begin{cases} w_1 & \text{if } z_1 \leq -\frac{r}{1-r} \\ w_2 & \text{if } z_1 \geq \frac{r}{1-r} \\ re^{i\theta}z + \lambda & \lambda = \frac{z_1(1-r)}{r} \quad \text{otherwise} \end{cases}$$

VII.1.7 This is essentially the general case of the preceding problem. Let $\{A; f\}$ be the associated shift dynamical system of a totally disconnected hyperbolic IFS on $\mathbb{R}^2$, where A is the attractor of the IFS. Then we must have

$$f(a) = w_i^{-1}(a) \qquad \text{for } a \in w_i(A).$$

Since the IFS is totally disconnected, we have

$$w_i(A) \cap w_j(A) = \emptyset \qquad \text{if } i \neq j$$

Furthermore, since the IFS is hyperbolic, $w_i(A)$ is compact for each i. Consequently, in $\mathbb{R}^2$ this means $d(w_i(A), w_j(A)) > 0$ if $i \neq j$. Then we may extend $\{A; f\}$ using any function $g(x)$ which smoothly connects the maps w_i^{-1} in the intervening distance. For example, since there are finitely many maps, we may define

$$\epsilon = \min_{i \neq j} d(w_i(A), w_j(A)) > 0.$$

Now construct functions which are equal to w_i^{-1} on $w_i(A)$, and decrease to zero at a distance less than $\epsilon/2$ from these sets, by taking functions

$$\tilde{w}_i(x) = \begin{cases} (1 - \frac{d(x, w_i(A))}{k\epsilon})w_i^{-1}(x) & \text{if } d(x, w_i(A)) \leq 1 \\ 0 & \text{otherwise} \end{cases}$$

with $k \geq 2$. Finally, sum these together to get

$$g_k(x) = \sum_{i=1}^{N} \tilde{w}_i(x)$$

Because the distance ϵ was chosen to be less than the distance between images of the attractor under w_i, these functions are continuous, agree with f on the attractor, and since k can be taken at any value greater than 2, there are infinitely many of them.

VII.1.9 We get the iteration formula by generating the real and imaginary parts of $f(z) = z^3 - \lambda$. Letting $z = x + iy$, we have

$$z^3 = (x + iy)^3 = x^3 + 3x^2 iy + 3xi^2 y^2 + i^3 y^3$$
$$= x^3 + i3x^2 y - 3xy^2 - iy^3$$
$$= (x^3 - 3xy^2) + i(3x^2 y - y^3)$$

so that the new formulas, letting $\lambda = \lambda_1 + i\lambda_2$ are

```
newx = x^3 - 3*x*y^2 - lambda1
newy = 3*x^2*y - y^3 - lambda2
```

The choice of R can be derived as follows: Writing $z = re^{i\theta}$, we have $z^3 = r^3 e^{i3\theta}$, which has a distance from the origin of r^3 compared to the distance r of z. Subtracting λ will have maximal effect on this distance when z^3 and λ lie on the same ray from the origin, namely when $\lambda = \rho e^{i\phi}$ and $\phi = 3\theta$. Consequently the problem reduces to this maximal case, namely that

$$r^3 - |\lambda| \geq r$$

or

$$r^3 - r - |\lambda| \geq 0$$

The minimum value for $|\lambda|$ is zero, and the inequality acheives equality for this λ value when $r = 1 + |\lambda|$. Choosing this as a candidate value for R, we find that

$$(1 + |\lambda|)^3 - (1 + |\lambda|) - |\lambda| = |\lambda|^3 + 3|\lambda|^2 + 3|\lambda| + 1 - 1 - 2|\lambda|$$
$$= |\lambda|^3 + 3|\lambda|^2 + |\lambda| \geq 0$$

so that for any $\lambda \in \mathbb{C}$, choosing $R > |\lambda| + 1$ will work in the escape time algorithm.

Answers to Chapter VII, section 2

VII.2.1 In this case $X = \hat{\mathbb{C}}$, we must show the presence of sets such that $W(A) = A$, where $W(A) = w_1(A) \cup w_2(A)$. Choose $A = \{\mathcal{O}\}$, then $w_1(A) = w_2(A) = \mathcal{O}$. Likewise, choose $A = \{\infty\}$, then $w_1(A) = w_2(A) = \{\infty\}$. Consequently, the unrestricted IFS in this case does not contain a unique invariant set. In point of fact, neither of these sets above are attractors of the IFS. The system is not contractive around these points, which are repellors. The point $z = 1$ is also a fixed point of the system, it is a repellor, and is part of $J(f)$.

VII.2.2 Let $\lambda \in (3/4, 5/4)$ and $f(z) = z^2 - \lambda$. We want to look at solutions to

$$z^2 + z + (1 - \lambda) = 0.$$

Using the quadratic equation we obtain

$$z = \frac{-1 \pm \sqrt{1 - 4 + 4\lambda}}{2} = \frac{-1 \pm \sqrt{4\lambda - 3}}{2}$$

We let

$$z_1 = \frac{-1 + \sqrt{4\lambda - 3}}{2}$$

$$z_2 = \frac{-1 - \sqrt{4\lambda - 3}}{2}$$

The easiest way to examine cycles of period 2 here is to check to see whether $f(z_1) = z_2$ and $f(z_2) = z_1$ (different approaches are used for different problems, this avoids quardic equations here). We have

$$z_1^2 = \frac{1}{4} - \frac{\sqrt{4\lambda - 3}}{2} + \frac{4\lambda - 3}{4} = (\lambda - \frac{1}{2}) - \frac{\sqrt{4\lambda - 3}}{2}$$

$$z_2^2 = \frac{1}{4} + \frac{\sqrt{4\lambda - 3}}{2} + \frac{4\lambda - 3}{4} = (\lambda - \frac{1}{2}) + \frac{\sqrt{4\lambda - 3}}{2}$$

Subtracting λ from each of these yields the desired values $f(z_1) = z_2$, and $f(z_2) = z_1$, hence this is a cycle of period 2. To check whether the cycle is attractive, we have

$$f^{2'}(z) = 4z^3 - 4\lambda z$$

We can simplify this for each z_i by making use of the period 2 properties:

$$z_1^3 = (z_2 + \lambda)z_1 = z_1 z_2 + \lambda z_1$$
$$z_2^3 = (z_1 + \lambda)z_2 = z_1 z_2 + \lambda z_2$$

giving us

$$f^{2'}(z_1) = f^{2'}(z_2) = 4z_1 z_2$$

and

$$4z_1 z_2 = 4 \frac{1 - (4\lambda - 3)}{4} = \frac{4 - 4\lambda}{4} = 1 - \lambda$$

For $\lambda \in (3/4, 5/4)$ we have $(1 - \lambda) \in (-1/4, 1/4)$ so that

$$|f^{2'}(z_i)| \leq \frac{1}{4} < 1$$

Hence this is an attractive cycle of period 2.

VII.2.3 Let λ be specified and suppose $a \cdot i \in J_\lambda$. Then $-a^2 \in J_\lambda$ and $a^4 \in J_\lambda$ are real line points. The set J_λ is connected, therefore there is an extreme point of this form which maps to the left most extremity of J_λ, on both sides of the real axis. The pair of IFS cut J_λ at the negative extremity and map this image to the left and right side, depending on the definition of the square root in the two IFS equations.

VII.2.5 For the dynamical system $\{\hat{\mathbb{C}}; f(z) = z^2 - \lambda : \lambda > 2\}$, we wish to show that the orbit of the critical point $\{\mathcal{O}\}$ converges to the Point at Infinity. We know that the basin of attraction for ∞ contains at least the points for which $r > 1/2 + \sqrt{1/4 + \lambda}$ so that if $\lambda > 2$, we have $r > 1/2 + 3/2 = 2$. $f^2(\mathcal{O}) = f(-\lambda)$ which is 2 when $\lambda = 2$ and increases for increasing λ. Hence $\lim_{n \to \infty} f^n(\mathcal{O}) = \infty$. By Theorem 2.1 in Chapter VII, the IFS $\{X = \hat{\mathbb{C}} \setminus B(\infty, \epsilon); +\sqrt{z + \lambda}, -\sqrt{z + \lambda}\}$ possesses a unique attractor.

Answers to Chapter VII, section 3

VII.3.1 The Newton's transformation associated with $F(z)$ is given by

$$f(z) = z - \frac{F(z)}{F'(z)} = z - \frac{z^4 - 1}{4z^3}$$

To see that 1 is indeed fixed, we compute

$$f(1) = 1 - 0 = 1$$

Taking the derivative with respect to z evaluated at 1, we have

$$f'(z) = 1 - \frac{(4z^3)^2 - 12z^2(z^4 - 1)}{(4z^3)^2} = 1 - \frac{16z^6 - 12z^6 + 12z^2}{16z^6}$$

$$= \frac{12z^6 - 12z^2}{16z^6} = \frac{3}{4} - \frac{3}{4z^4}$$

evaluated at 1 we have $f'(z) = 3/4 - 3/4 = 0$ so that 1 is an attractive (sometimes called super-attractive because $f'(z) = 0$) fixed point of the Newton's transformation.

VII.3.2 The Newton's transformation associated with $F(z) = z^2 + 1$ is

$$f(z) = z - \frac{z^2 + 1}{2z} = z - \frac{1}{2}(z + \frac{1}{z})$$

$$= \frac{1}{2}z - \frac{1}{2z} = 1/2(z - \frac{1}{z})$$

so that the corresponding IFS is got by taking inverses of $f(z)$ to yield

$$w = 1/2(z - \frac{1}{z})$$

$$2wz = z^2 - 1$$

$$z^2 - 2wz - 1 = 0$$

$$z = \frac{2w \pm \sqrt{4w^2 + 4}}{2} = w \pm \sqrt{w^2 + 1}$$

Consequently, the appropriate IFS is

$$w_1(z) = z + \sqrt{z^2 + 1} \qquad \text{and} \qquad w_2(z) = z - \sqrt{z^2 + 1}$$

For any $z \in \mathbb{R}$, $z^2 + 1 > 0$, hence $w_i(z) \in \mathbb{R}$. For $z = \infty$, we have $w_1(z) \to \infty$, and $w_2(z) \to 0$, consequently $\mathbb{R} \cup \{\infty\} \mapsto \mathbb{R} \cup \{\infty\}$, and the map is onto, by virtue of the behavior of $w_2(z)$ for large z. Then if $A = \mathbb{R} \cup \{\infty\}$, $W(A) = A$. We must make sure we can exclude the fixed points of the transformation. These are given by

$$\frac{1}{2}(z - \frac{1}{z}) = z$$

or $z_f = \pm i$. The derivative of $f(z)$ at these points is $1/2(1 + 1/(z^2)) = 0$ so these are attractive. Hence removal of a small ϵ-ball around $\pm i$ results in a space $\mathbf{X} \subset \hat{\mathbb{C}}$ on which A is a unique invariant set. Therefore A is the attractor of the IFS.

VII.3.5 The Newton's transformation associated with $F(z) = e^z - 1$ is given by

$$f(z) = z - \frac{e^z - 1}{e^z} = (z - 1) + e^{-z}$$

It has two fixed points in $\hat{\mathbb{C}}$, for we have

$$z = (z - 1) + e^{-z} \to e^{-z} = 1 \to z = 0$$

indicating that 0 is a fixed point. We also have

$$\lim_{z \to \infty} (z - 1) + e^{-z} = \infty$$

so that ∞ is also fixed. The derivative of $f(z)$ is given by

$$f'(z) = 1 - e^{-z}$$

which is zero at $z = 0$, and approaches 1 at infinity. Hence the attractive fixed point is 0, ∞ is a neutral fixed point.

VII.3.7 We wish to find a Möbius transformation $h(z)$ which establishes a conjugacy between $f(z)$ and $g(z)$. This means finding $h(z)$ such that

$$f \circ h(z) = h \circ g(z)$$

where

$$f(z) = \tfrac{1}{2}(z - \tfrac{1}{z}) \quad g(z) = z^2 \quad h(z) = \tfrac{az+b}{cz+d}$$

We mention at the outset that such a transformation is indeterminate with respect to scaling factors, so there will be choices among a, b, c, and d. Writing out the two terms of the conjugacy we have

$$h \circ g(z) = \frac{az^2 + b}{cz^2 + d}$$

and

$$
\begin{aligned}
f \circ h(z) &= \frac{1}{2}\left(\frac{az+b}{cz+d} - \frac{cz+d}{az+b}\right) = \frac{1}{2}\left(\frac{(az+b)^2 - (cz+d)^2}{(cz+d)(az+b)}\right) \\
&= \frac{1}{2}\left(\frac{a^2z^2 + 2abz + b^2 - c^2z^2 - 2cd - d^2}{acz^2 + bcz + adz + bd}\right) \\
&= \frac{1}{2}\left(\frac{(a^2 - c^2)z^2 + 2(ab - cd)z + (b^2 - d^2)}{acz^2 + (ad + bc)z + bd}\right)
\end{aligned}
$$

so that

$$\frac{2az^2 + 2b}{cz + d} = \frac{(a^2 - c^2)z^2 + 2(ab - cd)z + (b^2 - d^2)}{acz^2 + (ad + bc)z + bd}$$

We may eliminate the linear terms in z by setting

$$ad + bc = 0 \quad ab - cd = 0$$

which give us

$$a = -\frac{bc}{d} \Rightarrow -\frac{b^2c}{d} - cd = 0 \Rightarrow d^2 = -b^2$$

$$d = -\frac{bc}{a} \Rightarrow ab + \frac{c^2b}{a} = 0 \Rightarrow a^2 = -c^2$$

With these equalities assumed, we have

$$\frac{2az^2 + 2b}{cz^2 + d} = \frac{(a^2 - c^2)z^2 + (b^2 + d^2)}{acz^2 + bd}$$

Since, as mentioned, we have choices of scale, we set $a = 1$. This gives us

$$\frac{2z^2 + 2}{cz^2 + d} = \frac{(1 - c^2)z^2 + (b^2 + d^2)}{cz^2 + bd}$$

and comparing denominators leads to $b = 1$. Then we have

$$2 = 1 - c^2 \quad \text{and} \quad 2 = 1 - d^2$$
$$c = \pm i \quad \text{and} \quad d = \pm i$$

and we again have a choice (this time the sign of the scale factor). We choose $c = i$. Then $ab - cd = 0$ or $1 - id = 0$, and $d = -i$. Consequently we have

$$h(z) = \frac{z+1}{iz-i}$$

Checking, we have

$$f(h(z)) = \frac{1}{2}\left(\frac{z+1}{iz-i} - \frac{iz-i}{z+1}\right) = \frac{1}{2}\frac{z^2+2z+1-(-z^2+2z-1)}{iz^2-iz+iz-i}$$

$$= \frac{1}{2}\frac{2z^2+2}{iz^2-i} = \frac{z^2+1}{iz^2-i} = h(g(z))$$

as desired.

Answers to Chapter VII, section 4

VII.4.1 We analyze this map by first noting that this map is equivalent to the map

$$f(z) = \begin{cases} z^2 - 1 & \text{for } z_1 > 0 \\ z^2 - 1 + \lambda z_1 & \text{for } z_1 \le 0 \end{cases}$$

where z_1 is the real part of $z = z_1 + iz_2$. Consequently, for $\lambda = 0$, and for any open set lying entirely in the right half plane, the mapping is open. For $\lambda \in [-1, 0]$, a look at a small ϵ-ball about the origin shows why the map is still continuous, namely, the action of the term λx is to shift the points within the ball over towards the right. This produces a closed oval, with points to the interior mapping to the interior and the boundary mapping to the boundary. Consequently an interior point is still surrounded by an open ball of the image of interior points, and the map remains open. Specifically, given (x, y), if we have a point (x_1, y) to the left of (x, y), then since near the origin, $x^2 \ll x$, the λx term dominates, and the disk is reversed, $f(x_1, y)$ is to the right of $f(x, y)$, and points with the same x value map to the same x value, creating an oval around the center (x, y). Then an open set maps to an open set.

The mapping is not open for $\lambda = 1$ however. To see this, choose first $\epsilon_0 < 1/2$, so that for points in an ϵ_0-ball about the origin, $2x^2 < x$. Then for points on the boundary of the left half ball $x^2 + y^2 = \epsilon_0^2$ and the intersection with the y-axis, the portion on the axis maps to an interval on the real axis $[-\epsilon_0^2 - 1, -1]$. The rest of the boundary meets this interval at $-\epsilon_0^2 - 1$, but lies entirely to the left of it:

$$x^2 + y^2 = \epsilon_0^2$$
$$f(x, y) = (x^2 - y^2 - 1 + x, 2xy)$$
$$= x^2 - (\epsilon_0^2 - x^2) - 1 + x, 2xy)$$
$$= (-\epsilon_0^2 - 1) + 2x^2 + x, 2xy)$$

Since we have chosen ϵ_0 such that the x term dominates the $2x^2$ term this indeed lies to the left. If we look at this figure from $\epsilon = \epsilon_0$ as $\epsilon \to 0$, then the envelope of the images of the boundary form the boundary of the image of the ball $B_{\epsilon_0}(\mathcal{O})$. In particular, there are points interior to this ball which map to the boundary of the image. For any such point, a smaller ball around that point folds interior to this envelope, exposing it, hence there is no open set around such a point which maps to an open set, and the map is not open.

Finally, by choosing a ball around the origin sufficiently large, for any fixed λ, the x^2 and y^2 terms will dominate the λx term so that $f(B_R(\mathcal{O}) \supset B_R(\mathcal{O})$. In fact, as $(x, y) \to \infty$, the map approaches the analytic map $f(z) = z^2 - 1$.

Answers to Chapter VIII, section 1

VIII.1.3 We must first show that we satisfy the Lipshitz continuity requirement of Theorem 11.1 in Chapter III, namely for each w_i there is a k_i, independent of p, x, such that

$$d(w_{i_p}(x), w_{i_q}(x)) \leq k_i \cdot d_P(p, q)$$

For $w_1(x) = \lambda_1 x$, we have

$$d(\lambda_{1_p} x, \lambda_{1_q} x) = |\lambda_{1_p} x - \lambda_{1_q} x| = |x||\lambda_{1_p} - \lambda_{1_q}| = |x| d_P(p, q)$$

which is not independent of x. Similarly, for $w_2(x) = \lambda_2 x + 1 - \lambda_2$, we have

$$d(\lambda_{2_p} x + 1 - \lambda_{2_p}, \lambda_{2_q} x + 1 - \lambda_{2_q}) = |\lambda_{2_p} x + 1 - \lambda_{2_p} - \lambda_{2_q} x - 1 + \lambda_{2_q}| = |x - 1| d_P(p, q)$$

also not independent of x. We must restrict the domain as in exercise VIII.1.2. We have $-.9 \leq \lambda_1, \lambda_2 \leq .9$, and we look for an interval $[a, b]$ such that $W([a, b]) \subset [a, b]$ for all $p \in P$.

For $\lambda_1, \lambda_2 > 0$, $[0, 1]$ contains all the attractors. To see this, we have

$$w_1([0, 1]) = [0, \lambda_1] \quad w_2([0, 1]) = [\lambda_2, 1] \quad w_3([0, 1]) = [1/2, 1]$$

When $\lambda_1 < 0$, we have $w_1([0, 1]) = [\lambda_1, 0] \not\subset [0, 1]$, and likewise $w_2([0, 1]) = [1, 1 - \lambda_2] \not\subset [0, 1]$. One pushes the value for the necessary lower bound for the interval down, the other pushes the upper bound up. These values reach their minimum and maximum when $\lambda_1 = \lambda_2 = -.9$. Consequently, we will have found the proper interval $[a, b]$ when

$$W_{(-.9, -.9)}([a, b]) \subset [a, b].$$

Writing this out explicitly, we have

$$w_1([a, b]) = [-.9b, -.9a]$$
$$w_2([a, b]) = [-.9b + 1.9, -.9a + 1.9]$$
$$w_3([a, b]) = [.5a + .5, .5b + .5]$$

which means that a, b must satisfy

$$a \leq \min(-.9b, -.9b + 1.9, .5a + .5)$$
$$b \geq \max(-.9a, -.9a + 1.9, .5b + .5)$$

Now in order to include the interval $[0, 1]$ needed by the positive values for λ_1, λ_2, $a \leq 0$, and $b \geq 1$, which means that

$$\min(-.9b, -.9b + 1.9, .5a + .5) = -.9b$$
$$\max(-.9a, -.9a + 1.9, .5b + .5) = -.9a + 1.9$$

Setting the inequalities above to equality to solve we have

$$a = -.9b \quad b = -.9a + 1.9 \quad b = .18b + 1.9$$

which yields

$$b = \tfrac{1.9}{.82} \approx 2.3 \quad a \approx -2.1$$

so that, for convenience, the interval $[-3, 3]$ contains all the attractors, and we may prove the continuity on the IFS

$$\{[-3, 3]; w_1, w_2, w_3\}$$

to get it for all of $\mathbb{R}$. Looking back to our original x dependencies we now see that by substituting the endpoints of this interval for x in the distance expressions, we have the following Lipshitz conditions:

$$d(w_{1_p}(x), w_{1_q}(x)) \leq 3 \cdot d(p, q)$$
$$d(w_{2_p}(x), w_{2_q}(x)) \leq 4 \cdot d(p, q)$$

Consequently, the attractor varies continuously with $\lambda \in P$.

Answers to Chapter VIII, section 2

VIII.2.2 We first prove the hint. Let N be the set of all points $B \in \mathcal{H}(\mathbf{X})$ such that B is disconnected. Suppose $B \in N$. Then by definition, there are sets $B_1, B_2 \subset B$, such that

$$B_1 \cup B_2 = B; \; B_1 \cap B_2 = \emptyset.$$

Furthermore, since these sets are closed in B, they are compact. Then if $\delta = h(B_1, B_2)$, $\delta > 0$. Suppose $B' \in \mathcal{H}(\mathbf{X})$ such that $h(B, B') < \delta/3$. Then for any $x \in B'$, we have $d(x, B) < \delta/3$, and for any $y \in B$, we have $d(y, B') < \delta/3$. Let

$$B_1' = \{b \in B' : d(b, B_1) < \delta/3\}$$
$$B_2' = \{b \in B' : d(b, B_2) < \delta/3\}$$

Since each point in B' must be close to a point in B, we must have

$$B_1' \cup B_2' = B'.$$

Now suppose that $x \in B_1' \cap B_2' \neq \emptyset$. Then

$$d(x, B_1) + d(x, B_2) \leq \frac{2\delta}{3} < \delta$$

and by the triangle inequality, we would have, for some $b_1 \in B_1$ and $b_2 \in B_2$,

$$d(b_1, b_2) \leq d(x, b_1) + d(x, b_2) < \delta$$

which is a contradiction. Hence

$$B_1' \cap B_2' = \emptyset$$

and B' is also disconnected. But B' was *any* set within $\epsilon = \delta/3$ of B, so there is an open ball $B_\epsilon(B) \subset N$, and N is open in $\mathcal{H}(\mathbf{X})$.

Now let p be a parameter value such that $A(p) \in N$, that is, $A(p)$ is disconnected. Then there is an $\epsilon > 0$ such that $B_\epsilon(A(p)) \subset N$. This ball is open, and by the assumptions of the problem, the map $A : P \to \mathcal{H}(\mathbf{X})$ is continuous, so $A^{-1}(B_\epsilon(A(p)))$ is an open set in P containing p such that every element corresponds to a disconnected attractor. Hence the set of points in P for which $A(p)$ is disconnected is open in P. The complement is $\mathcal{M}$, which must therefore be closed.

VIII.2.4 For any value of λ in the parameter space, the attractor A can be written as the set of points of the form

$$z = \sum_{k=0}^{\infty} \pm \lambda^k$$

Since if $z \in A$, then the point

$$-z = \sum_{k=0}^{\infty} \mp \lambda^k$$

is also in A, we have

$$A = -A = \{z : -z \in A\}$$

(in other words the attractor of an element of this family of IFS is itself radially symmetric). Defining

$$\lambda A = \{\lambda z : z \in A\}$$

to make the notation simple, we have

$$A = (\lambda A + 1) \cup (\lambda A - 1).$$

And

$$-\lambda A + 1 = \lambda(-A) + 1 = \lambda A + 1$$
$$-\lambda A - 1 = \lambda(-A) - 1 = \lambda A - 1$$

in other words, $A(\lambda) = A(-\lambda)$. It then follows that if the attractor of λ is connected, so is the attractor of $-\lambda$ since they are identical, so $\mathcal{M}$ is symmetric about the origin.

VIII.2.5 We have the IFS

$$\{\mathbb{C}; \lambda z + 1, \lambda z - 1\}$$

where $\lambda = (0, 1/\sqrt{2})$. We may rewrite this as the IFS

$$\{\mathbb{R}^2; w_1(x, y), w_2(x, y)\}$$

where

$$w_1\begin{pmatrix} x \\ y \end{pmatrix} = \begin{pmatrix} 0 & -1/\sqrt{2} \\ 1/\sqrt{2} & 0 \end{pmatrix}\begin{pmatrix} x \\ y \end{pmatrix} + \begin{pmatrix} 1 \\ 0 \end{pmatrix} = 1/\sqrt{2}\begin{pmatrix} -y \\ x \end{pmatrix} + \begin{pmatrix} 1 \\ 0 \end{pmatrix}$$

$$w_2\begin{pmatrix} x \\ y \end{pmatrix} = \begin{pmatrix} 0 & -1/\sqrt{2} \\ 1/\sqrt{2} & 0 \end{pmatrix}\begin{pmatrix} x \\ y \end{pmatrix} - \begin{pmatrix} 1 \\ 0 \end{pmatrix} = 1/\sqrt{2}\begin{pmatrix} -y \\ x \end{pmatrix} - \begin{pmatrix} 1 \\ 0 \end{pmatrix}$$

Suppose we look at the action of these two maps on a rectangle centered at the origin. Each map flips the rectangle along the corners, reduces each side by $1/\sqrt{2}$, and then translates it either one unit to the left or to the right. We may solve for the position of a corner of such a rectangle, symmetric about the origin, such that it went to half the original rectangle. To do this, we must have (in absolute value)

$$x = 1/\sqrt{2}(y) + 1$$
$$y = 1/\sqrt{2}(x)$$

which has the solution $x = 2$, and $y = \sqrt{2}$.

Consequently, if we have the rectangle with corners $(\pm 2, \pm\sqrt{2})$, Then each map takes this rectangle to one of the horizontal halves of the original rectangle, just-touching along $x = 0$. The attractor is thus the rectangle, which tiles the plane.

VIII.2.7 Suppose that $w_1(A) \cap w_2(A) = \{z_0\}$, a single point. We wish to argue that if I is a path which intersects itself, then I is a point. Certainly such a path which is not trivial must be contained in either $w_1(A)$ or $w_2(A)$ or, since it must pass through z_0 this is a point of intersection. Then we may argue the same about the points $w_1(z_0)$ and $w_2(z_0)$, so these are points of self intersection. By induction, such a non-trivial non-self-intersecting path can not exist.

VIII.2.8 Let $e = e_1 e_2 \cdots e_n \cdots$ be a point in Σ a code space on the symbols $\{-1, 1\}$. For $\lambda \in \mathbb{C}$, define

$$f(\lambda) = e_1 + e_2\lambda + e_3\lambda^2 + \cdots$$

We want the radius of convergence of $f(\lambda)$. $f(\lambda)$ is bounded above by $1 + \lambda + \lambda^2 + \cdots = \sum \lambda^n$ (corresponding to $e = 1\bar{1}$). It is bounded below by $-1 - \lambda - \lambda^2 \cdots = -\sum \lambda^n$, corresponding to $e = -1\bar{1}$. These two series have a radius of convergence of $|\lambda| < 1$ and diverge at the endpoints. Consequently $f(\lambda)$ has a radius of convergence of 1. We show that the attractor for each λ is the set of points

$$\pm 1 \pm \lambda \pm \lambda^2 \pm \cdots$$

and that the map $f(\lambda)(e) = \phi(e)$. The attractor is the limit of the sequence in $\mathcal{H}(\mathbb{C})$ given by $\{W^n(\{0\})\}$. This sequence yields

$$0 \mapsto \pm 1 \mapsto \pm\lambda \pm 1 \mapsto \pm\lambda^2 \pm \lambda \pm 1 \mapsto \cdots$$

By induction, $A(\lambda)$ is the set of points

$$\pm 1 \pm \lambda \pm \lambda^2 \cdots$$

which is the image of Σ under both $f(\lambda)$ and $\phi : \Sigma \to A(\lambda)$.

Answers to Chapter VIII, section 3

VIII.3.3 The attractive orbit here is not a cycle, but rather aperiodic. There are other bubbles with periods equal to powers of 2 multiplied by the various odd numbers out past this point however. Try experimenting with the numbers around $\lambda = 1.75$ for instance.

Answers to Chapter VIII, section 4

All exercises in this section are programming experiments.

Answers to Chapter IX, section 1

IX.1.3 $p_1 \cdot p_3 = 0.03$ since the probabilities are independent.

IX.1.4 The attractor is a Sierpinski triangle with vertices at $0, i, 1$. The ball B_4 is located at $2 + i$ and has radius $\sqrt{2}$ and therefore intersects the attractor only at the point 1. The probability of getting only one string in code space is 0.

IX.1.5 $p1 = .275$, $p_2 = .125$, and $p_3 = .5$. We want to know about the ball B_1. B_1 includes only $w_1(A)$, hence one expects a probability of .275.

IX.1.6 There is a set of random strings (of measure 0 in code space) for which this number will not occur, e.g. B_2 in the above problem with the string $1\bar{1}$ has incidence 0.

Answers to Chapter IX, section 2

IX.2.2 X is the set of leaves, which is finite. $\mathcal{F} = 2^X$ the collection of all subsets of X. $\mathcal{G}$ is the collection of sets consisting of one leaf. Any set in $\mathcal{F}$ contains a finite number of leaves, therefore it is the finite union of sets in $\mathcal{G}$ so by definition, $\mathcal{G}$ generates $\mathcal{F}$.

IX.2.3 In order to prove that S is not in the field $\mathcal{F}$ we must show that it can not be written as a finite union of sets in $\mathcal{F}$. Suppose that $S = \bigcup_{n=1}^{N} A_n$ for some collection of $A_n \in \mathcal{F}$ where each A_n is the union of finitely many subintervals. Then at least one of the A_n must contain the disjoint union of an infinite number of intervals. But this contradicts the hypothesis about A_n. Hence $S \notin \mathcal{F}$.

IX.2.4 Let $X = [0, 1] \times [0, 1] \subset \mathbb{R}^2$, and $\mathcal{G}$ be the set of rectangles of the form $[p/q, r/s] \times [t/u, v/w]$ that is subrectangles with rational corners. Let $\mathcal{F}$ denote the field generated by $\mathcal{G}$. Let $S \in \mathcal{F}$. We need to show that S has rational area. i) $\bigcup_{n=1}^{N} A_n$ has rational area, where the A_n are subrectangles since the area of the union is given by $\sum_{n=1}^{N} \text{area}(A_n) - \sum_{i \neq j} \text{area}(A_i \cap A_j)$. ii) Similarly for any set of A_n as above, the complement $X \setminus \bigcup_{n=1}^{N} A_n = \bigcap_{n=1}^{N} X \setminus A_n$ has rational area, because $X \setminus A_n$ can be written as the finite union of open subrectangles. Consequently any set $S \in \mathcal{F}$ has rational area. The area of $B(\mathcal{O}, 1)$ is π which is irrational, consequently $B(\mathcal{O}, 1)$ is not in the field.

IX.2.5 Call the sets in the figure

a	G_1	G_2	b
c	$\emptyset$	e	f
g	h	i	j
l	m	X	o

Then we have the following list of equivalences, where those generated earlier in the list are used to generate the rest (once they have been established to be in the field).

$$a = X \setminus G_1$$
$$X = G_1 \cup a$$
$$\emptyset = X \setminus X$$
$$c = G_1 \setminus G_2$$
$$e = c \setminus G_1$$
$$j = X \setminus e$$
$$i = G_1 \setminus e$$
$$f = c \cup i$$
$$l = X \setminus f$$
$$m = i \cup l$$
$$h = m \setminus e$$
$$o = X \setminus h$$
$$g = G_1 \setminus o$$
$$b = X \setminus g$$

IX.2.6 Let e be the image of the binary number $a = .a_1 a_2 \cdots = \sum \frac{a_i}{2^i}$ under the map π which takes $a_i + 1 = e_i$, in other words $\pi(a) = e$. The complement of $e = \pi(a)$, a irrational, in Σ is not the union of a finite number of cylinder sets, which under the inverse of this map map to intervals of rational width around rational points in $[0, 1]$.

IX.2.7 X is a space, let $\mathcal{F} = 2^X$, the power set of X. Let $A \in \mathcal{F}$. Then since $X \setminus A \subset X$, we have $X \setminus A \in \mathcal{F}$. Since $X, \emptyset \subset X, \emptyset \in \mathcal{F}$. Compliments of subsets of X, and finite unions of subsets of X are subsets of X and are therefore elements of $\mathcal{F}$. Hence $\mathcal{F}$ is a field (in fact the biggest field over X).

IX.2.8 $\mathcal{B}$ is the σ-field generated by rectangles with rational corners on the unit square. These rectangles form a countable base in the metric topology induced by the Manhattan metric, as evidenced by the fact that the σ-field contains all balls of radius ϵ for all ϵ in the statement, and by the equivalence of metrics. To show this is the Borel field, the proof of Theorem 2.3 shows that the σ-field generated by a countable base for the topology of X is the Borel field.

IX.2.9 Σ is the codespace on the symbols $\{0, 1\}$. The cylinder sets are defined as all the sets, for all n, of the form

$$C(e_1 \cdots e_n) = \{e_\sigma \in \Sigma : e_{\sigma_1} = e_1, \ldots, e_{\sigma_n} = e_n\}$$

Let e be a point in Σ. Then the cylinder set $C_n(e)$ is precisely the set of elements $x \in \Sigma$ such that $x_i = e_i$ for $i \leq n$. Consequently,

$$d(e, x) = \sum_{i=1}^{\infty} \frac{|e_i - x_i|}{3^i} = \sum_{i=n+1}^{\infty} \frac{|e_i - x_i|}{3^i} \leq 1/3^n$$

By choosing $\tilde{e}$ such that $\tilde{e}_i = e_i$ for $i \leq n$ and $\tilde{e}_i \neq e_i$ for $i > n$, we have that $C_n(e) = B_{\frac{1}{3^n}}(e)$. The balls of radius 3^{-n} form a countable base for the metric topology on Σ, hence, by the proof of theorem 3, they generate the Borel field $\mathcal{B}$ on Σ.

IX.2.10 Let $\Delta \subset \mathbb{R}^2$ be the Sierpinski triangle, and $\mathcal{G}$ be the set of connected components of $\mathbb{R}^2 \setminus \Delta$. Notice that since by definition of component, this generating set is disjoint. Consequently, one cannot get a subset of a component (other than the component itself or the empty set) by union or compliment of the elements in $\mathcal{G}$. If $\mathcal{F}$ is the σ-field generated by $\mathcal{G}$ then $\mathcal{F}$ likewise contains no nonempty proper subsets of any component of $\mathbb{R}^2 \setminus \Delta$. Let $\epsilon > 0$ be chosen so small that if x is the center of the largest triangle inside the Sierpinski set, $B_\epsilon(x)$ fits entirely inside this component. Then $B_\epsilon(x)$ is not in $\mathcal{F}$, but it is in $\mathcal{B}$. Hence $\mathcal{F} \neq \mathcal{B}$. $\mathcal{F} \subset \mathcal{B}$ however, since any half plane is in $\mathcal{B}$ and any triangle is the union of the compliments of 3 half-planes.

IX.2.11 Let X be a metric space, $\mathcal{G}$ a generating set for the field $\mathcal{F}_1$, and for the σ-field $\mathcal{F}_2$. Let $\mathcal{F}_3$ be the σ-field generated by $\mathcal{F}_1$. Then

$$\mathcal{F}_3 = \mathcal{F}_1 \cup \{\bigcup_{n=1}^{\infty} A_n : A_n \in \mathcal{F}_1\}$$

We must show that this is the smallest σ-field containing $\mathcal{G}$. Suppose not. Then there is an $A \in \mathcal{F}_3$ such that $A \notin \mathcal{F}_2$. But $\mathcal{F}_1$ is the smallest σ-field containing $\mathcal{G}$ so the extra entry must be of the form

$$\bigcup_{n=1}^{\infty} A_n; \{A_n\} \subset \mathcal{F}_1.$$

But the smallest σ-field containing $\mathcal{G}$ must contain all of these sets, since it contains $\mathcal{F}_1$. Hence $\mathcal{F}_2 = \mathcal{F}_3$.

IX.2.12 Let $\mathcal{F}$ be any field of subsets of X. Choose $A \in \mathcal{F}$. Then $X \setminus A \in \mathcal{F}$ and the union of these two is in $\mathcal{F}$. The union is $A \cup X \setminus A = X$.

Answers to Chapter IX, section 3

IX.3.3 Let $X = [0, 1] \subset \mathbb{R}$. $\mathcal{F}$ is generated by subintervals, and for $a \le b$ we define $\mu((a, b)) = \mu([a, b]) = b - a$, and $\mu(A) =$ sum of lengths of subintervals of A.

(i) $\mu : \mathcal{F} \to [0, \infty)$. By definition $\mu(A) \ge 0$. Since the sum of lengths of disjoint intervals is less than or equal to $\mu([0, 1]) = 1$ it follows that $\mu(A) < \infty$.

(ii) Let $[a, b]$ be an interval, and let $b = lim_{n \to \infty} a_n$ for some sequence $\{a_n\}$ such that $a_1 \in [a, b]$. Then the sum of the lengths of intervals in the union

$$[a, a_1] \cup [a_1, a_2] \cup \cdots = [a, b]$$

is

$$\mu([a, a_1]) + \sum_{n=1}^{\infty}(a_{n+1} - a_n)$$

This is a telescoping sum and is therefore equal to

$$\mu([a, a_1]) + \mu([a_1, lim_{n \to \infty} a_n]) = [a, b]$$

Thus since any element in the field which can be written as the union of countably many elements must be unions of this form together with finitely many other elements, μ is countably additive.

(iii) Let S be the set of rationals in $[0, 1]$. Then S is not in the field $\mathcal{F}$. To see this, we note that neither S nor its compliment can be written as the union of finitely many intervals, since both contain countably many components, thus any finite union contains an element with countably many components.

(iv) Let $\hat{\mathcal{F}}$ be the σ-field generated by $\mathcal{F}$. For any field or σ-field, since

$$A \cap B = X \setminus (X \setminus A \cup X \setminus B)$$

the finite intersection of elements of a field are also in the field. Thus for any $x = p/q$ where p, q are integers, the set $\{x\} = [a, x] \cap [x, b] \in \mathcal{F}$ and we have $\mu(\{x\}) = 0$. The rationals (and therefore S) are a countable set, consequently we can write $S = \bigcup_{n=1}^{\infty} A_n$ where $\{A_n\}$ are the elements of S. Consequently, $S \in \mathcal{F}$. We have

$$\hat{\mu}(S) = \hat{\mu}(\bigcup_{n=1}^{\infty} A_n = \sum_{n=1}^{\infty} \hat{\mu}(A_n) = \sum_{x=p/q} \hat{\mu}(\{x\}) = 0$$

IX.3.4 Let Σ be the code space on symbols $\{0, 1\}$, and define $0 \le p_1 \le 1$, $p_2 = 1 - p_1$. We define μ on cylinder sets by

$$\mu(C(e_1 \cdots e_n)) = p_{e_1} \cdots p_{e_n}$$

To define μ on other sets in the field $\mathcal{F}$ there is really only one way to be consistent with this definition, and satisfy countable additivity. We describe this:

$$\mu(X) = \mu(C(1) \cup C(2)) = p_1 + p_2 = 1$$

$$\mu(A \cup B) = \mu(A) + \mu(B) - \mu(A \cap B)$$

where $A \cap B$ is in the field, since it is equal to $X \setminus ((X \setminus A) \cup (X \setminus B))$ by de Morgan's rules. The easiest way to see how this defines a countably additive measure on the field $\mathcal{F}$ is to note that for any finite set of codes $\{e_1, \ldots e_n\}$ the measure defined on the cylinder sets of these codes *taken in any order* is the same. Notice also that two cylinder sets are disjoint if they disagree on any specified symbol. Taken together, the measure of a union of these sets is the product of the codes on which they agree, if they are the same length, and this product times the product of the 'tail' of the longer string if they are different lengths. There are no elements of the field $\mathcal{F}$ which can be formed which specify values for an infinite number of places, since this is the disjoint union of an infinite number of components. Consequently, $S = \{x : x \text{ is odd}\}$ is not in $\mathcal{F}$, and $\mu(\{x : x_7 = 1\}) = p_1$. The set $S \in \hat{\mathcal{F}}$ since it can be written as a countable intersection (and therefore a countable union) of finite unions of cylinder sets. Its measure, as above, is the product of the probabilities on which the elements agree, namely all odd places, and is therefore $\lim_{n \to \infty} p_1^n = 0$. Notice that since the cylinder sets form a countable basis for the metric topology on Σ, $\hat{\mathcal{F}}$ is the Borel field on Σ.

IX.3.5 Since multiplication is commutative, it is easier to define the indices in reverse order. Thus we want the intervals

$$w_{e_n} \circ \cdots \circ w_{e_1}([0, 1])$$

as a generator for the field $\mathcal{F}$. Notice that the interval $w_1([0, 1])$ has measure p_1 and the interval $w_2([0, 1])$ has measure p_2. Define $\mu(X) = 1$. Then we must have

$$\mu(w_1 \circ w_{e_n} \circ \cdots \circ w_{e_1}([0, 1]) \cup w_2 \circ w_{e_n} \circ \cdots \circ w_{e_1}) = p_1 p_{e_n} \cdots p_{e_1} + p_2 p_{e_n} \cdots p_{e_1}$$

in order to have even finite additivity, but

$$p_1 p_{e_n} \cdots p_{e_1} + p_2 p_{e_n} \cdots p_{e_1} = (p_1 + 1 - p_1) p_{e_n} \cdots p_{e_1} = p_{e_n} \cdots p_{e_n}$$

so that, since these are the subintervals of $w_{e_n} \circ \cdots \circ w_{e_1}([0, 1])$ the 'middle third' has measure 0. Hence we have $\mu(B) = 0$ for any $B \in \mathcal{F}$ such that $B \subset X \setminus A$. This implies countable additivity, since every element of $\mathcal{F}$ must be the union of closed or open intervals beginning or ending on the boundary of a subinterval of the above form, and each subinterval is the countable union of subintervals, and satisfies countable additivity. Hence μ is a measure on $\mathcal{F}$. This provides $\mu(A) = 1, \mu(X \setminus A) = 0, \mu([1/3, 2/3]) = 0$.

IX.3.6 We define $\mathcal{F}$ on $[0, 1]$ by using the IFS

$$\{[0, 1], w_1 = \frac{1}{2}x, w_2 = \frac{1}{2}x + \frac{1}{2}, p_1, p_2\}$$

as in the previous problem. Here the subintervals touch, but the definition from problem 4 of $\mu(X) = 1$ yields that the intersection of two intervals at their boundaries has measure zero. In this case the field is the set of finite unions of subintervals which have boundaries of the form $k/2^n$. By expressing the numbers on the interval in binary, since any element of the interval is the countable sum $\sum_{n=1}^{\infty}(e_n - 1)/2^n$ for $e_i \in \{1, 2\}$ we generate all the subintervals of the interval, hence the extension of $\mathcal{F}$ is $\mathcal{B}([0, 1])$. In order to generate the measure which takes the measure of an interval as the length of an interval, it is necessary and sufficient to have translation independence. This requires $p_1 = 1/2$.

IX.3.7 Let (X, d) be a compact metric space, μ a Borel measure on X such that $\mu(X) \neq 0$. Let S be the support of μ on X. Then $\mu(S) = \mu(X) \neq 0$, so that $S \neq \emptyset$. S is closed which implies that S is compact. Hence $S \in \mathcal{H}(X)$.

IX.3.8 Let μ be a measure on σ-field $\mathcal{F}$, and let $\overline{\mathcal{F}}$ be the class of sets of the form $A \cup B$ where $A \in \mathcal{F}$ and $B \subset C \in \mathcal{F}$ such that $\mu(C) = 0$. $\overline{\mathcal{F}}$ is a σ-field since clearly $X \setminus B = X \setminus C \cup C \setminus B$ and $C \setminus B$ is a subset of an element of $\mathcal{F}$ of measure zero. The same implies with unions of this form. Define $\overline{\mu}(A \cup B) = \overline{\mu}(A) = \mu(A)$. Then if we have a countable union of sets of this form, we have

$$\overline{\mu}(\bigcup_{n=1}^{\infty}(A_n \cup B_n)) = \overline{\mu}((\bigcup_{n=1}^{\infty} A_n) \cup (\bigcup_{n=1}^{\infty} B_n))$$

which is the union of an element of $\mathcal{F}$ with a subset of an element of $\mathcal{F}$ of measure zero. Hence

$$\overline{\mu}(\bigcup_{n=1}^{\infty}(A_n \cup B_n)) = \overline{\mu}(\bigcup_{n=1}^{\infty} A_n)$$

and this latter obeys countable additivity by assumption. Hence $\overline{\mathcal{F}}$ is a σ-field, and $\overline{\mu}$ is a measure on it.

Answers to Chapter IX, section 4

IX.4.1 Let $f : [0, 1] \to \mathbb{R}$ be a piecewise constant function with finitely many discontinuities, located at $\{x_1, \ldots x_n \subset \mathbb{R}\}$. Let $x_0 = 0$, $x_{n+1} = 1$. Then f has a single value on each of the intervals $I_0 = [x_0, x_1], \ldots I_n = [x_n, x_{n+1}]$, and we let $y_i = f(x), x \in [x_i, x_{i+1}]$. Then if $\chi_i = \chi_{[x_i, x_{i+1}]}$ is the characteristic function on the interval I_i, we may rewrite

$$f(x) = \sum_{i=0}^{n} y_i \chi_{I_i}$$

If μ is the Borel measure on $[0, 1]$ given by $\mu([a, b]) = b - a$, we have, by definition

$$\int_{[0,1]} f d\mu = \sum_{n=0}^{n} y_i \mu(I_i) = \sum_{n=0}^{n} y_i(x_{i+1} - x_i).$$

For the Riemann integral of f on $[0, 1]$ we have

$$\int_0^1 f(x)dx = \int_{x_0}^{x_1} f(x)dx + \cdots + \int_{x_n}^{x_{n+1}} f(x)dx$$

$$= \int_{x_0}^{x_1} y_0 dx + \cdots + \int_{x_n}^{x_{n+1}} y_n dx$$

$$= y_0(x_1 - x_0) + \cdots y_n(x_{n+1} - x_n)$$

$$= \int_{[0,1]} f d\mu$$

IX.4.2 We want a measure on (■,Euclidean), defined on the field $\mathcal{G} = \{A : A \text{ is a rectangle }\}$, which corresponds to area. This means (assuming $a < b, c < d$),if $A = [a, b] \times [c, d]$ or the equivalent expression using open or half open intervals, we define $\mu(A) = (b - a)(d - c)$, and for unions of rectangles define $\mu(A \cup B) = \mu(A) + \mu(B) - \mu(A \cap B)$. Let $A \in \mathcal{F}$ such that $A = \bigcup_{n=1}^{\infty} A_n$, $A_n \in \mathcal{F}$ with the A_n disjoint. Then for the union to be in $\mathcal{F}$ we must be able to express it as a finite union of rectangles or the compliments of rectangles. Notice that the compliment of a finite union of rectangles is expressible on X as a finite union of rectangles. The measure of this region is then finitely additive by hypothesis, therefore μ is countably additive. Suppose there were another measure μ' on $\mathcal{F}$ which yielded the area of rectangles in X. Then since both measures need to be countably additive, and since they agree on each subrectangle of X, the two measures agree on $\mathcal{F}$. Hence $\mu' = \mu$. The extension of μ to $\mathcal{B}$ is also a measure which yields the area of sets $A \in \mathcal{B}$, because $\mathcal{B}$ contains all open sets in X, the infimum represented in the Caratheodory extension theorem can be replaced by an infimum on open sets, and is therefore the area of the closure of A.

To show that $\triangle$ is in $\mathcal{B}$ it suffices to note that it is closed. This follows since the compliment is then open, and can be covered by open sets in $\mathcal{F}$. This cover has a finite subcover, which implies that $X \setminus \triangle$ is a countable union of sets in $\mathcal{F}$, hence $\triangle \in \mathcal{B}$.

IX.4.3 The integral is defined on the intersection of two strips with $\triangle$. To calculate it, we first determine the measure of an arbitrary strip $[a, b] \subset [0, 1]$ intersected with $\triangle$. The measure is countably additive, so note that given $A \subset \triangle$, with the measure of A being $\mu(A)$, that we have

$$\mu(A) = \mu(w_1(A) \cup w_2(A)) + \mu(w_3(A)) = (p_1 + p_2)\mu(A) + p_3\mu(A) = .5\mu(A) + .5\mu(A)$$

Suppose that for $n < N$ $\mu([k/2^n, (k + 1)/2^n] \times \mathbb{R} \cap \triangle) = 1/2^n$ for $0 \leq k \leq 2^n$. That is, the measure of $\triangle$ falling within a strip of length $1/2^n$ positioned on a boundary of $[0, 1]$ subdivided into equal $1/2^n$ intervals is its length. We note that each such strip intersected with Δ is a vertical stack of disjoint Sierpinski triangles. We now subdivide the strip into two parts. The left-hand parts contains a stack of twice as many triangles, half the size, while the right-hand is a stack of as many triangles, half the size. Each triangle in the original strip is produced by transforming Δ by some sequence $(w_i \circ \cdots w_k)$ of transformations, those on the stack in the left-hand portion are either of the form $w_i \circ \cdots w_k \circ w_1$ or $w_i \circ \cdots \circ w_k \circ w_2$ while those on the right have an included transformation by w_3. Consequently the measure of the $1/2^n$ width strip is divided equally on its two halves. By induction, and finite additivity, the measure of any strip beginning and ending on boundaries of the form $k/2^n$ is its length. By countable additivity, the measure concentrated on any strip $[a, b] \times \mathbb{R} \cap \Delta$ is its length.

Having demonstrated this, the integral is $1 \cdot 2/3 + (-1) \cdot 1/3 = 1/3$. We will use the above result in further exercises.

IX.4.4 Let $\alpha, \beta \in \mathbb{R}$, and f, g simple functions. Then we have

$$\alpha f = \alpha \sum y_i \chi_{I_i}(x) = \sum (\alpha y_i) \chi_{I_i}(x)$$
$$\beta g = \beta \sum z_i \chi_{J_j}(x) = \sum (\beta z_i) \chi_{J_j}(x)$$

are simple functions. By definition, $\alpha f + \beta g$ is αf where $g = 0$, and βg where $f = 0$ and the sum elsewhere. We need to write the collection of sets $\bigcup I_i \cup \bigcup J_j$ as the union of disjoint sets. This is done by writing

$$K_k = I_k \setminus \bigcup J_j, \ K_{m+j} = J_j \setminus \bigcup I_i, \ K_{m+n+(i+j)} = I_i \cap J_j, \ K_k = \emptyset \text{ elsewhere.}$$

Define $w_i = \alpha y_i, \ w_j = \beta z_j, \ w_k = \alpha y_i + \beta z_j$ on the sets defined above. Then $w(x) = \sum w_k \chi_{K_k}$ is a simple function, and equal to $\alpha f + \beta g$. The integral of the sum is the sum of the integrals by the above definition of $w(x)$.

IX.4.6 We need to rewrite the function, as in the above problem, as a simple function with disjoint sets. Writing $C = A \setminus B, D = B \setminus A, E = A \cap B$, we have

$$f(x) = \chi_C + (2.3)\chi_D + (1 + 2.3)\chi_E$$

To evaluate the integral, we need the measures on each of these sets. We define them in terms of strings: C is the set of strings beginning with 1, and not having a 2 in the second place, i.e., $C = C(1, 1)$ the cylinder set beginning in two 1's. D is the set of strings not beginning with 1 which have a 2 in the second place, or, $D = C(2, 2)$. E is the set beginning with 1 with a 2 in the second place, or, $E = C(1, 2)$. By the definition of μ we have $\mu(C) = p_1 p_1 = (.4)^2$, $\mu(D) = p_2 p_2 = (.6)^2$, $\mu(E) = p_1 p_2 = (.4)(.6)$. The integral is given by

$$\int_\Sigma f d\mu = (.4)^2 + (2.3)(.6)^2 + (1 + 2.3)(.4)(.6) = 1.78$$

IX.4.7 Define the Borel measure δ_a by $\mu(B) = 1$ if $a \in B$ and 0 otherwise. Let f be continuous $f : (X, d) \to \mathbb{R}$. For each n, denote by $\{B_i\}_{i=1}^{k_n}$ a Borel partition of X into disjoint sets B_i where $\text{diam}(B_i \leq 1/n$. Define $\{f_n\}$ to be a sequence of simple functions defined on each partition by $f_n = \sum y_i \chi_{B_i}$ with $y_i = f(x_i)$ on the sets which do not include a and $y_{i_a} = a$ on the set in each partition $a \in B_{i_a}$. Then since the measure of all the sets is zero with the exception of B_{i_a} on which it is 1, we have

$$\int_X f_n d\delta_a = f(a) \forall n.$$

Hence the limit as $n \to \infty$ is $f(a)$. Our partition is independent of the choice of x_i with the exception of the choice of y_{i_a}, but the function is continuous, so that as $n \to \infty$ another choice of x_{i_a} approaches a as the diameter of B_{i_a} goes to zero.

IX.4.8 We proceed by first deducing that since $f(x, y)$ is continuous, and the measure on $\blacksquare$ is defined on rectangles by their area, the value of the integral is the Riemann integral of $f(x, y)$. By definition, the Riemann integral is done by taking a sequence of rectangular partitions of $\blacksquare$, and constructing two sequences of simple functions, choosing y_i, y_i' by

$$y_i = \inf\{f(x, y) : (x, y) \in B_i\}; \ y_i' = \sup\{f(x, y) : (x, y) \in B_i\}$$

By Theorem 4.1, since $f(x, y)$ is continuous, the two sequences converge to the same limit, namely $\int_\blacksquare f d\mu$, since they converge and converge to the same limit, the Riemann integral is defined and is equal to this limit. The value of the integral, by taking the Riemann integral, is $1/2$.

IX.4.10 Refering to exercise 3.4, the measure of any strip intersecting $\triangle$ is simply it's width. The function $f(x, y) = x^2$ is independent of y so we partition the space into strips, their intersection with $\triangle$ produces decreasing diameters. $f(x)$ is continuous, hence this partition and the choice of x_i are immaterial. Choose x_i to be the left-hand boundary of each strip, intersecting some point on $\triangle$. Then the value of the integral is just $\int_0^1 x^2 dx = 1/3$. Let $\mathbf{X}$ be the set of pixels on a computer display. For simplicity, define the distance between the pixel (i, j) on the ith row and in the jth column and the pixel (k, ℓ) as $|i - k| + |j - \ell|$, which is a Manhattan metric. Then $\mathbf{X}$ may be related to the subset of $\mathbb{R}^2$ consisting of integer coordinate points in the first quadrant of the same size. This set is closed and totally bounded. Hence $\mathbf{X}$ is compact. Since there are finitely many pixels, we may define the generator for the Borel subsets to be all the subsets consisting of one pixel each. This defines the Borel field on $\mathbf{X}$ to be the power set of $\mathbf{X}$. Thus any set of pixels is a Borel subset. Suppose there are J rows and K columns so that the total number of pixels is $N = J \times K$. Define a Borel measure by $\mu(A) = $ (number of pixels in $A)/N$. Any function from $\mathbf{X} \to \mathbb{R}$ is continuous because the pixels themselves are open sets, hence the inverse image of an open set in $\mathbb{R}$ is always open.

We will choose the function $f(i, j) = i + j$. Any function on the finite set of pixels can be described by its value on each pixel, hence its value at a pixel times the characteristic function at that pixel, hence the sum of such terms makes any function simple. Then the integral of f over $\mathbf{X}$ is

$$\int_X f d\mu = \sum_X \frac{i + j}{N} = \frac{1}{N} \sum_X (i + j)$$

where the summation notation indicates summing over all pixels in $\mathbf{X}$. Since there are K pixels with first value i, and J pixels with second value j, we have

$$\frac{1}{N} \sum_X (i + j) = \frac{1}{N} \left(\sum_{n=1}^{K} {}^{J} n + \sum_{n=1}^{J} {}^{K} n \right)$$

$$= \frac{1}{N} \left(\frac{K(J(J + 1))}{2} + \frac{J(K(K + 1))}{2} \right)$$

$$= \frac{1}{N} \frac{N(J + 1) + N(K + 1)}{2}$$

so the integral is

$$\int_X f d\mu = \frac{J + K + 2}{2}$$

Answers to Chapter IX, section 5

IX.5.1 Let

$$\mathbf{X} = \{(i, j) : i, j \in \{1, 2, 3, \ldots, K\}\}$$

and define a metric $d : \mathbf{X} \to \mathbb{R}$ by

$$d((i_1, j_1), (i_2, j_2)) = |i_1 - i_2| + |j_1 - j_2|$$

Then $(\mathbf{X}, d)$ is a compact metric space. Define μ to be an atomic measure with

$$\mu((i, j)) = \frac{i + j}{K^3 + K^2}$$

and let ν be another atomic measure with

$$\nu((i, j)) = \frac{1}{K^2}$$

for every $(i, j) \in \mathbf{X}$.

We wish to calculate the distance in the Hutchinson metric between μ and ν, given by

$$d_H(\mu, \nu) = \sup\{\int_X f d\mu - \int_X f d\nu : f \in C(\mathbf{X}, \mathbb{R}), |f(x) - f(y)| \le d(x, y) \forall x, y \in \mathbf{X}\}$$

where $C(\mathbf{X}, \mathbb{R})$ is the set of continuous functions from $\mathbf{X}$ to $\mathbb{R}$. Now $\mathbf{X}$ is a discrete space under a metric which is the restriction of the Manhattan metric on $\mathbb{R}^2$, consequently $f : \mathbf{X} \to \mathbb{R}$ is always continuous. As well, the integrals are the integrals of simple functions. Hence for every f, we the integrals can be expressed

$$\int_X f d\mu = \sum_X f(i, j) \frac{i + j}{K^3 + K^2}$$

$$\int_X f d\nu = \sum_X f(i, j) \frac{1}{K^2}$$

so the difference is, for any function f,

$$\sum_X f(i, j) \left(\frac{i + j}{K^3 + K^2} - \frac{1}{K^2} \right) = \frac{1}{K^2} \sum_X f(i, j) \left(\frac{i + j}{K + 1} - 1 \right)$$

Intuitively, one is tempted to pick the function $f(i, j) = i + j$, which among the set of valid functions is the largest of the form

$$f(i, j) = \alpha i + \beta j$$

However, the range of the term $(i + j)/(K + 1) - 1$ is from $2/(K + 1) - 1 = -(K - 1)/(K + 1)$ to $2K/(K + 1) - 1 = (K - 1)/(K + 1)$ which gives a negative coefficient to $f(i, j)$ over almost half of $\mathbf{X}$ detracting from the value of the sum. We must therefore endeavor to pick $f(i, j)$ as large as we can in absolute value, negative when the measure subtraction term is negative, and positive when it is positive. That is, we maximize by choosing the largest value for $|f|$ allowable under the distance constraint where $|(i + j)/(K + 1) - 1| = |C(i, j)|$ is largest. The interested reader can confirm that this exceeds the intuitive result. The value of $C(i, j)$ is constant along each diagonal from $(1, n)$ to $(n, 1)$ and is equal to zero along the major diagonal from $(1, K)$ to $(K, 1)$. It is anti-symmetric with respect to this diagonal, negative above it and positive below.

We thus place the largest values for $|f|$ at $(1, 1)$ and (K, K), and since the distance between these is $2K - 2$, we set them to $\pm(K - 1)$. We then proceed outward from these points in Manhattan metric circles. The distance between such circles (which are the diagonals where they intersect $\mathbf{X}$) is always the horizontal distance between the extensions of the diagonals, thus we end up satisfying the distance constraint by putting f to

$$
\begin{array}{cccccc}
-(K - 1) & -(K - 2) & \cdots & -1 & 0 \\
-(K - 2) & -(K - 3) & \cdots & 0 & 1 \\
\vdots & \vdots & \cdots & \vdots & \vdots \\
-1 & 0 & \cdots & (K - 3) & (K - 2) \\
0 & 1 & \cdots & (K - 2) & (K - 1)
\end{array}
$$

This may be summed by taking twice the sum above the diagonal, and noticing that there are n entries where $(K - n)$ is the absolute value of the entry, yielding

$$\frac{2}{K^2(K+1)} \sum_{n=1}^{K} n(K-n)^2$$

$$= \frac{2}{K^2(K+1)} \sum_{n=1}^{K} (nK^2 - 2n^2K + n^3)$$

$$= \frac{2K^2}{K^2(K+1)} \sum_{n=1}^{K} n - \frac{4K}{K^2(K+1)} \sum_{n=1}^{K} n^2 + \frac{2}{K^2(K+1)} \sum_{n=1}^{K} n^3$$

$$= \frac{2K^2 K(K+1)}{2K^2(K+1)} - \frac{4K K(K+1)(2K+1)}{K^2(K+1)} + \frac{2K^2(K+1)^2}{4K^2(K+1)}$$

$$= \frac{K-1}{6} = d_H(\mu, \nu)$$

IX.5.2 Since $\mu \in \mathcal{P}(\mathbf{X})$ the measure of the entire space is $\mu(\mathbf{X}) = 1$. This value is identical with the value on the support of μ which we will call S, that is, this implies $\mu(S) = 1$. In particular it is non-zero, and since, by Theorem 3.1 in Chapter IX statement (3) we have $\mu(\emptyset) = 0$, $S \neq \emptyset$. Then by Theorem 3.4 in Chapter IX, the support of a Borel measure on a metric space is closed. Since $\mathbf{X}$ is compact, a closed subset of $\mathbf{X}$ is also compact, hence S is compact. Since S is compact and non-empty, S, the support of μ is an element of $\mathcal{H}(\mathbf{X})$.

Answers to Chapter IX, section 6

IX.6.1 We want to show that the Markov operator is indeed a transformation on $\mathcal{P}(\mathbf{X})$, that is, $M(\nu) : \mathcal{P}(\mathbf{X}) \to \mathcal{P}(\mathbf{X})$. It suffices to show that

$$\nu \circ w_i^{-1} : \mathcal{P}(\mathbf{X}) \to \mathcal{P}(\mathbf{X})$$

since a sum of these operators is a Borel measure if they are, and $\sum_i p_i = 1$ indicates that they are normalized if each entry is.

(i) $w_i^{-1} : \mathbf{X} \to \mathbf{X}$, since for hyperbolic IFS these are invertible maps. In addition, if we define $w_i^{-1}(\{x\}) = \emptyset$ whenever $x \notin w_i(\mathbf{X})$ then $w_i^{-1}(\mathbf{X}) = \mathbf{X}$. Hence $\nu \circ w_i^{-1}(\mathbf{X}) = 1$, and this is a normalized Borel measure. Hence $M(\nu)(\mathbf{X}) = 1$.

(ii) $\nu \circ w_i^{-1}$ is a map $\mathbf{X} \to [0, \infty)$ (actually $\mathbf{X} \to [0, 1]$). Furthermore, by definition, if $\{A_j\}_{j=1}^{\infty}$ are mutually disjoint,

$$\nu \circ w_i^{-1}(\bigcup_{j=1}^{\infty} A_j) = \nu(\bigcup_{j=1}^{\infty} w_i^{-1}(A_j))$$

$$= \sum_{i=1}^{\infty} \nu(w_i^{-1}(A_i)) = \sum_{j=1}^{\infty} (\nu \circ w_i^{-1})(A_j)$$

so that $\nu \circ w_i^{-1}$ is countably additive. Since w_i^{-1} maps open sets to open sets (because w_i is a contraction, it is continuous), and since $w_i^{-1}(\mathbf{X} \setminus A) = w_i^{-1}(\mathbf{X}) \setminus w_i^{-1}(A)$, w_i^{-1} takes the generators for the Borel subsets into themselves, and preserves the operation of set complementation. Hence $w_i^{-1} : \mathcal{B} \to \mathcal{B}$, and the expression $\nu \circ w_i^{-1}$ is well defined. Hence we have $\nu \circ w_i^{-1} : \mathcal{P}(\mathbf{X}) \to \mathcal{P}(\mathbf{X})$ and therefore $M(\nu) : \mathcal{P}(\mathbf{X}) \to \mathcal{P}(\mathbf{X})$, as desired.

IX.6.2 We define L by

$$L = \{f \in C^0(\mathbf{X}, \mathbb{R}) : |f(x) - f(y)| < d(x, y) \forall x, y \in \mathbf{X}\}$$

and let $f \in L$. Define $\tilde{f} = s^{-1} \sum_i p_i f_i \circ w_i$, and we wish to show that $\tilde{f} \in L$. Since $\tilde{f}$ is a scalar times the sum of compositions of continuous functions, $\tilde{f}$ is clearly in $C^0(\mathbf{X}, \mathbb{R})$. We must therefore only verify the distance condition. For all $x, y \in \mathbf{X}$ we have

$$|\tilde{f}(x) - \tilde{f}(y)| = |s^{-1} \sum_i |f \circ w_i(x) - f \circ w_i(y)|$$

$$\leq s^{-1} \sum_i p_i |f \circ w_i(x) - f \circ w_i(y)|$$

$$\leq s^{-1} \sum_i p_i |w_i(x) - w_i(y)| \text{ since } f \in L$$

$$\leq s^{-1} \sum_i p_i s |x - y| \text{ since } s = \max_i s_i$$

$$\leq s^{-1} \cdot s |x - y| \cdot \sum_i p_i$$

$$\leq |x - y| = d(x, y) \text{ since } \sum_i p_i = 1$$

Hence $\tilde{f} \in L$.

IX.6.7 Let **X** consist of three points: $\mathbf{X} = \{A, B, C\}$, and let an IFS be defined by

$$\{\mathbf{X}; w_1, w_2; p_1 = .6, p_2 = .4\}$$

where

$$w_1(A) = B \quad w_1(B) = B \quad w_1(C) = B$$
$$w_2(A) = C \quad w_2(B) = A \quad w_2(C) = C$$

Furthermore, let $\mu_0(A) = \mu_0(B) = \mu_0(C) = 1/3$. We now let $\mu_n = M^n(\mu_0)$ and find constant entries for the *Markov transition matrix* $\tilde{M}, a, b, \ldots, i$, such that

$$\begin{pmatrix} \mu_n(A) \\ \mu_n(B) \\ \mu_n(C) \end{pmatrix} = \begin{pmatrix} a & b & c \\ d & e & f \\ g & h & i \end{pmatrix} \begin{pmatrix} \mu_{n-1}(A) \\ \mu_{n-1}(B) \\ \mu_{n-1}(C) \end{pmatrix}$$

By definition,

$$M(\mu) = \sum_i p_i \mu \circ w_i^{-1} = .6(\mu \circ w_1^{-1}) + .4(\mu \circ w_2^{-1})$$

so that

$$\mu_n = .6(\mu_{n-1} \circ w_1^{-1}) + .4(\mu_{n-1} \circ w_2^{-1})$$

and we have

$$\mu_n(A) = .6(\mu_{n-1} \circ w_1^{-1}(A)) + .4(\mu_{n-1} \circ w_2^{-1}(A)) = 0 + .4\mu_{n-1}(B)$$
$$\mu_n(B) = .6(\mu_{n-1}(A) + \mu_{n-1}(B) + \mu_{n-1}(C)) + 0$$
$$\mu_n(C) = 0 + .4(\mu_{n-1}(A) + \mu_{n-1}(C))$$

giving us

$$\tilde{M} = \begin{pmatrix} 0 & .4 & 0 \\ .6 & .6 & .6 \\ 0 & .4 & .4 \end{pmatrix}$$

For a discrete space like this, $\tilde{M}$ completely determines $M : \mathcal{P}(\mathbf{X}) \to \mathcal{P}(\mathbf{X})$, by defining μ_n at each point in the space. In the limit, $\tilde{M}^n \to \overline{M}$, an operator which takes every point in $\mathcal{P}(\mathbf{X})$ to some measure μ_f, and which is therefore necessarily singular. By definition $\mu_f = \tilde{M}(\mu_f)$. This is an eigenvalue equation which means that μ_f is an eigenvector of the eigenvalue 1.

IX.6.8 We first look at the sequence of measures $\{\mu_n\}_{n=0}^{\infty}$. We have

$$\mu_1(A) = \sum_i p_i \mu_0(w_i^{-1}(A))$$

$$\mu_2(A) = \sum_i p_i \mu_1(w_i^{-1}(A))$$

$$= \sum_i p_i \left(\sum_j p_j \mu_0 w_j^{-1}(w_i^{-1}(A)) \right)$$

$$= \sum_i p_i \left(\sum_j p_j \mu_0(w_j^{-1} \circ w_i^{-1}(A)) \right)$$

Since the sum after n such iterations is finite, the order is not important and we can write

$$\mu_n(A) = \sum_{i,j,\ldots,k} p_i p_j \cdots p_k \mu_0(w_k^{-1} \circ \cdots \circ w_j^{-1} \circ w_i^{-1}(A))$$

where the sum indicates a sum over all the indices $i, j, \ldots, k$.

μ_0 is concentrated on A since, by assumption, $\mu_0 \in \mathcal{P}(\mathbf{X})$, and $\mu_0(A) = 1$ so the support of μ_0 is a subset of A. If $x \notin A$, all the terms in the above sum are zero because of this, hence the support of μ_n is again a subset of A, and we must have $\mu_n(A) = 1$, because $M : \mathcal{P}(\mathbf{X}) \to \mathcal{P}(\mathbf{X})$. Now let μ be the unique invariant measure of the operator M, that is $M(\mu) = \mu$. From the formula just derived with respect to μ_0 for $M^n(\mu_0)$, and from the invariance of μ we must have, for any set B,

$$\mu(B) = \sum_{i,j,\ldots,k} p_i p_j \cdots p_k \mu(w_k^{-1} \circ \cdots \circ w_j^{-1} \circ w_i^{-1}(B))$$

Now choose $B = w_i \circ w_j \circ \cdots \circ w_k(A)$. Then we have

$$\mu(w_i \circ w_j \circ \cdots \circ w_k(A)) = \sum_{m,n,\ldots,l} p_m p_n \cdots p_l (w_m^{-1} \circ w_n^{-1} \circ \cdots w_l^{-1}(w_i \circ w_j \circ \cdots \circ w_k(A))$$

where the sum is over the same number of indices as in the composition for B. Since each element of the sum is positive or zero, the sum is greater than or equal to each of its entries, and we have

$$\mu(w_i \circ w_j \circ \cdots \circ w_k(A))$$

$$\geq p_i p_j \cdots p_k(w_k^{-1} \circ \cdots \circ w_j^{-1} \circ w_i^{-1}(w_i \circ w_j \circ \cdots \circ w_k(A))$$

$$\geq p_i p_j \cdots p_k \mu(A) = p_i p_j \cdots p_k$$

Equality holds precisely when this is the only contributing term of the sum, that is

$$\mu(w_m^{-1} \circ w_n^{-1} \circ \cdots \circ w_l^{-1}(w_i \circ w_j \circ \cdots \circ w_k(A)) = 0$$

whenever the ordered set of indices $(m, n, \ldots, l) \neq (i, j, \ldots, k)$, which says for all of these ordered sets the set

$$w_m^{-1} \circ w_n^{-1} \circ \cdots \circ w_l^{-1}(w_i \circ w_j \circ \cdots \circ w_k(A))$$

is at least of measure zero, if not empty. It is empty for the totally disconnected case, hence equality holds when the IFS is totally disconnected.

IX.6.11 The measure μ has support on A, consequently it suffices to restrict to Borel subsets of A. We first let the IFS be totally disconnected. Then the sets of the form

$$W_i = w_{\sigma_n} \circ w_{\sigma_{n-1}} \circ \cdots \circ w_{\sigma_1}(A)$$

for all n form a countable basis on the space A. Hence since for the totally disconnected case, these are all both closed and open, for any Borel subset B of A we have

$$B = \bigcup_{i=1}^{\infty} W_i$$

for some disjoint choice of W_i. But

$$\phi^{-1}(B) = \phi^{-1}(\bigcup_{i=1}^{\infty} W_i) = \bigcup_{i=1}^{\infty} \phi^{-1}(W_i) = \bigcup_{i=1}^{\infty} C_i$$

where C_i are disjoint cylinder sets in Σ. The measure ρ on Σ by observation gives the measure on any cylinder set as

$$\rho(C(\sigma_1, \ldots, \sigma_k)) = p_1 \cdots p_k$$

which is precisely the measure μ applied to the corresponding basis set W_i on A. Consequently, the measures agree, and we have

$$\rho(\phi^{-1}(B)) = \mu(B)$$

When the attractor is not totally disconnected, we have for each W_i finitely many cylinder sets in the image $\phi^{-1}(W_i)$, which are disjoint, and correspond to terms in the sum given in exercise IX.6.8. Consequently the measures agree on Borel subsets of A in this case as well. Since the Borel subsets of A are precisely the subsets $A \cap B$ where B is a Borel subset of $\mathbf{X}$, and since the support of the measure is A, the equality holds for any Borel subset of $\mathbf{X}$.

Answers to Chapter IX, section 7

IX.7.1 We want to prove that $\mu([x, x + \delta]) = \delta$, where μ is the invariant measure of the IFS

$$\{[0, 1]; w_1 = 1/2x, w_2 = 1/2x + 1/2; 1/2, 1/2\}$$

To do this it suffices to show that this expression holds on any closed interval of the form $[k/2^n, (k + 1)/2^n] \subset [0, 1]$, since we may then build an approximation of any interval from these, good to any error, by the disjoint union of subsets of this form which cover the interval. Since μ is defined as the infimum of the measures of such covers, we let the error go to zero.

To show that $\mu([k/2^n, (k + 1)/2^n]) = 1/2^n$ we note that for each iteration of $W(A)$ the image is 2 times the number of subintervals of A of half the length of those of the iteration before. Each of these subintervals is of the form

$$w_{\sigma_1} \circ \cdots \circ w_{\sigma_k}(A)$$

where there are k terms in the composition for $W^k(A)$. Consequently, these intervals of length $1/2^k$ have measure

$$\mu(w_{\sigma_1} \circ \cdots \circ w_{\sigma k}(A)) = p_{\sigma_1} \cdots p_{\sigma_k}$$

Since each term in the product is $1/2$, the right hand side is $1/2^k$ as desired.

IX.7.2 We must prove the μ applied to any rectangle in $\blacksquare$ gives the area of the rectangle, where μ is the invariant measure of the IFS

$$\{\blacksquare; w_1, w_2, w_3, w_4; 1/4, 1/4, 1/4, 1/4\}$$

and the w_i are given by

$$w_1 \begin{pmatrix} x \\ y \end{pmatrix} = \begin{pmatrix} \frac{1}{2} & 0 \\ 0 & \frac{1}{2} \end{pmatrix} \begin{pmatrix} x \\ y \end{pmatrix}$$

$$w_2 \begin{pmatrix} x \\ y \end{pmatrix} = \begin{pmatrix} \frac{1}{2} & 0 \\ 0 & \frac{1}{2} \end{pmatrix} \begin{pmatrix} x \\ y \end{pmatrix} + \begin{pmatrix} \frac{1}{2} \\ 0 \end{pmatrix}$$

$$w_3 \begin{pmatrix} x \\ y \end{pmatrix} = \begin{pmatrix} \frac{1}{2} & 0 \\ 0 & \frac{1}{2} \end{pmatrix} \begin{pmatrix} x \\ y \end{pmatrix} + \begin{pmatrix} \frac{1}{2} \\ \frac{1}{2} \end{pmatrix}$$

$$w_4 \begin{pmatrix} x \\ y \end{pmatrix} = \begin{pmatrix} \frac{1}{2} & 0 \\ 0 & \frac{1}{2} \end{pmatrix} \begin{pmatrix} x \\ y \end{pmatrix} + \begin{pmatrix} 0 \\ \frac{1}{2} \end{pmatrix}$$

that the invariant measure μ is uniform, that is, that for any rectangle, we have $\mu(([a, b], [c, d])) = (c - a)(d - b)$. We argue as in the previous problem: For any square with a side of length $\frac{1}{2^n}$ of the form

$$S = ([\frac{k}{2^n}, \frac{k+1}{2^n}], [\frac{\ell}{2^n}, \frac{\ell+1}{2^n}])$$

we have some sequence $i, j, \ldots, k$ of digits, of length n such that this square is equal to

$$S = w_i \circ w_j \circ \cdots \circ w_k(A)$$

and hence its measure is

$$\mu(S) = p_i p_j \cdots p_k = (\frac{1}{2^n})^2$$

so that the measure has the uniform property on such squares. These squares form a countable base for the Borel subsets of A, consequently any rectangle may be written as

$$([a, b], [c, d]) = \bigcup_{i=1}^{\infty} S_i$$

where the S_i have been chosen such that $\mu(S_i \cap S_j) = 0$, that is, they meet only on the edges. Consequently, by countable additivity, we have $\mu(([a, b], [c, d])) = (c - a)(d - b)$, for this rectangle, consequently the measure of any rectangle in A is its area.

IX.7.4 The distance between two points in $([0, 1]_p, d_p)$ can be written

$$d_p(a, b) = |d_p(a, 0) - d_p(b, 0)|$$

We start with the sets in each space corresponding to addresses in a cylinder set in code space, namely

$$\phi(C(\sigma_1, \sigma_2, \ldots, \sigma_n)) \text{ and } \phi_p(C(\sigma_1, \sigma_2, \ldots, \sigma_n)).$$

If the measures agree on the size of these sets, which form a base for the Borel field in each space then since we can rewrite f on these as

$$f(C(\sigma_1, \sigma_2, \ldots, \sigma_n)) = \phi_p(\phi^{-1}(C(\sigma_1, \sigma_2, \ldots \sigma_n)))$$

we will have demonstrated that the measures are equal. On $[0, 1]$, almost by definition, the invariant measure ν applied to a cylinder set yields

$$\nu(C(\sigma_1, \sigma_2, \ldots, \sigma_n)) = p_1 p_2 \cdots p_n,$$

that is, the probability of landing in the cylinder set on n iterations.

On $[0, 1]_p$, the cylinder set corresponds to an interval with endpoints

$$.\sigma_1 \sigma_2 \cdots \sigma_n \overline{0} \quad \text{and} \quad .\sigma_1 \sigma_2 \ldots \sigma_n \overline{(N - 1)} = .\sigma_1 \sigma_2 \ldots (\sigma_n + 1) \overline{0}$$

How wide is this interval? Start with $n = 1$. The interval $(.\sigma_1, .(\sigma_1 + 1))$ can be read from the IFS we used to construct $[0, 1]_p$, to be of width p_{σ_1}. A contraction of the same relative subdivision of $[0, 1]$ is placed inside it to make our "tick marks" by the IFS, so that the portion of this subinterval which is also in the interval $(.\sigma_1 \sigma_2, .\sigma_1(\sigma_2 + 1))$ is thus $p_{\sigma_1} p_{\sigma_2}$. If this proceeds in this fashion up to σ_k, then the subinterval which is again divided by the action of the IFS is of length $p_{\sigma_1} p_{\sigma_2} \cdots p_{\sigma_k}$, and we take a fraction of this equal to $p_{\sigma_{k+1}}$ times its length. Consequently by induction, the length of the image of a cylinder set is

$$\mu(\phi_p(C(\sigma_1, \sigma_2, \ldots, \sigma_n))) = d(.\sigma_1 \sigma_2 \cdots \sigma_n, .\sigma_1 \sigma_2 \cdots (\sigma_n + 1)) = p_{\sigma_1} p_{\sigma_2} \cdots p_{\sigma_n}$$

which agrees with the value on $[0, 1]$ for the image of the same set under the invariant measure ν. Hence $\nu(A) = \mu(f(A))$, and f is measure preserving.

IX.7.5 The diagonal of the square forms right triangles with the vertical lines generated by the tick marks on the x-axis, and with the horizontal grid lines generated by tick marks on the y-axis. The slope of this line is 1, hence the distance between horizontal grid lines on the diagonal is $\sqrt{2}$ times the distance on the x-axis. Consequently, a point picked at random on the diagonal has a p_i chance of falling between the $i - 1^{\text{th}}$ grid mark and the ith grid mark, and so forth. With probability 1 then, a point on the diagonal will have each digit occuring in proportion to the p_i spacing on the x-axis in its x coordinate, and the digits occur with equal frequency in its y coordinate.

Answers to Chapter IX, section 8

There are no mathematical exercises in section 8.

Index

Credits for Figures and Color Plates

Dr. John Herndon collaborated with the author on the computation of many of black and white figures in Chapters I–IX.

The Orchard Subset of $\mathbb{R}^2$, Figure III.46 was computed by Henry Strickland.

Figure VI.163 was produced by Peter Massopust.

The figures in Chapter X were produced by Louisa Anson.

The following list gives the author of each color image. The italics indicate who did most of the work.

Plate	Author
1	Michael Barnsley
2	Michael Barnsley, Alan Sloan
3	Michael Barnsley
4	Louisa Anson
5	Michael Barnsley
6	Michael Barnsley, *John Herndon*
7	Michael Barnsley, *John Herndon*
8	Michael Barnsley, *John Herndon*
9	Michael Barnsley, *John Herndon*
10	Michael Barnsley, *John Herndon*
11	Michael Barnsley, *John Herndon*
12	Michael Barnsley, *John Herndon*
13	Michael Barnsley, *John Herndon*
14	Michael Barnsley, *John Herndon*
15	Michael Barnsley, *John Herndon*
16	Michael Barnsley, *John Herndon*
17	Michael Barnsley, *John Herndon*
18	Michael Barnsley, *John Herndon*
19	Michael Barnsley, *John Herndon*
20	Michael Barnsley, *John Herndon*
21	*Michael Barnsley, Laurie Reuter*, Alan Sloan
22	*Michael Barnsley, Laurie Reuter*, Alan Sloan
23	Michael Barnsley
24	Michael Barnsley